The Wild Horse Dilemma

The Wild Horse Dilemma

Conflicts and Controversies
of the Atlantic Coast Herds

Written and Illustrated by
Bonnie U. Gruenberg

QUAGGA PRESS

ISBN 13: 978-0979002038 • ISBN 10: 0979002036 Library of Congress Control Number: 2014952179

Published by Quagga Press, an imprint of Synclitic Media, LLC • 1646 White Oak Road • Strasburg, PA 17579 • www.quaggapress.com

Also by the author
  Hoofprints in the Sand (as Bonnie S. Urquhart; Eclipse, 2002)
  Essentials of Prehospital Maternity Care (Prentice Hall, 2005)
  Birth Emergency Skills Training (Birth Guru/Birth Muse, 2008)
  The Midwife's Journal (Birth Guru/Birth Muse, 2009)
  Hoofprints in the Sand, Kindle Edition (Quagga Press, 2014)
  Wild Horses of the Atlantic Coast: An Intimate Portrait, Kindle Edition (Quagga Press, 2014)

Forthcoming
  Wild Horse Vacations: Where To See the East Coast Herds and What Else To Do While You're Visiting (Quagga Press, 2015)
  Discover! Wild Horses. A Kids' Guide to the East Coast Herds (Quagga Press, 2015)
  Birth Emergency Skills Training, 2nd Edition (Synclitic Press, 2015)

To my sons, Mark Bryan Scianna and
Keith J. Scianna

*Find your passion, and bring it to
everything you do.*

Reviewers

Jay F. Kirkpatrick, PhD • Reproductive Specialist, Senior Scientist • The Science and Conservation Center • Billings, MT

E. Gus Cothran, PhD • Director, Animal Genetics Lab. • Texas A&M University • College Station, TX

D. Phillip Sponenberg, DVM, PhD • Professor, Pathology and Genetics • Virginia- Maryland Regional College of Veterinary Medicine • Blacksburg, VA

Don Höglund MS, DVM • North Carolina State University College of Veterinary Medicine • Raleigh, NC

Karen Dalke, PhD • Anthropology; University of Wisconsin-Green Bay • Green Bay, WI

Sue Stuska, Ed.D. • Wildlife Biologist • Cape Lookout National Seashore • Harkers Island, NC

Doug M. Hoffman • Biologist, Resource Management Division • Cumberland Island National Seashore • St. Marys, GA

Karen McCalpin • Executive Director, Corolla Wild Horse Fund • Corolla, NC

Steve Edwards • Assistant Commonwealth's Attorney • Member, Board of Directors, Corolla Wild Horse Fund • Proprietor, Mill Swamp Indian Horses • Smithfield, VA

Pam Emge • Co-author, *Chincoteague Ponies: Untold Tails* • Tetonia, ID

And special thanks to the proofreaders:

Tabetha Fenton
Tim Ferry
Joyce Urquhart
Alex Gruenberg

Contents

Foreword 11

Author's Note... 17

Introduction 21

1 The Dilemma..25
   Wild Horses, Feral Horses

2 The Mustang Predicament.......................................63

3 Assateague, Maryland..115
   Camping with Horses

4 Chincoteague–Assateague, Virginia 171
   The Wild, Wild East

5 Currituck Banks, North Carolina 261
   The End of the Road

6 Ocracoke, North Carolina 341
   Bankers, Blackbeard, and Boy Scouts

7 Down East...413
   Shackleford Banks, Carrot Island, and Cedar Island, North Carolina

8 Cumberland Island, Georgia 493
   Island of Contradictions

Conclusion and Author's Commentary...................547

Appendix: Equine Evolution..................................555

Acknowledgments 565

Index...569

Tables

Colonial Spanish Horse Type Matrix....................369

Henneke Body Condition Scoring System 462

Lactational Condition Index.................................479

Foreword

Don Höglund, MS, DVM
Author of *Nobody's Horses*

*T*he *Wild Horse Dilemma* has better chronicled the life and times of the contemporary free-roaming horse than any previous resource. This study examines the various Atlantic free-roaming horse populations and echoes the observation that "live and let live" no longer applies to the environment—we affect everything, including our oceans. Human population growth, urban sprawl, and natural resource development have caused us to rethink how humans interact with the other elements of life in time and space, and these influences are further complicated by the presence of human emotion.

From my perspective, and building from the wonderful work of Dr. Dan Simons at the University of Illinois' Vic-Cog Laboratory, prominent free-roaming horse issues meld into a potpourri of illusion and emotion. Humans tend to believe that they are capable of accurately seeing what is right in front of them and of precisely remembering what they have seen and heard. They believe they can clearly determine the cause and effect of "wild" horse management plans. A concerned public often listens to expert authority. They believe the words of confident-sounding "wild-horse experts." As caring citizens, many believe that being merely familiar with free-roaming horse issues is the same as being knowledgeable. But familiarity with the horse quandary does not equal factual knowledge. *The Wild Horse Dilemma* helps the reader to gain objective knowledge of the free-roaming horse predicament and to challenge every statistic and claim made by any side of the free-roaming horse debate.

The illusion of awareness creates a misperception that all free-roaming horses are subject to the same laws. There are private populations of free-roaming horses owned by independent people or living on military installations, there are horses on public and private lands and there are horses gone-loose. Some of them are somebody's horses, some are nobody's horses, and some are everybody's horses. Any discussion about horses should begin with determining what population is being discussed, who is legally responsible for the horses and what issues are at hand.

Keep in mind two things as you read the enlightening prose between these covers. Confident sounding people are not necessarily accurate, but we tend to believe them, even though as Dan Simons' work has shown, experts are documented to be incorrect at least thirty percent of the time. For example, the fact that some horses were herded for many miles does not necessarily mean physical exhaustion caused any subsequent spontaneous abortions. Take into consideration the close proximity to human presence and being handled immediately after capture. Realize that contact with humans is unfamiliar to most free-roaming horses and it is a stressor. Just climb into an elevator and sense the distress among the occupants competing for space. Then, imagine releasing a harmless snake and imagine the resulting distress. Corralling wild horses yields similar distress. We must ponder all relevant factors when considering the horses' welfare.

No discussion of horse issues would be complete without a treatment of the emotions evoked by the thought of wild horses. This informative book addresses the emotional issues clearly, and readers should be cautious not to be overwhelmed by them. That humans are capable of

introspective awareness is obvious. At the same time, keep in mind, as Dr. Joseph LeDoux at NYU Center for Neural Sciences educates, "Thinking will not control emotion." To make matters more complicated, emotion—like stress—is not well defined, and if it is not clearly explained it is difficult to understand and study. One thing is for sure, however: we will never know what animals are thinking, feeling, or wanting. What we can do is focus on what the horses are doing.

We can certainly protect their welfare, measure their wellbeing, and manage them within the principles of humane handling. It seems to me that we should fund, study and manage free-roaming horses where they roam.

Many people have dedicated their lives and resources to the plight of the free- roaming horse. In *The Wild Horse Dilemma*, Bonnie Gruenberg has encircled the breadth of free-roaming horse difficulties. This work is a chronicle of human and horse history wrapped around the emotions of humans and the mystery of why horses have captured human interest.

You must teach your children that the ground beneath their feet is the ashes of their grandfathers. So that they will respect the land, tell your children that the earth is rich with the lives of our kin. Teach your children what we have taught our children, that the earth is our mother. Whatever befalls the earth befalls the sons of the earth. If men spit upon the ground, they spit upon themselves. This we know. The earth does not belong to man; man belongs to the earth. This we know. All things are connected. Like the blood which unites one family, all things are connected.

ATTRIBUTED TO CHIEF SEATTLE, 1780–1866
(But taken from *Home*, 1972, teleplay by Ted Perry)
Framed and hanging in the office of Lou Hinds, former Chincoteague National Wildlife Refuge manager

Author's Note

The wild filly watched me with gentle, intelligent eyes as I followed her movements with my camera. Pretending to ignore me, she grazed closer and closer until she was only few yards away. Then, unable to maintain her nonchalance any longer, she extended an outstretched neck and inquiring muzzle in my direction. She blew softly at me, filling flared nostrils with my scent. *How strange. A human. And it's just sitting there.*

Curiosity satisfied, she lifted her head and whickered softly to her herdmates. Silhouetted against the fading light, she was the essence of wildness, in tune with her own natural rhythms and in harmony with the fundamental forces that shape her world. In her momentary connection with me, she revived something in my own civilized soul that is as wild as she is.

When I first visited Cape Hatteras, North Carolina, in 1993, I was surprised to find that wild horses lived on several nearby islands and had once ranged along much of the Atlantic coast. They made their first hoofprints there not long after the arrival of early European settlers, and in time they ran free on innumerable North American islands from the Caribbean to Canada. I learned that small herds remained on the coast of Virginia, North Carolina, Maryland, and Georgia; on Sable Island, off Nova Scotia, Canada; and on Great Abaco Island in the Bahamas.

I looked for books about them, but credible information was scarce. Newspaper and magazine articles often appeared incorrect or fanciful, drawn from the statements of misinformed people, or based on superficial research. Once an inaccuracy appears in print, other writers mistake it for fact and perpetuate it. Frustrated by the lack of solid information, I set about writing an accurate, comprehensive book useful to other interested people.

I spent long hours in libraries and salt marshes separating fact from myth. I learned that each population of horses has its own character, its own history, and its own set of problems and concerns and that in most cases these animals have made a unique contribution to local history. I learned that most of these herds are controversial and that each has its own detractors and defenders.

My first book, *Hoofprints in the Sand: Wild Horses of the Atlantic Coast,* was published in 2002 by Eclipse Press after six years of research. This full-color hardcover volume presented an overview of the history and present situation of the wild horses. After a few years in bookstores, it went out of print.

Horses remain on the islands, but their stories and prospects have changed. Genetic testing has revealed secrets. Management conflicts have evolved, encompassing political, economic, and cultural issues as well as purely scientific ones. On many fronts, the role of horses in the wild is being redefined.

Picking up where *Hoofprints* left off, *Wild Horse Dilemma* lays out facts garnered from reliable sources and from my own observations. Whereas *Hoofprints* was an overview written for a general readership, *Dilemma* is more substantial and more thoroughly referenced to

satisfy both the layman and the academic. I have tried to present all sides of the issues fairly so that readers might weigh information and develop their own opinions. I researched storms and shipwrecks, equine behavior and genetics, history, epidemiology, barrier-island dynamics, sea-level rise, beach development, and the perpetual clash of viewpoints. I studied hundreds of documents, from historical papers to scholarly journals to court transcripts, so that I might accurately present the pertinent issues.

I visited each of the herds numerous times, meeting site managers and wild-horse advocates and learning what has changed over the years. Many people took time out of their busy lives to help me understand the core issues and to acquaint me with each herd.

It seems that most people find beauty in horses, and, like me, thrill to the sight of a wild-born filly grazing on the dunes at sunset. Yet free-roaming horses are also considered pests, forced into inhospitable habitats, or eradicated.

As the earth's dominant species, we have the power to preserve or destroy the wildlife of the world and the ecosystems in which they live. The choices we make regarding wild horses are far-reaching. We alter their destiny whether we act or choose to do nothing.

At this writing, the United States is in a stubborn economic downturn. The federal government is trillions of dollars in debt. Struggling state and local governments and charitable organizations find it difficult to justify funding for horses that survived centuries unaided. Moreover, wild horses are an intellectually, emotionally, and politically volatile topic. Horses inspire passionate sentiment both for and against their continued residence in the wild, and emotion threatens to override reason in decision making.

Wild horse management is necessarily influenced by economics, finance, politics, nostalgia, the genetics of the horses themselves, and the personalities and agendas of people involved with them. We must work together to do right by these animals. To this end, we must keep these influences in perspective, anticipate consequences, and understand how unexamined attitudes affect our actions.

We must learn to respect wild creatures and avoid interactions that inevitably erode their wildness. The other day I talked with a co-worker, a sensitive soul who cares deeply about animals and people alike. She had recently visited Assateague and was excited to learn that I was writing a book about the wild horses that live on barrier islands. "Oh, Bonnie, did you go up and pet them? Did you bring them treats?" she exclaimed, envious of my imagined close contact with the horses. She seemed puzzled when I explained that I had been careful NOT to approach and certainly not to touch or feed the wild horses I studied. All my photography was done though telephoto lenses that allowed me to keep my distance. When horses approached closer than a few yards away, I retreated. My goal was to remain so peripheral to their lives, in time they nearly forgot I was nearby.

People who brave traffic, expense, inclement weather, and insatiable pests to see wild horses are frequently disappointed when the horses stick their heads in car windows or raid coolers. "Those horses aren't wild. I could walk right up to them!" they complain. In fact, the horses are probably as wild as their prehistoric forbears, but their behavior often crosses some murky divide in our expectations between wild and domesticated. Because countless people have stroked them, fed them, and lured them, some can be momentarily docile, occasionally indifferent, or routinely bold and pushy in the presence of people. As anyone bitten or trampled can attest, they are no less wild than horses that avoid human contact.

When we impose ourselves and our desires on their lives, when we habituate them to our presence, when we teach them to approach us for food and attention, we rob them of their wildness. When we treat them as we would their domestic counterparts, we miss the opportunity to observe them in a natural state, that is, to appreciate the things that make them irresistibly attractive. We miss the very point of driving past thousands of their tame kin to seek them out. We create something like a petting zoo hazardous to us and to them. If we truly love and respect wild creatures, we must learn to stand back and enjoy watching them from afar. Only then can they—and we—know the real meaning of wildness.

We can begin to deal wisely with wild horses by understanding the facts and discovering how the threads of their existence are woven into the tapestry of life. Only through understanding can we hope to make rational, educated decisions about the welfare of these unique, free-roaming animals.

Bonnie U. Gruenberg
Strasburg, Pennsylvania
November 30, 2014

Introduction
Jay F. Kirkpatrick, PhD
Senior Scientist, The Science and Conservation Center

Wild horses first came to the consciousness of the American public in a big way in December 1971, when the Wild and Free-Roaming Horses and Burros Act was signed into law by Richard Nixon. Since that time the issues of wild horses and their management have assumed massive dimensions and coalesced a whole spectrum of interested parties, including wild horse advocacy groups, federal agencies, conservation groups, animal welfare organizations, the livestock industry, various outdoor recreational groups, evolutionary biologists, geneticists, and public land advocates, among others. Because of the huge economic—taxpayer and otherwise—implications of the wild horse issue, even Congress managed to include wild horses in their collective consciousness. Although I can't confirm the accuracy of the claim, it has been stated that the single issue that has produced the second-largest numbers of letters to the U.S. Congress is wild horses (only the Vietnam War produced more).

The issues that have arisen are the direct result of the aforementioned well-intentioned but poorly conceived piece of legislation, and the magnitude of the public interest is directly proportional to the vast areas of Western public lands involved in the controversy (some 31 million acres in ten Western states). Ask anyone on the street about wild horses and they will immediately speak of the west, prairies, desert, mountains and sagebrush. Even if they have not followed the politics, they have seen movies like *Electric Cowboy*, and PBS documentaries like *Cloud, Wild Stallion of the Rockies*. They know about Western wild horses. But you will seldom hear anything about the thousand or so horses inhabiting the barrier islands of the East Coast. Most Americans are not even aware that wild horses roam these wind-swept masses of sand and marsh.

Despite the lack of knowledge about the island horses, there are parallels between them and the issues of Western wild horses. Legislation of one kind or another can usually be found at the root of the barrier island horses' very existence today. Some of it was wisely conceived and some not so wisely. Economics, as with most else in our world today, rules supreme in driving the issues, although it is often disguised as animal welfare or conservation issues. And, as it did with the Western horses, Congress has found its way into the morass.

Historically, the barrier island horses, which range from Assateague Island National Seashore in Maryland over 700 miles to the south on Cumberland Island National Seashore, transcend most of the Western horses, having arrived much earlier. Some populations approach 500 years in age. Biologically, these horses are no wimps lounging on sand and in surf, but suffer some of the harshest and most inhospitable environments to which a horse can be exposed. And while they don't have perhaps the same poignant cultural history as the Western horses, which gave rise to an entire—but tragically transient— culture, they have their own cultural stories, which reach back hundreds of years.

Another inevitable parallel between Western wild horses and barrier island horses is the political Cuisinart® that works its way around the edges of the issue of management. It matters not whether horses live on a barrier island or a desert—they must be managed in some manner. We continue to refer to Western horses as free-roaming, but they are not. They are confined by fences, and our barrier island horses are confined by sea and bay. Civilized behavior requires that we manage them in some responsible and preferably humane manner, and hopefully without compromising their wildness. To do otherwise would be the same as placing a given number of dogs in a large kennel, providing them with some finite supply of food and water and then walking away.

Aside from the thinly disguised economic issues surrounding barrier island horses, their management, like Western horses, has become an issue of contention. Some, ill-advised to be sure, would have no management and let nature take its course (as it would with the dogs in the kennel!). Others, ignorant of the economic and genetic costs and trauma to the horses, would prefer to see them periodically removed. These folks have learned nothing from the massive failure of Western wild horse management, its brutality and the unsustainable logistics and costs. Still others would manage population growth through fertility control. As *The Wild Horse Dilemma* makes clear, no consensus has emerged. What is most remarkable, however, is that no one agency or organization can seem to learn anything from the other barrier island horse experiences. Each more or less operates in a vacuum.

Unlike Western horses roaming their sagebrush flats, barrier island horses were probably never indigenous to barrier islands, although from a continental standpoint, they are a reintroduced native species (at least on the basis of the "best available science" if not strident opinion). Herein lies one of the most contentious issues surrounding the saltwater equids. But law often trumps science, or history, and most of the horses reside on their seaside habitats with the force of law behind them. And *The Wild Horse Dilemma*, in an extraordinary manner, reveals the various issues surrounding each population.

Perhaps an introduction should not become personal, but after 28 years living and working with barrier island horses, I can attest to their struggles, their unique biological aspects, their cultural impacts on local folks, and yes, their romance. There is something very special about standing in a bayside marsh at sunset, with the sounds of gulls and geese, watching wild horses wading cautiously through guts, and grazing in secret (not to them) meadows on the edges of loblolly pine forests. It is just short of awesome to watch them running across the sand flats of the interdune areas on their way somewhere important to them. It is even humorous to watch them eyeing humans warily from the edges of seemingly impenetrable greenbrier jungles and then disappear into them without a sound. And for those fortunate enough to actually study them, it is inspiring to know who is related to whom, and even who their great-grandparents—now gone—were. And last but not least, it is satisfying to know that I am looking at a continuous string of generations that reaches back to before the country was born.

Not everyone understands the poignancy of extinction of a species. Horses per se are not threatened with extinction, but a long-standing part of American history and culture, and some unique biology hang in the balance on our eastern shores. Most of the managing agencies—and primarily, but not exclusively the National Park Service—must look beyond horses and protect entire delicate ecosystems. But there is still a place on these islands for what can only be described as one of the world's most adaptable large mammalian species.

Bonnie Gruenberg has brought to life the amazing biology, history, culture and, yes, conflicts associated with these horses. Without rancor, she also sheds light on the human arrogance and misguided agendas that threaten some of these populations more than any hurricane could. Let us hope that some balance is found and that in another 500 years, the descendants of today's horses are still wandering about these wonderful islands.

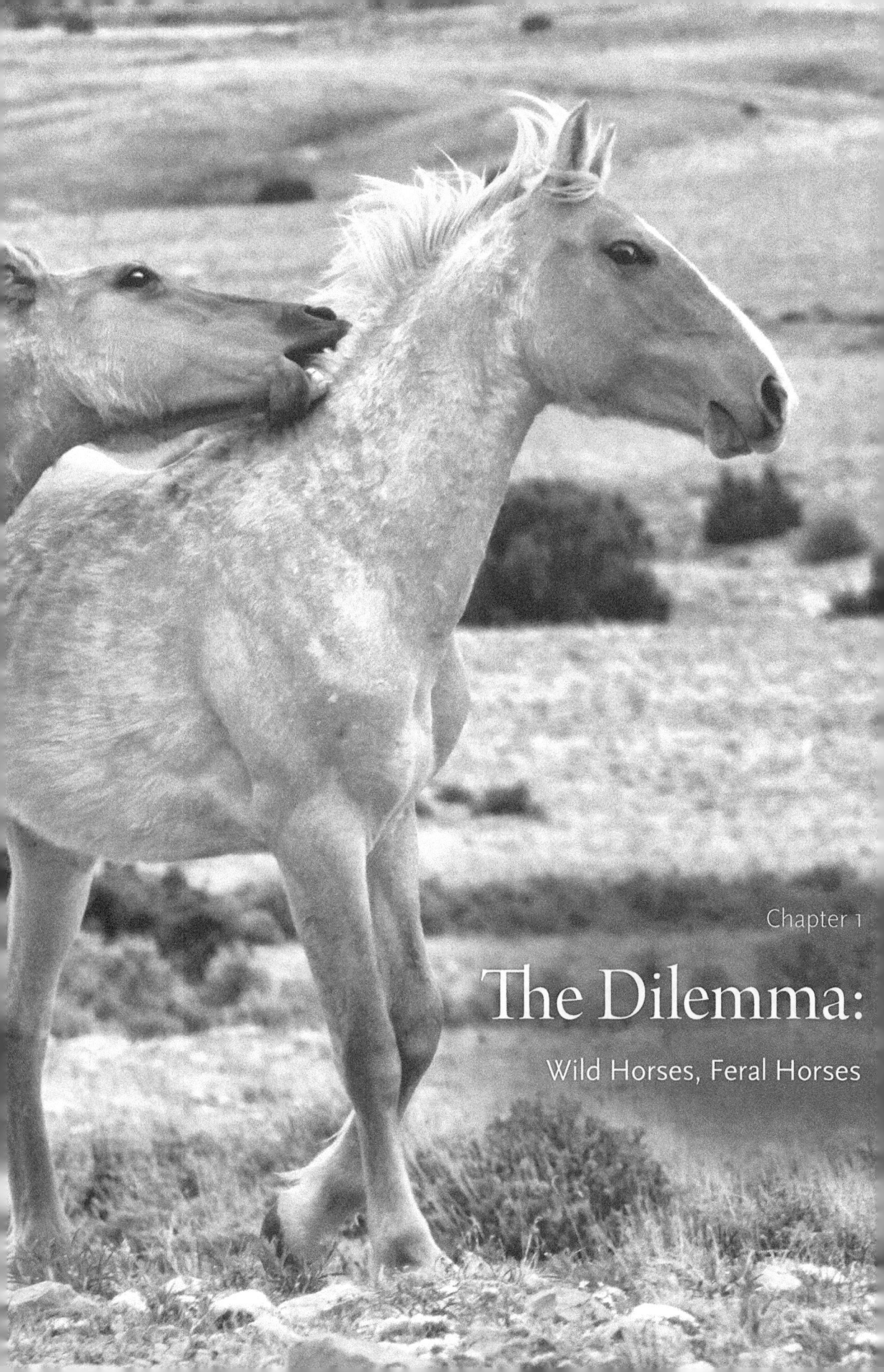

Chapter 1

The Dilemma:

Wild Horses, Feral Horses

Herds *of the* Atlantic Coast

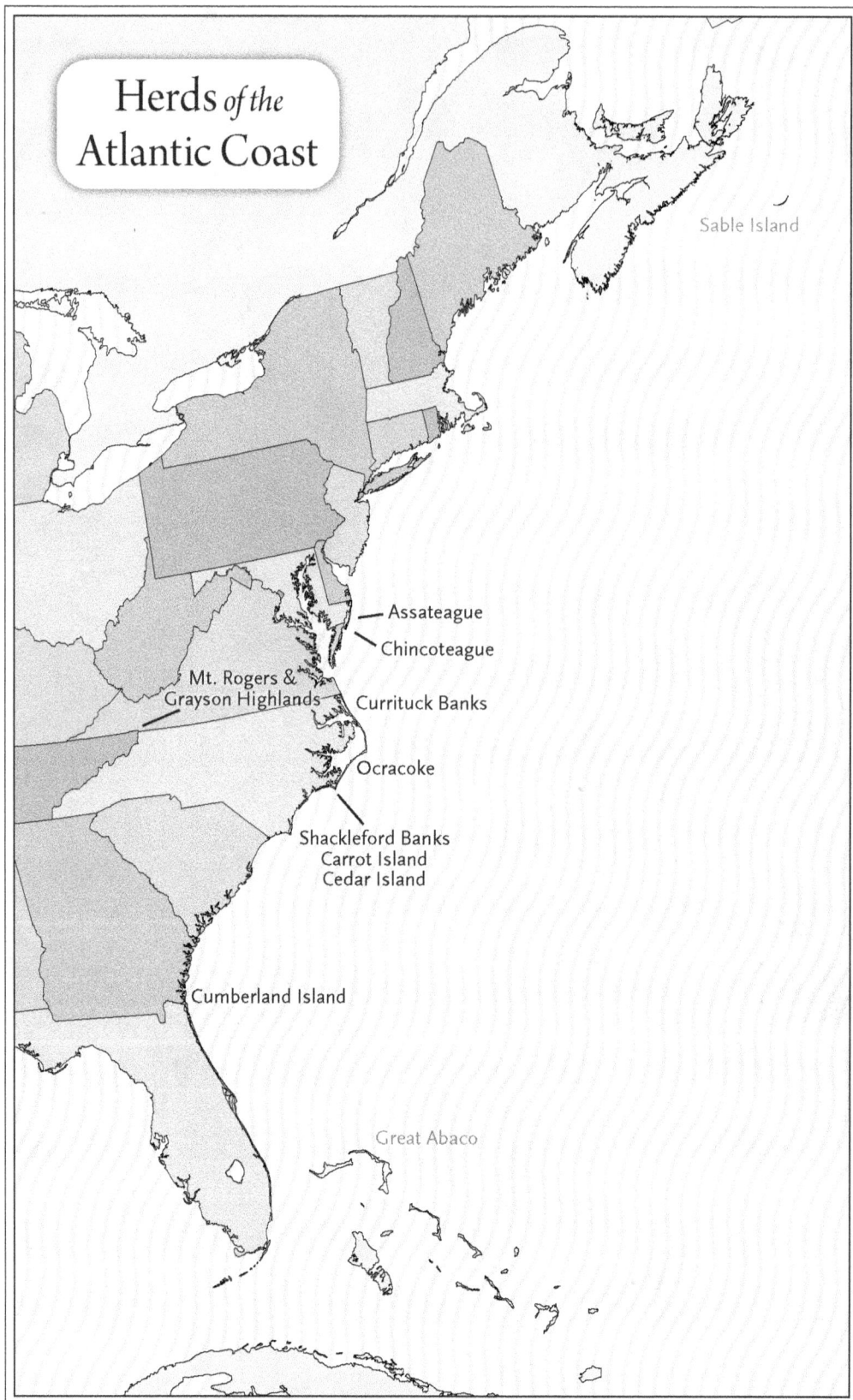

Sable Island

Assateague

Chincoteague

Mt. Rogers & Grayson Highlands — Currituck Banks

Ocracoke

Shackleford Banks
Carrot Island
Cedar Island

Cumberland Island

Great Abaco

Mystery surrounds the wild horse herds of the Atlantic coast. Their origins are enigmatic and raise many questions. Did they swim to the islands from the wreckage of ships, or were they placed there by early European colonists, seafarers, or pirates? Do their bloodlines represent the most rare and ancient of American breeds, or are they a random mixture of many genetic flavors? Do they descend from the proud steeds that helped conquistadors subjugate much of a hemisphere? How have they managed to survive in the wild for over 400 years with little assistance despite hurricanes, droughts, floods, epidemics, temperature extremes, disease, and the scourge of biting insects?

On closer examination, however, the true perplexity is found not in the wild horses themselves, but in the domesticated humans who assign meanings and labels to them. Because we cannot agree whether wild horses are wildlife, nuisances, historical symbols, or embodiments of abstractions such as beauty, our thinking is often confused, so our management plans are burdened with contradictory policies and incongruous goals. Boundaries blur between the plight of flesh-and-blood animals and the attitudes, agendas, and sentiments we project upon them.

Also enigmatic is the pull wild horses have on our souls. We resonate with horses at a level much deeper than rationality. Many people who have never been close to a living horse identify with its power, grace, and independence and appreciate its beauty in film and in art. When cats and dogs—also companion species—escape to the wild and roam free, they elicit primarily censure, repugnance, and even fear. Yet millions of people iconize the wild horse as a symbol of freedom. They are inspired and comforted by the simple knowledge that somewhere on the planet, horses run wild in outposts where humanity seldom treads.

On the ferry to Cumberland Island in 1998, two elderly bird-watchers equipped with wide-brimmed hats, binoculars, and Peterson Field Guides® expressed their hope of seeing a painted bunting. When the subject of the free-roaming Cumberland horses came up, the women stiffened slightly. This was a sore subject for them.

"The horses shouldn't be there at all!" asserted one birder, a retired pediatrician. "They aren't native wildlife, and I don't understand why the Park Service allows them to remain!"

"But they are pretty!" her companion chimed in helpfully. The pediatrician raised an eyebrow at her friend, and then conceded, "Yes, they are beautiful animals. But they have no business roaming on a national seashore."

A line is drawn in the sand, and adversaries separate. On one side are those who embrace the wild horses, and on the other stand those who would reduce their numbers or eliminate them altogether. Disagreement smolders between those who regard the feral horse as an exotic pest and those who regard it as a cultural and historical icon. Some truly believe that free-roaming horses are native wildlife and that putting them back on the North American landscape constitutes a restoration of the natural order, the righting of a wrong perpetrated

Wild horses share strong emotional bonds. They also stir strong emotions in us.

by prehistoric people who probably contributed to their extinction. The argument is not only biological, but also cultural, economic, and political (Linklater, Stafford, Minot, & Cameron, 2002).

"If you ask anyone vaguely familiar with horses about free-roaming American equine issues, be prepared to spend considerable time in debate," says Don Höglund, DVM, author of *Nobody's Horses.*

> There are plenty of valid viewpoints, several of which are quite emotional. The multitude of positions has created a deafening whirl in the halls of Congress as well as out on the western rangelands. Concerns for and about wild horses pit urbanites against ranchers, public servants, and horsemen. For instance, just referring to free-roaming horses as wildlife is a great way to get bogged down in arguments, even law suits, about definitions. (Höglund, 2010)

The biological status of free-roaming horses is also enigmatic. Because they are descendants of introduced livestock, many scientists consider them exotic. Others, such as Jay Kirkpatrick and Patricia Fazio (2010) make a scientifically valid case for accepting them as a reintroduced native species.

When large numbers regard wild horses as wildlife, lawmakers and others are inclined to grant them protection. When large numbers label them exotic, laws exclude them, and public opinion supports their removal. Nuances of terminology affect management policies and influence the destiny of these herds, so we must begin by considering how exactly we classify an animal as native wildlife.

Perhaps an animal is truly wild only if its ancestors were untouched by human intervention. In the modern world, few animals could meet this standard. Consider the dingoes of Australia. Dingoes may descend from dogs brought in by Aborigines or from semi-domesticated dogs introduced much later by Asian seafarers. Dingoes have held a well-defined ecological niche in Australia for thousands of years. Evolving in relative isolation from people and domestic dogs, they have developed unique features and instincts that set them apart from both ancestral gray wolves (*Canis lupus*) and domestic dogs (*C. lupus familiaris*) and earned them their own subspecies (*C. lupus dingo*). Should we consider the dingo wild or feral? Widespread hybridization with domestic dogs complicates the question.

Wildness can be influenced by circumstances of birth. A domestic kitten born to a stray may grow up as wild and wary as a bobcat. A bobcat bottle-fed in a zoo may become as gentle as a domestic kitten. Some wild species such as zebras and mink are difficult to tame even if socialized by humans for generations. Others, such as elephants and ruffed grouse, can be domesticated easily if caught young and raised by a knowledgeable handler. Wild chipmunks take seed from human hands, and wild ducks commingle with their domestic counterparts on farm ponds.

Human activities can shape the evolution of a wild species. Wild birds that have access to seed and suet provided by humans are more likely to forgo migration and winter in their breeding grounds. The Eurasian blackcap (*Sylvia atricapilla*), a small songbird, split into two distinct lines in response to the availability of birdfeeders. Each fall, one group migrates southwest to Spain, the traditional wintering ground. The other group flies northwest toward Britain, to spend the winter feasting on the bounty of birdfeeders. In the spring, both sets of birds return to their German breeding grounds. The British birds arrive first and select mates from other British migrants. The Spanish birds arrive later and find mates among themselves. Over time, the British winterers have diverged from the Spanish, developing rounder wings; longer, narrower beaks; and an imperative to return to Britain for the winter. The presence of bird feeders shifted migration habits and appears to have set the two populations of blackcaps on different evolutionary trajectories that may eventually result in specific divergence.

Migratory Canada geese (*Branta canadensis*) are a desirable native wild species, but where lush lawns and agriculture have created a winter food supply, many cease to migrate. They remain in their breeding grounds year-'round, sometimes moving into farms and feeding alongside domestic geese. They foul ponds, golf courses, and community pools and create health hazards with their droppings (Clark, 2003). Are geese that have chosen to live with people still wild?

We must also ponder bird species such as the brown-headed cowbird, barred owl, wood thrush, and Inca dove, which extended their ranges into new territory unbidden. The ranges of wild animals expanded and contracted long before humans appeared on the scene. And when people accidentally introduced house sparrows, hogs, zebra mussels, and Norway rats into new habitats, these species experienced such success that they pushed aside their native competitors. How long must an animal live in a new location before one can consider it indigenous?

Coyotes (*C. latrans*) have expanded their range because humans have killed off their biggest competitors, wolves. Now there are more coyotes than ever before, living in every state except Hawaii (Editors of *E Magazine*, 2002).

Some wild creatures are untamable, but still owe their very existence to the conservation efforts of humans. The Przewalski Horses of Mongolia are nearly impossible to train, even if gentled for generations. Humanity pressed them to the brink of extinction, but humanity also saved them. The current wild-living population has been reestablished from zoo stock that was selectively bred for generations. Are they wild?

Sometimes the introduction of a non-native animal actually benefits certain desirable wild species. Golden eagles (*Aquila chrysaetos*) came to the Northern Channel Islands of California to hunt feral hogs (*Sus scrofa*). When human hunting and trapping reduced the hog population, the eagles preyed more on the endangered island fox, *Urocyon littoralis* (Lowney, Schoenfeld, Haglan, & Witmer, 2005).

Zoologist Ronald Keiper is of the opinion that wildness is not about whether a species is unafraid of people or tolerant of human activity. To him, wildness is about self-determination. A wild horse is one who is free to make its own choices about where and when to find food and water, select its own mates, and encounter natural processes such as injuries and disease (Frydenborg, 2012).

Why *do* we deem horses more worthy of preservation than the feral hogs that share parts of their range and may have similar origins? Although some people are fond of free-roaming hogs, nobody has yet formed a feral-hog protection league, nor do hogs elicit the emotional firestorm sparked by free-roaming horses. Unlike wild horses, wild hogs are hunted for meat. Domestic swine provide substitutes for faulty human anatomy such as heart valves and insulin. But utility does not always equate with value. Songbirds are not particularly useful to us, yet we see fit to protect them.

Emotions cloud facts when horses are involved. People have admired and romanticized horses for thousands of years. The graceful, noble horse speaks to the emotions in a way that hogs and goats do not.

Wherever horses run wild in America, their history is cloaked in confusion, and their right to continued residency is disputed. Should we allow sentimentality or prejudices to influence our management plans? Should we be swayed instead by an incomplete or unexamined view of our own rationality?

> Unlike most other wild animals, free-roaming horses may not be hunted as game in the United States. Activists demand protection for horses and their habitat. Ranchers complain that they compete with livestock for forage and often consider them a nuisance. If we cannot hunt them, eat them, ride them, or in most cases even get close enough to watch them, some argue that free-roaming horses are not useful to humanity. Other people—some of them scientists—argue that the horses are culturally, biologically and historically significant wildlife and have a right to remain. If we accept them as wildlife, should we label them exotics that have no right to the habitat that they have claimed for themselves? Does our choice of terminology in any way alter the essence of what they are?

Officially, free-roaming horses in the United States are considered feral, exotic and non-native. *Merriam-Webster's Collegiate Dictionary* defines *feral* as "having escaped from domestication and become wild."

The U.S. National Park Service defines native species as "all species that have occurred or now occur as a result of natural processes," and defines exotic species as those that have arrived "as the result of deliberate or accidental human activities" (Zimmerman, Sturm, Ballou, & Traylor-Holzer, 2006, p. 47). The Park Service generally uses *exotic, non-native, alien,* and *invasive* synonymously. On the Cape Lookout National Seashore Web site (*Frequently Asked Questions—Horses,* 2011), the Park Service says of the horses,

To many, a wild horse is the embodiment of freedom and independence.

Feral means that the horse or its ancestors were once domesticated, which is the case with all free-roaming horses in North America. Wild implies that the horse lives its life in nature without being tame or domesticated. Either term can be used.

Exotic species are not considered natural components of the ecosystems in which they live because they did not evolve in concert with the species native to the place. Exotics frequently gain a tenacious foothold on the local biota; *The Vascular Flora of Pennsylvania* (Fowler & Klein, 1993) says that 37% of plant species in the state were not present when Europeans arrived. Decisions about which species to manage and how are complex, and sometimes arbitrary. While the management of exotic or feral species usually focuses on reducing their numbers or eliminating them entirely, some exotic animals, such as pheasants, are encouraged to proliferate.

Exotic is sometimes defined as not present when European explorers first set foot in North America. But what would Europeans have found had they sailed to the New World a few thousand years earlier? Native horses whose distant ancestors first appeared in North America more than 55 million years ago. Until the last Ice Age, modern horses, *Equus caballus*, roamed their continent of origin in great numbers.

The earliest equids were small, about 8–12 lb, and had multiple toes on each limb. Over the millennia, the lines that gave rise to the modern horse shifted from browsing forest-dwellers to grazers. They gained features that increased their success in the grasslands—long legs balanced on a single hoof for swifter flight, panoramic vision and increased height to spot predators, and a gut adapted to eating continuously on the move. (For a comprehensive discussion of horse evolution, see Appendix.)

Then suddenly, over a few thousand years starting about 13,000 years ago, horses vanished from North and South America. The last native North American horse died about 8,000 years ago (Franzen, 2010). For the first time in 55 million years, the lands where equids once numbered in the tens of millions no longer bore the hoofprints of wild horses.

Horses were not the only species to disappear during this period. North America lost 33 of 45 large mammal species, South America lost 46 of 58, Australia 15 of 16, Europe lost 7 of 23, and sub-Saharan Africa only 2 of 44. One hominid species or subspecies vanished in this time frame—*Homo floresiensis* ("Flores Man" or, more fancifully, "Hobbit") of Indonesia—evidently a few thousand years after *H. sapiens neanderthalensis* (Neanderthal Man) of Eurasia, and the recently discovered Denisovans. Researchers have long sought an underlying pattern that could explain why some species survived into the Holocene, and some did not. Why did grizzly and Kodiak bears persist, while short-faced bears and cave bears succumbed? The North American death toll included woolly mammoths, dire wolves, saber-toothed cats, lions, the stag-moose, several species of musk oxen and bison, the automobile-sized armadillo glyptodon, giant ground sloths, and camels, as well as several species of horses and the Yukon wild ass. The last mammoths survived on Wrangel Island, north of the Siberian coast, and St. Paul Island, in the Bering Sea, until around 4,000 years ago (Franzen, 2010).

The cheetah, abundant in North America, Europe, Asia and Africa through the Pleistocene, apparently suffered a nearly complete population collapse about this time. Because of this tight genetic bottleneck, cheetahs are so genetically similar that unrelated animals will accept skin grafts from one another. Whereas the coat patterns of tigers and leopards differ among individuals, all cheetahs have the same arrangement of spots.

There is no clear reason for the extinction of the ancient horses. Our Paleolithic ancestors hunted them heavily, apparently driving them into traps and pitfalls or over cliffs—although more recent research disputes this idea (Franzen, 2010). Some believe that hunting pressures brought them to the brink of extinction, and environmental causes, such as climatic changes, administered the death blow (Luis, Bastos-Silveira, Cothran, & Do Mar Oom, 2006; McGreevy, 2004). Others maintain that horses were declining in body size before humans arrived, and if humans had a role in horse extinction, it followed major ecological change that was already well under way (Guthrie, 2006). It is hard to conceive of an environmental shift that could bring about the destruction of the highly adaptable horse family, which had prospered though dramatic ecological upheavals over tens of millions of years, including a previous series of ice ages and interglacial periods. Why did horses vanish while other ungulates remained, including mighty herds of bison?

Until recently, the time of equine extinction was calculated by dating fossil bones and teeth. Fossilization, however, is a rare event. When most animals die, they simply rot, or their bodies are scavenged by other species. Recently, researchers have turned to recovering DNA from core samples of Alaskan permafrost. These molecules can reveal traces of animals that lived there but did not leave bones or teeth behind. James Haile's research team showed that woolly mammoths and horses survived in interior Alaska until at least 10,500 to 7,600 years ago, thousands of years later than previously demonstrated (Haile et al., 2009). This study indicated that mammoths and horses were contemporaneous with humans for several thousand years, proving that the incursion of people did not cause the immediate extinction of these species. Horses may have persisted longer in warmer climates, but in the absence of permafrost, no DNA remains in the soil.

Despite these findings, it seems likely that mankind played some role in their disappearance. Horses were abundant in Europe, but by the close of the last Ice Age, human overpredation had greatly reduced their numbers (Kelekna, 2009). By the Neolithic (8000–4500 BCE) the vast herds had vanished, leaving only small populations in central Europe and Iberia, while they remained abundant on the grassy steppes of Russia and the Ukraine into Central Asia.

Over the past 2,000 years, more than 100 mammal species have become extinct, including the Tasmanian wolf and the Caspian tiger, along with countless species of bird such as the passenger pigeon, the Carolina parakeet, and the dodo. On the islands of Oceania, more than 2,000 bird species have become extinct in the 3,000 years since the arrival of human beings—more than 20% of the world total (Donlan et al., 2006). Other species cling to the edge of the eternal abyss—mountain gorillas, Cuban crocodiles, Sumatran orangutans, European sturgeons, black rhinoceroses, Siberian tigers, Ethiopian wolves, Saiga antelope, Rothschild's giraffes . . . the list goes on. The rate of extinction has accelerated, largely because of an explosion of modern human activity (Traill, Brook, Frankham, & Bradshaw, 2010).

The quagga was a South African subspecies of the plains zebra that survived into the 1800s. (The name is onomatopoeic, derived from a Khoikhoi word for "zebra" that is a phonetic representation of the plains zebra's barking call—and presumably the quagga's.)

Quaggas were striped like zebras in the front, but with brown upper parts to the rear, and appeared to be half horse, half zebra. Adults were about 13.2 hands at the withers (54 in./1.37 m) and built like rugged ponies. They were docile and easily domesticated. Sir John Barrow (1801, p. 93) commented, "It is . . . well shaped, strong limbed, not in the least vicious, but, on the contrary, is soon rendered by domestication mild and tractable: yet, abundant as they are in the country, few have given themselves the trouble of turning them to any kind of use." William Burchell described a quagga foal that curiously followed a horse into town and allowed itself to be freely handled (Renshaw, 1904). Sadly, colonial stockmen saw quaggas as competitors for grazing resources and hunted them to extinction.

Renshaw (1904, pp. 179–180) wrote,

> The extermination of the quagga is all the more to be regretted because it seems possible that it might have been systematically broken in to bit and harness like a horse . . . The first example which [Swedish naturalist Anders] Sparrman noticed was a very tame specimen, pleased when anyone stroked its sleek sides and delighting to be caressed. On a subsequent occasion the doctor actually saw a quagga driven in the street harnessed with five horses: and with a rare foresight far in advance of his age he urged the domestication of the species, pointing out that in his day it could be even more easily obtained than the horse, that it would of course eat the harsh grass of the country—its natural food—and that it would probably be immune from the horse-sickness. . . . The relations of white men with the true quagga during the long years that have elapsed between 1772 and 1879 are summed up in one word—extermination!

He added (1904, p. 195),

Immense enthusiasm marked the commencement of ostrich-farming at the Cape in 1865–70: had similar zeal been exerted, even at that late hour, to save the true quagga, it might have been profitably domesticated, and surely a Colony which has saved the ostrich for the sake of its feathers—mere ornaments at best—might have bestirred herself to preserve the quagga for far more practical purposes.

The last wild quaggas were probably caught in 1870, but a small population may have survived until 1878. The last known quagga died in captivity in the Amsterdam Zoo in 1883. She is mounted and on display in the Amsterdam Zoological Museum.

Until the 1980s, most scientists believed the quagga more closely related to the horse than to the zebra. But Higuchi et al. (1987) analyzed mitochondrial DNA from a museum specimen and found it virtually identical to that of the plains zebra. A subsequent immunological study corroborated the results. Still, many scientists were skeptical and questioned the validity of the methods used.

Roughly two decades later, Orlando et al. (2009) used modern equipment to examine DNA samples from 13 quagga specimens as well as that from the Pleistocene "giant" Cape zebra, previously *E. capensis*. (The latter may have weighed 400 kg/882 lb and stood nearly 5 ft /1.5 m/15 hands high at the withers, about the size of an average riding horse, and was native to the same general area as the quagga.) Orlando's team concluded that all specimens of both species represented variations of the extant plains zebras. The quagga was a subspecies descended from a population that had been isolated during the Pleistocene about 120,000–290,000 years ago (Leonard et al., 2005). Apparently the plains zebra is morphologically plastic, giving rise to strains so physically dissimilar that scientists long believed that they were different species. Following this analysis, academics concluded that the plains zebra should no longer bear the name *E. burchelli* and renamed the species *E. quagga*. The quagga, formerly *E. quagga*, became a subspecies, *E. q. quagga*.

The confirmation of this lineage is good news for the Quagga Project, a venture founded by Dr. Reinhold Rau of Iziko Museums of Cape Town to reconstruct the quagga (Harley, Knight, Lardner, Wooding, & Gregor, 2009). The Quagga Project selectively bred plains zebras with quaggalike characteristics from nine foundation animals, captured in Namibia's Etosha National Park in 1987, with other wild-caught zebras introduced to increase genetic variability. By 2005, many foals, bearing English names such as Henry, Ziggi, Mike, Albert, Duncan, George, and Sebastian, showed minimal striping and darker coloration on their hindparts. By 2008 there were 89 "quaggas" in 12 locations. Because the quagga arose from the plains zebra, the original genes that produced the quagga are still obtainable. Critics say that these animals cannot represent the extinct subspecies, since they probably differed from the plains zebra in more characteristics than coloring—such as clearly more tractable dispositions. But even if the modern quagga is only superficially similar to its extinct ancestor, one can appreciate the effort to right the wrong that Europeans inflicted on the species. Yet the question remains—if repatriated to their ancestral homeland, should we consider these quagga wild or feral? Reintroduced or re-engineered?

Besides the quagga, in the last few centuries humankind has exterminated other equids— the prolific, successful Tarpan horse, the Persian onager (the "wild ass" mentioned in the Bible)—and very nearly killed off the Przewalski as well. It is fortunate the horse achieved such success as a domesticated species, or they might have been forever lost to us.

With keen senses and long legs engineered for speed, the horse is well equipped to recognize predators at great distances and take flight at a hint of danger.

E. caballus, our native horse, evolved in North America, but became extinct here as recently as 8,000–10,000 years ago, to be reintroduced by Spanish conquistadors in the 1500s. Their Iberian horses were bred for war and conquest and were renowned for intelligence, endurance, and tractability.

Horses have lived on the Iberian Peninsula since the Pleistocene and have contributed to the development of many modern European and New World breeds (Luis, Oom, & Cothran, 2007). Around 1100 BCE mariners from the eastern Mediterranean ventured west and north as far as the British Isles. On the Iberian Peninsula their fine-boned horses crossed with rugged, deep-chested, short-legged native breeds, producing robust offspring with great stamina and heretofore unseen colors and markings (Bennett, 2008b; Bennett & Hoffman, 1999). Some were gaited; that is, they covered ground quickly in a smooth, fluid singlefoot that horse and rider could maintain for great distances.

Over the following centuries, traders and conquerors of many ethnicities—Celtic, Carthaginian, Roman, Germanic—took Iberian crossbreeds all over Europe and North Africa. In the latter region, mating Iberian imports with their distant desert kin produced a light, maneuverable new breed, the Barbary or Barb, especially suitable for close combat with swords (Bennett, 2008b).

In the 7th century, armies mobilized by a new religion, Islam, poured out of the Arabian Peninsula. Like earlier Byzantine tacticians, they relied heavily on cavalry and emphasized mobility and use of projectile weapons, e.g., bows and lances (Bennett, 2008b; Leonhard, 2004), often against numerically superior forces. Consequently, they put a premium on speed. As their sphere expanded, another new breed of horse arose: the Arabian.

African, Libyan, and Numidian horses were swift, tough, intelligent, and agile, with the long, arched neck, dished face, short back, and high tail carriage seen in the modern Arabian horse (Kelekna, 2009). Breeders guarded the Arabian's bloodlines strictly, crossing purebred mares with purebred stallions to the point of consanguinity. A chance mating with an outside horse would disqualify the mare from future breeding, because of fear that any future foals would carry impure blood (Kelekna, 2009). There was minimal forage in the desert, but warriors and nomadic pastoralists somehow sustained their horses on a diet of camel milk and dates (Kelekna, 2009).

Bedouin warriors treasured their beautiful war horses. They preferred to ride mares into battle, did their best to protect them, and rushed to recapture them if they were unhorsed during the melee. Sometimes the warriors sewed their mares' vulvas closed before the conflict to avoid accidental mating with an inferior stallion (Kelekna, 2009).

Islamic invaders from North Africa overran the Iberian Peninsula beginning in 711 CE. During a presence lasting nearly 800 years, they crossed swift desert Barbs and probably Arabians with local horses, with an infusion of Norse Dun, giving rise to yet another type, the agile, durable Jennet (or Genet or Jennetta) of Andalusia and Seville.

Bred for medieval combat, the Jennet stood only about 13–14 hands (52–56 in./1.32–1.42 m) in height until the 14th century, when there was a concerted effort to breed larger horses (Mills & McDonnell, 2005). Jennets were maneuverable in battle, yet enviably handsome and practical for farm work and general use (Ryden, 2005). They fared well on poor forage, could tolerate extremes of weather, were smooth-gaited, and could carry heavy riders long distances without fatigue.

Jacques de Solleysel's 1664 book *Le Parfait Maréchal* (first published in English in 1696 as *The Compleat Horseman*) was for more than 100 years the authoritative work on horses. Quoted by Thornton Chard (1940, pp. 90–91), it describes the Spanish horse as follows:

> I have seen the Spanish horse and have owned some of them; they are extremely beautiful and the breediest of all, being portraits by a careful brush or fit for the mount of a king when he wishes to show the people his glory and majesty; though they are not as slender as the Barbs, nor as thick as the Neapolitans, still they have the perfection of the two. The Genet has a superb and bold walk, a lofty trot, an admirable gallop and a very fast run; in general they are not very large, nor excessively broad; if they are well selected it would be hard to find any horse more noble than they. I have heard accounts of their remarkable courage; they have been seen with their entrails hanging outside of their belly and while all stabbed and wounded, losing all their blood, yet they bring back the rider safe and sound, with the same ardour and spirit with which they carried him out, and finally, having less life than courage, collapse. The best breeds are in Andalusia; and especially the breed which the Spanish King has in Cordova is the best; that of Cardonne is very excellent; also the Molina, as well.

In 1492, Spain had good horses whose ancestors had been foaled in northern Africa and Arabia (Denhardt, 1951). By the time of Columbus's voyages, the Spanish equine population had been depleted by almost 800 years of warfare, and it was difficult for the country to part with large numbers of superior horses to facilitate exploration. King Ferdinand II was so determined to reverse the chronic shortage of horses that in 1494 he made riding a mule a capital crime. His apparent reasoning was that if he forced the people to choose between walking and riding horses, they would find a way to breed or acquire more horses.

By the 1400s, the Spanish Jennet had three distinct body types. In the north and west regions, the "Celtic" type was more strongly influenced by Tarpan and draft bloodlines. In the south and east, the "Spanish" type evidenced major contributions by the African Barb. In the central area, Spaniards preferred a cross between the two types. "There were three separate "castes" in Spanish horses, likely from Roman times onward" says D. Phillip Sponenberg, DVM, PhD, professor of pathology and genetics at Virginia-Maryland Regional College of Veterinary Medicine. "These were *jaca serrana*, the small pony/draft type from the north and east, *española* from the center and west (and often gaited, likely the source of the Jennet), and *fina* from the south, ancestral to the modern Andalusian" (personal communication, June 5, 2014). Because most of the early Spanish voyages to America originated in Seville, Cadiz, and other southern ports, most horses shipped across the Atlantic were probably of the "Spanish" type (Mirol, Peral Garcia, Vega-Pla, & Dulout, 2002). It is likely, however, that other types were also sent.

Columbus took 15 stallions and 10 broodmares to the Caribbean on his second voyage in 1493. These first New World horses were short-backed, compact, and sure-footed, with sloping pasterns and comfortable gaits (Denhardt, 1951). In 1508, Spain authorized the transport of 40 small, resilient horses of the Celtic type from Castile to mount the expedition to Panama organized by Alonso Ojeda and Diego de Nicuesa. Records indicate that Columbus and others took Marismeño mares from Guadalquivir River salt marshes to the New World (Luis et al., 2006). Spanish horses exported to the New World traveled from the ports of Spain to their final destinations, with a stop in the Canary Islands or the Antilles (Mirol et al., 2002).

At first it was unclear whether exploration of the Americas would yield riches, so Spain sent lower-quality animals on the earliest voyages. Columbus complained that the high-quality horses shown to him before his second expedition had been exchanged for cheaper animals. Some later explorers, however, chose to sell their finest horses in Spain and replace them with less expensive animals, pocketing the difference in price (Sponenberg, 1992).

The Spanish vessels that transported these horses were usually low-waisted and well-armed with cannon (Denhardt, 1951). The horses stood in narrow compartments with their heads tied short, a board in front and a board behind, unable to lie down for the duration of the 2–3-month voyage. To minimize colic and ration supplies, the horses were fed as little as would sustain them.

Because the stall was too narrow to let a horse lie down, sailors could fasten a sling under his body like a hammock when the sea was fair, so that he could use the support to rest his feet. A 19th-century British cavalry officer gave the following advice on using the sling:

> A horse should be slung in calm or moderate weather only, and allowed to stand on his legs in a rough and stormy sea; and for this reason: that in smooth weather he will rest his legs and feet by throwing his whole weight into the hammock ... whereas to sling a horse in rough weather ... would only have the effect of knocking him about, from head to tail, according as the ship rolls to and fro, and would, moreover, cause him to be severely chafed either by the friction of the side bails, or by the buttock-board, or breast-beam. ...

[N]o attempt should be made to raise him off his feet, which would only make him uneasy, and probably terrify a nervous horse into the bargain. When the horse has been standing on his legs for some days previously . . . he will . . . soon learn to throw his weight upon the horse hammock; indeed, with some horses, when they once find the support and relief which it affords them, it is necessary to use great quickness in making the ropes fast before they throw their whole weight on the canvass, relaxing all their muscles, and laying for some minutes as if dead. (Shirley, 1854, pp. 24–25)

The intrepid mounts destined for the New World stood a 50-50 chance of surviving the 3,000-mi/4,800-km voyage. Popular history holds that in the regions around 30° north and south of the equator, ships were often immobilized when the winds ceased to blow for weeks at a time. As the equine death toll mounted, horse carcasses or live horses were pushed overboard. Thus these regions came to be known as the Horse Latitudes. There is no evidence to support the legends of mariners on becalmed ships throwing live horses into the ocean.

By the early 1500s Spain had established breeding ranches on Puerto Rico, Jamaica, Hispaniola, Cuba, and the mainland to provide mounts for expeditions and other purposes. At least 250 Spanish horses found new homes in the New World between 1493 and 1540 (Mirol et al., 2002).

In addition to the Jennet foundation stock, breeders imported select Barbs from North Africa to meet the demand for horses and add more flavor to the genetic soup. New World horse breeders took pride in producing superior stock, and many made their fortunes selling horses to those who sought to settle or exploit the New World. The islands were ideally suited to ranching, with no large predators, abundant forage, and a mild climate ideal for not only horses, but also for cattle and other livestock.

In 1510, King Ferdinand encouraged colonists bound for Puerto Rico to take livestock with them and authorized Pedro Moreno and Geronimo de Bruselas to take mares to the new settlement to serve as breeding stock. Later that year, merchant ships began to carry mares and other animals from Spain to Puerto Rico. In 1511 the king permitted certain individuals on Hispaniola to transfer livestock without restriction to privately owned horse farms and government-owned estancias in Puerto Rico, Jamaica, and Cuba (Johnson, 1943). These four islands supplied horses and other livestock to Central America, Mexico, and Peru and later, illicitly, to the English colonies in North America (McKnight, 1959).

In 1519, Hernando Cortés disembarked in Mexico with 600 men, intending to attack the native Aztecs at Tenochtitlán (Mexico City), then one of the largest cities in the world, with a population of more than 200,000. The contingent brought 17 horses of chestnut, bay, pinto, sorrel, and roan, thus unintentionally returning the horse to the land of its ancestors. The horses were clearly important to Bernal Díaz del Castillo, a conquistador who wrote an eyewitness account of the conquest of Mexico. In his journal he carefully described each horse, for example, "a grey mare, a very good charger which Cortes bought . . . with his gold buttons," adding that "more horses were not taken, for there were none to be bought" (1632/1908, pp. 86–87). Castillo's attention to minute particulars is understandable. War horses were not only essential to conquistadors' survival and victory, but also huge investments. Some may have cost more than $500,000 in modern currency (Bank of England, n.d.; Chard, 1940; *Purchasing Power*, 2013; Turner, n.d.).

The native Aztecs of Mexico had never seen a horse before and were initially terrified, believing this apparition to be supernatural. They were even more amazed when they watched the creature divide into man and beast, then rejoin to form a single entity, unharmed by the separation. Was this creature a magical deer, or a large medicine dog that only man-gods could ride? Cortés was quick to use their awe and wonder to his advantage.

According to Spanish accounts, Montezuma II, the ruler of the Aztecs, proclaimed Cortés the prophesied remanifestation of their god Quetzalcoatl and presented him with gold and silver gifts. But Cortés wanted wealth, not worship, conquest, not coexistence. Other tribes, historic enemies of the mighty Aztecs, joined forces with Cortés, and the Aztecs fell. Cortés looted the Aztecs' gold and gemstones, and his success inspired further conquests.

Similarly, Pizarro devastated the Incas at Cajamarca, where 168 Spaniards slaughtered an army that outnumbered them 500:1. The Incas had been fighting a civil war that left them depleted and divided, and Pizarro exploited their vulnerability. Mail-clad Spaniards, many of them mounted, sliced through the Indians' quilted armor with steel blades and rapidly slaughtered the majority of them. Not a single Spaniard was lost. Following Cajamarca, Pizarro bested the Indians at Jauja, Vilcashuaman, Vilcaconga, and Cuzco, pitting a few dozen Spanish against thousands or tens of thousands of Indians (Diamond, 2009).

Native tribes soon realized that the Spaniards were not magical and they could kill either a horse or a rider. Nonetheless, horses gave Europeans a remarkable advantage in battle. Mounted warriors could slash at foot soldiers from the protection of a high, mobile platform. They could cover great distances quickly, attack, and retreat out of harm's way before native sentries could warn their troops. A horse could press through crowds or dense forest, climb mountains, and ford rivers. By 1536, the Incas were ambushing horsemen in narrow passes, but had no defense in open country (Diamond, 2009). The Europeans also brought diseases with them—smallpox, measles, influenza, typhus, bubonic plague, and others to which the native people had no immunity. These diseases spread from tribe to tribe, sometimes in advance of the European explorers, and ultimately killed about 95% of the pre-Columbian Native American population (Diamond, 2009).

Horses were lost from early expeditions and escaped to the wild to multiply. By the 17th century, Indian tribes in present-day Texas and New Mexico—notably Comanche and Apache—began to acquire horses. Initially, like their prehistoric ancestors, they probably used horses as a food source. Francis Haines (1938, p. 429) wrote,

> The initial obstacle to be overcome in converting the Indian to the use of the horse was his ignorance in the care and use of the strange animal. This was overcome by the constant contact between the Indians and the horse-using Spaniard, rather than by the chance acquisition of a stray animal by some tribe. To such a tribe the stray would have suggested a dinner rather than a servant.

Any horses straying from the earliest Spanish expeditions were probably male, and therefore unable to form a breeding herd in the wild. The conquistadors preferred to ride stallions and geldings; only five of the 17 horses that Cortés took to Mexico were female, and on the Coronado expedition, two of 558 horses were mares (Haines, 1938).

In the early 1600s, it appears the Spanish traded mares and breeding stock to northern Rio Grande and Plains tribes. In 1680, the successful revolt of the Pueblo Peoples caused Spanish invaders to withdraw from the region, leaving thousands of their horses behind.

Native Americans quickly and deeply integrated the horse into their cultures. They used it for transportation and warfare. They honored the horse in art and song. Tribes traded and stole horses from one another, and horses escaped to the wild from the loosely tended unfenced herds.

Cortés repatriated the horse to its native North American grazing grounds, the place where horses and their pre-equine ancestors thrived for over 55 million years. The earliest equid fossils in North America were found in the Bighorn Basin of Wyoming and Montana—the very same place where the Pryor Mountain Mustangs now range. Reintroduced horses slipped into an ecologic niche where they thrived and multiplied without displacing wild species that shared their habitat. How is it, then, that many do not consider them a native species?

Donlan et al. (2006) wrote that it is cultural convention that decides which creatures are regarded as native and which are foreign, "usually irrespective of ecological and historical insights" (p. 664). Ecologist Daniel Janzen described feral horses as "a Spanish gift from the Pleistocene—invented here, then extinguished here by people, but surviving in the Old World 'refuge' and brought back here by people" (Barlow, 2000, pp. 32–33).

The fossil record and mitochondrial DNA analysis confirm that *E. caballus* is native to North America (Kirkpatrick & Fazio, 2010). Although domestication and other forces created a wide range of appearances on other continents, the horses that returned to North America had genomes identical to those last in residence. "The fact that horses were domesticated before they were reintroduced matters little from a biological viewpoint," write Kirkpatrick and Fazio (2010). "They are the same species that originated here, and whether or not they were domesticated is quite irrelevant."

E. Gus Cothran, PhD, a prominent equine geneticist at Texas A&M University, disagrees. "Yes, the species that evolved here is that of the domestic horse but there have been a great many changes genetically due to domestication, he said. "It is not the same animal as what left the continent ages ago, and it is not genetically identical" (personal communication, September 9, 2014).

During the Late Pleistocene, caballoid horses were genetically diverse as they spread from Alaska into Eurasia and Africa. Most of these ancient lines were lost as native horses became extinct and the horse relied on domestic breeding to survive as a species (Cieslak et al., 2010). In the early Holocene and during the Copper Age, Iberian and Eurasian steppe horses became distinct.

Horses were the last of the five main livestock mammals (after goats, sheep, cattle, and swine) to be domesticated, and they have undergone less manipulation than these other species (Proops & McComb, 2010). Recent evidence suggests that horses may have been domesticated much earlier than was once believed. In 2009, a team led by Alan Outram of the University of Exeter discovered artifacts indicating that the Botai Culture of the steppes east of the Ural Mountains in northern Kazakhstan was the first to domesticate horses, circa 5,500 years ago (Outram et al., 2009). Bones found at the site had more in common with Bronze Age domestic horses from many centuries later than with indigenous wild horses, indicating that they had been selectively bred for some time. Most were large enough to ride—13–14 hands (52–56 in./1.32–1.42 m), larger than the average Roman cavalry mount (Kelekna, 2009). Horses were also apparently used in ritual sacrifices. The Botai were unique among contemporaneous ancient peoples in that they had

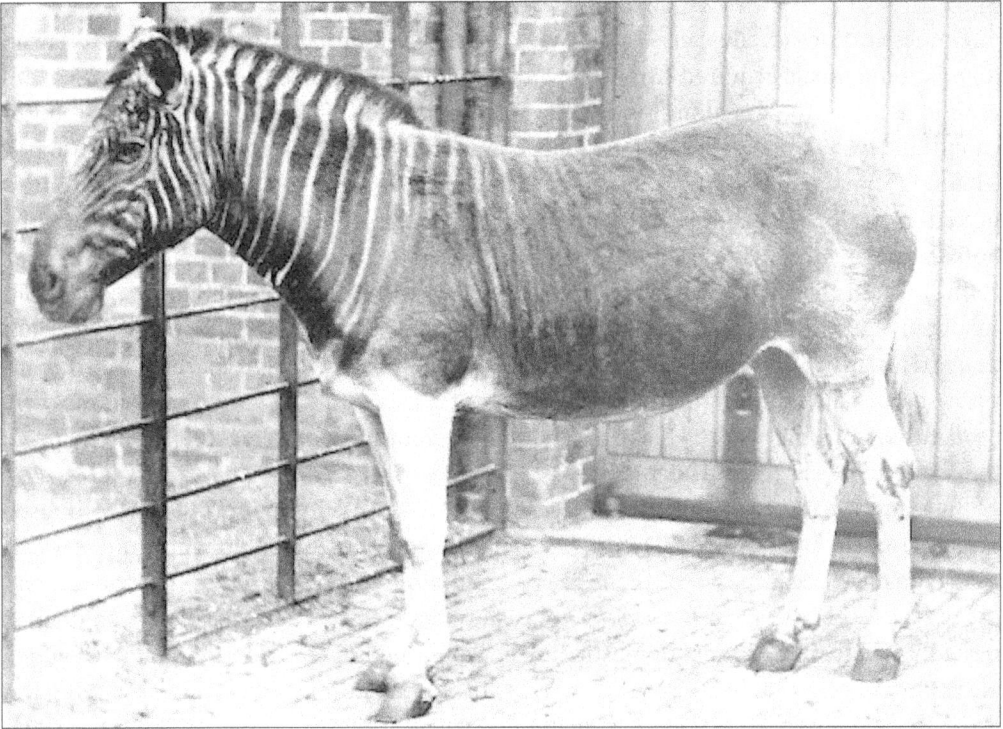

The extinct Quagga (*Equus quagga quagga*) was probably a subspecies of plains zebra. This mare, photographed in the Regent's Park Zoo in London by F. York in 1870, was one of the last of her kind and the only live quagga ever photographed. Wikimedia, public domain image.

tools specifically designed to manufacture smooth rawhide thongs, which are important in horse cultures that use reins, whips, hobbles, and ropes (Kelekna, 2009). Large quantities of manure found at the site indicate captive horses and may have been used to insulate homes and fuel fires.

Many of the horse remains found at the site showed abrasion of the teeth consistent with bits fashioned from leather, bone, and hemp or horsehair rope. Researchers found that a horse would have had to be ridden for more than 150 hours for such wear to occur (Kelekna, 2009).

The Botai evidently used horses for meat and milk as well. When researchers analyzed a fatty residue on Botai pottery, they found evidence of lipids from mare's milk. In Kazakhstan, horse milk is a traditional beverage, and it is often fermented into koumiss, an effervescent, alcoholic kefir-like drink. Although it was known that koumiss was an ancient beverage, this study shows the practice dates back to the very earliest horse herders.

Warmuth et. al. (2012) subsequently found corroborating evidence in the genome of the domestic horse. They concluded that horses were initially domesticated on the steppes of the Ukraine, southwest Russia, and west Kazakhstan—precisely where the earliest archeological evidence of horse husbandry was found. A second center of equine domestication may have arisen on the Iberian plains, either through an independent episode or through wild Iberian mares used to augment the bloodlines of domesticated horses brought in from eastern Europe (Lira et al., 2010). After the first domestication event, it appears that domestic horses spread quickly to other cultures, probably because a horseman could more easily make long journeys

Domestication usually causes a genetic bottleneck, limiting alleles to those present in foundation animals acquired from the wild. Uniquely, the horse has a limited amount of genetic material to pass along from male to male via the Y-chromosome, but the mitochondrial DNA inherited though the female is uncommonly diverse (Cieslak et al., 2010). Although roughly 30% of the original maternal lineages present in the Bronze age domesticated horses have been lost through 5,500 years of breeding (Cieslak et al., 2010), the horse remains more genetically diverse than most other domestic animals. Jansen et al. (2002) found that living domestic horses descend from at least 77 different mares, but Lindgren et al. (2004) suggests a smaller number of ancestral stallions. Apparently ancient domestic herds were periodically augmented with local wild-caught mares from the Eurasian steppes and the Iberian Peninsula (Lira et al., 2010), but very few stallions were brought in from the wild to make a genetic contribution—or if more were introduced, they have no modern descendants.

Building on earlier research, Bennett (2008b; Bennett & Hoffman, 1999) has proposed seven variously named subspecies of E. caballus based on geographic range and physical characteristics. Because two of them disappeared around the end of the Pleistocene, and the Przewalski Horse (E. c. przewalskii) of eastern and central Asia has never been domesticated, the ancestry of most domestic horses would be limited to four lines, all of which are extinct in the wild: the Tarpan (E. c. ferus) of eastern Europe and central Asia; the desert-adapted Afro-Turkic horse (E. c. pumpelli), which gave rise to the Arabian, Barb, and other "hot-blooded" or "Oriental" breeds; the Central European or Mosbach Horse (E. c. mosbachensis), ancestor of warm-blooded breeds, including Friesians; and the West European horse (E. c. caballus), progenitor of the pony and draft breeds.

Interestingly, DNA studies have also revealed the coat colors of certain ancient horses. To pin down when horses were first domesticated, Ludwig et al. (2010) examined the genes for coat color in ancient DNA to determine when they first diverged from the uniform coloration of the original wild equids. They discovered that the Pleistocene horses of Siberia and Europe were bay-based, and Iberian horses from the early Holocene were bay-based or black-based. Modifying genes often alter coat color, and Sponenberg theorizes that the actual color of these horses was dun or grulla (personal communication, June 5, 2014). Starting about 3000 BCE, a rapid, significant increase in the number of color variants occurred in the horses of Siberia and Eastern Europe. The first evidence for chestnut coat color was found in the remains of a Siberian horse dating to 3000 BCE, and through the Bronze Age chestnut horses became common. The first tobiano spotting was identified in a Chinese horse dating from 1200–800 BCE. Overo pinto apparently is a more recent color; it does not appear in the ancient samples.

Then Melanie Pruvost and a team of researchers (Pruvost et al., 2011) analyzed DNA of 31 prehistoric horses from Siberia, eastern and western Europe, and the Iberian Peninsula that lived from about 20,000 to 2,200 years ago. Eighteen horses were bay-based, seven were black-based, and six had the "Lp" gene for white with dark spots, like today's leopard Appaloosa. All these color variants appear in ancient cave murals, suggesting that prehistoric artists painted from life. Sponenberg points out, however, that because two genes work together to produce Appaloosa spotting, some solid-colored horses also have the Lp gene, and therefore researchers cannot conclusively identify the coat color of horses with the gene (personal communication, June 5, 2014).

The wild horses of Shackleford Banks, N.C., can trace their bloodlines back to the Spanish Jennet of the 15th century, although just how they inherited this legacy remains enigmatic.

A white horse with leopard spots would have blended in well with irregular patches of snow on an Ice Age landscape. When the climate warmed, dun coats would have provided better camouflage, and spotted horses may have grown scarce in the wild. Modern Appaloosas are genetically predisposed to uveitis, which can lead to blindness; if this association was present in ancient spotted horses, it may have contributed to their decline in the wild.

Research intended to simulate the process of domestication using hundreds of generations of foxes and rats demonstrated that tractable temperaments and unusual coat colors tend to appear together (Ludwig et al., 2010). Even if the animals are selected exclusively for tameness and not for appearance, coat color variations increase rapidly as a species becomes more domesticated.

Since the domestication of that original long-forgotten horse, humanity has continually manipulated the genes of the original wild stock by selecting for body type, speed, disposition, and coloration. The value and significance of horses were decided within the context of culture. Horses were ascribed powers from the mystical to the mundane. The Scythian horse cultures were fierce nomads who lived intimately with horses, their most treasured possessions. In China, Ferghana horses were prized for producing blood-colored sweat. Blood-sweating horses were believed to have resulted from a mating between a dragon and a mare. More prosaically, the bloody coat was probably caused by the parasitic nematode *Parafilaria multipapillosa*, which burrows under the skin and causes blisters that burst and ooze blood (van Dierendonck & Goodwin, 2005).

The Spanish horses brought to North America by Cortés appear far removed from wild *Equus*, yet a foal born to a newly freed mare grows up with all its wild instincts intact, as if millennia of domestication had never occurred. In contrast, a domestic dog in a wild state remains distinct from its Pleistocene progenitor, the wolf. Domestic dogs have become their own subspecies with different behavior and physiology. The DNA of dogs is very similar to that of wolves; the differences appear attributable to differences in patterns of gene expression. Wolves usually produce one litter a year, but feral dogs may have two or three. Within a wolf pack, only the alpha pair may produce pups. Among feral dogs, any member of the group can reproduce. Wolves are more aggressive than dogs, inclined to destroy intruders in situations where a dog is likely to bark, growl, and snap. In other words, a dog may bite the mailman, but a wolf may try to kill him.

Although most domestic animals are subspecies of ancestral wild animals, the horse is the very same species it was in the Pleistocene. Domestication creates changes in behavior, form and gene expression— but apparently not in the underlying genome. The modifications caused by selective breeding are, in most cases, no more extreme than those caused naturally by genetic drift. There are 682 recognized horse breeds in the world today, ranging from the 1-ton/900-kg Shire to miniature horses no bigger than a large dog. All of them are the very same species as the *E. caballus* that lived in North America during the Pleistocene.

On the other hand, Sponenberg notes, "There are genetic changes under domestication, which is the main significance of the wild/domesticated split. That is, domestication 'is forever,' and you never get the original wild animal back, even after generations of feral existence" (personal communication, June 5, 2014). All over the world free-roaming horse populations descend from lost or released domestic stock genetically manipulated to serve the needs of man, or as in the case of the Takhi, reintroduced from genetically conserved zoo stock. There are wild horses in Japan, Argentina, Brazil, and Puerto Rico; on Sable Island off Nova Scotia; and until recently on Great Abaco Island in the Bahamas. In Australia, there are about 400,000 wild Brumby horses, descending (according to one account) from six horses that escaped from Irish and British settlers in 1788 (Gill, 1994). This is a huge number, almost 10 times the number of wild horses in the United States. Australia is only 80% as large as this country, though its human population is much less dense. The Kaimanawa horses of New Zealand, descended from escaped Welsh and Exmoor ponies and flavored by other breeds over the years, have been roaming since the 1840s.

The "white horses" of the Camargue have been free-roaming in the Rhône delta of southern France for hundreds or even thousands of years, but biologists doubt that they have been truly wild in historic times (Gill, 1994). Some believe they are descended from the extinct horses that roamed the Solutré region of France 17,000 years ago (Upper Paleolithic period), but so far there is no proof. The population is currently managed by removing inferior horses and by gelding colts to keep the breed true to type. Camargue horses probably contributed genes to the Spanish Jennet.

In the British Isles, free-roaming ponies include the Dartmoor, Exmoor, Welsh, New Forest, Fell, Highland, Shetland, Connemara, Lundy, and Eriskay. Most have run wild for centuries, but have been greatly altered by outside blood. A few herds of these ponies have been free roaming for thousands of years and are thought by many to carry blood from the original native animals.

If the terms *wild* and *feral* are vague, controversial, or emotionally loaded, they are too useful to abandon. Plausible substitutes such as *free-roaming* and *free-ranging* are not synonyms for either term. *Free-roaming* and *free-ranging* imply self-determination, and the ability to chose mates and wander at will. All the world's wild horses are naturally or artificially restricted to more or less fixed parcels of land. Dr. Jay Kirkpatrick, reproductive specialist and senior scientist at the Science and Conservation Center of Billings, Montana, describes them as "fenced, confined and literally trapped," much like dogs and cats

The climactic meeting of Hernando Cortés and Montezuma II in the Aztec capital, Tenochtitlan (Schijnvoet, 1724). After a cordial reception, relations soured and Montezuma was killed, either by Spaniards or by his own people. Aztecs eventually expelled the Spaniards in 1520, but Cortés returned the following year and rapidly destroyed the Aztec Empire and many of its architectural wonders. Although the Spanish were greatly outnumbered, superior weaponry and awe-inspiring horses were keys to their victory. Courtesy of Jay I. Kislak Collection, Rare Book and Special Collections Division, Library of Congress.

with limited food and water (personal communication, May 29, 2014). In a purely natural existence, horses migrate and reduce or expand their ranges when resources, climate, predation, mating prospects, or social hierarchies change, sometimes traveling hundreds of miles in reaction to new conditions. No wild population enjoys such freedom today. Says Kirkpatrick,

> I was always struck by the pathetic situation in that winter of 1977–78, when half the Pryor herd [in the Pryor Mountains Wild Horse Range of Montana and Wyoming] died. If they could have walked a mere quarter mile to the west of the fenced range they could have survived. They were but a quarter mile from the riparian zone of Crooked Creek, with grass and cottonwood. But they could not—because of a fence. (Personal communication, May 29, 2014)

Still, space to survive and freedom to move are the very essence of wildness. The 90–120 horses in the Pryor Mountain herd are restricted by artificial barriers to a marginal range of nearly 40,000 acres/16,200 ha. Their preserve is tiny in comparison with the expanse of scrub and prairie where their prehistoric ancestors roamed in the millions before human beings reached North America. But within this enclave the herd meets few unnatural

An Apsáalooke (Crow) man on horseback, probably in the Pryor Mountains of Montana (Curtis, 1908). The Crow probably did not acquire horses until the 18th century. Before long, however, they were renowned horse breeders and traders, and they had more horses than any other Plains tribe. The horse in this image shows strong Colonial Spanish characteristics: short-coupled, deep-bodied, narrow-chested, with sloping croup and low tail set. Courtesy of Library of Congress, Prints and Photographs Division, LC-USZ62-1206.

obstacles, finds nourishment, and exhibits a broad spectrum of wild behavior. Horses are free to choose mates—though their options have been limited by BLM culling—and reproduce at will when not constrained by contraceptive vaccines. They are free-roaming within the boundaries set.

Conversely, the 27 descendants of wild and domestic horses inhabiting Park Service property on Ocracoke Island, North Carolina, are confined to only 180 acres/73 ha, an area much too small to allow survival without supplemental feeding, veterinary care, and intensively managed reproduction. Though their ancestors had the run of the entire island, today's herd is not free-roaming by any definition. At times some of them have the run of their enclosure, at other times they reside in paddocks, nourished by twice-daily feedings of hay and grain. They weather severe storms in sturdy barns. Opportunities for natural interaction are virtually absent. Many are gelded, and all are tame. It is likely, though, that this herd would become self-sufficient and revert to a wild state within a few generations on a range as large as that afforded the Pryor Mountain herd.

The other East Coast herds fall somewhere between these extremes. They are not free-roaming by Pleistocene standards, but many of these horses experience self-determination on a daily basis, choosing mates, food sources, and daily activities without continual human interference. In this sense, they may reasonably be considered free-roaming. Yet despite

A wild Pryor Mountain adolescent, Kapitan, who was killed with his father, Admiral, by a drunk driver in July 2011. He has many Spanish characteristics seen in the preceding photograph, taken more than a century earlier.

their apparent freedom, they live under other human-imposed restrictions. We gather them, remove them, and limit their reproduction, skewing their genetic makeup and altering their behavior. They change their band dynamics and restructure their home ranges in response to our placement of artificial water sources, test plots, archaeological digs, and other intrusions. Many argue cogently that these horses are not free-roaming, but they are clearly freer than most of their domestic kin.

Free-roaming and *free ranging*, then, add new controversies to a contentious discussion without resolving any of the others. Differences in freedom of movement among wild populations and between wild horses and domestic are often differences of degree, not of kind. But these terms, like *wild* and *feral*, are too useful to exclude, even for the sake of temporary harmony. Although no wild horse population in the United States is free to travel great distances in response to environmental or other changes, many herds go where they please within the physical limits imposed by topography or civilization.

In its 2013 report, the National Research Council used the term *free-ranging* for populations "allowed to use spatially extensive habitats in ways that increase access to forage, improve their physiological condition, and increase the probability of their own and their population's viability." The committee chose "semi-free-ranging" to describe equids "confined to limited areas, for example, in fenced reserves or protected areas that are nevertheless expansive enough for the animals to move freely over larger areas than typical farms

The Takhi, or Przewalski Horse, is clearly equine, but is proportioned differently from domestic horses, with a large head, upright mane, and long ears. This species has 66 chromosomes, in contrast to the 64 found in domestic horses. Matings between the two species yield fertile hybrids with 65 chromosomes.

or ranches." The term *domestic* "describes an animal that is kept by humans, typically as a companion animal or as livestock" (NRC, 2013, p. 19).

For the purposes of this book, *free-roaming* and *free-ranging* are interchangeable and mean "able to move without significant restriction over a range large enough to allow natural subsistence, interaction, and reproduction." Whether a given population actually does subsist, interact, or reproduce naturally is another matter.

The Przewalski \shə-'väl-skē\ Horse, or Takhi (*E. c. przewalskii*), is the closest living relative to the domestic horse. Their common ancestor lived only about 500,000 years ago. This close relationship seems supported by the fact that domestic horses bred to Takhis yield fertile offspring. In contrast, zebra chromosomes differ from those of the domestic horse in many aspects.

While *Homo sapiens* was still living primitively in caves, the Przewalski Horse proliferated across the plains of central Asia. These animals are distinctively stocky and solid, with large heads and dun coloration. They possess an upright mane which, unlike that of *E. caballus*, sheds and regrows every spring. This species persisted largely untouched by humankind until the 1960s.

Ten thousand years ago, these horses probably migrated seasonally across the Siberian steppes, relocating in herds of thousands in the same fashion as reindeer. Eventually, mankind displaced the Przewalski from its preferred environment to more remote corners

Paleontologists find the bones of early equids in abundance in the Bighorn Basin of Wyoming, where the Pryor Mountain mustangs range today. The very dust kicked up by the hooves of this wild mare may contain the fossilized remains of her ancient ancestors—yet many people do not consider wild horses a native North American species.

where people were few. These horses, best suited for living in the cold of the steppe, found sanctuary in the furnace-like conditions of the Gobi desert.

Historically, Mongols revered and respected their indigenous horse. The name *Takhi* means "holy" or "spiritual" in Mongolian.

Johann (Hans) Schiltberger (1380–*c*. 1440), a German writer, described the horses in the journal he kept while held prisoner in Mongolia. In 1879, the celebrated Russian traveler Nikolai Przewalski presented a skull and hide brought from Central Asia to zoologist I.S. Poljakov, who described them as a new species in 1881 and named them to honor the explorer. Around 1900 Carl Hagenbeck, a German circus owner and exotic-animal merchant (and proponent of the "Hagenbeck revolution" in zoo design) captured a number of Takhis and placed them in preserves. To collect them, hunters searched vast tracts of wasteland for bands of the elusive wild horses, pursued them relentlessly using a number of remounts, shot the exhausted adults and captured the foals with an urak/arkan, a stick with a loop of rope at the end (Kelekna, 2009). They wrestled the foals into bags and slung them onto camels for the trek out of the wilderness, then loaded them into train cars. Their captors sustained them on the milk of sheep and domestic mares. Many died *en route*. One 1901 expedition exterminated the adult horses in at least 25 harems to capture 52 foals, 28 of which survived transport (Kelekna, 2009).

Przewalski Horses persisted in the wild through the 1940s, but by the 1950s, zoologists counted only about 100 Mongolian wild Takhis in several bands (Kelekna, 2009). Even in the Gobi wasteland, many stockmen viewed the untamable horses as a nuisance that competed with domestic livestock for resources and shot them on sight. Populations of people and livestock expanded and excluded Takhis from water holes. The last documented sighting of wild Takhis was in May 1968 on the border between Mongolia and China. Later reports could not be verified.

Zoos never acquired large numbers of the animals, and captivity-bred horses faced a serious risk of inbreeding (Wakefield, Knowles, Zimmermann, & van Dierendonck, 2002). One of the largest and most genetically valuable captive herds was maintained at Askania Nova, a nature preserve in the Ukraine. During the military occupation of the Ukraine in World War II, German soldiers methodically shot all the Takhis on the refuge. (Today Askania Nova once again holds the largest captive breeding program for Przewalski Horses.)

In 1945, there were only two captive breeding herds remaining, in Prague and in Munich. All Takhis alive today can be traced to 13 horses held in captivity in 1945, which in turn descended from 15 horses caught around 1900.

In 1973, only 206 Takhis existed worldwide, all in zoos. Twenty years later, the number had increased to 1,000. After the Chernobyl accident, several dozen Przewalski Horses were introduced to the evacuated area, putting it to use as a nature preserve.

Takhis returned to the wild in reserves in Xinjiang, China, in 1988 (Kelekna, 2009) and Mongolia in 1990. In 1993, the Mongolian Association for Conservation of Nature and the Environment and the Foundation Reserves for the Przewalski Horse introduced Takhis to the Hustain Nuruu Steppe Reserve in the rolling hills of central Mongolia, which was historically a khan hunting preserve, and managed it for a year while it adjusted to the native food sources and challenging climate. In June of 1994, the foundation successfully freed 19 horses where their ancestors had once ranged. The horses returned swiftly to their ancient ecological niche and social structure. By 2005, there were 248 individuals in the wild herd and 1,500 Takhis worldwide. The World Conservation Union listed the Przewalski Horse as extinct in the wild until 2008, when it changed the designation to critically endangered.

The winter of 2009–2010 was harsh for the herd of 137 Przewalskis living in the Dzungarian Gobi in southwest Mongolia. Severe cold and heavy snowstorms assaulted the region from December through March, killing a total of 104 Takhis despite heroic rescue efforts (Kaczensky, Ganbaatar, Altansukh, & Enkhsaikhan, 2010). Once again, the herd teeters on the brink of extinction.

Should these repatriated Przewalskis be considered wild or feral? They are the pureblooded descendants of horses taken from the wild and reestablished in their natural environment. Then again, they are the result of about 25 generations of matings carefully engineered by zoos to give the few survivors the greatest possible genetic diversity. Przewalski Horses will submit to captivity and breed in zoos, but they are willful and aggressive and will not reliably bend to the whims of a trainer. Most sources claim that the Przewalski Horse has never been trained to ride or drive, but Franzen (2010) features a photograph of a wild-caught stallion named Vaskia, captured in 1899, being ridden with saddle and bridle. (Other wild equine species, such as asses, onagers, quaggas, and zebras have been saddle- or harness-broken, some with great difficulty.) Takhis survived their years of confinement with their largely untamable spirits intact, and quickly reoccupied their original ecological niche when given the chance.

The Visionary, painting by the author. What was in the minds of primitive artists as they rendered horses on cave walls?

Sponenberg maintains that the Takhi is a true wild horse. "In the absence of selection for domesticated behavior and morphology," he said, "the captive propagation of Takhis would not qualify as domestication" (personal communication, June 5, 2014).

The Tarpan or Eurasian wild horse, *E. c. ferus*, roamed the vast forests and grasslands from eastern Poland to the Ural Mountains (Bennett, 2008a). This variety was the first wild horse to be domesticated. Many researchers believe that the Tarpan was the progenitor of the modern domestic horse. Tarpans were stocky, solid, mousy dun or grulla animals of 13–14 hands (52–56 in./1.32–1.42 m). Berenger (1771, p. 144) described them thus:

> The *Tarpans* are a kind of wild-horses, in the desert, east of the river Yaik [Ural].
> They are of a middling size, roundish, short, generally of a bluish-grey colour, with big heads, and ewe-necked. They are taken with a noose, and broken to saddle, by being coupled to a tame horse.

Here again, indigenous horses proved incompatible with agriculture and other human activities. Expanding human cultures domesticated a few Tarpans, ate others, and killed the rest as annoyances or for sport, pushing them into extinction. Until 1913, a herd of these animals was kept in the Bialovesh Forest Preserve in Poland (Bennett, 2008b). The last known wild Tarpan mare was accidentally killed during an attempt at capture in the Ukraine in 1851, and the last captive animal died in 1918 (USBLM, Wild Horse and Burro Program, 2005).

A number of scientists have tried to re-create the Tarpan by mating horses thought to have a high percentage of Tarpan blood. Over time, they developed a primitive-looking, steel-gray dun animal that can be seen in zoos today. These animals look like Tarpans and have been released into the wild in several western European reserves, but genetically they cannot be considered to represent the original subspecies.

Berenger provides a vivid description of the wild horses that were alive during his time. He tells us of the desert horses descended from cavalry mounts that escaped in 1697. They became uniform over time, "inclining to red, the hair of their skins being curled, and waved like lamb-skin; but when they grow old, it changes to a mouse-grey, their manes and tails being black, and having a black list along their backs" (1771, p. 143). He describes Cossacks driving them into deep valleys full of snow and catching them with nooses. Most were killed with spears, but the younger animals were kept for use and "found infinitely stronger than a common horse" (p. 144).

From the depths of prehistory, horses have been a preferred food for humankind. Horse bones are common at archeological sites and often bear the cut marks of human tools. Sometimes entire herds were trapped by chasing them into blind canyons or ambushing them in narrow passes. A site at Solutré in eastern France holds the remains of 32,000–100,000 horses killed year after year while seasonally migrating to their summer grazing habitat (Kelekna, 2009).

The African ass was probably domesticated before the horse. It was initially valued for its meat and milk, but proved valuable as a pack animal along ancient trade routes (Kelekna, 2009). Asses were ridden, but their conformation required the rider to sit over the rump, where the jolting of the hind legs made high speeds impractical. It appears that people domesticated the Persian onager east of the Tigris. Onagers were difficult to tame and bred poorly in captivity, but by 2800 BCE, it became common practice to breed them to donkeys to produce hardy, tractable hybrids (Kelekna, 2009).

Serendipitously, features that gave the modern horse evolutionary advantages are precisely what facilitated domestication. Though other animals have been ridden and driven—cattle, yaks, reindeer, elephants, camels, and even moose, bison, and ostriches—the horse alone is so well suited to the task it almost appears custom-made to mesh with human needs.

The same dominance/submission social patterns that gave a horse security within the hierarchy of his herd allowed him to submit to human domination. Many wild species never became trusting or trustworthy despite decades of captive breeding, but horses were easily tamable, even after escaping and running wild for generations. They learned quickly and had indelible memories, were large enough to ride, and had a bone structure able to hold a saddle and bear the weight of a rider. Few animals were as fast or could sustain speed over great distance while carrying a load.

Horses could survive on the poorest food in a wide range of climates. They were inquisitive and versatile. They had no horns or antlers to injure a rider. Nature even provided a convenient gap between their molars and incisors that allowed a bit to rest on the sensitive gums, permitting a greater degree of control.

Wear on horse teeth found at Dereivka, Ukraine, and evidence from the Botai investigations mentioned above indicates that people rode horses with bits before the invention of the wheel. One wonders how long before that people used leather or fiber halters, hackamores, and neck ropes that left no trace in the archaeological record. A curb bit made of bone and antler was in use as early as 1500 BCE in Switzerland, and a jointed snaffle bit surprisingly similar to its modern counterparts was in common use in central Eurasia as early as 1000 BCE.

Tribes that used horses for transportation and draft had significant advantages over those that did not. Each horse could pull several hundred pounds on a travois. A mounted warrior

Some mustang herds, such as the Pryor Mountain horses, have Spanish characteristics, the legacy of their Jennet ancestry.

was far more intimidating and effective than one on foot and could travel great distances quickly, attacking or retreating. The domestication of the horse brought greater mobility, innovations in hunting and warfare, increased leisure time, changes in social and economic values, and the sheer joy of racing the wind above the drumming of powerful hooves.

During this period of domestication, people used two basic approaches to training. One method, favored in early Rome and elsewhere, was to force the horse to submit through domination and punishment—"breaking" the horse (Goodwin, 2002). The Scythians of central Eurasia and the Greeks employed a cooperative approach based on an understanding of the behavior of the horse. Both methods are still used today; compare the rodeo-style bronco breaking with the round-pen join-up approach and desensitization training used in "natural horsemanship" circles.

It probably did not take long for early breeders to select animals with desirable traits such as speed, stamina, disposition, size, or eye-catching coloration. For the first time in history, horse phenotypes were determined more by human preferences than by environmental suitability and evolutionary advantages. Domestic horses gradually diverged from their wild cousins, and for thousands of years humanity continued to manipulate equine bloodlines, creating many diverse breeds within the species.

Many researchers believe that if not for domestication, the horse would have become extinct. The range and numbers of wild horses contracted dramatically as the last Ice Age drew to a close. Many of the wild populations from which the horse had been domesticated became extinct. Then suddenly, around 6000 BCE, horses resurged, becoming common wherever people lived. Other species followed a similar pattern. People hunted the abundant aurochs to extinction after it had given rise to domestic cattle. People nearly exterminated the wolf, but dogs thrive. Wild sheep and goats are scarce, but domestic descendants flourish.

The Spanish Jennet horse that Cortés returned to the land of its ancestors does not exist today. In the years that followed the discovery of the New World, Spanish horse breeders selected stock that best suited their needs and aesthetics. Some types were refined until they became distinct new breeds, such as the modern Andalusian. Some Spanish horses were outcrossed to other breeds to create new varieties. Thus the Jennet gave rise to countless other breeds.

In 1588, the English repelled the Spanish Armada. This victory eventually allowed England to plant settlements all along the Eastern Seaboard. Many English colonists were farmers, and they often bought livestock from Spanish ranches in the West Indies to avoid the difficulty and expense of trans-Atlantic shipping. Over time, American breeds such as the Narragansett Pacer were developed for smooth travel over poor roads. (Paul Revere was reportedly astride a Narragansett Pacer on his famous ride.) Later, Tennessee Walkers, Quarter Horses, and Morgans became common along the East Coast.

In 1670, the English Parliament levied a tax on fences in the American colonies, creating a hardship for subsistence farmers who were often on the razor edge of survival. Coastal settlers cleverly circumvented this tax (and the cost of fencing, with or without tax) by putting their stock out to graze on barrier islands and peninsulas where the animals could range largely unfenced. Periodically, stockmen held roundups for the purpose of branding foals and calves and collecting stock for sale or use. Otherwise, the herds were left largely to their own devices, migrating from island to island, grazing the marshes, and living the lives of wild horses. Sometimes livestock owners would die or move away, leaving the herds to a wild existence.

Native Americans also used natural barriers to confine animals for exploitation. Lawson (1709, p. 207) notes, "they go and fire the Woods for many Miles, and drive the Deer and other Game into small Necks of Land and Isthmus's, where they kill and destroy what they please."

In the Western reaches of the New World, hardy Jennets were instrumental in the success of the conquistadors. Indians who had learned to ride while working for the Spanish rancheros probably traded for—and stole—a number of these horses. When the Pueblo Peoples revolted against the Spanish in the 1680s, thousands of Spanish horses escaped into the wild from settlements in the Rio Grande valley (Kelekna, 2009). Competing tribes stole ponies from the unfenced herds of rivals, running off with some horses and scattering others. Within a century or two, free-roaming mustangs (so named from the Spanish *mesteño*, meaning "stray" or "untamed") numbered in the millions. These horses thrived in large numbers from Florida through the West. They spread up through Oregon and Idaho into Canada. They enjoyed such amazing success that Charles Darwin, pondering the extinction of earlier horses, wrote in *Origin of Species* (1902, p. 103),

> I was filled with astonishment; for, seeing that the horse, since its introduction by the Spaniards into South America, has run wild over the whole country and has increased in numbers at an unparalleled rate, I asked myself what could so recently have exterminated the former horse under conditions of life apparently so favorable.

Along the Appalachian Trail in southwestern Virginia, two feral ponies scratch themselves on a sign. Shetland ponies were deliberately introduced by the U.S. Forest Service in 1974 to reduce brushy overgrowth on Mt. Rogers and the Grayson Highlands.

Early North American mustangs descended entirely from Spanish horses, but many other sources have contributed genes over the years. The herds that have lived longest in the wild (Bookcliffs of Colorado, the Sulphur mustangs of Utah and Nevada, Cerbat mustangs of Arizona, and the Pryor Mountain mustangs of Wyoming and Montana) have strong Spanish characteristics, as do old Spanish breeds such as the Florida Cracker, the Cherokee horse, the Choctaw Horse, and others. All of these New World Spanish types are collectively known as the Spanish Colonial Horse. Many consider the Kiger horses of Oregon a Colonial Spanish herd, but Sponenberg disagrees. "They are not Colonial Spanish, at least most of them," he said. "These herds were actually assembled/constructed from dun horses from a variety of places, rather than being found right there in one place as a unique population. Subsequent selection for the dun colors made them relatively uniform, but that selection also took them away from the usual Colonial Spanish phenotype" (personal communication, June 5, 2014).

While the horses that repopulated the North American west were originally of Spanish lineage, most mustang herds today show influences many American and Old World horse breeds. European settlers took hold in the East, then pushed westward into the uncharted vastness of plains, forest, and mountains, bringing an array of horse breeds with them. Stock, saddle, draft, and carriage horses (largely of Thoroughbred, Morgan, and Quarter Horse lineage) helped the pioneers carve the wilderness into farms and towns and hauled goods to and from trading posts. Agile cavalry horses galloped beneath resolute soldiers intent on conquest. Many of these horses escaped or were set free. Farmers who tried to wrest a living from the land often failed and returned east, abandoning their livestock. Indians obtained horses from trappers, soldiers, farmers, and one another, and their own loosely tended stock sometimes escaped to the wild as well.

Some of today's mustangs still have a distinctly Spanish appearance, and some possess genetic markers suggesting Iberian lineage. The Spanish horses were used to improve many other breeds, such as Quarter Horses and Morgans, however, and Spanish markers can be found in most of them. Science cannot determine whether these genes were inherited directly from Spanish horses of the colonial era, or from other breeds. The herds at greatest risk for genetic erosion are the small, scattered groups of Colonial Spanish Horses. Sponenberg explains, "Among the feral horses only the Colonial Spanish horses are really irreplaceable genetic resources. The other herds have genetics that are duplicated in domestic herds, and could be reconstituted if needed" (personal communication, June 5, 2014).

Spanish or not, each mustang herd has its own characteristics, which many consider worthy of preservation. The United States government classifies free-roaming horses as feral livestock, but many scientists view them as a reintroduced native wildlife and believe that, as such, they are entitled to protection. Genetic analysis of Pleistocene fossils has demonstrated that the *E. caballus* repatriated by European explorers after a comparatively short absence is biologically identical to the ancient horse that roamed the ancestral grasslands.

How long should a North American herd run free to earn the designation of repatriated native wildlife? In 1974, the U.S. Forest Service deliberately introduced a breeding herd of mostly Shetland ponies to the Grayson Highlands of Virginia, to control the vegetation that blocked the scenic vistas. A group of volunteers loosely manages the herd, but they have roamed free for almost 40 years. Can we consider them wild? What status should we give domesticated horses released to "improve" a wild herd—or because they are no longer wanted?

For thousands of years, up until the recent past, horses were instrumental to the progress of humanity. Historically, people have valued them as highly as human beings—sometimes higher. Horse theft was often a capital offense through the 1800s in both the United Kingdom and in parts of the United States. Horses have provided humanity with transportation, muscle power, companionship, food, and sacrifices to various gods. They have increased the efficiency of farming methods and fertilized crops with manure. Lactating mares have provided milk, and that milk has been fermented into an intoxicant. Horse hair has been mixed into plaster to strengthen walls; stuffed into upholstery; and made into paintbrushes, bowstrings, and a durable fabric called haircloth. The hormone-replacement drug Premarin™ contains estrogen extracted from pregnant mares' urine. Horsehide is tanned into leather, horseflesh is ground for dog food, and horse bones and hooves are rendered into glue. Horsepower, which James Watt created for marketing his steam engine, is still a widely recognized unit of measure. Horsemeat was a dietary staple for primitive humans, and it remains a delicacy in some cultures, taboo in others.

We ride horses and drive them, harness their strength to do our work, keep them as pets, and eat them. They are status symbols, companions, and icons of freedom and beauty.

While we still value the domestic horse as a partner and companion animal, we cannot seem to agree on whether horses are important to us in the wild. We determine the value of a wild species, usually by its apparent usefulness. If we consider wild horses worth saving, we can find many good reasons to keep them roaming the rangelands and barrier islands of North America. If not, we can find many good reasons for eradicating them from the wild. As Benjamin Franklin (1771/2008, p. 51) observed, "So convenient a thing it is to be a reasonable creature, since it enables one to find or make a reason for everything one has a mind to do."

Although raw emotion should not overrule good judgment, how we feel about the facts is at least as important as the facts themselves. There is no purely rational basis for embracing feral horses while slaughtering feral cattle or for reintroducing genetically impure red wolves to wildlife refuges while removing reintroduced horses that are authentic *E. caballus*. The best any of us can do is to consider how emotions and biases affect our decision-making and bring both our minds and our hearts to the task.

References

Barlow, C. (2000). *The ghosts of evolution: Nonsensical fruit, missing partners, and other ecological anachronisms.* New York, NY: Basic Books.

Barrow, J. (1801). *An account of travels into the interior of southern Africa, in the years 1797 and 1798.* London, United Kingdom: For T. Cadell and W. Davies.

Bennett, D. (2008a). *Introduction to horse evolution: Anatomical characteristics, classification, and the stratigraphic record.* Retrieved from http://www.equinestudies.org/evolution_ horse_2008/intro_to_horse_evolution_2008_pdf2.pdf

Bennett, D. (2008b). *The origin and relationships of the Mustang, Barb, and Arabian horse.* Retrieved from http://www.equinestudies.org/origin_mustang_arab/origin_mus-tang_arab_barb_pdf1.pdf

Bennett, D., & Hoffman, R.S. (1999). *Equus caballus. Mammalian Species,* 628, 1–14.

Berenger, R. (1771). *The history and art of horsemanship.* London, United Kingdom: For T. Davies and T. Cadell.

Castillo, B.D. (1908). *The true history of the conquest of New Spain. By Bernal Díaz Del Castillo, one of its conquerors.* London, United Kingdom: Hakluyt Society (Original work published 1632).

Chard, T. (1940). Did the first Spanish horses landed in Florida and Carolina leave progeny? *American Anthropologist,* 42(1), 90–106. doi: 10.1525/aa.1940.42.1.02a00060

Cieslak, M., Pruvost, M., Benecke, N., Hofreiter, M., Morales, A., Reissmann, M., & Ludwig, A. (2010). Origin and history of mitochondrial DNA lineages in domestic horses. *PLoS ONE,* 5(12): e15311. doi: 10.1371/journal.pone.0015311

Clark, L. (2003). A review of pathogens of agricultural and human health interest found in Canada geese. In K.A. Fagerstone & G.W. Witmer (Eds.), *Proceedings of the 10th Wildlife Damage Management Conference* (pp. 326–334). Fort Collins, CO: Wildlife Damage Management Working Group of The Wildlife Society. Retrieved from http:// www.aphis.usda.gov/wildlife_damage/nwrc/publications/03pubs/clar034.pdf

Curtis, E.S. (Photographer). (1908). *The scout in winter—Apsaroke.* Retrieved from http:// www.loc.gov/pictures/item/2002722308/

Darwin, C. (1902). *Origin of species by means of natural selection, or the preservation of favored races in the struggle for life* (6th ed.). New York: P.F. Collier & Son.

Denhardt, R. (1951). The horse in New Spain and the borderlands. *Agricultural History,* 25(4), 145–150.

Diamond, J. (2009). *Guns, germs, and steel: The fates of human societies* (Kindle ed.). New York, NY: W.W. Norton.

Donlan, C.J., Berger, J., Bock, C.E., Bock, J.H., Burney, D.A., Estes, J.A., . . . & Greene, H.W. (2006). Pleistocene rewilding: An optimistic agenda for twenty-first century conservation. *American Naturalist, 168*(5), 660–681. doi: 10.1086/508027

Editors of E Magazine. (2002, April 30). Going, going . . . exotic species are decimating the United States' native wildlife. *E/The Environmental Magazine, 13*(3), 1–8. Retrieved from http://www.emagazine.com/magazine-archive/going-going—-exotic-species-are-decimating-americas-native-wildlife

Fowler, A.R., & Klein, W.M., Jr. (1993). *The vascular flora of Pennsylvania: Annotated checklist and atlas*. Philadelphia, PA : American Philosophical Society.

Franklin, B. (2008). *The autobiography of Benjamin Franklin, 1706–1757* (Facsimile). Bedford, MA: Applewood Books (Original work published 1771).

Franzen, J.L. (2010). *The rise of horses: 55 million years of evolution*. Baltimore, MD: Johns Hopkins University Press.

Frequently asked questions—horses. (2011, October 31). Retrieved from http://www.nps.gov/calo/naturescience/horse-faqs.htm

Frydenborg, K. (2012). *The wild horse scientists*. Boston, MA: Houghton Mifflin Harcourt.

Gill, E. (1994). *Ponies in the wild*. London, United Kingdom: Whittet Books.

Goodwin, D. (2002). Horse behaviour: Evolution, domestication and feralisation. In N. Waran (Ed.), *The welfare of horses* (pp. 1–18). Dordrecht, Netherlands: Kluwer Academic Publishers.

Guthrie, R. (2006). New carbon dates link climatic change with human colonization and Pleistocene extinctions. *Nature, 441*, 207–209. doi: 10.1038/nature04604

Haile, J., Froese, D.G., MacPhee, R.D.E., Roberts, R.G., Arnold, L.J., Reyes, A.V., . . . Willerslev, E. (2009). Ancient DNA reveals late survival of mammoth and horse in interior Alaska. *Proceedings of the National Academy of Sciences of the United States of America, 106*(52), 22352–22357. doi: 10.1073/pnas.0912510106

Haines, F. (1938). The northward spread of horses among the Plains Indians. *American Anthropologist, 40*(3), pp. 429–437.

Harley, E.H., Knight, M.H., Lardner, C., Wooding, B., & Gregor, M. (2009). The Quagga Project: Progress over 20 years of selective breeding. *South African Journal of Wildlife Research, 39*(2), 155–163.

Higuchi, R.G., Wrischnik, L.A., Oakes, E., George, M., Tong, B. & Wilson, A.C. (1987). Mitochondrial DNA of the extinct quagga: Relatedness and extent of postmortem change. *Journal of Molecular Evolution, 25*(4), 283–287. doi: 10.1007/BF02603111

Höglund, D. (2010, March 23). *Management of the free-roaming horse*. Retrieved from http://pryorwild.wordpress.com/2010/02/14/february-15-2010-pzps-reversibility/

Jansen, T., Forster, P., Levine, M.A., Oelke, H., Hurles, M., Renfrew, C., . . . Olek, K. (2002). Mitochondrial DNA and the origins of the domestic horse. *Proceedings of the National Academy of Sciences of the United States of America, 99*(16), 10905–10910. doi: 10.1073/pnas.152330099

Johnson, J.J. (1943). The introduction of the horse into the Western Hemisphere. *Hispanic American Historical Review, 23*(4), 587–610.

Kaczensky, P., Ganbaatar, O., Altansukh, N., & Enkhsaikhan, N. (2010, August). *Winter disaster in the Dzungarian Gobi—Crash of the Przewalski's horse population in Takhin Tal 2009/2010*. Retrieved from http://www.takhi.org/media/

forschung/2010_Winter- disaster-in-Dzungarian-Gobi-2009_10.pdf

Kelekna, P. (2009). *The horse in human history*. Cambridge, United Kingdom: Cambridge University Press.

Kirkpatrick, J., & Fazio, P. (2010, January). *Wild horses as native North American wildlife*. Retrieved from http://awionline.org/content/wild-horses-native-north-american-wildlife

Lawson, J. (1709). *A new voyage to Carolina; Containing the exact description and natural history of that country....* London, United Kingdom. Retrieved from http://docsouth. unc.edu/nc/lawson/lawson.html

Leonard, J.A., Rohland, N., Glaberman, S., Fleischer, R.C., Caccone, A., & Hofreiter, M. (2005). A rapid loss of stripes: The evolutionary history of the extinct quagga. *Biology Letters, 1*(3), 291–295. doi: 10.1098/rsbl.2005.0323

Leonhard, R.R. (2004). Belisarius and the small force theory. *Armchair General, 1*(1), 26. Retrieved from http://www.jhuapl.edu/ourwork/nsa/papers/SmallForceTheory.pdf

Lindgren, G., Backström, N., Swinburne, J., Hellborg, L., Einarsson, A., Sandberg, K., ... Ellegren, H. (2004). Limited number of patrilines in horse domestication. *Nature Genetics, 36*, 335–336. doi: 10.1038/ng1326

Linklater, W.L., Stafford, K.J., Minot, E.O., & Cameron, E.Z. (2002). Researching feral horse ecology and behaviour: Turning political debate into opportunity. *Wildlife Society Bulletin, 30*(2), 644–650.

Lira, J., Linderholm, A., Olaria, C., Durling, M.B., Gilbert, M.T.P., Ellegren, H., ... Götherström, A. (2010, January). Ancient DNA reveals traces of Iberian Neolithic and Bronze Age lineages in modern Iberian horses. *Molecular Ecology, 19*(1), 64–78. doi: 10.1111/j.1365-294X.2009.04430.x

Lowney, M., Schoenfeld, P., Haglan, W., & Witmer, G. (2005). Overview of impacts of feral and introduced ungulates on the environment in the eastern United States and Caribbean. In D.L. Nolte & K.A. Fagerstone (Eds.), *Proceedings of the 11th Wildlife Damage Management Conference* (pp. 64–81). Fort Collins, CO: National Wildlife Research Center.

Ludwig, A., Pruvost, M., Reissmann, M., Benecke, N., Brockmann, G.A., Cieslak, M., ... Hofreiter, M. (2010). *Variation of coat color genotypes pinpoints down the roots of horse domestication*. Presented at the 9th World Congress on Genetics Applied to Livestock Production in Leipzig, Germany. Retrieved from http://www.kongressband.de/wcgalp2010/assets/pdf/0180.pdf

Luis, C., Bastos-Silveira, C., Cothran, E., & Do Mar Oom, M. (2006). Iberian origins of New World horse breeds. *Journal of Heredity, 97*(2), 107–113.

Luis, C., Juras, R., Oom, M.M., & Cothran, E.G. (2007). Genetic diversity and relationships of Portuguese and other horse breeds based on protein and microsatellite loci variation. *Animal Genetics, 38*(1), 20–27. doi: 10.1111/j.1365-2052.2006.01545.x

McGreevy, P. (2004). *Equine behavior: A guide for veterinarians and equine scientists*. London, United Kingdom: W.B. Saunders.

McKnight, T.L. (1959). The feral horse in Anglo-America. *Geographical Review, 49*(4), 506–525.

Mills, D.S., & McDonnell, S.M. (Eds.). (2005). *The domestic horse: The evolution, development and management of its behaviour*. Cambridge, United Kingdom: Cambridge University Press.

Mirol, P., Peral Garcia, P., Vega-Pla, J., & Dulout, F. (2002). Phylogenetic relationships of Argentinean Creole horses and other South American and Spanish breeds inferred from mitochondrial DNA sequences. *Animal Genetics, 33*(5), 356–363. doi: 10.1046/j.1365-2052.2002.00884.x

National Research Council. (2013). *Using Science to Improve the BLM Wild Horse and Burro Program: A way forward.* Washington, DC: The National Academies Press.

Orlando, L., Metcalf, J.L., Alberdi, M.T., Telles-Antunes, M., Bonjean, D., Otte, M., . . . Cooper, A. (2009). Revising the recent evolutionary history of equids using ancient DNA. *Proceedings of the National Academy of Sciences of the United States of America.* Advance online publication. doi: 10.1073/pnas.0903672106

Outram, A.K., Stear, N.A., Bendrey, R., Olsen, S., Kasparov, A., Zaibert, V., . . . Evershed, R.P. (2009). The earliest horse harnessing and milking. *Science, 323*(5919), 1332–1335. doi: 10.1126/science.1168594

Proops, L., & McComb, K. (2010). Attributing attention: The use of human-given cues by domestic horses (*Equus caballus*). *Animal Cognition, 13*(2), 197–205. doi: 10.1007/s10071-009-0257-5

Pruvost, M., Bellone, R., Benecke, N., Sandoval-Castellanos, E., Cieslak, M., Kuznetskova, T., . . . Ludwig, A. (2011). Genotypes of predomestic horses match phenotypes painted in Paleolithic works of cave art. *Proceedings of the National Academy of Sciences of the United States of America, 108*(46), 18626–18630. doi: 10.1073/pnas.1108982108

Renshaw, G. (1904). *Natural history essays.* London, United Kingdom: Sherratt & Hughes.

Ryden, H. (2005). *America's last wild horses* (Revised ed.). New York: Lyons Press.

Schijnvoet, J. (Engraver). (1724). [Cortés received by Montezuma II]. In A. Solis, *History of the conquest of Mexico by the Spaniards* (T. Townsend, Trans.). London, United Kingdom: Printed for T. Woodward, J. Hooke, and J. Peele. (Original work published 1684). Retrieved from http://memory.loc.gov/master/ipo/qcdata/qcdata5/early_americas/tifs/ea0086_03.tif

Shirley, A. (1854). *Remarks on the transport of cavalry and artillery: With hints for the management of horses, before, during, and after a long sea voyage.* London, United Kingdom: Parker, Furnivall, and Parker.

Sponenberg, D.P. (1992). The colonial Spanish horse in the USA: History and current status. *Archivos de Zootecnia, 41*(154/extra), 335–348. Retrieved from http://dialnet.unirioja.es/servlet/fichero_articulo?codigo=278710&orden=90526

Traill, L., Brook, B., Frankham, R., & Bradshaw, C. (2010). Pragmatic population viability targets in a rapidly changing world. *Biological Conservation, 143*(1), 28–34. doi: 10.1016/j.biocon.2009.09.001

van Dierendonck, M., & Goodwin, D. (2005). Social contact in horses: Implications for human-horse interactions. In F. de Jonge & R. van den Bos (Eds.), *The human-animal relationship: Forever and a day* (pp. 65–82). Assen, Netherlands: Royal Van Gorcum BV.

Wakefield, S., Knowles, J., Zimmermann, W., & van Dierendonck., M. (2002). Status and action plan for the Przewalski's Horse (*Equus ferus przewalskii*). In P.D. Moehlman (Ed.), *Equids: Zebras, asses and horses. Status survey and conservation action plan* (pp. 82–89). Gland, Switzerland, and Cambridge, United Kingdom: IUCN/SSC Equid Specialist Group.

Warmuth, V., Eriksson, A., Bower, M.A., Barker, G., Barrett, E., Hanks, B.K., . . . Manica, A. (2012, May 22), Reconstructing the origin and spread of horse domestication in the Eurasian steppe. *Proceedings of the National Academy of Sciences of the United States of America, 109*(21), 8202–8206. doi: 10.1073/pnas.1111122109

Zimmerman, C., Sturm, M., Ballou, J., & Traylor-Holzer, K. (Eds.). (2006). *Horses of Assateague Island population and habitat viability assessment workshop: Final report.* Apple Valley, MN: IUCN/SSC Conservation Breeding Specialist Group.

Chapter 2

The Mustang Predicament

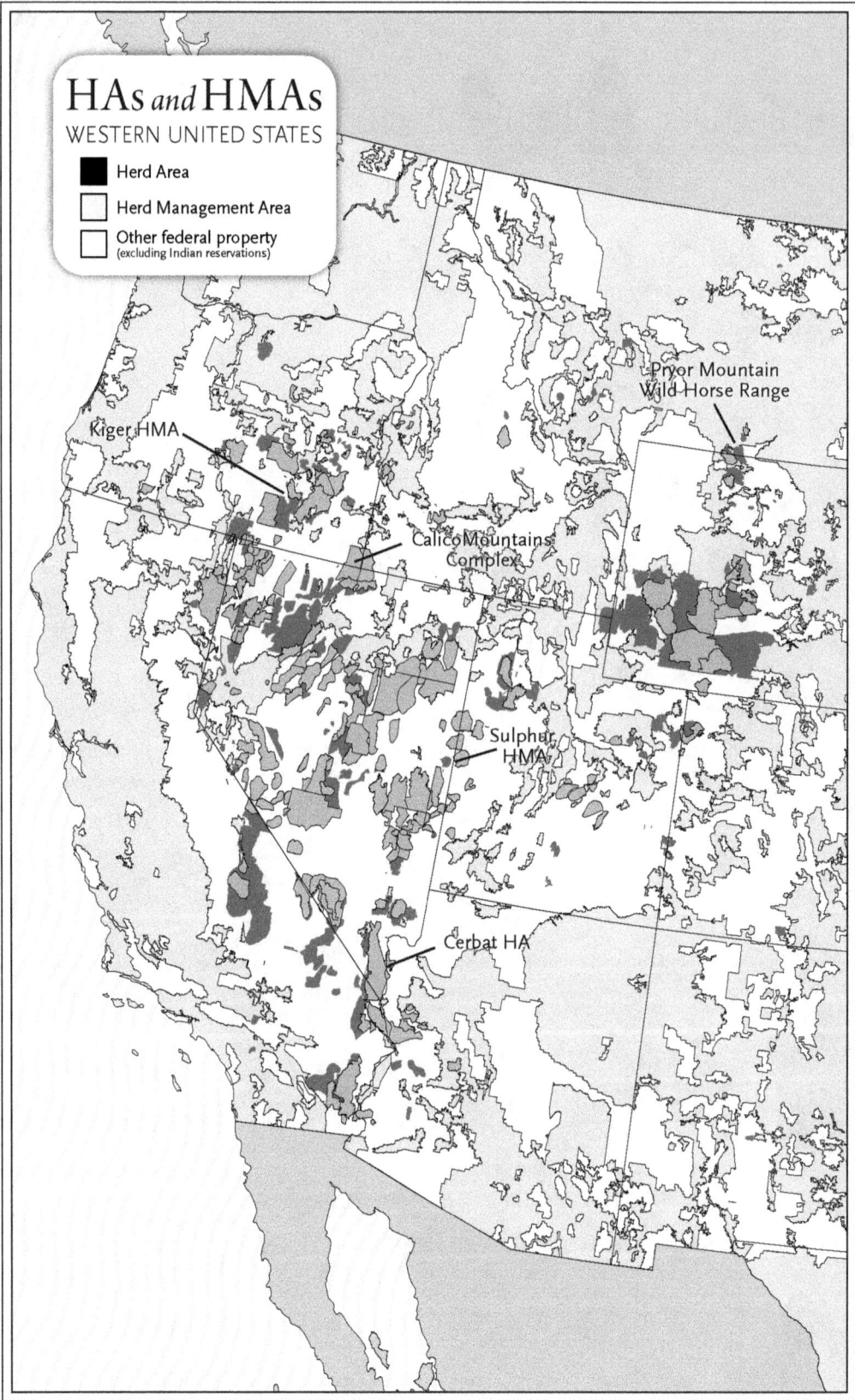

HAs *and* HMAs
WESTERN UNITED STATES

- ■ Herd Area
- ▨ Herd Management Area
- ☐ Other federal property
  (excluding Indian reservations)

Kiger HMA

Pryor Mountain
Wild Horse Range

Calico Mountains
Complex

Sulphur
HMA

Cerbat HA

"You do that to me again, I'm gonna punch you out" (Glionna, 2012; Philipps, 2012, November 6; Cloud Foundation, 2012). Those were the words Interior Secretary Ken Salazar, the man who administered almost one seventh of the United States and stood eighth in the line of presidential succession, used when local reporter Dave Philipps publicly asked questions about wild horse policy. Visiting his native Colorado on Election Day, 2012, Salazar stopped by an Obama campaign office near Colorado Springs. After addressing volunteers, he took questions and seemed to tense at the mention of wild horses. Seconds later, he threatened Philipps.

Wild horses are a politically volatile subject, and when Salazar took office in 2008, he stepped onto a hornets' nest. Detractors saw Salazar, a Colorado rancher, as yet another politician biased toward cattle-raising interests and positioned to determine the fate of America's wild horses. The U.S. Bureau of Land Management, an Interior Department agency, had amassed a record of questionable stewardship over wild horses and burros sharing public range with privately owned livestock.

Although the 1971 Wild and Free-Roaming Horses and Burros Act made it illegal to slaughter wild horses, thousands of mustangs wound up at the abattoir during the Reagan, G.W. Bush, and Obama administrations. Shortly before the Election Day blowup, Philipps (2012, September 28) reported that from the start of Salazar's term in 2009, one livestock hauler—Salazar's Colorado neighbor and longtime associate Tom Davis—had bought 70% of the wild horses sold by the bureau for $10 each. Davis will not and the bureau cannot account for 1,730 horses that changed hands.

To understand the conflicts affecting the small, high-profile herds in the East, we should look carefully at passions aroused by, precedents set for, and assumptions made about the mostly larger, lower-profile herds in the West. Wherever they roam free on either side of the continent, wild horses are subject to complex environmental, cultural, economic, legal, administrative, political, and personal forces. Often they end up like the rope in a tug of war.

Anthropologist Karen Dalke wrote of the mustang (2010, p. 100),

> In many respects it is an icon of national freedom produced through media images, but fed to an urban public relatively isolated from the public lands where the mustang roams. This dissonance between the actual experience of the mustang and the barrage of historical images it evokes allows a variety of beliefs about the wild horse to emerge and gain a following. The exchange of information results in the wild horse becoming a product of a collective imagination that views it as an icon of freedom.

Some are inclined to dismiss many of the activists campaigning on behalf of free-roaming horses as emotional, unrealistic "horse-huggers." But these activists are often the first to see the gestalt of the mustang situation, to alert us to trouble, and to harness public sentiment for the good of the free-roaming horse.

America's mustangs have suffered greatly at our hands. They have dwindled from robust abundance in the 19th century to small, scattered, fenced populations ranging from a few dozen to several thousand.

The Bureau of Land Management is responsible for the welfare of these iconic animals, and most of its employees say they want to do right by them. Government agencies in general, however, have a long history of using deception, betrayal, and shaky science to advance their objectives, and they have not always acted in the best interest of the mustangs. "The BLM was a livestock and oil and gas extractive user management organization; then they had a wildlife species foisted upon them," says Ginger Kathrens, volunteer executive director of the Cloud Foundation (Rhodes, 2010). The ranching industry exerts strong influence on BLM policy and has been largely successful in pressuring the government to remove free-roaming horses from public lands (Stillman, 2005).

Yet the majority of American citizens stand behind the mustang and believe that wild horses should continue to roam their cradle of origin in significant numbers. In a democratic republic, strong public sentiment should count for something.

By some accounts, millions of wild horses roamed the American West through much of the 19th century. Nobody really knows how many there were at peak population; indeed, there were no formal large-scale mustang counts in the Western states until the BLM conducted them. Wild horses were plentiful in the Old West, though, and the tally of mustangs removed over the years to be sold as saddle horses, meat, or hides suggests that the population could well have numbered in the millions.

According to Wyman (1945/1962), when Zebulon Pike explored the southwestern part of the Louisiana Territory in 1806, wild horses were so abundant that he sent scouts ahead to frighten them away, lest they run off with his party's domestic horses. During Texas cattle drives, cowboys would sometimes startle a wild herd into flight, and then "as many as 200 head would sweep right through the cattle" (Wyman, 1945/1962, p. 133), scattering them and necessitating as much as an entire day to reconsolidate the herd.

Rufus Steele, a mustanger who was intimately familiar with the wild horses of the West, wrote (1910, p. 198),

> In Nevada there are to-day not less than fifty thousand wild horses. There may be one hundred thousand, for their habits are such as to make any exact count impossible. . . . There are bays, albinos, chestnuts, red and blue roans, pintos, sorrels, buckskins, and milk-whites. The mares average eight hundred pounds in weight, and the stallions frequently weigh three hundred pounds more than that; they stand from thirteen to fourteen hands high. Their endurance is phenomenal, and as for agility, the marks of their unshod hoofs are found at the summits of monumental boulderpiles which even a mountain goat might reasonably be expected to cut out of his itinerary. They keep to an elevation of from six to nine thousand feet, descending to the plains hardly at all. The water-holes are from twenty to fifty miles apart, but when the taint of man is upon a drinking-place, they will turn aside from it, even in midsummer, and wander on until instinct leads them to a spring that man has not defiled. In winter the waterholes may be solid ice, but the horses are not inconvenienced—they eat the snow. Bunchgrass is their sustenance in summer; then the first frosts cure the white sage, and that becomes palatable; they paw through the snow to reach it, and keep fat throughout the winter.

GAS MASKING THE HORSE

PROTECTION FOR MAN AND BEAST

If the deadlock on the Western Front is broken and trench warfare is abandoned for open fighting, the cavalry horse will again become an important factor. The supply of horses is not so great that the Allies can afford to sacrifice them and therefore many animals at the front are now equipped with gas masks.

Countless mustangs served in foreign wars, not always with American soldiers. As this item from 1917 suggests, horses and men faced the same horrors, though horses' survival rate was much lower.

Farther north, wild horses had difficulty surviving the bitter winters without human assistance. Bison fared well through the harsh winters of the northern Great Plains by virtue of insulating body fat, thick woolly coats that trap body heat, and the warmth of other herd members. Bison can survive in deep snow for days without food or water. Horses, mules, and cattle, however, sometimes froze solid in their tracks and "toppled to the ground like statues" (De Steiguer, 2011, p. 80). On August 24, 1805, Meriweather Lewis commented in his diary on the dearth of wild horses in what is now western Montana (De Steiguer, 2011).

As Europeans and their descendants domesticated the Wild West for farming and ranching, they came to view the wild horse as an enemy, a rival for resources on the range, and a threat to crops. Wild stallions would raid ranches and claim domestic horses, and cattlemen systematically exterminated them as pests. Theodore Roosevelt wrote, "the wild stallions are, whenever possible, shot; both because of their propensity for driving off the ranch mares, and because their incurable viciousness makes them always unsafe companions for other horses, still more than for men. A wild stallion fears no beast except the grizzly, and will not always flinch from an encounter with it" (1896, p. 33).

By 1824, ranchers and government agents were building corrals at the edge of town where whole communities would join forces to gather roaming horses. Once he had corralled them, the rancher would release the horses one at a time—branded domestic horses would be reclaimed by their owners, and mustangs would be shot (Wyman, 1945/1962). Until the 1920s, noted big-game hunters would shoot them for sport, as they did the buffalo (Wyman, 1945/1962). Wild burros were similarly hunted in Arizona. Emerson Hough (1898, p. 328) paraphrases one hunter who described them as "about as hard game to hunt as any he ever saw. . . . wilder than deer and will run at sight of man at any distance."

C.F. Davis (1887, p. 447) related the experiences of his friend Walker, who spent 3 years raising cattle in southwestern Kansas. Walker told Davis,

> We are much annoyed by wild stallions, which come into our stock herds and steal away our mares. . . .
>
> Nearly every ranchman in my section of the country has at least one horse trained for hunting wild stallions. . . .
>
> [Y]ou can sometimes manage to kill five or six in a single run. . . .
>
> You have to use a revolver, for, going at such breakneck speed, it would be impossible to use a rifle with any accuracy.
>
> Every wild herd has a 'king stud'—usually a four-year-old—and it is desirable, if possible, to kill him first; for, deprived of a leader, a panic seizes the herd, it scatters somewhat, runs with less energy, and you have a much better chance to get in your work. If you do not kill him, you may not be able to get a single one—of course, you don't care to kill the mares—and if you wound him, you want to look out for him.

Walker described his encounter with one stallion,

> He was a beauty—black as jet. . . .
>
> He darted hither and thither with lightning rapidity; now urging on those that showed any disposition to lag behind, and now restraining the foremost ones. . . .
>
> I had ridden fully ten miles—in the meantime killing two 'spikes' (yearling studs)—before I got a shot at the king. . . .

I had never seen anything in the line of horseflesh move with such nimbleness as that same mustang. But . . . the bullet struck him, as I afterward learned, in the neck. . . .

I noticed that they were running less rapidly—evidently holding up for their king; and when I caught a glimpse of the latter, I saw that he showed unmistakable signs of distress. . . .

I soon succeeded in getting another shot at him, and this time I planted a bullet just back of his fore shoulder, and, with a wild scream, he plunged heavily forward, and rolled over quite dead. His race was run, and nobly, too. . . .

I cut off the king's tail as a trophy, and have it now.

From the late 1800s through the 1930s, if not later, the U.S. Army Remount Service captured many mustangs (U.S. Bureau of Land Management, Billings Field Office, 2009) to "remount" soldiers who repeatedly had their horses shot from beneath them. The Second Boer War in South Africa (1899–1902) found the British Empire scrambling to supply its soldiers with horses. Mustangs were abundant and inexpensive to gather; more than 190,000 horses, most of them mustangs, were shipped from New Orleans to South Africa (Campbell, 1908). This was unfortunate for the horses; three quarters of them perished *en route*, and most of the rest died soon after. Sometimes the fallen horses provided food for soldiers after rations were exhausted. The British and their allies lost more than 300,000 horses in two and a half years and as many as 500 a day. The equine carnage was so devastating that the British dedicated a Horse Memorial in Port Elizabeth, South Africa, in 1905.

Then, as now, the Nevada wilderness remained the last major holdout of the American mustang. In 1900, under pressure from the Nevada Live Stock Association, the state of Nevada passed a law that allowed people to shoot mustangs on sight. Professional horse hunters would position themselves by springs and shoot horses as they came to water (Wyman, 1945/1962). Over the next two years, 15,000 mustangs were gunned down, their hides sold to manufacture conveyer belts, upholstery, and baseballs. As the wild population dwindled, hunters began shooting free-range domestic horses that wore the brands of local ranchers. The stockmen hastily lobbied for the repeal of the law and then quietly continued paying bounties on mustangs.

Others profited by capturing mustangs, breaking them to saddle, and selling them as riding horses. Steele wrote "To many men the catching of these horses is a source of livelihood. They live among mustangs, they think mustangs, they measure in mustangs" (1910, pp. 199–202). Because greater volume equaled greater profit, the emphasis was on capturing large numbers of horses and breaking them as quickly as possible using violent, aggressive techniques that typically killed about 25% of them. Steele wrote admiringly about one prominent South Dakota man who shipped "seven thousand splendid horses" from Nevada to Middle Western markets over the course of 6 years. Steele estimated that an additional 2,000 horses were killed in the capture and saddle training process.

North America was not alone in its exploitation of wild horses. In South America, an astounding number were exterminated. In 1891 alone, 700,000 wild horses were slaughtered for their hides in Argentina, and a comparable number were killed in Brazil (Wyman, 1945/1962).

In 1908, the U.S. Forest Service made plans to shoot 15,000 mustangs that roamed Lander County, Nevada. The press picked up the story, and officials were inundated by letters from outraged Americans protesting their destruction—as well as inquiries from hunters seeking employment and from knackers hoping to process the fallen animals. The Forest Service quickly abandoned the plan, not wishing to associate itself publicly with such a controversial practice as horse extermination.

World War I generated a greater need for pack and saddle animals. Horses captured from the wild earned mustangers $30–40 each, roughly $500–700 today (McKnight, 1959). The Remount Service processed around 571,000 horses and mules during World War I (Born, 2007). Around 68,000 were killed in combat, many of the remainder were apparently left in Europe, and some of them were probably eaten. The United Kingdom bought another 618,000 from North America (Jurga, 2012). At the end of the war, there were 39 remount depots in the United States with a capacity of about 229,000 animals (Born, 2007). Only seven remained by 1941. In World War II the United States used far fewer horses than it had in World War I, but Germany, despite its advanced technology, used more (Keegan, 1993).

To meet the demand for military horses, the Army Remount Service methodically shot wild stallions, then introduced Arabians and Thoroughbreds into the herds to take their places and mate with wild mares (USBLM, Billings Field Office, 2009). Mustangers sold young horses to the government, which hastily broke them and pressed them into service. Sponenberg (2011) says that "only a very small minority of feral horses (mustangs) in North America qualifies as being Spanish in type and breeding."

Mustangs' numbers dwindled under the pressures of hunting and capture through the early 20th century, but then the population rebounded once more. The mustang population in the American West was estimated at about 1 million by 1925 (Wyman, 1945/1962). Wild horses had no federal protection, so mustangers were free to capture unbranded horses for commercial gain. California chicken-feed processors required enormous quantities of horsemeat and pressured the railroads to allow a special "chicken-feed rate" on live horse shipments. Wyman (1945/1962, p. 204) writes, "By designating a carload of horses as 'chicken feed,' the railroad was under no legal obligations to give humane treatment to the cargo. Under this rate, thousands of horses purchased for one cent a pound or less on ranges as far east as the Dakotas were transported to California."

The 1920s also saw an unprecedented demand for canned dog food, marketed as a "balanced ration" (Wyman, 1945/1962, p. 205) that was superior to table scraps. P. M. Chappel of Rockford, Illinois, and his two brothers sold their first cans of pet food under the label of Ken-L Ration. Initially, the company canned the meat of mostly aged domestic horses, but as demand increased, the brothers turned to raising their own herd of meat horses (Cruise & Griffiths, 2010). Mustangs proved a cheap and abundant source of flesh, comprising most of the 1,446 horses processed for the pet food industry in 1923. By 1933, the industry was canning almost 30 million pounds of horsemeat for dog food, most of it from mustangs (Wyman, 1945/1962).

When soldiers returned from World War II, attended college under the GI Bill, and set up homes in the suburbs, they acquired dogs and cats that were treated like members of the family. To meet the demand, mustangers captured more horses and sent them to processing plants (De Steiguer, 2011).

The choicest cuts of horsemeat were shipped to Europe for human consumption, where they brought a premium price. Manufacturers processed scraps into glue, and fashioned hides—especially the tender skins of unborn foals—into pony-fur jackets for women.

If your dog doesn't like Hill's Horse Meat with Gravy,

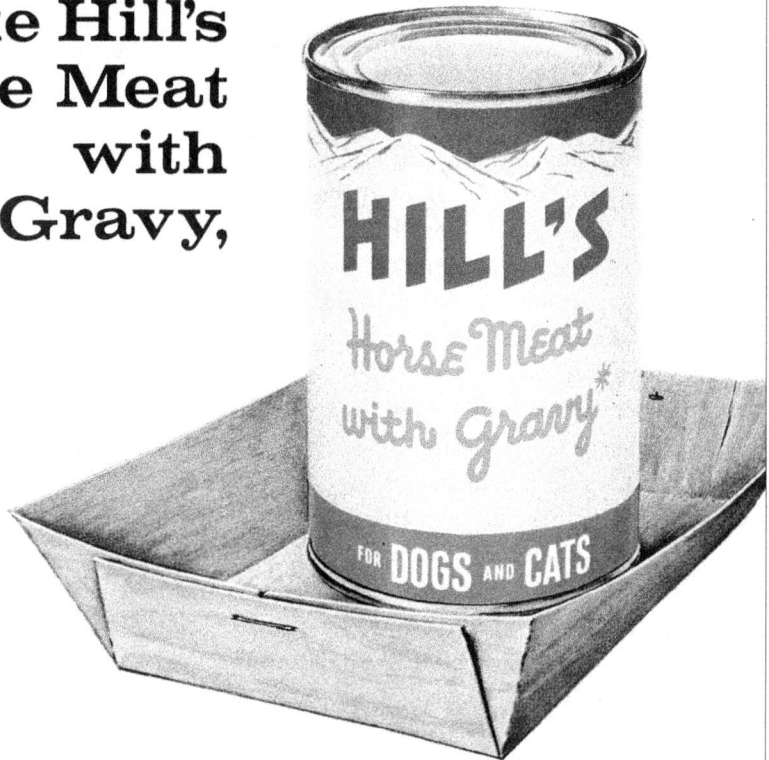

it's because he's a vegetarian.

Hill's Horse Meat with Gravy is for meat-lovers. It's real meat. All meat. And real gravy, made from natural juices. Gives your dog a 24-hour supply of vitamins and minerals. Deliciously. If your dog is not a vegetarian, get him Hill's Horse Meat with Gravy. After all, if you ate just one meal a day, you'd want it to be a good one, too.

Mustangs not turned into cannon fodder often ended up as dog food. Advertisements like this one, published regionally in *Life Magazine* July 24, 1964, barely 3 months after Ford introduced its popular Mustang sports car, cause discomfort today, illustrating how public opinion has changed.

Wild horses were seen as a nuisance competing with cattle for grazing land. Labeled an exotic species, the horses had no protection. Paradoxically, food crops such as wheat and rice and livestock such as cattle and sheep are non-indigenous species, but are considered desirable because they bring profit.

Through the 1920s, wild horses were gathered by the thousands in Montana and Wyoming. In 1933 and 1934, a single buyer shipped 22,000 wild horses out of the Bighorn Basin from the vicinity of the modern Pryor Mountain herd.

Wild horses were packed crippled and bleeding into trailers and trucked to slaughterhouses, barely alive. Because they were classified as feral livestock, none of the laws protecting wildlife applied. As Hope Ryden observed, "Therefore, in a sense, it was open season on wild horses twelve months of every year, and a hunting license was not required to shoot, gather, trap, or torment them" (2005, p. 16).

With the passage of the Taylor Grazing Act in 1934, the Bureau was empowered to manage livestock density levels on these previously unregulated lands. Grazing rights were rented to stockmen, and the number of grazing animals allowed on each tract was limited to the carrying capacity of the range. A wild horse eats and drinks about as much as a domestic horse, a cow, or five sheep. Stockmen saw wild horses as competition with their own stock for the limited resources of the range. BLM grazing officials helped stockmen gather the horses in cooperative "county roundups" throughout the West, and the animals were generally sold for slaughter (Johnson, 1943; McKnight, 1959).

Anti-mustang sentiment among stockmen was by no means universal. Dayton Hyde, founder of the Black Hills Wild Horse Sanctuary, wrote (2009, p. 115), "for every rancher who hated wild horses there were many more who tolerated them or even loved them." Fences kept mustangs off the prime grazing lands and pressed them to forage in mountains and deserts unable to support cattle or sheep.

Abattoirs would not accept dead animals, and wild horses invariably resisted slaughter to their last breath. So mustangers subdued and weakened horses bound for slaughter to minimize their resistance when they were packed into trucks and trains. Their methods invariably caused great suffering while keeping the horses alive long enough to collect payment. Sometimes the men would lasso the animal around the neck and allow it to drag a tire until it was exhausted and nearly asphyxiated. One technique was to "crease" the horse—that is, shoot him in the upper part of the neck, in front of the withers, near the spine (Hough, 1897, pp. 85–86). Hough wrote that this practice produced a dozen dead horses for every live one. When the horse was stunned by the gunshot, the mustanger would tie the horse's feet together. Other captors would cut a ligament behind the horse's knees with a sharp knife and drain the joint fluid, then force him to stumble for over 100 miles to a railway depot (Wyman, 1945/1962). Sometimes mustangers tied the neck of one horse to the tail of the next in bunches of four or five. Alternatively, the captor tied a rope from the neck of the horse to a foreleg, or through the front and back legs to the tail, making it impossible for him to lift his head, and leave him standing without food or water for up to 24 hours while they collected other horses (Cruise & Griffiths, 2010). Ropes were sometimes tied to barbed-wire loops threaded through the cartilage between the nostrils. Sometimes the mustangers deprived the horses of air by throwing them down and sewing their nostrils partially closed or gouged out their eyes to blind them (Ryden, 2005). The men had an infinite capacity for devising tortures to immobilize a horse through pain and fear. During the 1940s and 1950s, mustangers gathered wild horses using airplanes and trucks. They sold some to ranchers as saddle horses and others as rodeo bucking stock. They shipped the majority to processing plants to be canned as pet and fur-farm food (McKnight, 1959). More than a million wild horses were killed between 1900 and 1950, their destruction often approved and funded by the BLM (Dalke, 2010). In the four years following WWII, the BLM authorized the removal of more than 100,000 mustangs from Nevada alone, virtually all of which went to slaughter (De Steiguer, 2011).

The laws were tightened chiefly because of the life work of one individual, Velma Johnston, better known as "Wild Horse Annie." Johnston, who worked as a bank secretary in Reno, Nevada, was driving to work one day in 1950 when she noticed a livestock truck leaking blood (Ryden, 2005). She followed it to its destination and was appalled to find it packed with severely injured mustangs. Many were wounded with buckshot, with eyes and lips shot out and hooves worn to bloody stumps from being chased over rocks to the point of collapse (Cruise & Griffiths, 2010). A colt that had lost his footing had been trampled to a pulp by the others. This cruelty was perfectly legal.

Johnston investigated and found that the treatment of mustangs bound for slaughter was "barbaric beyond belief" (De Steiguer, 2011, p. 154). She spent the rest of her life fighting for legal protection for wild horses, and her efforts resulted in the 1959 Wild Horse Annie Act (An Act To Amend Chapter 3 of Title 18 . . . 1959), which prohibited hunting or harassment of wild horses on public lands using motorized vehicles or aircraft.

But abuses and even full-scale massacres continued. Over several months in 1989, at least 500 wild horses were deliberately shot in the backcountry outside Lovelock, Nevada. The shooter(s) were never convicted, and the community was inexplicably divided about whether the perpetrators should be considered villains or heroes (Stillman, 2009).

In 1998, Reno construction worker Anthony Merlino and two U.S. Marines, Lance Corporal Scott Brendle and Lance Corporal Darien Brock, shot 34 wild horses in Nevada's Virginia Range. Evidence included a trophy photograph of Brock standing on a dead horse. Merlino, the ringleader, allegedly had been shooting wild horses since adolescence and bragged about it to friends. At their trial, State Veterinarian David Thain presented evidence from the autopsies performed on each carcass—ruptured organs, collapsed lungs, and shredded hearts—one horse had even been sprayed in the face with a fire extinguisher. Many lay dying for days, digging hollows in the desert earth in a vain effort to stand and escape. The veterinarian worked his way down the list, and by the time he reached the seventh horse, he broke into tears (Stillman, 2009). As the press revealed the details to the public, at least 50,000 letters to the media expressed outrage and horror at the shootings and called for justice. Yet when the dust settled, all three received light sentences. After this highly publicized trial, the Nevada legislature made killing a wild horse a felony.

To many who made a living from the resources of the West, the mustang was a rival to exploit, dominate, or eradicate. Yet to a much larger faction, the wild horse came to represent freedom, independence, and self-determination, attributes prized by Americans. Monty Roberts wrote in *Shy Boy* (1999, p. 4), "To some extent we attach to the mustang the same characteristics we would like to see in ourselves: strong, wild at heart, sensitive, graceful—and above all, free." Dalke (2010, p. 102) observed,

> Television began to reach more people than any other mode of communication by the mid-twentieth century, creating mythology about cowboys and mustangs, and

viewers believed what they saw. Even cowboys began to conform to the televised images of themselves. . . . The mustang becomes an object of beauty and wonder as it moves across an idyllic, natural landscape. The mustang is no longer an animal, but an image of an animal contextualized in romanticism.

The Ford Mustang, a fixture in American popular culture since 1964, owes much of its enduring appeal to its four-legged namesake. Ford's marketing strategy attracts individuals who fancy that they are free-spirited and nonconformist.

At the same time, it influences how people who have never seen a living wild horse understand mustangs and "know" what they are like. Although there is no direct contact with the animal, people imagine what it must be like based on its association with a car, and resonate with the ideals attributed to it. Thus the mustang becomes a powerful, independent creature that savors its freedom (Dalke, 2010).

On December 15, 1971, the U.S. Congress passed the Wild and Free-Roaming Horses and Burros Act of 1971, which reads,

Congress finds and declares that wild free-roaming horses and burros are living symbols of the historic and pioneer spirit of the West; that they contribute to the diversity of life forms within the Nation and enrich the lives of the American people; and that these horses and burros are fast disappearing from the American scene. It is the policy of Congress that wild free-roaming horses and burros shall be protected from capture, branding, harassment, or death; and to accomplish this they are to be considered in the area where presently found, as an integral part of the natural system of the public lands.

The law charged the Department of the Interior to "manage wild free-roaming horses and burros in a manner that is designed to achieve and maintain a thriving natural ecological balance on the public lands" and to "consider the recommendations of qualified scientists in the field of biology and ecology, some of whom shall be independent of both Federal and State agencies." After passage of the Public Rangelands Improvement Act of 1978, the Bureau of Land Management created 303 herd management areas on 65,780 mi^2/170,369 km^2 of range throughout 10 Western states, an area roughly the size of Florida (Behar, 2010).

The Bureau of Land Management estimates that the free-roaming equid population totaled about 17,000 horses and 8,045 burros in 1971 (USBLM, Wild Horse and Burro Program, 2007) and rose to 65,000–80,000 in 1980 (USBLM, 2012, August)—although another federal document reports 9,500 in 1971 (USBLM, Wild Horse and Burro Program, 2005). In 2008, the agency claimed that Nevada HMAs supported about 6,000 more horses than the recommended population level, and other ranges had 18,000 mustangs where the BLM said there should be no more than 12,000, a number that jumped by about 20% with each crop of foals (Lewis, 2009, February 1). According to the Bureau, as of February, 2012, an estimated 31,500 wild horses and 5,800 wild burros roamed on BLM-managed rangelands in 10 Western states, mostly in Nevada (74%) and Wyoming (10%) (USBLM, 2012, September 28).

Wild horse advocates question these numbers because they disagree not only with one another, but also with those put forward by other authorities. Many mustang advocates claim that the BLM is overestimating the number of horses in the wild. As Ginger Kathrens said in a blog interview, "The BLM would have us believe there are over 30,000, and yet the best statistical review that's been done would indicate as few as 15,000" (Rhodes,

2010). Further, round numbers are naturally suspect because they are rare in nature. Tallies, projections, and population targets evenly divisible by a power of 10 may denote anything from unavoidable imprecision to haphazard guesswork to outright deception.

The term *population census* is misleading. A census implies that every animal is counted. Yet it is difficult for even the most conscientious scientists to count fast-moving wild animals over large areas of forbidding terrain. Horses are often well camouflaged or hidden by trees or vegetation, and weather conditions, topography, and technique affect accuracy. In a New Zealand survey of marked horses, the helicopter census overestimated herd size from 15% to perhaps 32% (Linklater & Cameron, 2002).

In North America, it appears that wild horse populations are often undercounted. A few herds can be counted using ground surveys with identification based on photographs or unique markings of individual horses. Most wild-horse censuses are conducted by helicopter, but in rugged terrain and wooded areas it is likely that only about half the animals are sighted. The 1971 equid count was based on an educated guess by census takers flying over the herds in fixed-wing aircraft. The National Research Council (1982, p. 45) concluded, "The 17,000 figure is undoubtedly low to an unknown, but perhaps substantial degree." The panel noted that spotters probably identified 93% of horses in open landscapes, but only about 40% in wooded, mountainous areas, while a helicopter census in the same area counted 48% (NRC, 1982, p. 4).

Likewise, the 2013 NRC report asserted, "the statistics on the national population size cannot be considered scientifically rigorous" (p. 5). The Wild Horse and Burro Program conducts a population census or estimate on every herd area every four years at a minimum (USBLM, Wild Horse and Burro Program, 2005). The committee proposes that reported annual counts for many HMAs probably underestimate the actual horse population by 10–50% (NRC, 2013).

Scientists studying bears, mountain lions, tigers, wolves, coyotes, and mountain gorillas have successfully monitored populations using DNA from feces, hair, and scent marks. DNA samples serve as genetic identification tags that reveal not only individual characteristics, but also population size, sex ratios, kinship, and levels of genetic diversity. Eggert et al. (2010) successfully genotyped horses from manure at Assateague Island National Seashore, and checked accuracy against the known roster of individuals. If researchers develop a quick, inexpensive method for genotyping and data analysis, non-invasive DNA surveys could gather accurate information without disturbing natural behavior (NRC, 2013).

A possible solution may be to utilize Department of Defense satellite imagery for identification of horses. Computer modeling could be employed to evaluate fertility management and genetics. Alternatively, drones (unmanned aerial vehicles/UAVs) may cause less disturbance than larger piloted aircraft, may gather more detailed information than satellites, and are likely more cost-effective than either. The FAA has issued several hundred permits to fly drones in domestic airspace for everything from research to fisheries management to law enforcement (Lynch, 2012; Martin et al., 2012). The World Wildlife Fund uses small battery -powered drones to study and protect rhinoceroses and tigers in Nepal (McGivering, 2012), and the U.S. Geological Survey is using decommissioned U.S. Army drones to monitor sandhill cranes, count moose, and study mountain pine beetles ("Cranes and Drones: Strange Airfellows?" n.d.).

The NRC also looked at wild horse population estimates and demographics, reproductive rates, competition, and impacts on range habitat. A population's ability to grow maximally under optimal conditions is called its biotic potential, represented by the letter r in formulas. Usually environmental conditions are not optimal, and population increase is limited by food, water, climate, habitat, and other factors. Carrying capacity, the maximum number of individuals that a habitat can support, is represented by the letter K. The BLM defines equine carrying capacity more narrowly as the maximum number of horses that can live from year to year in a setting without causing damage to plants or related resources (USBLM, n.d.).

Reproductive physiologist Jay Kirkpatrick, senior scientist at the Science and Conservation Center of Billings, Montana, suggests that "acceptable impact" is a better term (personal communication, May 28, 2014). A single horse, he says, will damage plants and related species, and land managers must determine what level of impact is acceptable within each unique range.

Species whose populations increase until they reach the carrying capacity of their environment and then keep a balance are referred to as K-selected. "Horses fall into the K-selected species description, where a species produces relatively low numbers of offspring annually and invests a lot of effort in the survival of the offspring," said wildlife biologist Doug Hoffman (personal communication, March 18, 2011). "The K-selected species is more vulnerable to environmental factors or events, resulting in relatively low or no population growth once established." Conversely, when the population is low and resources are abundant, K-selected species increase quickly until they reach carrying capacity. Human populations are also K-selected, along with eagles, apes, deer, elephants, and bats.

At the other end of the continuum are r-selected species, such as rabbits, mice, frogs, and mosquitoes. These species usually have short lives, mature quickly, and produce large numbers of young that require minimal or no parental care. Their populations fluctuate with intervals of rapid growth, mostly in response to physical conditions unrelated to the population density, such as rainfall and temperature.

The anti-mustang contingent—as well as a good many mustang advocates—voice concerns about the rapid reproduction of free-roaming horses. The Bureau proclaims that feral herds "grow at an average rate of 20 percent a year and can double in size every four years" (USBLM, 2013, March 8).

It points to studies for corroboration, such as Garrott and Taylor's (1990) which calculated an 18% growth rate for the herd at Pryor Mountain Wild Horse Range (31,000 acres/12,545 ha) using BLM data from 1970–1986. They wrote, "The highest foaling rates occurred after a 51% reduction in the population, suggesting a density dependent response.... The adult sex ratio was consistently skewed approximately 2 to 1 in favor of females and was attributed primarily to the removal of disproportionate numbers of males during herd reduction programs. The apparent mean growth rate of the population was 18% annually; however, by removing animals, the population was maintained at approximately 120–150" (1990, p. 603).

Eberhardt, Majorowicz, and Wilcox (1982, p. 367) say clearly that their growth rates, for two herds, were "estimated ... to be about 20% a year," but noted that such a rate is possible "only if survival rates are high and reproduction exceeds that normally expected from horses." They add, "We do not propose that the observed rates in these herds are necessarily

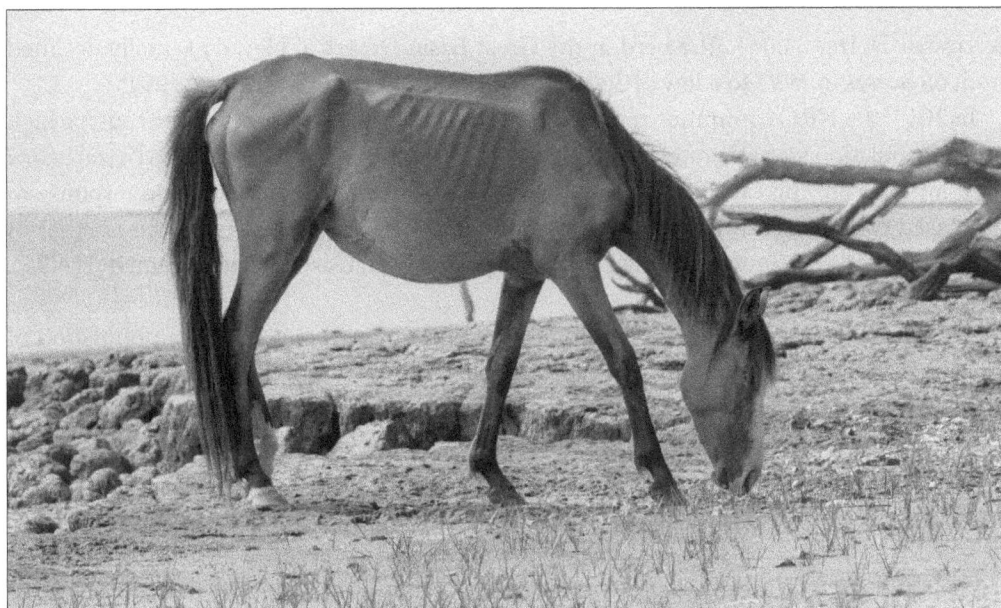

A goal of wild horse management is to keep the herd census in balance with environmental resources. When there is not enough food, horses die of starvation and disease. On Cumberland Island National Seashore, the Park Service does not manage the wild horse herd, and there is not enough forage to sustain all the horses. For some, life is a daily struggle to consume enough calories to stay alive, and mortality is high. This mare is about 4 or 5 years old, lactating, and perhaps pregnant again. The photograph was taken in June, when food is most abundant. The broken hairs at her tailhead also suggest anal itching due to a heavy parasite load, another consequence of overcrowding.

typical of feral horses in general," and they say again that "the observed rates are appreciably higher than might be expected on the basis of other experience with large animals" (p. 367).

Garrott, Siniff, and Eberhardt studied numerous herds, but concluded (1991, p. 641) that "many feral horse populations are currently being maintained at levels below which density-dependent responses operate and are, therefore, increasing at or near their biological maximum."

One of these studies was done by a federal contractor and two BLM employees, another was done by a professor and a BLM employee using BLM data, all are outdated, and the agency has misinterpreted or misrepresented all three.

The NRC made note of "several biases in the census data" (1982, p. 1) and concluded that "increase rates are undoubtedly subject to a variety of environmental pressures," including "variations in forage conditions associated with annual weather variations... variations in forage conditions associated with different equid population densities; and ... variations in forage conditions associated with use by other herbivores, both wild and domestic"(1982, p. 14).

Typically, when large herbivores are introduced to a new range or rebound from a population reduction, they quickly increase to peak abundance, crash to a lower abundance, and then increase to a carrying capacity lower than peak abundance (Forsyth & Caley, 2006). Consequently, it is uncommon for horse herds to grow by 18–20% annually unless the population density is low in relation to available resources, and then for only a short interval. Between 1967 and 1988, before population control was implemented on Assateague Island National Seashore, the horse herd grew by about 9% annually (Zimmerman Sturm, Ballou,

& Traylor-Holzer, 2006). The herd in the Great Basin Desert of Nevada actually declined from 65 horses in 1992 to a low of 36 horses in 1998 (Greger & Romney, 1999).

In 2013, the NRC committee reported that BLM management practices encourage high horse population growth rates (NRC, 2013). Periodic removals keep herd size below food-limited carrying capacity, causing the population growth to accelerate in response to decreased competition for forage. The committee concluded, "As a result, the number of animals processed through holding facilities is probably increased by management" (NRC, 2013, p. 6).

Carrying capacity fluctuates in response to environmental conditions; a population of large herbivores can exceed carrying capacity following a dry season, and fall well below K after a few weeks of rain without a change in density (Georgiadis, Hack, & Turpin, 2003). During long periods of drought, ungulate populations naturally decline, then rebound during wet years.

If allowed to reach carrying capacity, numbers remain more or less stable from one decade to the next, increasing in a series of good years and declining in bad. The unmanaged Sable Island herd, off the coast of Nova Scotia, has stabilized around 200–350 and stayed in that range for more than 200 years, proliferating in good years and dying off in large numbers when resources are scarce. The wild herd at Cumberland Island National Seashore has no current management plan in place, and consequently the population has equilibrated at carrying capacity. Between 1986 and 1990, the Cumberland Island herd grew by only 4.3% (Goodloe, Warren, Osborn, & Hall, 2000).

While herd growth slows as population density increases, it appears that horses self-limit their numbers only after they reach the carrying capacity of the range (NRC, 1982). Before herd growth ceases, overgrazing can cause environmental damage and reduce diversity of species, and starvation can occur.

Lou Hinds was manager of the Chincoteague National Wildlife Refuge, which is home to a famous herd of free-roaming ponies. He explained (personal communication, May 21, 2010),

> If you manage for carrying capacity, the ponies will eat all the good stuff first, then the secondary stuff, until they reach down and get the stuff that just supports their population. At that point, you have reached what most managers would call carrying capacity. The number of animals living on the island is supported by the island. Any more than that would starve to death. But this selective grazing affects natural plant diversity. And we are managing our wild lands for the total diversity of plants and animals, not for just one species or another.

While overgrazing is detrimental to both horses, plant life, and competing species, light to moderate grazing can actually benefit the environment. Numerous peer-reviewed studies have demonstrated that controlled grazing can create beneficial disturbance and promote biological diversity, accelerate succession, promote the growth of desired vegetation, and make forage more nutritious (Hobbs, 1996; Severson & Urness, 1994; Vavra, 2005). Despite these data, the BLM has primarily managed wild horses by restricting range, limiting fertility, and removing them from federal lands.

"Each herd management area (HMA) has an estimated appropriate management level (AML), or the number of horses that can be living on that land in a healthy manner without degrading the range," says Jay Kirkpatrick (2010, pp. 12–13).

A mustang stallion drives a mare and her yearling away from the author at a rapid pace. The Pryor Mountain herd ranges along a major Crow and Shoshone migration route, and tribal horses probably contributed to the herd genetics. Notice the barbed-wire fence delineating the boundary of the horse management area. In the modern world, even wild and free horses are, in a sense, confined.

Exactly how AML is determined remains a mystery on many HMAs but clearly differs from HMA to HMA, and how that AML relates to domestic livestock on the same land also remains unclear. The advocacy groups, in general, always dispute the horse AML as too low, and the agricultural groups dispute the numbers as too high. Both groups often see the AML number as some sort of conspiracy. But what is the answer? The wild horse and burro program reached a "critical crossroads" (USBLM, 2009, October 7) recently when pressures from the economic recession caused holding costs to skyrocket and adoptions to plummet, rendering the current mode of management unsustainable for the animals, the environment, and the taxpayers. In 2007, the program cost Americans $38.8 million. By 2011, its budget reached $75.8 million, $8.7 million of which was spent on "gathers" and removals, $7.3 million on adoption events, and a whopping $35.7 million (47%) to maintain captured horses, which can live for 25–30 years (USBLM, 2012, September 28).

Decades before this crisis, the Bureau had recognized the need for an alternative to slaughter, dispossession, and neglect, so it enlisted the help of reproductive specialists Jay Kirkpatrick and John Turner. Their solution was to adapt a method of birth control developed by Alex Shivers at the University of Tennessee in 1972. It took them nearly 16 years, but in 1988 they implemented an agent that literally vaccinates a mare against pregnancy.

The requirements for a successful contraceptive vaccine were many. Ideally it should curb reproduction without changing any aspect of equine behavior. It should be reversible to allow the herd to recover from a natural disaster or disease. One dose should render a horse infertile for at least one breeding season. And the treatment should be capable of remote administration by dart gun or bio-bullet, to minimize the handling of these wild animals.

For 15 years Kirkpatrick and Turner focused on steroids to check reproduction. These were successful in the laboratory, but in actual use they were inefficient and produced inconsistent results. In separate studies, other researchers tried various hormonal preparations.

Hormonal implants, successful in women and in skunks, were impractical in mustangs. Copper intrauterine contraceptive devices (IUDs) designed for 10-year use in women (Paragard®) were about 80% effective the first year, but ineffective thereafter, probably because they were comparatively small and easily expelled. Castration and hormonal sterilization of either sex alters behavior, and castration is irreversible. A wild herd's social dynamics hinge on sexual behavior. Disrupting these natural interactions with surgery or hormones degrades a wild herd to sexless domestic stock. Treating harem stallions with large doses of testosterone necessitated gathering, capturing, immobilizing, and injecting them on an annual basis and tended to shift foaling times to summer or fall (USBLM, Wild Horse and Burro Program, 2005).

Disappointed with the poor field results, Kirkpatrick and Turner tried a different approach. They shifted their focus from endocrinology to immunology. This was new territory for them. In 1988 they collaborated with Irwin Liu of the University of California at Davis, who suggested that they work on an immunocontraceptive derived from porcine zona pellucida, the transparent, noncellular protein layer surrounding the egg cells and early embryos of mammals, pigs in this case. Sperm must pass through the zona pellucida in order to fertilize the egg, so it is an ideal target for immunocontraception. Liu, a specialist in equine reproduction, was excited by the possibilities.

This technology actually allowed researchers to vaccinate the mares against pregnancy. The PZP injection works by persuading the recipient's body to block the attachment of sperm to ova. Complementary proteins on the surface of the sperm and the zona pellucida of the ovum allow the sperm to attach to the ovum and fertilize the egg. When researchers injected a mare with zona pellucida from pigs, the mare's immune system produced antibodies against this foreign protein, which attached to the sperm receptors on the zona pellucida of the mare's ova and blocked fertilization (USBLM, Wild Horse and Burro Program, 2005).

A three-person staff at the Science and Conservation Center of Billings, Montana—where Kirkpatrick is senior scientist—produces most of the PZP used to contracept wild horse populations (Frydenborg, 2012). Scientists take pig ovaries from sows that have been slaughtered for their meat, and slice, filter, homogenize, and cook them to produce a pure, injectable liquid. The final product is administered to wild mares via dart gun; two injections the first year, and an annual booster thereafter.

A vaccinated mare comes into heat as usual, every 21 days from March to September. She exhibits all the usual estrus and mating behaviors. But as her ova develop, antibodies attack them and spackle the surface, filling in specialized receptor sites in the zona pellucida. Sperm cells from the stallion swim up to fertilize the eggs, carrying molecules on their surfaces that fit those receptors like a hand in a glove. They are thwarted. The antibodies block fertilization, and no foal is conceived.

By April 1988 the vaccine was ready for field testing. None of the vaccinated mares became pregnant that season.

The vaccine appears fully reversible in the short term—about 21% of mares conceive the year after contraception is discontinued, 55% by 2 years, and 76% by 3 years; after 8 years only 3% are still infertile. The vaccine typically disrupts long-term ovarian function if contraception is repeated for more than 5 consecutive years on a given mare (USBLM, Wild Horse and Burro Program, 2005).

The McCullough Peaks HMA is located 70 mi/113 km east of Yellowstone National Park and includes 109,814 acres/44,440 ha of land. The BLM has set a limit of 100 mustangs on this wild-horse reserve, or one horse for every 1.7 mi²/4.4 km². The current number is 167, after 92 horses were removed in 2009. Before the roundup, horses were in good flesh, indicating that the HMA was meeting their needs. When the author spent a day on the refuge in 2011, she encountered only six horses in the vastness of the grassland. It would be easy to argue that McCullough Peaks can support many more wild horses than BLM permits.

In a natural herd, about half the mares will produce a foal in any year. The effects of the vaccine decline over time, and the mare may be able to reproduce again the next year if not contracepted. After 7 years of treatment, the mare may become permanently infertile (USBLM, Wild Horse and Burro Program, 2005). One-dose annual booster shots display 90% effectiveness in preventing pregnancy. Occasionally, a contracepted mare will produce a "surprise" foal.

The PZP vaccine is relatively noninvasive, nontraumatic, and well controlled. It does not damage the environment and will not pass into the food chain after a mare's death. The majority of animal welfare organizations in the United States support its use. It is also inexpensive, at about $20 per dose. The mares do not appear to be harmed in any way by the vaccine. During the initial trials, a few pregnant mares were accidentally darted as well, but all went on to deliver healthy foals, which have had healthy foals themselves.

A number of behavioral studies demonstrated that contracepted and uncontracepted mares had virtually identical activity patterns, related to the harem stallion in the same way, and were not more likely to leave their bands (Ransom, Cade, & Hobbs, 2010). Another study, however, suggested that contracepted mares are more likely to leave their harem stallions and join with other mates (Madosky, Rubenstein, Howard, & Stuska, 2010). However, these studies were not controlled for band stallion age and tenure, which have been previously shown to be the dominant factor in maintaining harem stability (Rutberg,1990).

In July 2009 the House of Representatives passed the Restore Our American Mustangs Act (ROAM), which proponents said would revolutionize the management of wild mustangs on the range. Three months later, Interior Secretary Ken Salazar introduced a new comprehensive plan intended to make the Wild Horse and Burro Program sustainable long-term. The plan involved removal of excess horses from the range, aggressive contraception of mares, and managing the sex ratio at 40% mares and 60% stallions (USDI, 2010, p. DH-47). Young mares would be allowed to reproduce and contribute to the gene pool and then would be contracepted. The Bureau also considered the release of nonreproducing mustangs into existing HMAs. Under the initiative, Congress or the secretary of the Interior could recognize certain "treasured herds" on public lands to increase public awareness and appreciation, increase interest in environmental education, and stimulate ecotourism that could bring jobs to nearby rural communities.

Salazar's plan met with active opposition from many organizations, especially wild-horse-advocacy, animal-welfare, and public-lands-reform groups. Mustang removals from the Pryor Mountain Wild Horse Range (Montana and Wyoming) and the Calico Mountains Complex (Nevada) elicited harsh criticism and protests in person and on the Web (Harty, 2010). Most resisted the large-scale elimination of horses from the range; some protested the use of immunocontraception and the skewing of sex ratios, which can create unnatural social structures in herds. These detractors have been effective in focusing media, public, and congressional attention on their issues and preferred solutions.

Some wild horse advocates would like to grant horses the freedom to reproduce at will until they reach the carrying capacity of their ranges. When the equine population reaches the maximum number the land can support, the herds will stop growing. Kirkpatrick (2010, p. 5) contends, however,

> While that has some biological logic behind it, the theory does not address the massive deterioration of the range on which these horses live while their numbers explode, nor does it acknowledge the disease and starvation that ultimately cause that population to collapse. One advocacy group holds to the idea that we should just let them starve to death because it's "nature's way."

Death by starvation *is* nature's way when population outstrips resources, or where people have introduced horses, but limited their range, exterminated large predators, or both. But death by starvation is lingering and cruel, causing much more misery to the animal than the stress of gathers and the disruption of fertility control. Skeletally thin mares scrounge for weeks to keep their milk flowing for hungry foals which, like their mothers, eventually succumb when the food runs out. Dalke (2010, p. 108) wrote,

> At a national BLM meeting, an animal advocacy group member said, "If we are going to consider them wild, then they need to fend for themselves and there is going to be starvation!" . . . Some gasped at the statement, but it was obvious that wildness and strength were defining elements of her imagery. Upon further discussion, it became clear that the speaker believed that these mustangs could overcome anything, and if starvation did occur it would be quick and only a few would be impacted. Being from an urban center in the East, she had not experienced the animals or the public lands directly; she did not understand how a mustang could

not find enough to eat on several thousand acres, nor had she seen them starve. In this scenario, the image was so strong it blocked the realities of the local conditions. Believing that these strong, beautiful, independent creatures can survive in any condition, she advocated a laissez-faire management style.

Some horse advocates oppose fertility control, arguing that suppressing reproduction is unnatural. But very little of the environment inhabited by our North American wild horses can be considered natural. The Wild West has been domesticated. Nature has not taken its course with these animals in the last 9,000 years. On this continent the repatriated horse has been displaced from expansive, lush ancestral prairies to resource-poor deserts and mountaintops. The wide-open spaces have been fragmented. Instead of roaming at will, wild horses are contained within fenced management areas.

Clearly there is nothing natural about manipulating fertility, but contraception effectively keeps herd size in balance with the environment while significantly extending the lifespan of treated mares and increasing foal survival. "How does fertility control, which allows almost every mare to breed at some point in her life, compare with gathers and removals, where the vast majority of horses taken off the range have never bred and will never make a genetic contribution to the herd?" asks Kirkpatrick. "In the 2009 Pryor Mountain gather and removal, 37 of 57 horses removed never had the opportunity to breed there—and they never will" (Kirkpatrick, 2010, p. 9).

In addition to liquid PZP, GonaCon™, an immunocontraceptive vaccine that interferes with gonadotropin-releasing hormone, is approved by the U.S. Environmental Protection Agency to prevent pregnancy in wild horses (NRC, 2013).

At the onset of the reproductive cycle, the hypothalamus, at the base of the brain, produces gonadotropin-releasing hormone, or GnRH, which prompts the pituitary gland to release luteinizing hormone and follicle-stimulating hormone. LH and FSH induce sperm production in males and follicular development and ovulation in females, resulting in release of steroid hormones such as estrogen and testosterone. These hormones, in turn, drive sexual function and behavior. GonaCon causes the animal to produce antibodies against GnRH, which bind with the hormone and render it ineffective. (National Wildlife Research Center, 2008). Without GnRH signaling the reproductive hormones, both males and females remain in an asexual state and no longer exhibit natural reproductive behavior.

PZP vaccine is the agent most extensively tested on free-ranging horses, and the National Research Council says that it may be the most promising option for fertility control (NRC, 2013). The vaccine is available in three forms. The most tested and most often used is liquid PZP, prepared by finely slicing pig ovaries to release oocytes from surrounding tissues. It is effective for 1 year.

Turner, Liu, Flanagan, Rutberg, and Kirkpatrick (2007) developed a pelleted form of controlled-release PZP known as PZP 22 because it remains about 85% effective after 22 months. When used on 96 free-roaming mares in Nevada, it produced fertility rates of 5.2%, 14.9%, 31.6%, and 46.2%, respectively, in post-vaccination years 1–4 (NRC, 2013). The rate for untreated mares was 53.8%.

The PZP in SpayVac™ is prepared by grinding and homogenizing whole ovaries, then encapsulating the PZP in liposomes to extend the period of release. SpayVac™ seems to contain more extraneous ovarian material than liquid PZP. This foreign material seems to increase the immune response, producing greater efficacy, but perhaps greater autoimmune reactions. Although SpayVac™ has been used on other wildlife, only one published study (Killian, Thain, Diehl, Rhyan, & Miller, 2008) has assessed the efficacy of SpayVacTM in horses (NRC, 2013). In a small sample of 12 domestic mares, the efficacy rate was 100% in year 1 and 83% in years 2–4. There was evidence of uterine edema in the mares treated with SpayVac™, but the researchers could not determine whether these changes were pathological (NRC, 2013). Another study by Bartell in 2011 found that the ovaries of mares treated with SpayVac™ appeared damaged—lighter in weight than those of control mares with smaller oocytes and thinner zonae pellucidae (NRC, 2013).

Whereas liquid PZP is well studied, not much is known about pathological effects or long term reversibility of SpayVac™. SpayVac™ can be given by hand or dart (NRC, 2013). Liquid PZP can be administered by dart or bio-bullet, but pelleted PZP must be injected by hand. This involves gathering and other expensive, traumatic activity.

Fertility control programs limit the growth of the herds to what the environment can support, allowing BLM to gather fewer horses less frequently. Fewer unadoptable animals live out their lives in holding facilities at public expense, and a larger percentage of gathered animals may find good homes through adoption.

"There are of course concerns that must be addressed when using contraceptives to control overpopulation in free-roaming horses," says Höglund (2010):

> Any program will need to be funded, easy to administer, cost-effective, effective for multiple years, and have few or no known contraindications or adverse effects. Currently, since the vaccines are injectable, horses must be captured and processed or shot from a remote location with a dart or bio-bullet. Aerial-facilitated darting or use of bio-bullets can be inefficient, and these types of vaccination methods make research study of the health and biomedical effects of the contraceptive difficult. Further, annual capture of free-roaming horses for booster vaccination . . . usually becomes more difficult with each subsequent attempt.

Gathers increase expense, stress on the horses, and the potential for injury to horses and personnel. Bartholow (2007) found that the most cost-effective option is to use 2-year or 3-year contraceptives combined with removals that keep the population more than 50% male.

Kirkpatrick prefers to use the original annual vaccine that he researched and popularized. "The longer acting form of the vaccine is very experimental, and costs quite a bit more than the native PZP we routinely use at Assateague Island National Seashore, Cape Lookout National Seashore, Corolla, and Carrot Island. We do not recommend its use on any horses, western or eastern, where the horses can be treated remotely. We firmly believe in not handling the animals, to reduce stress" (personal communication, April 12, 2011).

PZP is not a commercial product, and nobody makes money from its production or use. The nonprofit Science and Conservation Center in Billings, Montana, produces the vaccine and provides it at 60% of the cost of production, about $24 for a standard, one-year, 100-mcg dose. Researchers developed the vaccine with public funding and consider it in the public domain (Kirkpatrick, Rutberg, & Coates-Markle, 2012). Consequently, they kept it an "experimental" drug under rules of the U.S. Food and Drug administration long after it had ceased to be experimental. In 2012, however, the U.S. Environmental Protection Agency assumed regulatory control over PZP and registered it for use in horses under the name ZonaStat-H. Because it is no longer officially experimental, private companies and individuals heretofore uninvolved with the development and testing of the agent may eventually produce and sell it with expectation of profit. But the SCC maintains that it "has no intention of commercialization" (Kirkpatrick, Rutberg, & Coates-Markle, 2012, p. 6).

Population control becomes even more crucial as horse ranges shrink. Since 1971, about one third of the designated mustang habitat—*20,000 square miles* (51,800 km²)—has been sold for private development, oil and gas exploration, and renewable-energy projects. Of the original 303 HMAs, only 180 remain, fragmented rangelands encompassing 45,150 mi²/116,938 km². The area deducted from the horses' range since 1971 is about the size of Massachusetts, Vermont, and New Hampshire combined.

"My overall analysis reveals an effective displacement of the wild equids from at least three fourths of the public lands to which they are legally entitled as the 'principal' presences for which they are to be managed," says wildlife ecologist Craig Downer (2008).

> These HA's were supposed to be determined by where these equids were found at the passage of Public Law 92-195 [the Wild and Free-Roaming Horses and Burros Act]—and I take this to mean not just the tiny portion of ground they stood on at the exact hour and date of the Act's passage, but rather the home ranges of all the bands of every herd throughout the BLM/USFS [U.S. Forest Service] West that was then occupied on a year-round basis.

The Wild and Free-Roaming Horses and Burros Act transformed the management of mustangs on public lands and gave the horses some protection, but it generated new tensions among the bureau and horse advocates, users of public lands, and citizens of states with wild horse and burro herds. The bureau was required to manage the herds at the 1971 population levels, and it began conducting periodic gathers to remove horses from unauthorized areas and restore a thriving ecological balance between free-roaming horses and the rangeland (USBLM, Billings Field Office, 2009). After the horses are gathered, some (usually young, colorful horses with good conformation) are put up for adoption, others are assigned to a holding facility, and some are released back to the range. This selection process is largely arbitrary.

Dalke wrote (2005, p. 108),

> When reading the policy associated with rounding up wild horses it seems like management is based on factual and scientific decision-making void of any human emotion. This could not be further from the truth. The horses are measured against standards that groups have created for being more or less appropriate for adoption. Everyone has an opinion. Opinions can be tied to a larger community of sentiment, but most people do not seem cognizant of this . . . at this event it is understood as acting in the best interest of the horses.

With roundups came the need for government holding facilities, federally funded private pastures, and staff to supervise public adoptions. Höglund (2010) wrote, "The traumatic capture and adoption of free-roaming horses have been so successful, that six of the original sixteen western states that were home to wild horses in 1971 no longer have herds."

Wild horses on public lands are entitled only to the ranges that their forebears occupied in 1971. With one third of their rightful land given away to other uses, the horses escaped the HMAs and extended their ranges into unauthorized areas.

As the HMAs contracted, the horses multiplied. Wolves, brown bears, and big cats are natural predators, but they do not threaten the horses in most of the ranges. Under the Act, "excess" horses are subject to "removal" to achieve "appropriate management levels," but it is unclear how *excess* is defined (Harty, 2010). Equine advocacy groups claim that the bureau brashly decided that there were too many horses in the HMAs, then removed them despite lack of scientific evidence to support these actions (Harty, 2010).

Downer observed (2011),

> Certain herds have been reduced to unsustainable numbers. In addition to reducing the original 300+ Herd Management Areas (HMA) to 180, federal officials have proceeded to approve of management levels that are, in nearly all cases, genetically non-viable within each given HMA.... Of the 180 greatly reduced HMAs throughout the West, a glaring 130, or 72%, have AML of less than 150 horses or burros, and many of these are much less than 100 equids, even numbering in the teens, e.g. Ely BLM's Diamond Hills South AML of 22 horses, or Carson City BLM's Dogskin Mountain herd with AML of 15 horses, and Lahontan herd with AML of 10 wild horses to remain. According to BLM's own typically cited standard of 150 equids, in California 19 out of 22 HMAs have non-viable AMLs; in Utah 17 out of 21; in Idaho 5 out of 6; in Montana, one out of one (6 of the original 7 HA's having been zeroed out); and in Nevada, 67 out of 90 of the scant remaining herds are similarly non-viable.

Studying a small wild herd in southern Spain, Vega-Pla, Calderón, Rodríguez-Gallardo, Martínez, and Rico (2006) considered "the longstanding debate on minimum viable population size" and concluded (p. 576) that "there is no single number that tips a species into extinction or survival but that a number of demographic, stochastic, and genetic variance factors play fundamental roles in the fate of small populations. The present Retuertas horse population shows no signs of inbreeding depression and has recovered from only some tens of horses."

Kirkpatrick explained,

> This issue of genetics and sustainable populations should not be discounted, but it has become a scientifically distorted subject, in order to advance agendas. The MVP number for ASIS has been determined to be 48 and 52 for the Pryors, but these are only theoretical and have no empirical data to support them. Unless reproduction is failing, for physiological (genetic inbreeding depression) reasons, there is no genetic issue. (Personal communication, May 28, 2014)

There is, however, evidence to the contrary. Wright (1931) determined that inbreeding depression in small, isolated populations could result in loss of fitness and increased risk of extinction, and even modest interbreeding between small groups could reduce genetic drift and inbreeding before genetic compromise occurred. In the 1970s, a number of studies concurred that one immigrant per generation could offset inbreeding depression in small populations (NRC, 2013). The 2013 NRC report, drawing on more recent research, concluded that one immigrant per generation is an absolute minimum and recommended that the BLM exchange 10 animals, preferably groups of post-pubescent mares, between genetically vulnerable populations every 10 years (NRC, 2013).

Translocation is not without risk. Newcomers can bring diseases to which the recipient herd has no immunity, and an influx of new genes can shift bloodlines away from historic ancestry. Still the NRC evidently considers these risks acceptable:

At the Jackson Mountains wild horse gather in 2012, a helicopter drives a band of mustangs into the wings of a trap as wranglers stand ready to close the gates behind them. Photograph courtesy of U.S. Bureau of Land Management (USBLM, 2012, June 18).

For herds that have strong associations with Spanish bloodlines—such as those of the Cerbat Mountain, AZ; Pryor Mountains, MT; and Sulphur, UT—or herds that contain unique morphological traits—such as the Kiger, OR, herd—BLM will need to balance concerns about maintaining breed ancestry with the need to maintain optimal genetic diversity. Herds that remain isolated over the long term will inevitably lose genetic diversity inasmuch as maintaining or slightly increasing herd sizes will not offset the effects of genetic drift. . . .

The committee recommends that BLM consider some groups of HMAs to constitute a single population and manage them by using natural or assisted migration (translocation) whenever necessary to maintain or supplement genetic diversity. Although there is no magic number above which a population can be considered forever viable, studies suggest that thousands of animals will be needed for long-term viability and maintenance of genetic diversity. Very few of the HMAs are large enough to be buffered against the effects of genetic drift, and herd sizes must be maintained at prescribed AMLs, so managing the HMAs as a metapopulation will reduce the rate of reduction of genetic diversity in the long term. (NRC, 2013, pp. 169–170).

Reproductive failure is a late sign of a shrinking gene pool. The herd north of Corolla, N.C., is still fecund, but genetic analysis shows genetic variability to be dangerously low. Its inbreeding expresses itself with an increase in recessive disorders such as locking stifles, which requires surgical correction.

Horse advocates, scientists, lawmakers, land managers, accountants, ranchers, hunters, and members of the general public have widely divergent ideas of how wild horses should be managed, and all determinedly promote their opinions and agendas. Kirkpatrick (2010, p. 2) writes,

> In reality the debate points are many and varied and they cross the two fundamental sides, creating a foggy miasma rather than clarity. Horses versus no horses; many horses versus fewer horses; status as a reintroduced native species or as a non-native exotic species; "natural" versus "unnatural" management schemes; gathers and removals versus fertility control; domestic livestock on public lands versus no livestock. Most everyone, on both (or the many) sides of the issue cannot seem to come to grips with what is real, or remotely possible, while at the same time infusing the debate with strong emotions, unfounded opinions and an astounding resistance to established facts.

By the early 1980s the Bureau of Land Management was giving horses away to large-scale adopters. Many of these animals were promptly shipped to slaughterhouses and subjected to inhumane treatment along the way. When the public protested this abuse, the government put an end to large-scale adoptions (Symanski, 1996). But even before the Salazar blowup, journalists, bloggers, and others accused the bureau of putting down healthy horses and continuing to sell unlimited numbers of horses without due diligence, sometimes clearly for slaughter—practices that horse advocates say have gone on for four decades.

In 2004, Montana Republican Senator Conrad Burns, acting in stockmen's interests, inserted a rider into the Consolidated Appropriations Act of 2005 (a 3,000-page omnibus appropriations bill) that granted the BLM the authority to sell "unadoptable" mustangs more than 10 years of age or passed over at adoption events at least three times without limitation. This amendment to the Wild and Free-Roaming Horses and Burros Act passed without hearings or public review and overturned the federal policy that nominally safeguarded wild horses from slaughter. President George W. Bush signed the Burns Amendment into law and in so doing negated more than 30 years of lobbying and legislative action in behalf of the horses.

The first casualties of the Burns Amendment were six horses gathered from the Antelope Hills Herd Management Area, Wyoming, in October 2004. Dustin Herbert of Meeker, Oklahoma, a former rodeo clown, bought the horses in April 2005 for $50 each. He told the BLM that they would be used for a church youth program, but sold them for slaughter less than three days later (Humane Society of the United States, 2007). A Sioux Indian group bought 83 horses and resold 35 to a meat broker, who in turn sold them for slaughter. An average 1,000-pound/450-kg horse in good condition recently sold for about C$0.35–C$0.40/lb ($0.77–$0.80/kg) live weight, or C$350– C$400 (roughly $329–$376 at 2008 exchange rates) (Alberta Equine Welfare Group, 2008).

On its Web site, the Bureau flatly denies that it is selling or sending wild horses to slaughter:

> This charge is absolutely false. The Department of the Interior and the Bureau of Land Management care deeply about the well-being of wild horses, both on and off the range, and the BLM does not and has not sold or sent horses or burros to slaughter. (USBLM, 2012, August)

In May, 2005, the BLM changed requirements to make it difficult for mustang buyers to resell to an abattoir and was cited by the Government Accountability Office for noncompliance with the Burns Amendment, which calls for sales without limitation. Yet as the Davis case shows, BLM horses are sometimes sold for slaughter, and the agency's claims to the contrary are either grossly uninformed or dishonest—and inexcusable.

"Some captured free-roaming horses are not federally protected from commercial processing," says Höglund (2010). "Domestic horses, including some of the refugees from the wild life, are not protected from human consumption."

In the United States, domestic horses are commonly sold at auction if a fast sale is needed. Many summer camps buy a string of lesson horses in the spring and resell them in the fall to circumvent the cost of maintaining them through the winter. Sometimes divorce or the death of an owner forces immediate sale. Any horse that does not sell quickly is bid on by "kill buyers." Although they usually bid on the old, lame, or behaviorally challenged, kill buyers often take away useful animals such as outgrown children's ponies, horses recovering from minor injuries, and unbroken pets or broodmares that could serve a rider well with a few months of saddle training.

Until 2007, the majority of horses slaughtered in the United States were processed at three facilities in Illinois and Texas. Almost all of their meat was exported to France, Belgium, Switzerland, Italy, Japan, and Mexico. In 2006 the United States exported about 18,740 tons/17,000 metric tons of horsemeat valued at $65 million. Most of these animals were domestic horses raised for recreational purposes, such as riding.

In the United States, unwanted horses present a divisive problem. Increasingly, all over the country, recreational and work horses are being abandoned like unwanted kittens, left tied to posts, released into farmers' fields, or freed to run with feral bands. Reports of abuse and starvation are numerous, and horse rescue operations are filled to capacity. Some people believe that slaughter is an acceptable end for unwanted horses because using their meat and hides benefits people more than lethal injection and burial.

Veterinarians for Equine Welfare (2008) adamantly maintains, however, that "horse slaughter **is not** and **should not** be equated with humane euthanasia [emphasis in original]. Rather, the slaughtering of horses is a brutal and predatory business that promotes cruelty and neglect." In 2007, the three equine slaughter facilities in the United States were closed by state statutes. Horse dealers circumvented the restriction by shipping slaughter horses to plants in Canada and Mexico. In 2008, almost 107,000 horses were shipped from the United States for slaughter in foreign plants—about the number of horses that had been slaughtered annually in American plants (U.S. Department of Agriculture, 2010).

The Slaughter Horse Transport Program, administered by the USDA, monitors the treatment of horses bound for slaughter. There is little legal incentive to provide humane treatment *en route* to the processing plants. Between 2005 and 2009, 43 shippers owed almost $174,000 in unpaid fines imposed for inhumane treatment, yet were permitted to continue shipping slaughter horses (USDA, 2010).

Horses bound for slaughter are typically packed flank to flank with many others for up to 24 hours at a stretch without food, water, or rest. Some of them are very young, pregnant, or seriously injured, and aggressive horses often attack and terrorize the weaker animals.

The slaughter process can be, and often is, inhumane. As the horses approach the kill floor, they can see, hear, and smell the carnage, and they usually panic, resisting death in every possible way. Horses are typically rendered unconscious by a captive-bolt gun that causes blunt trauma to the brain. Often the bolt is poorly aimed as the animal fights for its life, and workers must shoot the horse as many as five or six times to subdue it (Veterinarians for Equine Welfare, 2008). When the animal is insensible, a chain is placed around its leg, which suspends it upside down so that workers can cut its throat and remove its innards.

Footage from hidden cameras in Mexican slaughterhouses shows that horses are not always unconscious when they are bled and gutted. In Mexico, workers paralyze horses by stabbing them in the neck, often repeatedly until the knife severs the spinal cord. Whereas the captive bolt ideally renders the horse insensible, the paralyzing knife only immobilizes them. The horses are fully aware as their throats are cut and their internal organs are removed. Young foals who follow their dams onto the kill floor watch their mothers meet a terrifying end and then are themselves systematically killed by workers and cast into the gut pile, too small to be profitable.

Horses shipped to Canada and Mexico must travel greater distances and are more likely to suffer inhumane treatment in transit. With the closing of plants in the United States, the number of horses transported to Alberta for slaughter increased by 40%, prompting processors to open a new plant in Saskatchewan to meet demand (Alberta Equine Welfare Group, 2008). Bjerga and Crawford (2013) calculated that the United States exported 197,442 live horses to neighboring countries in 2012, almost twice as many as in 2008.

About 1% of Canadian horses are raised solely as meat animals, bred for maximal yield and raised as food on the hoof like steers or lambs. At any given time, about 20,000 "feeder" horses are fattening up in one of four Alberta equine feedlots before live shipment to Japan or export as boxed meat to satisfy European appetites.

Three herds of wild horses, thought to be Spanish-blooded, roam the forests of Alberta and British Columbia. The Alberta herd numbered roughly 800–1,000 animals in 2011, and the provincial government issues a number of permits for their capture each year. In recent years, about 30 horses have been taken annually; but between December 2011, and February 2012, 216 horses were gathered and shipped directly to slaughter. The horses have no legal protection, and people emotionally back them or bash them, as in the United States.

At this writing, Valley Meat Company of Roswell, N.M., is seeking permission from the U.S. Department of Agriculture to turn an idle cattle-processing plant into an abattoir that can slaughter about 100 horses a day. Despite numerous legal and regulatory difficulties (Bjerga & Crawford, 2013; Front Range Equine Rescue, 2013), the company applied for inclusion in the federal inspection program, a necessary first step (Associated Press, 2012), and sued the USDA in October 2012 to force the matter. (It also sued three humane groups for defamation.) Despite apparent progress toward approval, proposed opening dates have been repeatedly delayed (Clausing, 2013; *Paulick Report* staff, 2013).

On March 12, 2013, the Safeguard American Food Exports Act of 2013 (H.R. 1094/S. 541), which would prohibit both slaughtering horses in this country and exporting them for slaughter, began its journey through Congress. Perhaps in reaction, Tim Sappington, a Valley Meat employee/contractor, posted a video on YouTube wherein he cursed at "animal activists," shot a docile, healthy-looking horse in the head, and walked away as it died (Coffey, 2013; Gillian, 2013; Karlin, 2013). It is unclear whether his actions are criminal. If the shooting is deemed a malicious act, Sappington could be charged with cruelty. If, as Sappington claims, he was killing the animal for food, his actions were entirely legal.

Horsemeat has been food for humans for tens of thousands of years. Our prehistoric ancestors often preferred horses over all other game, and until recently most civilizations that had access to horses ate them. Horsemeat has 20% more protein than high-quality beef, 25% less fat, nearly 20% less sodium, double the iron, and less cholesterol (AEWG, 2008). Today more than 1 billion people, or 16% of the world population, consume horsemeat (AEWG, 2008), 27.6% more than in 1990.

Necessity made horsemeat more popular than usual in the United States and the United Kingdom during and after World War II, when other meat was scarce, and again in the United States during the 1970s, when the price of beef soared. Few people in either country eat horsemeat now, but some remain willing to sell it to those who do. In 2005, France imported 32,000 tons/29,000 metric tons of horsemeat, Belgium and Russia about 30,000/27,200 each, and Italy about 24,000/21,800, all for human consumption. Italians like a fattier meat, Belgians choose leaner cuts, and the French and Quebecois prefer chewier meat from older animals. The Japanese prefer their horseflesh served raw within 3 days of slaughter, so they tend to import live horses for local processing.

In April 2007 the U.S. House of Representatives passed H.R. 249, An Act To Restore the Prohibition on the Commercial Sale and Slaughter of Wild Free-Roaming Horses and Burros, intended to overturn the Burns Amendment and restore the prohibition against commercial sale and processing of captured horses and burros. Höglund writes: (2010),

> Sessions of Congress last two years, and at the end of each session all proposed bills and resolutions that haven't passed both houses of congress [sic] are cleared from the books. H.R. 249 later evolved into H.R. 1018 that passed the House in July, 2009 and hibernates peacefully in the Senate—Committee on Energy and Natural Resources—as S.R. 1579, a companion bill to the House version. Like it or not, as of February, 2010, the fate of the free-roaming horse is regulated under the protections of legislation formed and passed on or before fiscal year 2005.

As of 2012, the BLM has sold more than 5,400 horses and burros since the passing of the amendment. The current laws do not prevent buyers from shipping them outside the United States to slaughter.

A wild horse removed to a BLM pasture enters a world preferable to the slaughterhouse, but still quite different from the open spaces of his homeland. BLM personnel move captured mustangs from the temporary pens to the short-term holding corrals in stock trailers. On arrival at the facility, wranglers place them in holding pens that allow a minimum of 700 ft^2/65 m^2 per animal and access to high-quality hay and water. A veterinarian examines each horse; treats health concerns and injuries; and sometimes euthanizes those affected by chronic or incurable disease, injury, lameness, or serious physical defect. The animals are freeze-branded with unique identification numbers, tested for equine infectious anemia (swamp fever), vaccinated, and dewormed, and males are castrated.

The bureau puts young, desirable horses up for adoption to carefully screened individuals who meet stringent requirements. For example, the agency requires adopters to have a sturdy corral with 6-ft/1.8-m) fences; anything less would allow the skittish animal to escape. The government retains title to the horse for the first year and evaluates its treatment and condition, and the adopter becomes the legal owner if he or she has upheld the adoption contract and has taken good care of the horse. But the BLM will pass title immediately with the sale of a mustang greater than 10 years of age or of younger horses that have been passed over for adoption at least three times. Despite the sale and adoption programs, most horses gathered by the BLM are fated to live out their lives in holding facilities.

Many who would like to adopt a mustang do not have the space or the know-how to train a panicky 1,000-lb/450-kg mustang. While a wild mustang fends for itself quite well, horse owners are obliged to provide food, water, and health care to their livestock, and gentle them until at minimum they will accept veterinary care without struggle. Dalke (2010, p. 104) observed,

> While adopting a historical icon is enticing, horsemanship skills that have been disappearing since the introduction of the car are needed to assure success. . . . As a result, too often people without these skills are adopting an image that they find difficult to manage . . . The responsibility is not placed on the human trainer, but rather on the "wild" animal.

A number of agencies and individuals champion the domesticated mustang, and sponsor events to encourage adoption. The Mustang Heritage Foundation (MHF) sways potential adopters through exciting training programs and competitions, which include Extreme Mustang Makeovers, Mustang Million, Youth and Yearling Mustang Challenges, Youth "Camp Wildfire," and the Trainer Incentive Program. Since 2007, Extreme Mustang Makeovers pair horse trainers of all ages and abilities with untouched mustangs, who work with the horses for approximately 120 days before competing. After the trusting mounts willingly perform maneuvers before a rowdy crowd and complete an obstacle course, they are offered to the public for adoption. The Mustang Million awards $1 million in cash and prizes, and highlights the trainability, athleticism and versatility of the trained mustang.

Beginning in the 1980s, natural horsemanship became increasingly mainstream. Many mustang adopters use training methods taught by clinicians such as Buck Brannaman, Monty Roberts, Clinton Anderson, Craig Cameron, and Pat and Linda Parelli, all of whom have made humane training techniques accessible to many inexperienced horse owners directly and through books and videos.

State penitentiaries in Wyoming, Colorado, Nevada, Kansas, and Utah have developed innovative programs pairing mustangs with convicts for gentling and basic training. The horses learn to trust and obey a rider while the inmates develop skills that can lead to rewarding careers as wranglers or trainers. Almost 95% of these trained horses find homes. At one such prison, the Wyoming State Honor Farm, about 900 inmates gentled about 3,600 mustangs in the first two decades of the program (Behar, 2010).

Another program actually pays adopters in New Mexico, Texas, Oklahoma, and Kansas $500 to take home an adult mustang (USBLM, Wild Horse and Burro Program, 2009; 2011, January 13). The cash incentive is designed to ease the financial burden for the adopter, and the cost to the government is much less than that of keeping the horse in a long-term holding facility.

When the author perused the BLM adoption Web site in 2012, she saw several horses that qualified for this incentive. All were 4-year-old geldings: a 15-hand (60-in./1.52-m) bay with a white snip on his nose, captured July 2008 from the Nevada Wild Horse Range HMA; a sorrel with a broad white blaze, captured outside the Conant Creek Reservation in Wyoming; a dapple-gray with a broad, well-muscled chest, captured at the Sandwash Basin HMA in Colorado. Potential adopters of eligible horses register with the bureau and place bids. The high bidder gets the horse, and a year later, the check is in the mail. If the program is not carefully monitored, however, kill buyers and other exploiters could profit twice, once for taking horses off the government's hands and a second time through sales to abattoirs.

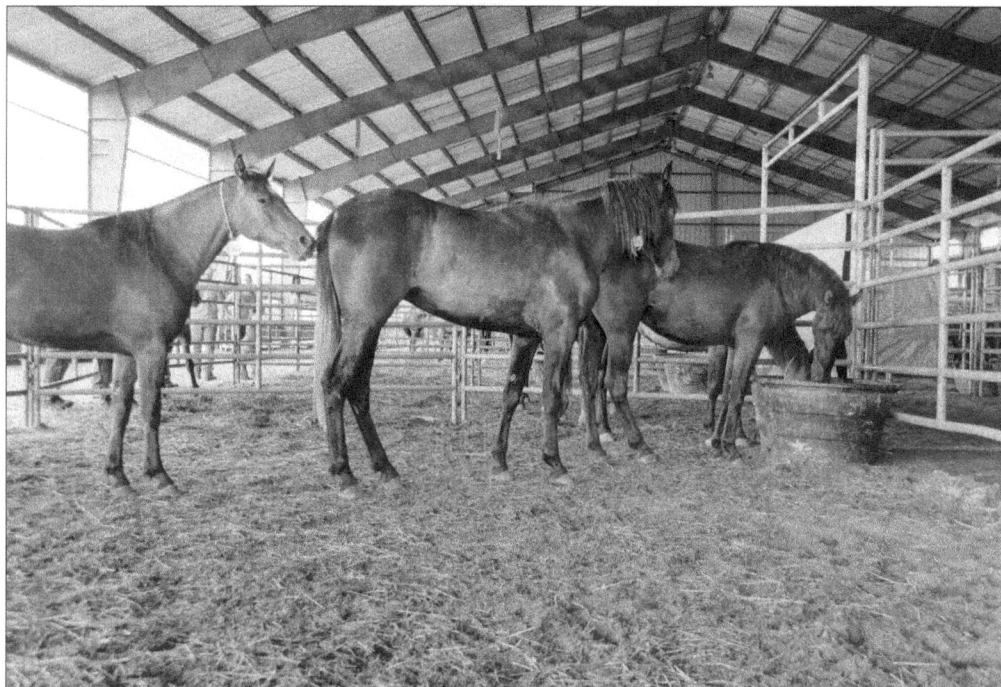

The Bureau of Land Management periodically trailers Western mustangs to adoption sites in the east. This event in Annville, Pennsylvania, in July 2012 offered about 40 horses, including yearlings, 2- and 3-year-olds, four older horses that had been returned by previous adopters for health or behavior issues, a very friendly unbroken 5-year-old gelding, and several burros. Note the Thoroughbred-like conformation of some of these 2-year-old fillies. The tallest was almost 16 hands (64 in./1.63 m) already and promised to grow uncommonly large for a mustang. Adopted mustangs have successfully competed in nearly every discipline alongside domestic breeds and won championships in reining, dressage, team penning, endurance, jumping, gymkhana, and numerous other activities. They are rugged and seldom need shoes.

Most mustangs more than 5 years old are not adopted. Instead, they are transported to long-term pastures. The horses are maintained in grassland large enough to allow free-roaming behavior and provided the forage, water, and shelter necessary to sustain them in good condition (USBLM, 2010). Mares are kept separate from the gelded stallions in all facilities but one. Horses in the LTPs remain available for adoption. These horses are handled minimally, though they are counted weekly and observed for illness and other problems.

In the early 2000s, the bureau gathered more than 10,000 mustangs annually, straining the capacity of the holding facilities and far outpacing the ability of the public to adopt them (USBLM, Wild Horse and Burro Program, 2005). Between 2007 and 2009, about 62% of gathered mustangs and burros were adopted and about 8% were sold to qualified individuals (USBLM, 2010). In the current economic downturn, horse owners find it increasingly difficult to maintain the animals they have and are hesitant to take on another mouth to feed.

The Bureau has two problems: it does not know what to do with excess horses running free on federal lands, and it does not know what to do with excess horses in federal holding facilities. The public is opposed to sending the horses to slaughter, shooting them, or euthanizing them. They are expensive to maintain; taxpayer dollars subsidize feed, cleanup, facility maintenance, vaccinations, veterinary care, hoof trimming, and deworming. As of

June 2012 there are about 45,000 captive mustangs in BLM facilities—12,400 in short-term corrals and 33,400 in Midwestern pastures (USBLM, 2012, September 28).

Many private landowners and nonprofit organizations have created small wild-horse sanctuaries on their own properties. In 2012, the BLM solicited proposals for partnerships with landowners with an eye to creating large, publicly accessible eco-sanctuaries for captured mustangs. Sanctuaries offer a natural environment where horses can roam free, engage in normal social behaviors, provide opportunities for visitors to observe the behavior of wild horses in a natural setting, and promote local tourism and economic stimulus (USDI, 2010, p. DH-48). Conducted with sensitivity, ecotourism causes minimal disruption to the horses, but garners more public appreciation, support, and protection for the horses.

The 4,000-acre/1,620-ha, family-owned Deerwood Ranch, about 30 mi/48 km west of Laramie, Wyo., was recently approved as an eco-sanctuary for up to 300 nonreproducing wild horses. Landowners Richard and Jana Wilson will develop educational programs and tours, and the BLM will pay the Wilsons $1.30 per horse per day—the same amount paid to landowners who provide long-term pasture for captured mustangs.

As of this writing, the most ambitious eco-sanctuary project on the table is headed by wild horse advocate Madeleine Pickens, businesswoman, philanthropist, and wife of oil billionaire T. Boone Pickens. The BLM is working with Pickens and the Saving America's Mustangs Foundation to establish a colossal wild horse eco-preserve, Mustang Monument, in Elko County, Nevada—about half the size of the state of Rhode Island (SAM Foundation, n.d.). The projected sanctuary would encompass most of the existing Spruce Grazing Allotment, including about 14,000 acres/5,670 ha of private land and 508,000 acres/205,580 ha of public land—with a boundary fence about 155 mi/250 km long! The vast tract includes 10,262-ft/3,128-m Spruce Mountain and expanses of remote wilderness areas "essentially untouched by man" (USBLM, Elko District, Wells Field Office, n.d., p. 12).

The BLM intends to return up to 10,000 wild horses from confinement in holding facilities to a semiwild existence at Mustang Monument. These horses will be unable to reproduce (males have been gelded) but will be protected from starvation, drought, and harassment by people. The Spruce/Pequop, Goshute, and Antelope Valley HMAs and the 1,384 wild horses within them will become part of Mustang Monument, and the SAM Foundation would like to maintain these groups separately as intact breeding herds. The horses will all remain under federal ownership, but SAM will oversee their management.

SAM proposes to build a modern, cost-effective short-term holding facility to replace that used by the BLM. Annually, the government will pay SAM the equivalent of the average long-term holding costs currently incurred by the agency, and $500 per year per horse indexed to inflation. Whereas private contractors now receive public money to take care of the horses, the money paid to SAM would remain within the foundation to be used solely for maintaining the land and the horses.

Pickens estimates that the Mustang Monument could save taxpayers hundreds of millions of dollars over the next 20 years and draw 1.5 million visitors each year. Mustang Monument would include a cutting-edge learning center for youth and adults, educational programs, creative pursuits, Western and Native American cultural activities, a modern horse-training and clinical facility, trails, campgrounds, and lodging in teepees and eco-friendly log cabins.

These horses were gathered from the Pancake Complex in eastern Nevada on an emergency basis due to drought. The BLM wrote an assessment: "Most are in poor body condition" (2012, September 14). In this photograph, however, the animals look healthy, if somewhat lean. How thin is too thin? Is it better for them to be thin but free, or well-fed and confined? Are the horses better served by taking them from the range before they become emaciated or by waiting until removal is the only option? Wild horse advocates, ranchers, and BLM officials have different perspectives on this issue. Photograph courtesy of USBLM.

Some people, primarily stockmen, landowners, and others who would like to see mustangs removed entirely from the wild, oppose the concept of wild horse eco-sanctuaries. When the *Las Vegas Review Journal* published an article about Pickens's sanctuary (Myers, 2012), it received dozens of online comments, many of them anti-mustang.

One writer said,

> The liberal circle is a funny one. Refuse to control population = thousands of useless horses destroying the landscape = millions in transporting, capturing and housing costs = millions in bureaucratic BS all because we couldn't bare [*sic*] to see population control.

Another wrote,

> The government is seeking to help cut costs, Easy, make them into dog food, sell it cheap, thus no more poisen [*sic*] food for dogs from the Chi-Coms and no more taking care of worthless disease ridden animals!

Wild-horse refuges are expensive to establish and maintain, since animals require food and veterinary care, unless officials decide to adopt a selective hands-off policy. If the mustangs on the range continue to reproduce, more sanctuaries will be needed over time. Most of the existing sanctuaries throughout the United States are in critical financial straits (Kirkpatrick, 2010).

Some are concerned that once eco-preserves become a viable option, the government would find a way to remove all the wild horses from the HMAs and deposit them in the sanctuaries. The Bureau intends to reevaluate HMAs that lie outside of the proposed eco-sanctuary to determine their ability to sustain a viable horse population, but it is unclear why this is necessary. While 93% of the Spruce/Pequop HMA would lie within the sanctuary, about 86% of the Antelope Valley HMA and 73% of the Goshute HMA extend beyond its borders. Some detractors predict that BLM would seize the opportunity to eliminate all remaining horses from the parts of the HMAs outside the sanctuary, and then from the HMAs that overlap other sanctuaries. From there, might the agency relocate free-roaming horses from HMAs *near* a sanctuary, then from HMAs in the same BLM administrative district, then from those in the same state? Might it eventually truck horses to sanctuaries 1,000 miles away to eradicate them from HMAs?

Another concern is over-commercialization. A Web site, unfinished at this writing, proclaims the Mustang Monument "a wild horse eco-resort (Mustang Monument, 2012) which would seem at odds with the pristine solitude desirable in a nature preserve. A large complex with heavy visitation and vigorous development around the periphery could quickly become as urbanized as the south rim of the Grand Canyon.

The Bureau of Land Management says that it removes wild horses from their legal ranges to protect the land and the animals from the consequences of overgrazing. The agency is responsible for maintaining range vegetation in reasonable condition, and its management policy assumes that without intervention, horse populations will expand until range vegetation is irreversibly damaged and the animals die a cruelly lingering death from starvation. Equids are widely considered detrimental to the range, a position supported by reports such as Pimentel's (2002). Under the revealing heading "Vertebrate Pests," Pimentel et al. say, "large populations of introduced wild horses and burros cost the nation an estimated $5 million per year in forage losses" (p. 293). It is unclear how they estimate the loss of forage or determine to whom it is lost. They seem to imply that publicly owned grass has value only if eaten by privately owned livestock and that ranchers would make constructive use of the grass now being eaten by wild equids.

Additionally, the patches of bare ground created by horses and cattle provide important feeding, nesting, dusting, and display sites for upland and passerine birds, as well as nesting habitat for birds such as the mountain plover, which evolved in association with bison (Vavra, 2005). Grazers break up homogeneous grass stands, producing patchy, open cover with a diversity of forbs. They create trails and openings that provide nesting sites for birds (U.S. Fish and Wildlife Service, 1999). And the urine and feces of herbivores recycle nitrogen faster than litter alone, making the soil more hospitable for grasses, forbs, and shrubs by hastening litter decomposition (Hobbs, 1996).

The Fish and Wildlife Service uses carefully managed grazing rotations as a tool to maintain healthy habitat for prairie chickens. More than 1 million Attwater's prairie chickens populated the Texas and Louisiana Gulf coastal prairie at the time of first European contact, their strange booming mating calls resonating over the endless grasslands each spring. The prairie chicken evolved alongside the vast herds of bison, horses, and other large herbivores that ruled the primordial tall-grass prairie. When the large grazers vanished, so did the birds. Overgrazing by cattle compounded the problem. By the 1930s, the species had dwindled to about 1% of its former population. Today, rotational grazing has helped to reestablish the bird in its native ranges.

During a rigorous Rocky Mountain winter, even minimal improvements in forage quality benefit wildlife (Vavra, 2005). A number of studies involving cattle showed that grazed habitat provided sparser but more nutritious forage. Properly timed grazing during the active growth stage of bunchgrasses slows maturation, making the aboveground part of the plant more beneficial to elk (Vavra, 2005). In winter and spring, elk preferentially forage in areas that cattle had grazed during the previous summer.

Research in the Netherlands involving cattle found that free-range grazing was a useful management strategy for a heathland nature preserve (Bokdam, 2003). Another study showed that periodic heavy grazing on the winter ranges of deer increased shrub production and growth of seedlings (Vavra, 2005). Biologists showed that in northern Nevada timed cattle grazing stimulates regrowth of forbs that attracts sage grouse (Evans, 1986).

In the 19th century the dense forest at what is now Mt. Rogers National Recreation Area and Grayson Highlands State Park in western Virginia was heavily logged, leaving the mountains denuded. Huckleberry bushes and other brush, no longer in competition with old-growth trees for sunlight, proliferated and choked the landscape. The U.S. Forest Service took over the land in the 1960s and decided to preserve the open highlands for wildlife habitat and for the enjoyment of visitors. Left alone, brush and forest would eventually overwhelm the grasslands and block the panoramic vistas. Historically, farmers used brush fires to keep the land clear, but this method put people and property at risk and left unsightly charring.

Grazing proved the best method for maintaining the open areas. In the early 1970s, the Forest Service placed a flock of sheep in the Pine Mountain area, but the sheep consumed the toxic native mountain laurel, and many died. They were replaced by cattle, but when autumn arrived they began to feed on snakeroot, another toxic plant, and were also poisoned.

In 1974, the Forest Service contracted with Bill Pugh of Sugar Grove, Va., to release a small herd of ponies, mostly Shetlands, onto the Pine Mountain crest area. The ponies were immediately successful at their new career. They were hardy enough to survive bitter winters at 5,000-foot elevations, reproduced easily, and were popular with the public.

Then they began to die. First several succumbed over a harsh winter, and then more and more until 28 ponies had died. People assumed that the ponies had starved and blamed Pugh and the Forest Service for their deaths. Pugh maintained that the ponies had been healthy and thriving, and their deaths were not due to disease or starvation. When the American Society for the Prevention of Cruelty to Animals launched an investigation, Pugh, disgruntled, turned responsibility for the ponies over to the Forest Service. The agency then sold the herd to the Wilburn Ridge Pony Association, a nonprofit organization founded specifically for the management of the ponies. When the association investigated the mysterious pony deaths, they found that all the victims had been shot. The Forest Service launched its own investigation, and the shootings stopped—though nobody was ever convicted. A quarter century later, the ponies have effectively maintained the desired balance of trees, brush and grassland, to the benefit of visitors who hike the peaks to enjoy the scenic overlooks.

If grazing is a useful tool for range management, what species should be doing it? Two grazing animals in a given area could constitute biological excess. If one of these animals is a wild horse and one is a domesticated cow, the decision about which to remove becomes one of human values or preference. Put another way, if an area can comfortably support 500 cattle and 500 horses, numbers greater than 1,000 animals would constitute biological excess. If the public prefers 1,000 horses, the area can be said in this context to have an

At Mt. Rogers National Recreation Area and Grayson Highlands State Park in the mountains of southwestern Virginia, the U.S. Forest Service uses free-roaming ponies to manage the growth of vegetation so the panoramic vistas remain open. Scenic balds, however, are not a natural feature of these peaks, but a result of deforestation from logging and farming.

excess of 500 cattle (Harty, 2010).The term *excess* has both a scientific and a social connotation, influenced not only by biological impact of a species but also by management policies, legal issues, and prevailing public opinion (Harty, 2010).

"The proportions of wild burros and horses in relation to livestock and big game on the public lands is grossly weighted in favor of the later, especially livestock," says Downer (2011, p. 2):

> The 1971 Act authorized a small fraction—namely, one-sixth—of the public BLM and USFS lands for wild horses and burros. Here, in these relatively minor areas, these two splendid species are by law authorized to be the "principal," though not exclusive, presences. Perversely, this core legal intent has been ignored by authorities, since rarely do wild equids receive even one fourth of the available forage allocation within their legal areas.

Cattle and sheep are grazed in wild-horse ranges, but not to an economically significant degree. They also benefit from government subsidies. In 2011, the average monthly lease rate for grazing on private lands in 11 western states in 2011 was $16.80 per head (Vincent, 2012). Yet in 2013, livestock permittees pay only $1.35 per month to graze a cow and her calf on federal land whereas the state of Texas would charge $65 to $150 for the same use.

Interestingly, the BLM pays $1.30 to $1.40 per horse per day to support the mustangs in long-term holding facilities in the Midwest. Some find it difficult to understand why the Bureau would pay to remove tens of thousands of horses from their HMAs, only to replace them with hungry cattle that do not come close to offsetting the costs of maintaining the horses on private pasture.

The majority of the meat eaten by Americans originates in feedlots, not on rangeland. "Less than three percent of the nation's entire beef cattle herd grazes on public lands, and sheep ranching is a welfare industry, subsidized by the government to the tune of $123 million a year," says Kirkpatrick (2010, p. 3). "We could remove every sheep and cow from public lands tomorrow and the nation would not notice a thing."

The elimination of sheep and cattle from public lands would give wild horses more forage for a time, but not over the long term. "After the cows and sheep are gone, horses would one day overpopulate the land," says Kirkpatrick (2010, p. 6). "This view buys time, but doesn't purchase a solution."

Still, cattle and sheep are the primary grazers on public lands. The forage requirements of grazers is measured in animal unit months. An AUM is the amount of forage needed to sustain one cow and her calf, one horse, or five sheep or goats for a month. "In FY 2005, forage consumed on BLM lands by livestock summed up to 6,835,458 AUMs, contrasting with wild horse/burro consumption of only 381,120 AUMs, or 5.6% that of livestock," says Downer (2011, p. 3).

> And the percentage is much less now with the recent, draconian roundups having taken place. . . . According to 2008 Public Lands Statistics, on USFS lands, livestock devours 6.6 million AUMs worth of forage, much in vital headwaters where cattle camp, while wild horses and burros struggle to get by on a meager 32,592 AUMs—an outrageous ½ of 1% of available forage.

Ginger Kathrens said in an interview published on an animal-advocacy blog,

> One of the ways the BLM gets rid of horses is to say they are starving, and that they have to be rescued; they are devouring the land and destroying the landscape and resources is the claim. But there are millions of head of cattle, and there are only a few thousand mustangs. It's just a method of devaluing the horse and getting rid of a species native to North America. (Rhodes, 2010)

Michael Holbert, BLM Deputy State Director for Nevada, verified that the Bureau removed between 3,000 and 4,000 mustangs from their Nevada ranges each year in 2006–2008 (Johnston, 2009). The Bureau suspended a 2010 gather when 7 wild horses died of dehydration after a long-distance helicopter chase (AP, 2010).

In 2012, BLM conducted an "emergency" gather that it had planned at least a year earlier ("BLM to gather wild horses," 2011; Harding, 2011; Ruggles, 2011). In October 2011, the Bureau stated its intention to remove 1,435–1,540 horses from an estimated population of 1,653–2,200 over the upcoming year. During the first gather, in January and February 2012, contractors captured only 880 horses (or 828) and killed 9, leading many to wonder whether the original count was overestimated or the mustangs were distributed so sparsely across their enormous range that they had eluded capture.

In July 2012 the U.S. Department of Agriculture issued a drought emergency declaration for all of Nevada (USDA, 2013). Two months later, BLM conducted an "emergency" two-day gather that captured 124 additional horses. Some people speculate that the bureau used the drought to justify continuing the controversial Nevada gathers despite the protests of horse advocates and other citizens.

Wild horse management is an emotional, political, and very expensive issue. Horse advocates do not always trust the federal government to make decisions that are in the best interest of the horses. The U.S. Institute for Environmental Conflict Resolution described a "*mutual* lack of credibility and confidence" (emphasis in original) between Wild Horse and Burro Program personnel and some stakeholder groups, including various animal-welfare and wild-horse-protection groups (Harty, 2010, p. 19). This agency recommended increased collaboration between the public and the Bureau of Land Management in managing mustangs as well as increased opportunity for public comments and suggestions (Harty, 2010).

The National Research Council (1982, p. 55) reported that BLM staff showed a "broad range" of attitudes about the Wild Horse and Burro Program. It reported that many BLM employees were sincerely committed to managing mustangs "in the spirit of the 1971 Act," but others were resistant to the program. Some said that they felt pressured to "depict range, population, and other conditions in an antihorse and antiburro context."

Governments and private entities try to conserve mustangs because much of the public values these animals, but sentiments differ even among those who advocate conservation. Mustangs are conserved primarily because they are valued by the public, but public sentiment varies.

At one end of the spectrum are those who strongly support the protection and management of horses as wild animals with minimal human interference. Before settlers tamed the West, horses roamed by the millions on enormous unfenced tracts of high-quality grassland and thrived without assistance. Now wild horses live in fenced parcels with limited resources. Any kind or degree of management necessarily entails some interference. The less interference, the more starvation, disease, suffering, and death. The horses of Cumberland Island, Georgia, are currently unmanaged, and consequently many individuals are excessively thin, show signs of chronic disease such as persistent fungal infections, and die young.

At the opposite end are those who believe wild horses have no right to roam the rangelands and would like to see them removed. Somewhere in the middle of the spectrum are those who advocate intense management to maintain small populations of wild horses and burros.

Then there are those who advocate situational management to preserve the largest sustainable populations in harmony with their environment without needless domestication. Large numbers prevent genetic bottlenecks and increase the likelihood that a population will survive a disaster. If preserving the population is the goal, and the largest sustainable number is not harmful, there is no good reason to prefer a smaller number.

Kirkpatrick wrote (2010, p. 13),

> Thirty-five thousand fecund and protected horses are a lot of animals. Even horse advocates might ask the question—do we need that many? Can we afford that many? Might it be better to have fewer horses in healthier condition than many horses living on the edge? What if we had 5,000 (or 10,000, or 15,000) horses,

In December 2011 the Pryor Mountain wild horse population was approximately 150 horses with 17 foals, exceeding the AML of 90–120 plus the current year's foals. Fifty-three mares were treated with PZP in 2011. The BLM expects to contracept 70–85% of the mares through 2015.

living on lands set aside strictly for horses, with no livestock conflicts? What if we had one or two "national wild horse ranges" in each of the ten states with wild horses, where there was no conflict with anything else? Twenty national wild horse ranges totaling 5,000 animals would translate into an average of 250 horses per range, well beyond what is genetically viable. Would the wild horse be valued more if we had 5,000, or 10,000 with no conflicts rather than 35,000 with multiple conflicts? We once valued the white-tailed deer, but after they moved into the suburbs and started wrecking our cars and eating our shrubbery (conflicts) we started being annoyed by them and hired people to shoot them.

Over the last few years, mustangs have been taken from their ranges in large numbers, often more than 100 at a time. Ginger Kathrens pointed out that in 2009, 12 herds were being removed entirely from 1.4 million acres/567,000 ha near Ely, Nevada, "because these lands are suddenly not appropriate for wild horses. . . . However, no action has been made to reduce cattle grazing in these areas" (Cloud Foundation & Equine Welfare Alliance, 2009, p. 1).

Downer (2011, p. 2) observes,

> Within the Triple B, Maverick-Medicine and Antelope Valley Complex of HMAs in Nevada BLM's Ely District, BLM plans to reduce wild horses from 2,198 to 472 individuals, thus cutting the horses from a 23% realized forage allocation to slightly less than 6%. Even at the high end of 889, this management level represents only one-ninth, or 11% of available forage. . . . At present the complex' sparse wild horse population of 2,198 within 1,682,998 acres translates to one wild horse per 776 acres—hardly an overpopulation! But if BLM's proposal is adopted, the wild horse

population would be reduced to 472, meaning there would remain an immense 3,566 acres of legal habitat per individual wild horse. . . .

The wild horse populations from the Calico Complex of five HMAs containing ca. 600,000 acres, were gutted in the first months of 2010 leaving hardly any to be observed during the flyover I made in early April, 2010. Only 31 wild horses were spotted in a transection of each of the HMAs, while ca. 350 cattle were still seen out munching on the forage within the horses' legal areas, mostly congregated around more grassy water sources, unlike the horses themselves, who were seen far from these sources and much more evenly distributed.

In Wyoming, the appropriate management level for the Adobe Town and Salt Wells Creek HMAs is set at 610–800 adult horses. In early 2010, the BLM census estimated the population at 2,438—at least three times the target number. Ranchers who hold grazing permits are voluntarily limiting the numbers of livestock they graze to avoid competition for water resources in a drought. A large-scale roundup and removal was executed in the fall of 2010 in which 2,268 animals were gathered, and 318 were returned to their home range. Despite concerns about overpopulation, the vast majority of horses scored in the healthy range of body-condition assessments, and only a few were overly thin.

Höglund (2010) writes,

Due to recent federal legislation by politicians beholden to private enterprises, free-roaming horses could vanish from the American west. What will the future hold for these noble creatures who are direct descendants of horses that once hauled us across the frontiers of America and helped us build a nation?

Roundups are stressful for the horses; injuries sometimes occur—broken legs, wounds, and even deaths; and keeping horses in close proximity invites the spread of infectious disease. In the hours and days after strenuous roundups in the Calico Complex Mountain Range of Nevada in February 2010, approximately 25 captured mares spontaneously aborted their unborn foals (Höglund, 2010). According to the Cloud Foundation, 357 horses were killed in gathers from 2004 through 2009—but this figure includes horses that were euthanized for unrelated conditions such as birth defects (*Calico Gather Wild Horse Deaths*, 2010).

There are many who fervently hold that America's mustangs have the birthright to a wild existence. Many of our mustangs are in limbo, confined within government holding facilities, quite possibly for the rest of their lives. Downer's proposed solution is to restore these animals as the principal species in each of their legal herd areas. He estimates that a total of 39,114 wild equids could be reinstated on federal lands throughout the West. "I have performed a calculation of the numbers that could be restored based on the various sizes of the legal Herd Areas," he says (2008, p. 5). "Though just the empty Herd Areas in Nevada and Wyoming alone could accommodate the 30,000, I recommend that these equids be used to restore more viable herds throughout all the 10 Western states from which they have been unfairly depleted."

Under the present law, this relocation would be illegal. Section 1339 reads "Nothing in this Act shall be construed to authorize the [Interior] Secretary to relocate wild free-roaming horses or burros to areas of the public lands where they do not presently exist." Further, the bureau correctly maintains that "no specific amount of acreage was 'set aside' for the exclusive use of wild horses and burros under the 1971 Wild and Free-Roaming Horses and Burros Act. The Act directed the BLM to determine the areas where horses and

Horses are often gathered and removed from Western herds. Which horses should stay, which should be taken to holding facilities, and which should be offered for adoption? In well-visited herds like the one at Pryor Mountain, visitors grow attached to certain charismatic individuals, particularly stallions, and want to see these horses remain in the wild. This stallion is Admiral, who was one of the most popular horses on the refuge until he was killed by a drunk driver in 2011.

burros were found roaming and to manage them 'in a manner that is designed to achieve and maintain a thriving natural ecological balance on the public lands'" (USBLM, 2012, September 28). The law set a solid upper limit: horses and burros must live forever on no more acres than they occupied in 1971. But it set no lower limit; a total of 0 acres for wild equids is entirely possible under the law, and evidently favored by many in positions of power and influence.

Wild horses do not live independent of human beings. We played a role in both the eradication and reintroduction of the wild horse in North America. Now that they have returned, management decisions are unavoidable. We decide whether to manage populations to achieve specific goals, or we decide to let nature take its course as we understand it. We decide whether to intervene if a horse is injured or starving. We decide how to respond if a herd expands beyond its food supply. Should we bring food? Remove animals? Stand by while animals slowly die from starvation? Reintroduce predators that may someday eat a hiker? If we remove animals, will they be chosen randomly, or will we target the young, the old, or specific bloodlines or colors?

Kirkpatrick (2010, p. 14) asks,

> Is it reality to suggest that the existing horse herds—fenced, chased and torn asunder for 40 years—are in any sense of the word natural, genetically or behaviorally? Is it reality to believe—in 2011—that in coming years there will be larger federal budgets to deal with the growing problem? Those who engage in these debates must come to grips with the established facts that govern management of wild horses,

which include existing laws, empirical and published scientific data, the unnatural nature of our wild horse populations and the limited choices on the table.

The Bureau of Land Management must make decisions about how to manage each mustang population, and each herd is unique. In unmanaged herds, natural selection determines which animals reproduce over the long term; but in the short term, old, thin, sickly, and inbred mares do conceive and struggle to bear and raise foals.

Some wild horse herds are manipulated to retain desirable characteristics. Kirkpatrick (2010, p. 15) comments,

> Some geneticists (and advocacy groups) believe the horses should be managed for phenotype—their physical appearance, how they look—usually with the goal of having horses with certain colors, or conformation, or striping, and something resembling Spanish horses of the 1500s.... In the Pryor Mountains of Montana, the color composition of the herd is entirely different from what it was 40 years ago, largely because of selective removals. Today the dun color is common (>20% of the herd), but 40 years ago there were but three duns in the entire herd.

In the Pryor Mountain herd, the goal is to maintain the Colonial Spanish conformation within the 90–110-horse herd and to prevent the loss of "Spanish" characteristics. The horses are managed to express the full spectrum of Spanish coat colors, to prevent any one color from being eliminated or becoming dominant, and to preserve certain bloodlines (USBLM, Billings Field Office, 2009). The Pryor Mountain herd, though isolated and remote, is well-visited by tourists, however, and the public has come to favor individuals within the herd, especially those that range near the highway. Well-known horses or groups, especially stallions, are retained regardless of their genetic contribution or the overall health of the herd. Many descend from a non-Spanish horse introduced from the Rock Springs Herd Management Area in southern Wyoming (Sponenberg, 2011) who mated his daughters and founded a very successful line of horses. Sponenberg (2011) notes that horses of this lineage are among the most charismatic and popular with the public, and remain when others are removed. These favored stallions tend to acquire artificially large harems through lack of competition.

Although most of North America's wild horses live in the western United States, as well as Alberta and British Columbia, there are scattered populations on a number of barrier islands along the Atlantic coast. The largest herds live on Sable Island, off the coast of Nova Scotia; Assateague Island, off Virginia and Maryland; Currituck Banks, Ocracoke Island, Carrot Island, Cedar Island, and Shackleford Banks, North Carolina; and Cumberland Island, Georgia. In addition, small, unexpected pockets of feral horses exist in Missouri, Virginia, and other states. They also roam in the Bahamas, Puerto Rico, and the Virgin Islands. They live in Venezuela, Argentina, England, France, Poland, Namibia, Japan, Australia, and New Zealand.

America's wild horses are the free-roaming descendants of domesticated horses imported from Europe. Detractors argue that horses, having been residents for 400–500 years at most, are not a native species, and they cause profound environmental damage to indigenous plants and animals that evolved in balance with one another, not with horses.

But aside from an absence which may be as few as 7,500 years, horses were part of the ecological balance—as their ancestors and cousins were for more than 50 million years before that. During the Pleistocene, from roughly 2.5 million to 12,000 years ago, horses, bison, camels, and mammoths were part of the ecological balance—and their ancestors and cousins for 55 million years before that. "The key element in describing an animal as a native species is (1) where it originated; and (2) whether or not it co-evolved with its habitat," write Kirkpatrick and Fazio (2010). "Clearly, *E. caballus* did both, here in North America. There might be arguments about 'breeds,' but there are no scientific grounds for arguments about 'species.'"

Despite all they have suffered at the hands of man, these horses are survivors. People have pushed wild horses to the most inhospitable habitats, including deserts, swamps, mountaintops, and sandbars, where they usually manage to thrive, often reproducing at a rate that strains the local ecosystem.

In Downer's eyes, the horse is one of the most truly native species in North America. Humankind, with its quirky notions regarding livestock grazing, resource exploitation, and city planning, is the invasive species here. "Because they combine power and beauty in a remarkable way, these animals are inspirational," he says (2011). "In the wild and through respectful and moderate visits, they uplift us from the doldrums of a civilization too exclusively fixated upon mankind and its creations, too artificial and out-of-tune with the greater world of Nature. The wild horses and burros will show us a better way of life, when we learn by observing them in the wild and on their own terms, not our imposed ones."

Free-roaming horses—Western or eastern, foreign or domestic, of long tenure or short—instigate a thorny tangle of issues wherever they occur. The mustang predicament has no easy solutions, and any action taken for or against them is certain to stir the flames of controversy. They are not usually considered wildlife, like elk or bison, or livestock like sheep or cattle. As a "national heritage species," they are classed somewhere between wild and domestic, but without the desirable traits and therefore the advantages of either designation. People who have never seen mustangs in their natural environment determine their fate in urban centers far from the horses themselves.

Their problems are many. Herd size can increase rapidly under certain circumstances. Many believe that the grass in their mouths would be better used sustaining a cow and her calf. It is illegal to shoot them, most Americans oppose their slaughter, and not enough people adopt them. If we leave them in parcels of wilderness to breed unchecked, they will overwhelm the environment. The act of capturing them transforms them from free-roaming American icons to dependent livestock requiring expensive food, shelter, and veterinary attention. Tens of thousands of captured wild horses are eating through the government budget for their care. Thousands more will give birth in the wild this year or next. We want a healthy environment, a balanced budget, and ranchers earning a good living. Most of us want mustangs to thrive in the wild and receive humane treatment in captivity.

There are no easy solutions. Every decision that solves one problem will cause or worsen problems on another front. This challenge is so complex; it is unlikely that any solution will satisfy all concerned. Compromise and cooperation are our best hope.

References

An Act To Amend Chapter 3 of Title 18, United States Code, So As To Prohibit the Use of Aircraft or Motor Vehicles To Hunt Certain Wild Horses or Burros on Land Belonging to the United States, and for Other Purposes. Pub. L. No. 86-234 (1959).

An Act To Restore the Prohibition on the Commercial Sale and Slaughter of Wild Free-Roaming Horses and Burros, H.R. 249, 110th Cong. (2007).

Alberta Equine Welfare Group. (2008, February). *The Alberta horse welfare report: A report on horses as food producing animals aimed at addressing horse welfare and improving communication with the livestock industry and the public.* Calgary, Alberta, Canada: Alberta Farm Animal Care Association. Retrieved from http://equineenews.osu.edu/documents/HorseWelfareReport1-AFAC.pdf

Associated Press. (2010, July 12). *BLM suspends Nevada horse gather after 7 animals die.* Retrieved from http://klas.dsys1.worldnow.com/story/12795947 blm-suspends-nevada-horse-gather-after-7-animals-die

Associated Press. (2012, December 20). *New Mexico meat company sues feds, claiming inaction is delaying horse slaughterhouse opening.* Retrieved from http://www.foxnews.com/us/2012/12/20/new-mexico-meat-company-sues-feds-claiming-inaction-is-delaying-horse/

Bartell, J.A. (2011). *Porcine zona pellucida immuncontraception vaccine for horses* (Unpublished master's thesis). Oregon State University, Corvallis.

Bartholow, J. (2007). Economic benefit of fertility control in wild horse populations. *Journal of Wildlife Management, 71*(8), 2811–2819. doi: 10.2193/2007-064

Behar, M. (2010). The mustang redemption. *Mother Jones, 35*(1), 50–58.

Bjerga, A., & Crawford, A.J. (2013, March 19). *Horse-slaughter jobs embraced in state where cowboys roam.* Retrieved from http://www.bloomberg.com/news/2013-03-19/horse-slaughter-jobs-embraced-in-state-where-cowboys-roam.html

BLM to gather wild horses from Pancake Complex between Ely and Eureka. (2011, December 8). *Ely Times* (Ely, NV). Retrieved from http://www.elynews.com/news/article_fc5b1c24-21cc-11e1-9d03-001871e3ce6c.html

Bokdam, J. (2003). *Nature conservation and grazing management: Free-ranging cattle as a driving force for cyclic vegetation succession.* Wageningen, Netherlands: Ponsen & Looyen.

Born, K.M. (2007, January 8). *The Quartermaster Remount Service.* Retrieved from http://www.qmfound.com/remount.htm

Calico gather wild horse deaths (as of January 26, 2010). (2010). Retrieved from http://thecloudfoundation.files.wordpress.com/2010/02/wild_horse_deaths_summary_report- 012610.pdf

Chen, H. (2013, June 26). Ellis-van Creveld Syndrome. *Medscape.* Retrieved from http://emedicine.medscape.com/article/943684-overview#a0199

Clausing, J. (2013, April 23). *Lawyer: Inspectors clear NM horse slaughterhouse.* Retrieved from http://www.knoxnews.com/news/2013/apr/23/lawyer-inspectors-clear-nm-horse-slaughterhouse/

Cloud Foundation & Equine Welfare Alliance. (2009, August 28). *Circumventing the wishes of Congress and the American public: BLM moves forward with massive removals of wild horses.* Retrieved from http://www.equinewelfarealliance.org/uploads/PryorGatherFinal.doc

Cloud Foundation. *Salazar Election Day interview.* (2012, November 13). Retrieved from http://www.youtube.com/watch?v=yzv1cd_M59I&list=UUjxuLU5B_KLQDSuijIWA41w&index=1&feature=plcp

Coffey, D. (2013, March 21). Here is the face of "pro-horse slaughter" advocates. *PPJ*

Gazette. Retrieved from http://ppjg.me/2013/03/21/here-is-the-face-of-pro-horse-slaughter-advocates/

Cohen, A. (2011, August 11). The quiet war against Wyoming's wild horses. *The Atlantic*. Retrieved from http://www.theatlantic.com/national/archive/2011/08/the-quiet-war-against-wyomings-wild-horses/243286/

Consolidated Appropriations Act, 2005, Pub. L. No. 108-447, Div. E, Title I, § 142, 118 Stat. 3039, 3070 (2004) (amending 16 U.S.C. § 1333).

Cranes and drones: Strange airfellows. (n.d.). Retrieved from http://www.fort.usgs.gov/RavenA/

Cruise, D., & Griffiths, A. (2010). *Wild Horse Annie and the last of the mustangs: The life of Velma Johnston*. New York, NY: Scribner.

Dalke, K. (2010, spring). Mustang: The paradox of imagery. *Humanimalia, 1*(2), 97–117.

Dalke, K.K. (2005). *The real and the imagined: An ethnographic analysis of the wild horse in the American landscape* (Unpublished doctoral dissertation). University of Wisconsin- Milwaukee.

Davis, C.F. (1887). A wild stallion hunt. *Frank Leslie's Popular Monthly, 23*(4), 446–447.

De Steiguer, J.E. (2011). *Wild horses of the West: History and politics of America's mustangs*. Tucson: University of Arizona Press.

Downer, C. (2008, October). *Forever wild and free*. Presented at the Wild Horse Summit in Las Vegas, NV. Retrieved from http://www.saveourwildhorses.org/PDF/Articles/FOREVERWILDANDFREEspeech102008short.pdf

Downer, C. (2011, March 3). *[Letter to National Wild Horse and Burro Advisory Board][Web log post]*. Retrieved from http://tuesdayshorse.wordpress.com/2011/03/06/craig-downer-appeals-to-national-wild-horse-and-burro- advisory-board/

Eberhardt, L.L., Majorowicz, A.K. & Wilcox, J.A. (1982). Apparent rates of increase for two feral horse herds. *Journal of Wildlife Management, 46*(2), 367–374.

Eggert, L., Powell, D., Ballou, J., Malo, A., Turner, A., Kumer, J., . . . Maldonado, J.E. (2010). Pedigrees and the study of the wild horse population of Assateague Island National Seashore. *Journal of Wildlife Management, 74*(5), 963–973. doi: 10.2193/2009-231

Evans, C. (1986). *The relationship of cattle grazing to sage-grouse use of meadow habitat on the Sheldon National Wildlife Refuge* (Unpublished master's thesis). University of Nevada, Reno.

Forsyth, D.M., & Caley, P. (2006). Testing the irruptive paradigm of large-herbivore dynamics. *Ecology, 87*(2), 297-303. doi: 10.1890/05-0709

Fraker, M.A., Brown, R.G., Gaunt, G.E., Kerr, J.A., & Pohajdak, B. (2002). Long-lasting, single-dose immunocontraception of feral fallow deer in British Columbia. *Journal of Wildlife Management, 66*(4), 1141–1147.

Front Range Equine Rescue. (2013, April 22). *Front Range Equine Rescue discovers would-be horse slaughterer falsified federal application, has committed multiple felonies*. Retrieved from http://frontrangeequinerescue.org/documents/April 22 Press Release re Felonies. pdf?_r=1&

Garrott, R.A. & Taylor, L. (1990). Dynamics of a feral horse population in Montana. *Journal of Wildlife Management, 54*(4), 603–612.

Garrott, R.A., Siniff, D.B. & Eberhardt, L.L. (1991). Growth rates of feral horse populations. *Journal of Wildlife Management, 55*(4), 641–648.

Gas masking the horse. (1917, September). *Illustrated World, 28*(1), 680. Retrieved from http://books.google.com/books/download/Illustrated_world.pdf?id=9qtMAAAAY AAJ&output=pdf&sig=ACfU3U2EJjlpfwk2NR0HMGIN1jRailGYeQ

Georgiadis, N., Hack, M., & Turpin, K. (2003). The influence of rainfall on zebra population dynamics: Implications for management. *Journal of Applied Ecology, 40*(1), 125–136. doi: 10.1046/j.1365-2664.2003.00796.x

Gillian, C. (2013, March 21). *Valley Meat horse buyer kills horse on-camera to send a message.* Retrieved from http://www.examiner.com/article/valley-meat-horse-buyer-kills-horse-on-camera-to-send-a-message

Glionna, J.M. (2012, November 14). Interior's Ken Salazar, in video, threatens wild horses reporter. *Los Angeles Times.* Retrieved from http://www.latimes.com/news/nation/nationnow/la-na-nn-ken-salazar-colorado-wild-horses-20121114,0,3145414.story

Goodloe, R.B., Warren, R.J., Osborn, D.A., & Hall, C. (2000). Population characteristics of feral horses on Cumberland Island, Georgia and their management implications. *Journal of Wildlife Management, 64*(1), 114–121. doi: 10.2307/3802980

Greger, P.D., & Romney, E.M. (1999). High foal mortality limits growth of a desert feral horse population in Nevada. *Great Basin Naturalist, 59*(4), 374–379.

Gura, T. (2012). Genomics, plain and simple. Nature, 483(7387), 20–22. doi: 10.1038/483020a

Harding, A. (2011, November 30). BLM OKs Pancake Complex horse roundup. *Elko Daily Free Press* (Elko, NV). Retrieved from http://elkodaily.com/news/local/article_4cb94f62-1b72-11e1-8382-001cc4c03286.html

Harty, J.M. (2010). *The secretary's wild horse and burro initiative: A plan for public engagement.* U.S. Institute for Environmental Conflict Resolution. Retrieved from http://www.azgfd.gov/inside_azgfd/documents/PublicEngagementPlan.pdf

Hobbs, N.T. (1996). Modification of ecosystems by ungulates. *Journal of Wildlife Management, 60*(4), 695–713.

Höglund, D. (2010, March 23). *Management of the free-roaming horse* [Online forum comment]. Retrieved from http://pryorwild.wordpress.com/2010/02/14/february-15-2010-pzps-reversibility/

Hough, E. (1897). *The cowboy.* New York, NY: Brampton Society.

Hough, E. (1898). A new sort of big game [Chicago and the West column]. *Forest and Stream, 51*(17), 328.

Humane Society of the United States. [2007, June 21]. *Protect America's wild horses—support H.R. 249.* Retrieved from http://www.humanesociety.org/assets/pdfs/legislation/110_wildhorses_HR249.pdf

Hyde, D.O. (2009). *All the wild horses: Preserving the spirit and beauty of the world's wild horses.* St. Paul, MN: Voyageur Press.

Johnson, J.J. (1943). The introduction of the horse into the Western Hemisphere. *Hispanic American Historical Review, 23*(4), 587–610.

Johnston, C. (2009, January–February). Wild horses. *Nevada Magazine.* Retrieved from http://nevadamagazine.com/issues/read/wild_horses/

Jurga, F. (2012, February 5). Your horse is in the army now! But where did war horses go before they were sent to France? [Web log post]. *War Horse News.* Retrieved from http://blogs.equisearch.com/warhorseblog/2012/02/british-war-horse-remount-depots-556/

Karlin, M. (2013, March 22). *Man shoots horse dead in video to defy animal activists: Company he works for likely to become first US horse slaughterhouse since 2007*. Retrieved from http://www.truth-out.org/buzzflash/commentary/item/17876-man-shoots-horse-dead-in-video-to-defy-animal-activists-company-he-works-for-likely-to-become-first-us-horse-slaughterhouse-since-2007

Keegan, J. (1993). *A history of warfare*. New York, NY: Vintage Books.

Killian, G., Thain, D., Diehl, N.K., Rhyan, J., & Miller, L. (2008). Four-year contraception rates of mares treated with single-injection porcine zona pellucida and GnRH vaccines and intrauterine devices. *Wildlife Research, 35*(6), 531–539. doi: 10.1071/WR07134

Kirkpatrick, J. (2010, December 29). *A (wild) horse, a (wild) horse, my kingdom for a (wild) horse*. Billings, MT, author.

Kirkpatrick, J., & Fazio, P. (2010, January). *Wild horses as native North American wildlife*. Retrieved from http://awionline.org/content/wild-horses-native-north-american-wildlife

Kirkpatrick, J.F., Rutberg, A.T., & Coates-Markle, L. (2012, June 6). Immunocontraceptive reproductive control utilizing porcine zona pellucida (PZP) in federal wild horse populations (4th Ed). P.M. Fazio (Ed.). Billings, MT: Science and Conservation Center, ZooMontana. Retrieved from http://www.sccpzp.org/wp-content/uploads/PZP-QA-June-6-2012.pdf

Lewis, J.M. (2009, January 1). Equine welfare: Unwanted horses—an epidemic. *DVM Newsmagazine, 40*(1). Retrieved from http://veterinarynews.dvm360.com/dvm/ article/articleDetail.jsp?id=574367

Lewis, J.M. (2009, February 1). Problems mounting for wild-horse management. *DVM Newsmagazine, 40*(2). Retrieved from http://veterinarynews.dvm360.com/dvm/ Veterinary+Equine/Problems-mounting-for-wild-horse-management/Articl Standard/Article/detail/581081

Linklater, W.L., & Cameron, E.Z. (2002). Escape behaviour of feral horses during a helicopter count. *Wildlife Research, 29*(2), 221–224. doi: 10.1071/WR01063

Lowney, M., Schoenfeld, P., Haglan, W., & Witmer, G. (2005). Overview of impacts of feral and introduced ungulates on the environment in the eastern United States and Caribbean. In D.L. Nolte & K.A. Fagerstone (Eds.), *Proceedings of the 11th Wildlife Damage Management Conference* (pp. 64–81). Fort Collins, CO: National Wildlife Research Center.

Lynch, J. (2012, April 19). FAA releases lists of drone certificates—many questions left unanswered. Retrieved from https://www.eff.org/deeplinks/2012/04/faa-releases-its-list-drone-certificates-leaves-many-questions-unanswered

Madosky, J.M., Rubenstein, D.I., Howard, J.J., & Stuska, S. (2010). The effects of immuno-contraception on harem fidelity in a feral horse (*Equus caballus*) population. *Applied Animal Behaviour Science, 128*(1), 50–56. doi: 10.1016/j.applanim.2010.09.013

Martin, J.J., Edwards, H.H., Burgess, M.A., Percival, H.F., Fagan, D.E., Gardner, B.E., . . . Rambo, T.J. (2012). Estimating distribution of hidden objects with drones: From tennis balls to manatees. *PLoS ONE, 7*(6): e38882. doi: 10.1371/journal.pone.0038882

McGivering, J. (2012, June 12). Drones to protect Nepal's endangered species from poachers. Retrieved from http://www.bbc.co.uk/news/science-environment-18527119

McKnight, T.L. (1959). The feral horse in Anglo-America. *Geographical Review, 49*(4), 506–525.

Mustang Monument. (2012). *Mustang Monument wild horse eco-resort*. Retrieved from http://www.mustangmonument.com/

Myers, L. (2012, April 20). Plans for Nevada wild horse eco-sanctuary pick up speed. *Las Vegas Review-Journal* (NV). Retrieved from http://www.lvrj.com/news/pickens-nevada-wild-horse-eco-sanctuary-plans-advance-148133635.html?numComments=36

National Research Council. (2013). *Using Science to Improve the BLM Wild Horse and Burro Program: A way forward*. Washington, DC: The National Academies Press.

National Research Council, Board on Agriculture and Renewable Resources, Committee on Wild and Free-Roaming Horses and Burros. (1982). *Wild and free-roaming horses and burros: Final report*. Washington, DC: National Academy Press.

National Wildlife Research Center. (2008, December 23) *GonaCon immunocontraceptive for deer*. Fort Collins, CO: Author.

Paulick Report staff. (2013, March 16). Attorney: Horse slaughter plant ready to go in three weeks. *Paulick Report*. Retrieved from http://www.paulickreport.com/news/the-biz/attorney-horse-slaughter-plant-ready-to-go-in-three-weeks/#.UZwJYL-ZMZNY.email

Pimentel, D., Lach, L., Zuniga, R., & Morrison, D. (2002). Environmental and economic costs associated with nonindigenous species in the United States. In D. Pimentel (Ed.), *Biological invasions: Economic and environmental costs of alien plant, animal, and microbe species* (pp. 285–303). Boca Raton, FL: CRC Press.

Philipps, D. (2012, September 28). All the missing horses: What happened to the wild horses Tom Davis bought from the gov't? *Pro Publica*. Retrieved from http://www.propublica.org/article/missing-what-happened-to-wild-horses-tom-davis-bought-from-the-govt

Philipps, D. *Salazar interview 11/6/12*. (2012, November 6). Retrieved from http://chirbit/8zx4Dh

Ransom, J., Cade, B., & Hobbs, N. (2010). Influences of immunocontraception on time budgets, social behavior, and body condition in feral horses. *Applied Animal Behaviour Science, 124*(1), 51–60. doi: 10.1016/j.applanim.2010.01.015

Reiner, R.J., & Urness, P.J. (1982). Effect of grazing horses managed as manipulators of big game winter range. *Journal of Range Management, 35*(5), 567–571.

Restore Our American Mustangs Act, H.R. 1018/S. 1579, 111th Cong. (2009).

Rhodes, D.G. (2010, Summer). Saving wild horses: An interview with Ginger Kathrens. *Friends of Animals Act•ionLine*. Retrieved from http://friendsofanimals.org/actionline/summer-2010/wild_horses.php

Roberts, M. (1999). *Shy Boy: The horse that came in from the wild*. New York, NY: HarperCo.

Roelle, J.E., Singer, F.J., Zeigenfuss, L.C., Ransom, J.I., Coates-Markle, L., & Schoenecker, K.A. (2010). *Demography of the Pryor Mountain Wild Horses, 1993–2007* (Scientific Investigations Report 2010–5125). Reston, VA: U.S. Geological Survey. Retrieved from http://pubs.usgs.gov/sir/2010/5125/pdf/SIR10-5125.pdf

Roosevelt, T. (1896). *Ranch life and the hunting-trail*. New York, NY: The Century Co.

Ruggles, A.M. (2011, October 13). Nevada's Pancake Complex HMAs to be site of BLM wild horse science project. *Images of the Vanishing West*. Retrieved from http://photo-rover.blogspot.com/2011/10/2011oct13-nevadas-pancake-complex-hmas.html

Rutberg, A.T. 1990. Intergroup transfer in Assateague pony mares. *Animal Behaviour, 40*(5), 945–952. doi: 10.1016/S0003-3472(05)80996-0

Ryden, H. (2005). *America's last wild horses (Revised ed.)*. New York: Lyons Press.

Safeguard American Food Exports Act of 2013, H.R. 1094/S. 541, 113th Cong. (2013).

Saving America's Mustangs Foundation. (n.d.). *Saving America's Mustangs: A prospectus*. Retrieved from http://www.blm.gov/pgdata/etc/medialib/blm/wo/Communications_Directorate/public_affairs.Par.76646.File.dat/ SAM_pospectus.pdf

Severson, K.E., & Urness, P.J. (1994). Livestock grazing: A tool to improve wildlife habitat. In M. Vavra, W.A. Laycock, & R.D. Pieper (Eds.), *Ecological implications of livestock herbivory in the West* (pp. 232–249). Denver, CO: Society for Range Management.

Sponenberg, D.P. (2011). *North American Colonial Spanish Horse update, July 2011*. Retrieved from http://www.centerforamericasfirsthorse.org/north-american-colonial-spanish-horse.html

Steele, R. (1909). Trapping wild horses in Nevada. *McClure's Magazine, 34*(2), 198–209.

Stillman, D. (2005, February 16). Mustang Sallies: Can America's wild horses survive another four years of Bush? *Slate*. Retrieved from http://www.slate.com/articles/news_and_politics/the_best_policy/2005/02/mustang_ sallies.html

Stillman, D. (2009). *Mustang: The saga of the wild horse in the American West* (Kindle Ed.). New York, NY: Houghton Mifflin Harcourt.

Symanski, R. (1996). Dances with horses: Lessons for the environmental fringe. *Conservation Biology, 10*(3), 708–712. doi: 10.1046/j.1523-1739.1996.10030708.x

Turner, J.W., Jr., Liu, I.K.M., Flanagan, D.R., Rutberg, A.T. & Kirkpatrick, J.F. (2007). Immunocontraception in wild horses: One inoculation provides two years of infertility. *Journal of Wildlife Management, 71*(2), 662–667. doi: 10.2193/2005–779

U.S. Bureau of Land Management. [2009, August 28]. *Questions and answers about the Pryor Mountain wild horse herd gather*. Retrieved from http://www.blm.gov/pgdata/etc/medialib/blm/wo/Planning_and_Renewable_Resources/wild_horses_and_burros/pls_herd_area_statistics/pryor_mountain_fact.Par.68036.File.dat/factsheet Pryor Mountains.pdf

U.S. Bureau of Land Management. (2009, October 7). Secretary Salazar announces new wild horse and burro plan [Press release]. Retrieved from http://www.blm.gov/co/st/en/BLM_Information/newsroom/2009/sec__salazar_announces.html

U.S. Bureau of Land Management. (2010). *Environmental assessment: Adobe Town-Salt Wells Creek Herd Management Area Complex wild horse gather. Environmental Assessments (WY)*, Paper 25). Retrieved from http://digitalcommons.usu.edu/wyoming_enviroassess/25

U.S. Bureau of Land Management. (2012, June 18). JacksonMtnGather-06_18_12_003. BLM Nevada. Retrieved from http://www.flickr.com/photos/blmnevada/7397031842/

U.S. Bureau of Land Management. (2012, August). *Myths and facts*. Retrieved from http://www.blm.gov/wo/st/en/prog/whbprogram/history_and_facts/myths_and_facts.html

U.S. Bureau of Land Management. [2012, September 11]. *Questions and Answers, Pancake HMA Horse Gather, September 2012*. Retrieved from http://www.blm.gov/pgdata/etc/medialib/blm/nv/field_offices/ely_field_office/wild_horse___burro/eydowhgat_pancake0.Par.14237.File.dat/Questions and Answers_Pancake HMA Emergency Gather.pdf

U.S. Bureau of Land Management. (2012, September 14). *Pancake gather mares & colts gathered [sic]*. Retrieved from http://www.flickr.com/photos/blmnevada/7986119790/in/photostream

U.S. Bureau of Land Management. (2012, September 28). *Wild horse and burro quick facts.* Retrieved from http://www.blm.gov/wo/st/en/prog/whbprogram/history_and_facts/ quick_facts.print.html

U.S. Bureau of Land Management. (2013, February 5). *Fact Sheet on the BLM's Management of Livestock Grazing.* Retrieved from http://www.blm.gov/wo/st/en/prog/grazing.html

U.S. Bureau of Land Management. (2013, March 8). *Wild horse and burro quick facts.* Retrieved from http://www.blm.gov/wo/st/en/prog/whbprogram/history_and_facts/quick_facts.html

U.S. Bureau of Land Management, Billings Field Office. (2009, May). *Pryor Mountain Wild Horse Range/Territory environmental assessment MT-010-08-24 and herd management area plan.* Billings, MT: Bureau of Land Management, Billings Field Office. Retrieved from http://www.blm.gov/pgdata/etc/medialib/blm/mt/field_offices/billings/wild_horses.Par.30079.File.dat/pmwhrFINAL.pdf

U.S. Bureau of Land Management, Elko District, Wells Field Office. (n.d.). Proposed northeast Nevada wild horse eco-sanctuary. Retrieved from http://www.blm.gov/pgdata/etc/medialib/blm/nv/field_offices/elko_field_office/information/nepa/eiss/archives/nenvwh_ecosanctuary.Par.21472.File.dat/EcoSanctuaryScopingBrief.pdf

U.S. Bureau of Land Management, Wild Horse and Burro Program. (2005, March). *Strategic research plan: Wild horse and burro management.* Retrieved from http://www.blm.gov/pgdata/etc/medialib/blm/wo/Planning_and_Renewable_Resources/wild_horses_and_burros.Par.91906.File.dat/Strategic Research Plan.pdf

U.S. Bureau of Land Management, Wild Horse and Burro Program. [2007]. *Wild horse and burro removal, adoption, population, AML table.* Retrieved from http://www.wildhorseandburro.blm.gov/statistics/PopRemAdopStats71-05.pdf

U.S. Bureau of Land Management, Wild Horse and Burro Program. [2009]. *We'll pay you! $500 when you adopt an adult horse (4yrs +).* Retrieved from http://www.blm.gov/pgdata/etc/medialib/blm/nm/programs/whb/whb_flyers.Par.58470.File.dat/WHB Incentive flyer.pdf

U.S. Bureau of Land Management, Wild Horse and Burro Program. (2011, January 13). *$500 Wild Horse Adoption Incentive Program.* Retrieved from http://www.blm.gov/nm/st/en/prog/wild_horse_and_burro/_500_Wild_Horse_Adoption_Incentive.html

U.S. Department of Agriculture. (2013, February 27). *Disaster designations under the amended rule.* Retrieved from http://www.usda.gov/wps/portal/usda/usdahome?contentidonly=true&contentid=emergency_response_maps.html

U.S. Department of Agriculture, Office of the Inspector General. (2010, September). *Animal and Plant Health Inspection Service administration of the horse protection program and the slaughter horse transport program* (Audit Report 33601-2-KC). Retrieved from http://usda.gov/oig/webdocs/33601-02-KC.pdf

U.S. Department of the Interior. (2010). *Wild horse and burro initiative.* Retrieved from http://www.blm.gov/pgdata/etc/medialib/blm/wo/Planning_and_Renewable_Resources/wild_horses_and_burros/national_page.Par.21307.File.dat/2011_Budget_in_Brief_WHB.pdf

U.S. Fish and Wildlife Service. (2010, September). *Back Bay National Wildlife Refuge comprehensive conservation plan.* Retrieved from http://www.fws.gov/northeast/planning/Back Bay/pdf/FinalCCP/BACKBAYNWRFinalCCP9_2010.pdf

Vavra, M. (2005). Livestock grazing and wildlife: Developing compatibilities. *Rangeland Ecology & Management, 58*(2), 128–134. doi: 10.2111/1551-5028(2005)58<128:LGAWDC>2.0.CO;2

Vega-Pla, J.L., Calderón, J., Rodríguez-Gallardo, P.P., Martínez, A.M., & Rico, C. (2006). Saving feral horse populations: Does it really matter? A case study of wild horses from Doñana National Park in southern Spain. *Animal Genetics, 37,* 571–578. doi: 10.1111/j.1365-2052.2006.01533.x

Veterinarians for Equine Welfare. (2008, January 9). *Horse slaughter—its ethical impact and subsequent response of the veterinary profession: A white paper.* Retrieved from http://www.vetsforequinewelfare.org/white_paper.php/

Vincent, C.V. (2012). *Grazing fees: Overview and issues.* Retrieved from https://www.fas.org/sgp/crs/misc/RS21232.pdf

Wild and Free-Roaming Horses and Burros Act of 1971, 16 U.S.C. §1331–1340 (1976).

Wyman, W.D. (1962). *The wild horse of the West.* Lincoln: University of Nebraska Press. (Original work published 1945.)

Zimmerman, C., Sturm, M., Ballou, J., & Traylor-Holzer, K. (Eds.). (2006). *Horses of Assateague Island population and habitat viability assessment workshop: Final report.* Apple Valley, MN: IUCN/SSC Conservation Breeding Specialist Group.

Chapter 3

Assateague, Maryland

Camping with Horses

Assateague *and* Chincoteague

MARYLAND–VIRGINIA

■ Highway	▨ Beach	
▤ Road	▨ Marsh	
┄ ORV route	□ Maritime forest	

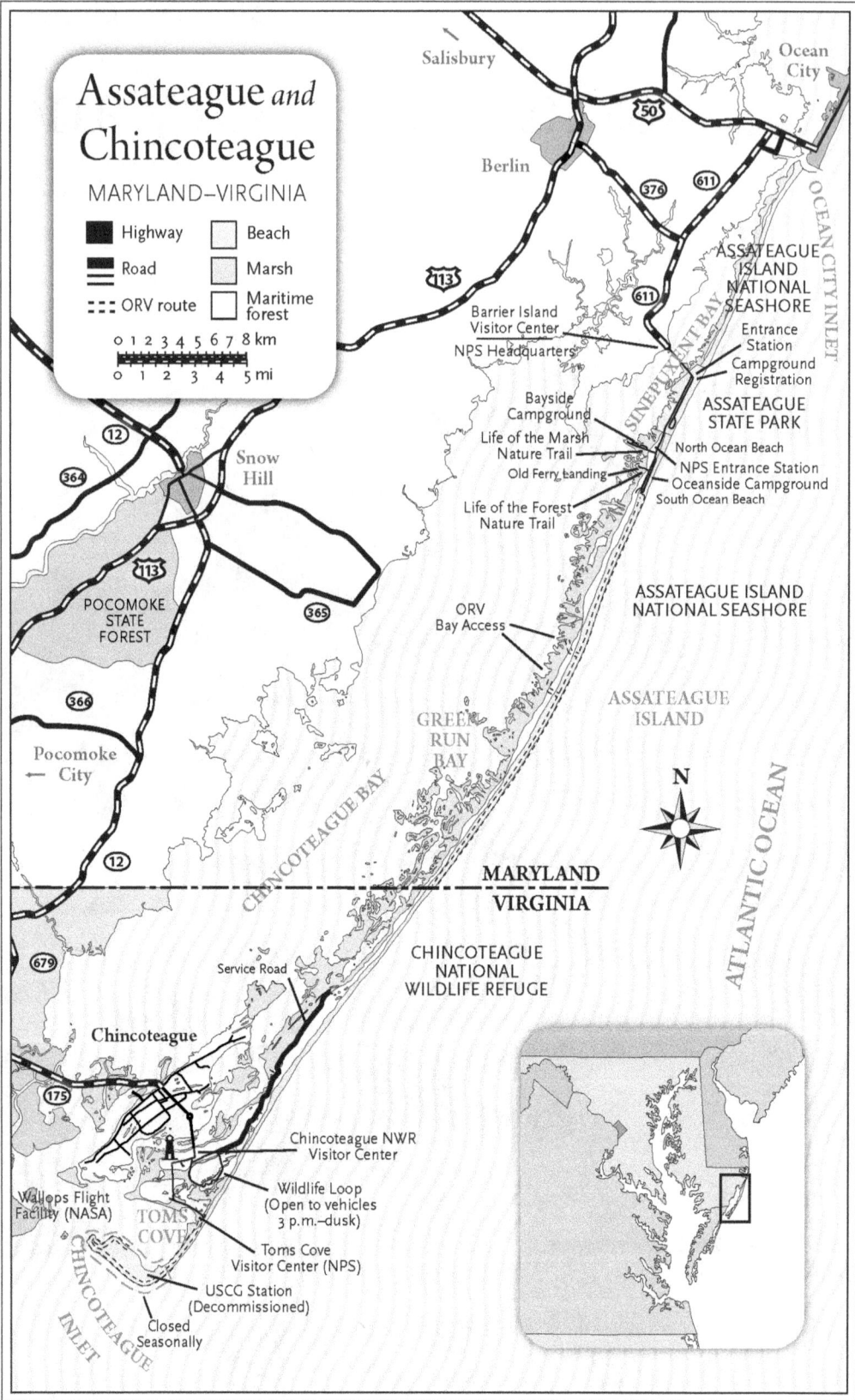

0 1 2 3 4 5 6 7 8 km
0 1 2 3 4 5 mi

Salisbury

Ocean City

Berlin

ASSATEAGUE ISLAND NATIONAL SEASHORE

Barrier Island Visitor Center
NPS Headquarters

Entrance Station
Campground Registration

Bayside Campground

ASSATEAGUE STATE PARK

Life of the Marsh Nature Trail
North Ocean Beach

Old Ferry Landing
NPS Entrance Station
Oceanside Campground
South Ocean Beach

Life of the Forest Nature Trail

Snow Hill

POCOMOKE STATE FOREST

ORV Bay Access

ASSATEAGUE ISLAND NATIONAL SEASHORE

Pocomoke
← City

GREEN RUN BAY

ASSATEAGUE ISLAND

CHINCOTEAGUE BAY

N

MARYLAND
VIRGINIA

ATLANTIC OCEAN

Service Road

CHINCOTEAGUE NATIONAL WILDLIFE REFUGE

Chincoteague

Chincoteague NWR Visitor Center

Wildlife Loop (Open to vehicles 3 p.m.–dusk)

Wallops Flight Facility (NASA)

TOMS COVE

Toms Cove Visitor Center (NPS)

USCG Station (Decommissioned)

Closed Seasonally

CHINCOTEAGUE INLET

SINEPUXENT BAY

OCEAN CITY INLET

At the Maryland end of Assateague Island, visitors to the state and national parks can get quite intimate with free-roaming horses. Too intimate. The 600-lb-plus (270-kg) animals troop across campsites, block access to bathhouses, and tramp over beach towels. Bands of ponies unpredictably cross the main road, causing unsuspecting motorists to hit the brakes fast or even hit the ponies themselves. Some ponies purposely block traffic to thrust searching muzzles into open car windows, hoping for a taste of human food.

Most of the Maryland horses have little fear of people. They brazenly wander into campsites in search of good grazing and any tidbits that they can beg from obliging humans. They walk under clotheslines and step on bathing suits. They track sand across tarpaulins while families work to pitch tents. They sniff and sometimes sample food cooking inches from open fires. Itchy foals view most human contraptions, from barbecue grills to truck bumpers, as potential scratching posts. They tear large holes in screen houses and walk on in, even if there is nothing inside. Campers shoo them away with the loud clanging of a spoon against a pot. It does not frighten them, but usually signals the animals that they are not welcome and should take their activities elsewhere. On hot days, when insects became intolerable, they cross the dune line to the open beach to stand at the waterline beside the bathers.

In his book *Into the Wind* (1994), Jay Kirkpatrick writes about an Assateague stallion "intelligent enough to take advantage of civilization's amenities" (p. 45). Slash, a pinto stallion "with long, white slashes down his side reminiscent of seagull droppings," would avoid the biting insects by keeping to the areas of the park that employees fogged with insecticide. When he was thirsty, he would stand by a water faucet until a camper came by, whereupon he would draw attention to himself by stamping one hoof. Eventually a camper would figure out what Slash wanted and turn on the water faucet so that he could drink. "As if to thank the camper," Kirkpatrick writes, "Slash would reward his trainee with an amazing array of facial expressions signifying his satisfaction, then drink his fill of sweet, fresh water" (p. 45).

Assateague is a 37-mile-long (60-km) island populated by two herds of feral horses separated by a sea-to-sound fence at the Maryland-Virginia line. Historical records document the presence of livestock on Assateague Island since the late 1600s. There are two versions of the story of their origins. For generations, local people grew up believing that the original horses swam to shore from a wrecked Spanish galleon. According to the U.S. National Park Service, they probably descended from domesticated stock owned by early settlers to the region. In colonial America, residents used grassy islands and necks as grazing commons that substituted bodies of water for fencing to contain horses, cattle, sheep, hogs, and goats. Periodically, stockmen conducted communal roundups, or "pennings," to brand, sell, or remove livestock to the mainland.

The village of Chincoteague, Va., revived Pony Penning in the 1920s after a brief hiatus, penning and selling young Assateague ponies to benefit the Chincoteague Volunteer Fire Company. When much of Assateague Island became a national seashore in 1965, most of

While visitors enjoy a day at the beach, a band of wild horses ransacks their campsite.

the free-roaming horses had been moved to the Chincoteague National Wildlife Refuge at the Virginia end of the island, which allows the firefighters to keep horses there by special agreement.

At the Maryland end of Assateague there remained a smaller group of native ponies. Authorities disagree on its size and growth rate. Barry Mackintosh (1982/2003), author of the official history of Assateague Island National Seashore, says that the Maryland herd descends from 10 horses that seasonal resident Paul Bradley donated to the Berlin, Md., Jaycees in 1965, and that the Jaycees donated the herd to the Park Service three years later. Dr. Ronald Keiper, a zoologist who has studied the horses extensively over the past 40 years, wrote that the Jaycees donated 21 horses in 1965—9 stallions and 12 mares (1985). Zimmerman et al. (2006, p. 45) claim that there were 9 horses in 1961, "10+" in 1966, and 21 in 1967.

Eggert et al. (2010) say there were 28 horses when the Park Service acquired the herd in 1968. The team analyzed the pedigrees and DNA of the Maryland ponies and traced the current population to 39 founders with 11 maternal lineages. The number of maternal lineages in a population cannot exceed, and often is much lower than, the number of individuals. If the Assateague herd ever shrank to 9 horses (Zimmerman et al., 2006) or 10 (Mackintosh, 1983/2003), it could not have preserved 11 matrilines. Either the founding herd was larger than reported, or there were undocumented introductions later.

In the 1960s, two agencies began establishing parks on the Maryland section of Assateague. The National Seashore owns most of it, but Assateague State Park occupies about 850 acres/344 ha. The horses roamed freely between federal and state property. Spencer P. Ellis, director of the Maryland Department of Forests and Parks, wanted the ponies removed, viewing them as destructive and a potential safety hazard for children (Mackintosh,1982/2003). Superintendent Bertrum C. Roberts of the national seashore preferred that the Park Service acquire them as a desirable exotic species, limit their reproduction, and manage them as wildlife.

In October, 1970, Roberts wrote to a colleague,

On the Maryland end of Assateague, wild ponies live more or less as wildlife. Park officials interfere with their lives only to administer contraception, which limits their numbers and extends the life expectancy of the mares. Here they share a tasty morsel of *Phragmites*, exotic animals eating an exotic plant.

The Service-owned herd of Assateague ponies have finally, after five heavy use seasons, become accustomed to human activity. On the surface this appears to be a great boon for the visitor. This year, however, we experienced our first cases of horse bites and kicks because of the "taming" of these wild little beasts. This according to our Solicitor must result in "do not feed, pet, or otherwise get involved with the pony" signs at the seashore entrances as well as in the appropriate park literature. It is difficult to conceive that this problem is with us at a seashore, but it is. (Mackintosh, 1982/2003, p. 105)

"All over the country, feral horses roam on federal lands managed by the National Park Service, the Fish and Wildlife Service, and the Bureau of Land Management," said Lou Hinds, former manager of the Chincoteague NWR (personal communication, May 21, 2010). "The American public often cannot see the distinctions between these federal agencies, and often do not realize that each agency might have a different way of looking at the animals." One thing unites all the agencies responsible for wild horses: they prefer to eliminate them from public lands.

For most parks, management plans hinge on whether the goal is to conserve species or to preserve natural processes (Houston & Schreiner, 1995). If the park prioritizes conservation of species that may be compromised by wildfire, grazing, or predation, officials manage the park to protect them from these pressures. If the goal is to allow nature to take its course and accept any consequences, park managers do not interfere with those processes. If the goal is to allow natural processes to occur unless they produce undesirable outcomes, park personnel intervene only when unacceptable results appear likely (Houston & Schreiner,

The horses of Assateague are small, only 12–13.2 hands (48–54 in./1.22–1.37 m) at the withers. Their short stature is partly due to generations of living in harsh conditions and partly due to genetics.

1995). Most larger national parks in the United States are managed in accordance with this third option with a strong emphasis on allowing natural processes free rein.

Two federal agencies manage the free-roaming horses of Assateague. The Park Service has total management control of the island north of the fence that runs from ocean to bay and keeps the herds mostly separate. In the past, the Park has sometimes transferred problem horses from the Maryland end of the island to join the Virginia population. Now, a wandering horse will stray around the fence occasionally, until the one of the managing agencies can return it to its home range.

The Maryland herd is managed differently from the Virginia herd. In Maryland, the Park Service owns the horses, manages them as wildlife, and has targeted a herd size of 80–100 animals. The horses north of the fence live with minimal human interference, as wild as the deer that share their habitat (Ingle, 2005). The horses in the Virginia herd reproduce at will, and the foals are sold annually during the Pony Penning festival. A few of the best foals each year are "turn-backs" purchased to be donated back to the herd as breeding stock that will spend their lives on Assateague. The herd is regularly vaccinated, wormed, and given veterinary and farrier attention. In many ways it is managed similarly to Western ranch horses.

The herds have a common origin, however, they have probably been present since the 17th century, and they have become a unique and relatively homogeneous breed. These horses are pony-sized to be sure; their average height is only 12–13.2 hands (48–54 in./1.22–1.37 m) at the withers. (Judged solely by height, any horse under 14.2 hands (58 in/1.47 m) is a pony.) They are built like ponies—short legs and backs, dense bones, and thick manes and tails. There are also genetic distinctions between ponies and horses. Centuries ago, the

The Assateague horses are in good condition overall and go about their lives with a relaxed, unwary attitude, unlike most Western mustangs. Even so, when a band lies down to rest, one horse remains standing
as sentry. At the time the author took this picture, it had rained heavily for 11 straight days, and with the
reappearance of sunshine, the ponies took the opportunity to nap on the drying grass.

animals were taller and built more like horses. Although it is possible that the foundation
stock was mostly Spanish, outside genes have been introduced over the years. According to
Keiper (1985) and others, Shetland Ponies were added to promote pinto coloration in the
1920s, and their genes might have also decreased the height of descendants. Despite outcrossing with ponies, both herds are still genetically horses, and foals sold to the mainland
from Chincoteague often outgrow their island brethren.

Genetics, then, is not solely responsible for their diminutive stature. Their small size
today probably results from the interplay of many influences. It is possible that tight space
on Assateague limits not only their numbers, but also their size. When large animals live in
small areas, particularly islands, their size often decreases over many generations in a phenomenon known as insular dwarfism. Harsh environmental conditions and low-nutrient
forage also restrict growth. The horses of Sable Island, off the coast of Nova Scotia, endure
a colder, stormier climate but remain larger because they have different bloodlines that predispose them to greater stature, even under adverse conditions.

Other breeds flavored the Assateague horses over the centuries, but no documented outside introductions of stock have been made to the Maryland herd since the Park Service
took over its management in 1965. Until that time, there was no real difference between the
Virginia and Maryland bloodlines. Since then, managers have added outside horses to the
Virginia herd to improve bloodlines. The current Park Service contraception program limits
fertility while maintaining maximal genetic diversity in this closed population.

Assateague horses, especially those in the prime of life, remain round and robust even
after a difficult winter. Carl Zimmerman, former resource manager for Assateague Island

A pregnant woman reviews her video footage while her toddler pets the muzzle of a wild mare. A man stands with a stroller, one hand stroking her glossy coat. These horses are unafraid of people, but they are not tame. Ponies shift gears quickly and often kick, bite, or trample people standing nearby.

NS, attributed their apparent good health to a combination of comparatively good nutrition and the use of birth control to limit reproduction (personal communication, 1998). Ponies in general tend to have efficient metabolisms and are usually "easy keepers" that stay plump on minimal forage.

From the time Assateague first became a national seashore, the horses have been free to live as wild horses, exhibiting natural behavior and subject to natural processes. Ponies may display lacerations, hoof overgrowth, or other injuries or signs of illness. Age takes its toll; elderly horses may develop prominent ribs, sharp spines, and rough coats. Because it regards the horses as wildlife, the Park Service summons a veterinarian only if human activities have caused injury.

Many domestic horses have a far greater fear of human implements and toys than these supposedly wild ponies do. Bright umbrellas flutter in the wind, screaming children race by them to the sea, and boogie boards wash up in the surf, yet the ponies seldom shy or spook. In a camping area, inline skaters swiftly blade down the pavement, passing so close to grazing horses that they could reach out and give them a pat. In fact, some do. Many domestic horses are at least somewhat intolerant of wheeled people whizzing by on bikes, skates, or other devices, yet these animals largely ignore the children on pedal toys that nearly collide with them at regular intervals. Dogs bark savagely at the equine interlopers visiting the campsites, but the ponies seem to know that dogs at the seashore must remain on a 6-ft/1.8-m leash—they are unimpressed with the bravado.

No matter how crowded the park may be, the horses attend to their activities unmindful of the audience. Stallions duel violently among parked vehicles and barbecue grills. Bachelor males chase one another across campsites at a mad gallop. Stallions enthusiastically mate with mares in the shade of the bath houses while bystanders pretend not to watch.

Only a small part of the Maryland section of the barrier island is tourist-friendly, and many of the horses prefer to live in the areas not frequented by people. These horses are more shy and reclusive than those that frequent the campgrounds. To the north and south

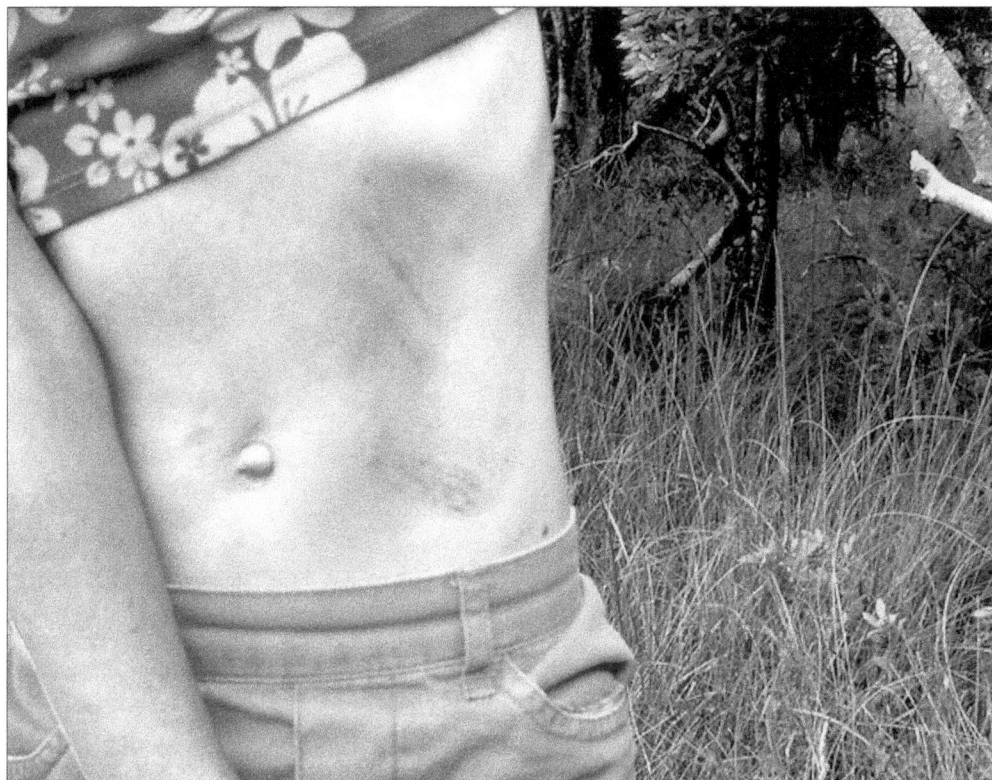

The hoofprint of a wild horse on the abdomen of a very lucky girl. This kick could have easily fractured her ribs or ruptured her spleen. Aimed slightly higher, a kick to the head or the sternum could have killed her. Photograph courtesy of the National Park Service.

beyond the camping area, Assateague is undeveloped and relatively unused. Only a small percentage of visitors ever leave the developed section. Consequently, much of this well-used park remains undisturbed and natural.

Park Service rangers impose fines of $175 per incident when they catch visitors feeding, petting, or approaching within 10 ft/3 m of horses or other wildlife. Many park visitors have trouble accepting that these friendly, curious animals pose any threat to them or vice versa, but they are not tame; they are just unafraid of people.

Unknowingly putting themselves in danger, tourists crowd around the horses when park personnel are absent, stroking them, braiding their manes and sharing bits of a picnic lunch. Unfortunately, this intimacy often has consequences unforeseen by many tourists: visitors are frequently kicked and bitten. The Park Service has a collection of photographs that illustrate the damage a pony can do to the human body. Often the perpetrator was docilely accepting a pat moments before the scene turned ugly.

One Park Service photograph shows a bare-midriffed child with a bright purple hoof-mark centered on her abdomen. She was very lucky. A random kick can dislodge teeth, blind an eye, rupture a spleen, crush a spine or chest, or cause brain injury or death.

Offering treats can result in broken fingers when the pony bites the hand that feeds it. Some visitors offer treats from their cars and, in effect, train them to stand in the road waiting for handouts. Drivers who did not expect to see horses on the pavement hit them at speed. One such incident claimed the life of a healthy 10-year-old mare. She was standing in

How accidents happen. Five drivers stopped on a narrow causeway to see the ponies and attracted them onto the road. Because other visitors have offered them treats through their windows, the horses often deliberately stop traffic and thrust their huge heads into cars looking for handouts. This horse is in an excellent position to inflict a serious bite. Note the car on the bridge, which was traveling at least 40 mph/64 kph. Its driver would not have had enough time to brake for ponies, people, or vehicles in the road.

the middle of the road hoping to stop traffic for a snack when she was stuck by a car. Her leg was shattered, she suffered internal injuries, and the car was badly damaged. Touchingly, her mate stood over her and would not leave. Park rangers humanely euthanized her.

Ticks, as tiny as pepper flecks, infest the ponies. Some of these harbor Lyme disease. Unseen, these ticks jump from pony to petter and can transmit a chronic disease that can cause debilitating fatigue, neurological damage, and muscular weakness. (The author contracted Lyme disease while following pony bands through the brush at a distance to obtain photographs for this book!) Many deer and white-footed mice on Assateague carry Lyme disease, but it is unknown how many ponies are infected. Mosquitoes can transmit encephalitis from horse to human, but only over short distances (Kirkpatrick, 1994).

The Park Service prohibits visitors from approaching within 10 feet of the ponies, but a bus length is a safer distance. A horse lashing out at another will often barge right over a person standing between them, causing bruises or broken bones. Even docile domestic horses that have been trained to inhibit aggressive impulses around people can show their irritability with teeth and hooves. Wild horses are all the more unpredictable and uninhibited.

Some visitors are downright foolhardy. One Park Service ranger told of a woman from New York who tried to ride one of the ponies. Thrown violently to the ground, she got up and remounted, only to be thrown again. When the rangers attempted to stop her, she insisted that it was okay for her to ride them because she knew what she was doing—she had horses of her own. As it turned out, she was a lawyer who should have known a thing or two about following regulations and about liability. A report in the Ocean City *Dispatch* described an incident in which two intoxicated men, one of them naked, were arrested for trying to ride the ponies and tackle sika deer ("Naked Rodeo on Assateague," 2007).

Horses will open containers and tear apart tents to find meals. Foals learn these techniques from their herdmates at a young age. Finding caches of human food reinforces

Busted! The horses know the Park Service vehicles and scatter when one appears on the scene. Like naughty children caught in the act, the ponies assume an air of exaggerated nonchalance: "It wasn't me—I didn't do it!" Although the tangle of vehicles dispersed when the ranger arrived, fraternizing with wildlife cost this visitor a $175 fine.

Visitors who feed horses from their vehicles are in effect training them to stand in the road waiting for handouts. This increases the odds that cars will hit them. A car striking a horse usually hits its legs, hurling its body over the hood and through the windshield, often causing serious injury to the occupants. The car is totaled, and the horse usually suffers a terrifying and painful death. Note the bridge in the background. This automobile struck its victim directly across the road from the spot where the car was feeding ponies in the preceding image. Photograph courtesy of the National Park Service.

Ponies and people bask on the beach at Assateague State Park on a hot July afternoon.

marauding behavior. Because horses knocked over trash cans and consumed everything from greasy paper towels to hot dogs, now the Park Service collects trash in horse-proof dumpsters. Rangers making the rounds of campsites will remove food left out where horses might try to get to it. Human food disrupts the balance of ponies' intestinal flora and is likely to cause colic in animals engineered as grass-eating machines. There is also little nutrition obtained from raids on human comestibles.

In the 1980s, the Park Service considered a pony a problem if it was involved in three or more documented incidents in one year that resulted in property damage or caused injury to a visitor. Before 1995, the Park Service removed a total of 39 horses from the national seashore when they became adept at raiding campsites or begging at the roadside. All those horses were moved south to the Chincoteague NWR. "Our strategy now is to place heavy emphasis on visitor education and viewing horses safely," says Allison Turner, a biological science technician at Assateague Island NS for over two decades (personal communication, February 17, 2011). "By eliminating inappropriate visitor behavior around horses, 'problem horses' should no longer be created."

In August 2011, however, the Park removed an 18-year-old stallion named Fabio from the island when his bold raids on campsites put visitors at risk. While most horses can be shooed away, Fabio believed that he was dominant over not only the members of his own band, but also the campers. When people tried to make him leave before he was ready, he asserted his dominance by kicking, biting, and charging, posing a significant risk to their safety. The Humane Society of the United States trailered him to the Doris Day Horse Rescue and Adoption Center in Texas, where he was to be trained and offered for adoption.

Sometimes ponies deliberately gather in the areas where people spend their time. For about a week during the summer of 2000, up to 75 ponies (which was then about half the Maryland herd) congregated in the state park day-use area, occupying a quarter-mile (400-m) stretch of beach alongside hundreds of bikini-clad bathers and screaming toddlers.

The stallion defends the mares in his band from the advances of other males, but he is not necessarily the most dominant horse in the group. In most bands, the alpha mare makes the day-to-day decisions, such as where to graze and when to water.

When asked why they took to the beach in such numbers, Maryland Park Ranger Rick Ward said,

> They're wild animals. They have minds of their own. Some think they go to the water to cool off and get away from the flies, and the day-use area is just the best source of food out there. . . . they're particularly fond of potato chips. They aren't dumb—they have even learned how to open coolers! (Personal communication, 1998)

The animals knocked over belongings, urinated on beach towels, and rolled in the sand beside sunbathers, but aside from begging food, were generally docile. The rangers concentrated on educating the visitors to avoid contact with the ponies and to keep food away from them. "Most of the time, the ponies and the visitors coexist peacefully. Very few people are a problem," Ward explained.

But he tells of one man who set out his family's lunch in a way that must have looked like a banquet to the ponies. The lifeguard warned him to put the food away, but he ignored her. One persistent pony would not take no for an answer. The man pushed and shoved at the hungry animal, then began to hit him with a shovel! After arguing with the lifeguard and lying to the ranger, the man and his family were evicted from the park.

Ward went on to relate other incidents involving clashes between park visitors and ponies:

> A drunk kid jumped on the back of one on a five dollar bet and was thrown into a bush. His friends took off and left him there. Then there were the two ladies trampled

Saltmarsh cordgrass is the preferred food of Assateague horses, but they will consume a diverse menu of herbage in small amounts even in the presence of abundant grass, presumably to gain trace nutrients or entertain the palate. This mare forced her way into rigid, unyielding undergrowth to nibble briefly.

by the ponies at the National Park. They were just in the way, I guess, lying on the beach. They required hospitalization. (Personal communication, 1998)

He also told of a couple who enthusiastically photographed their toddler walking underneath a stallion, unaware that if he had moved suddenly, the little girl could have been killed.

The Park Service is responsible for the feral horses, whether they are on state or federal property, and maintains them as a "desirable feral species" (Assateague Island NS, 2006). This arrangement necessitates balancing their needs with the park's other natural resources, and keeping ponies and people safe. The park's "pony patrol" is a group of volunteers that contributes more than 1,200 hours a year educating visitors and cautioning them to stay 10 ft/3 m away (Hayward, 2007).

The majority of visitors do not venture beyond the developed areas of the park. Their reluctance to hike into the isolated areas helps to preserve most of Assateague in its natural state. But the emphasis was not always on keeping Assateague wild.

In the 1950s, Leon Ackerman and a group of investors bought, surveyed and platted 15 miles of Assateague oceanfront property north of the Virginia line for residential and commercial lots. He advertised heavily in urban newspapers, tempting buyers to invest in his Ocean Beach development (Mackintosh, 1982/2003) with fantasies of idyllic vacation retreats and speculative profits from resale. By the early 1960s, about 3,200 investors had purchased 5,850 lots at Ocean Beach, and several dozen houses formed the nucleus of the community. Ackerman paved a road, Baltimore Boulevard, which ran to the Virginia line, and dug channels in the marshes for mosquito control. In 1957, Atlantic Ocean Estates, Inc., followed suit by subdividing the northern end of Assateague into 1,740 platted lots. The properties were promoted through radio advertisements that offered listeners "'down payments' of up to $1,000 if they could identify familiar 'mystery tunes' like 'You Are My Sunshine' and 'The Missouri Waltz'" (Mackintosh, 1982/2003, p. 14). Sales were brisk, even though there was no legal access to the property, and no streets, utilities, buildings, or other improvements ever existed. Moreover, the land in question was rapidly migrating westward—with every storm the ocean overwashed, eroding the beach and sweeping across the island to the bay. Ironically, many of the shorefront lots became quite literally Atlantic Ocean Estates as the sea claimed them (Mackintosh, 1982/2003). John T. Moton, the developer, was imprisoned in 1962 in an unrelated scandal (Associated Press, 1962).

Assateague seemed well on its way to becoming another Ocean City. In the 1930s, Assateague was surveyed as a potential site for a national seashore, but the plan never coalesced. By the early 1950s, the Park Service judged Assateague Island too developed for further consideration.

The only access to the island was by ferry, and many prospective property owners balked at buying homes in such an inaccessible location. Developers reasoned that a bridge across Sinepuxent Bay would boost sales and raise the value of the island homes, so they began construction near the ferry dock. Dredging up material from the marshes, they fashioned a causeway stub, but lack of funding forced them to discontinue the project.

Undaunted, the developers changed strategies. The state of Maryland had coveted Assateague as a potential state park, but neither its 1940 nor its 1952 proposal to acquire land there had borne fruit. In 1956, Leon Ackerman's North Ocean Beach, Inc., presented the state of Maryland with 540 prime Assateague acres (219 ha) to establish the park, fully expecting that the state would build a bridge to allow visitors and landowners easy access. The Maryland General Assembly quickly appropriated $750,000 to buy additional land for the park and authorized work on a bridge that ultimately cost the state nearly $2 million (Mackintosh, 1982/2003). Assateague appeared destined to become another bustling resort city—until the Ash Wednesday Storm (the Five-High Storm) of March 1962.

Although nor'easters, many of them severe, pound Assateague regularly, the Ash Wednesday Storm was powerful beyond anything in the memories of even the oldest coastal residents. This tempest wantonly destroyed almost every structure on Assateague. Sheets of seawater literally picked up houses and tossed them into the marsh. When the storm passed, the remains of 11 long-forgotten shipwrecks lay uncovered on the shore. Two new inlets sliced across the island. Twenty-two feral horses on the Virginia end of Assateague drowned in the storm. Baltimore Boulevard was severely damaged, and to this day visitors can observe large broken chunks of the roadway along the Life of the Dunes nature trail in the national seashore.

Other reminders of the storm persist. A freshwater pond stands on the Life of the Forest Trail, providing hydration for Assateague's fauna and a rich habitat for many species. The Ash Wednesday Storm created the pond when it demolished a house and whirled floodwaters around its foundation, scouring a depression in the sand to the level of the water table, where a lens of freshwater collects. The pond endures, and deep within it one can still find the remnants of the house.

After this reality check, developers and homeowners alike wondered whether the barrier island was too unstable to support a resort community. Two studies showed that for Assateague to support communities of any size, developers would need to construct a long line of large protective dunes, install an expensive sewer system, and raise the island to the

Ocean City, Md., is a densely populated commercial vacationland with high-rise hotels, boardwalks, restaurants, amusements, and heavy traffic. If the Ash Wednesday Storm of 1962 had bypassed Assateague, it might never have become a national seashore and would probably look very much like this.

minimum level recommended for permanent construction with 17 million cubic yards (13 million m³) of fill dredged from the floor of the bay (Mackintosh, 1982/2003). This is an astonishing amount of material. It is more than the volume of rock and soil moved to make the original 363-mile Erie Canal (roughly 10 million cubic yards/7.65 million m³) and more than the 16.5 million cubic yards/12.6 million m³ of asphalt used to pave 1,836 mi/2,955 km of Interstate 95 from Maine to Florida. Private and commercial development of the island was possible, but it would be an expensive and chancy undertaking.

Ackerman himself, having grossed about $4.5 million from Assateague real estate sales, declared Assateague unsuitable for private development. Profoundly depressed over the Ocean Beach fiasco and throttled by financial and legal problems related to other ventures, he committed suicide in April 1964 (Mackintosh, 1982/2003).

On the other hand, as the largest undeveloped beach between Cape Cod and Cape Hatteras, Assateague reemerged as an attractive candidate for a national seashore.

With abundant evidence that Assateague was too unstable for permanent development, one would expect lot owners to welcome federal acquisition of their property. Many, however, clung to their investments with fantasies of beach homes and profits. U.S. Representative Rogers C.B. Morton (MD 1) advocated continued residential development of Assateague. The "Morton Plan," advertised as "Assateague's reach for greatness," proposed three private villages about 10 miles apart, including a center for the fine and performing arts, sports facilities, a wildlife museum, and an auditorium (Mackintosh, 1982/2003). People who already owned lots on the island could trade their property for land in the new communities. When this plan drew insufficient support, Morton proposed another in 1964—a 600-acre/243-ha complex for commercial concessions and lodging. Even the Park Service

The remains of Baltimore Boulevard, the paved road built to support the development and commercialization of Assateague Island.

initially contemplated building two 100-room motels with restaurants, numerous concessions, hard-surface parking for 14,000 cars, and a 32-foot-wide (9.6-m) paved highway to extend from bridge to bridge through the refuge (Mackintosh, 1982/2003).

A report from the U.S. Department of the Interior issued in April 1963 recommended that the federal government acquire Assateague Island as a national seashore under the Park Service while letting Assateague SP and Chincoteague NWR retain their individual identities (Mackintosh, 1982/2003).

The three agencies administering the island often clashed. When the seashore was authorized, the Park Service tried to assimilate the state park, which clutched its holdings tenaciously. The Park Service considered plans to increase visitation, which the refuge opposed because its primary purpose is to provide habitat for birds and other wildlife.

In 1965, the areas of the Maryland end of Assateague not owned by the state park were designated a national seashore. The three agencies agreed to minimally develop parts of the island for both intensive (concentrated) and extensive (dispersed) day use and let visitors find food, lodging, and concessions on the mainland.

Now that the seashore existed on paper, the Park Service's next task was to acquire the balance of roughly 9,000 acres/3,600 ha from about 3,500 property owners, some of whom did not want to sell (Mackintosh, 1982/2003). On the northern part of the island was Atlantic Ocean Estates. Although most of its 3,657 lots were owned by Thomas B. McCabe, 195 had been sold (often sight unseen) to investors. The Park Service intended to assimilate this land into the national seashore, but deemed it a lower priority, concluding that lot owners were unlikely to develop this section because there was no land access.

McCabe blocked federal acquisition of Atlantic Ocean Estates at every turn, even though many lot holders wished to sell their rapidly eroding property. By 1970, the advancing sea had swallowed much of the original real estate, and the government filed a condemnation suit for what was left. Delays ensued, many owners who had rejected government offers hoping for a better price found their holdings completely submerged by a storm in 1974.

Like all other barrier islands, Assateague is fundamentally unstable, and it has changed dramatically over the centuries. Although its coordinates, the number of inlets, and the height of its hills have changed, it has remained part of a barrier chain separating ocean from estuary with beach, marsh, overwash flats, and maritime forest. Until recently, it maintained a fairly constant volume of land mass and remained roughly the same distance from the retreating mainland.

A barrier island is the product of complex interactions of wind, waves, sediment transport, and tides, changing contour and composition in response to erosion, storms, changing sea levels, and ocean currents (Hayward, 2007). Sand dunes build up, wash away, and form again in new places. Marshes are filled in, beach is redefined, and the entire island shifts westward.

In the early 1900s, Assateague was actually part of a peninsula extending through Ocean City, Maryland, and joining the mainland at Fenwick Island, Delaware. The Great Hurricane of August 23, 1933, opened Ocean City Inlet and transformed Assateague into an island again. Heavy rainfall pelted Assateague for days, engorging Sinepuxent Bay. When the winds shifted offshore and the tide ebbed, the waters of the bay surged seaward, bisecting the peninsula and forming present-day Assateague Island.

Southbound littoral currents, created by waves that hit the beach at an angle, attempted to refill the passage with sand; but by September 1933, jetties were under construction, and dredging kept the waterway open for the convenience of seagoing vessels. And because those jetties cut off the longshore currents, Assateague was starved for sand.

The Ash Wednesday storm of 1962 breached the northern part of the island in two places. Though the U.S. Army Corps of Engineers repaired them, the root cause of the difficulties remained: interruption of sediment transport, which narrowed the island and made it more vulnerable to inlet formation. Pushed by artificial erosion and accelerated shoreline migration since 1933, Assateague Island has retreated westward nearly 0.6 mi/1 km, creating changes in geography and habitat that would not have occurred otherwise (Pendleton, Theiler, & Williams, 2004).

Through the last century, engineers believed that a dynamic barrier island system could be immobilized with the right combination of interventions. Now, it is clear that

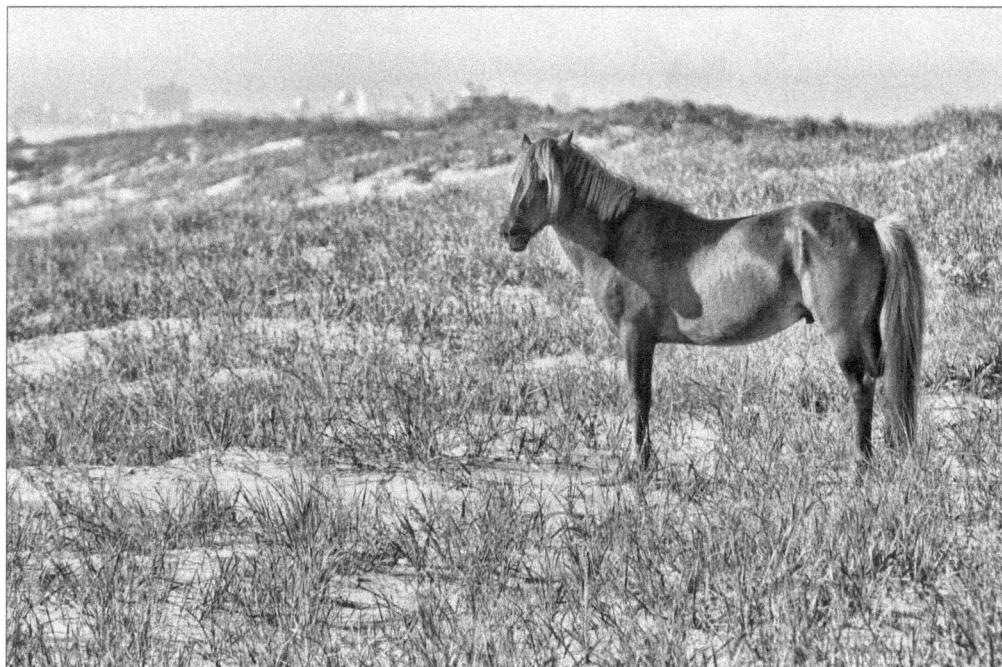

A study in contrasts. A wild stallion stands watch over a vast expanse of low, sandy dunes. Barely discernible on the horizon, Ocean City's geometric skyline marks the divide between the natural world and the modern one.

"improvements" to control one aspect of the system almost always result in undesirable effects elsewhere in the system. When people build dunes and disrupt currents to hold a barrier island in place, it usually begins to lose volume. When they temporarily stabilize both position and volume, they affect the position and volume of other islands in the chain. And in the long run, nature always wins. Despite the best efforts of clever minds and expensive stabilization projects, barrier islands remain mobile and untamable.

Shoreline armoring is the single greatest threat to island beaches (Pilkey, Rice, & Neal, 2004). Construction of seawalls and groins always leads to beach erosion over time. The Ocean City jetties disrupted the ocean currents, which dropped sand in shoals or on the north side of the inlet rather than on the northern end of Assateague. The south jetty eventually stood *in* the inlet until it was extended and reconnected to the shore using large amounts of trucked-in stone. So much sand built up along the north jetty, the Corps of Engineers had to raise the structure to keep that sand from overtopping it and filling the inlet. In 1980, the Corps recommended beach nourishment for the northern end of Assateague to combat erosion.

By this time, Ocean City was highly urbanized and wantonly sprawling northward to meet the development at the Delaware border. Engineers had supported its westward growth by filling in marshes until the Maryland Wetlands Act of 1970 limited this practice. The Clean Water Act of 1972 regulated dredging and filling activities, and the National Environmental Policy Act of 1969 (1970) required developers to conduct environmental impact assessments before beginning a project and to choose the least damaging option. In the 1980s, Ocean City widened its narrowing beaches, nourishing them with sand pumped from offshore shoals. Storms in the 1990s sucked away more of the beachfront,

A band of ponies rests on the beach, where strong breezes keep biting flies from alighting.

necessitating more sand replenishment. The ebb-tidal shoal became so large (more than 1 mi/1.6 km wide) that it merged with the northern end of Assateague. Without mitigation, the northern part of Assateague would become unstable and extremely vulnerable to inlet formation during powerful storms. Direct access to the ocean would destroy important habitat on Assateague and in nearby waters and would expose mainland communities to the full onslaught of storm-driven waves.

The Corps of Engineers and the Park Service collaborated on the North End Restoration Project to replace the sediment lost over the past 60 years to the construction of the Ocean City Inlet jetty system. The goal was to reestablish a "natural" sediment supply for northern Assateague that mimicked the way sand flowed before the jetties were built. The initial phase of the project, completed in 2002, transferred sand onto the beachfront from offshore shoals. Workers dredge the inlet twice a year to keep the channel open.

Ocean City alone hosts between 320,000 and 345,000 visitors on summer weekends, and the population of the surrounding areas has leaped prodigiously over the past 50 years (Ocean City Planning and Zoning Commission, 2006). Interestingly, Ocean City has arrived at these numbers using a mathematical formula called "demoflush," which estimates population based on flow amounts through the sewage treatment system. Demoflush provides a reliable tracking method that remains consistent from season to season and year to year (Goldschmidt & Dahl, 1976).

Probably as a result, water quality has plummeted in Sinepuxent and Chincoteague bays. Total phosphorus levels are 50 times the recommended threshold, ammonia and nitrogen are near the level toxic to fish, oxygen levels are suboptimal, and fecal coliform bacteria levels are elevated (Lambert, Ozbay, & Richards, 2009). If the situation worsens, water recreation and fishing may be affected, and marine life may suffer.

Birds gorge on ticks and insects plucked from the coats of horses and on invertebrates churned from the sand by their hooves.

Some blame Assateague's free-roaming horses for the decrease in water quality. As the logic goes, each adult pony on Assateague Island produces about 40 lb/18 kg of manure each day. The 150 adult horses and their foals living on the Chincoteague NWR collectively produce more than 3 tons/2,700 kg of manure each day, and the 80–100 Maryland horses produce almost 2 additional tons (1,800 kg) daily (Lambert, Ozbay, & Richards, 2009). Water containing urine and manure rich in nitrogen, phosphorus, potassium, magnesium, calcium, zinc, and copper drains off the marshes into the ocean and bays.

This equine pollution contribution is dwarfed, however, by agricultural runoff from fertilized fields and chicken barns on the mainland and by the sewage created by 300,000 or more people, about 14 million gallons/53 million L a day (Russo, 2009). About 36% of the watershed that feeds the bays is used for farming, and large-scale chicken raising and processing are prevalent all along the Delmarva Peninsula (Lambert, Ozbay, & Richards, 2009). The highest phosphorus levels occur in June, when farmers fertilize their fields.

On the other hand, solid horse waste offers potential benefits to island plant life. Any gardener can testify that manure builds organic reserves in soil and increases its ability to absorb and hold water. It stabilizes soil against the relentless winds, assists sprouting and root penetration of young plants, and encourages earthworms and other beneficial organisms (Buchanan, 2003).

The home ranges of these stallions overlap, and they encounter each other regularly. Rather than engaging in battle, they display ritual behaviors that proclaim, "I'm a tough guy, you don't want to mess with me!" Upon meeting, they sniff muzzles, necks, and flanks, their teeth and front legs facing the opponent in case bloodshed is necessary. After about 10–20 seconds of sniffing, both squeal. Another round of sniffing, another pair of ear-splitting squeals. The encounter might go on for 5–10 minutes. The one that yields is the subordinate.

Some alpha stallions allow a subordinate stallion to remain with the band. In this band, the pinto is the harem stallion, and he spends considerable energy affirming his dominance over the chestnut. In this picture, they were investigating manure left by other stallions. The chestnut then defecated on the pile, indicating "this is my home range." The pinto defecated on top of the chestnut's dung, signifying his dominance. Then they both turned to thoroughly sniff the pile again.

Assateague is one of the most visited parks on the Eastern Seaboard, and on the island, human waste is treated in septic fields that are sometimes breached by flooding. Likewise, most of Chincoteague's wastewater is discharged into seepage pits, cesspools, and drain fields. Water quality is an issue on barrier islands that do not support wild horses as well, mostly because of wastewater treatment outfall, public boat anchorages and marinas, and runoff from the mainland.

Mussels and other bivalve shellfish are sentinel species for water quality. Because they feed by filtering water, they tend to concentrate bacteria within their tissues. If those bacteria are harmful, consumption of raw or undercooked shellfish can cause illness. The microorganism content of bay water varies widely from one hour to the next, influenced by wind, currents, precipitation, and tides. Instead of analyzing water directly, ecologists assess water quality by examining shellfish. In one study of Chincoteague and Sinepuxent bays, levels of harmful bacteria in mussels were as high as 19 times the acceptable standard for edible bivalves (Lambert, Ozbay, & Richards, 2009).

Horses appear unaffected by the increasing water pollution. A more pressing issue is finding adequate freshwater on a sandy, salty island. Seawater is about 3% saline, blood is 0.9%. When a land mammal drinks seawater, the kidneys must excrete the extra salt; but they have a very limited ability to do so. Without a source of freshwater, horses will become fatally dehydrated.

Ponies often drink the lower-salinity bay water. Assateague has small freshwater ponds, but even these have some salt content. A certain amount of salt is beneficial—anyone who has seen the well-worn salt licks in the stalls of domestic horses knows that the equine craving for salt is a strong one. Ponies eliminate some of this salt through sweat, some through urine.

Lost body water concentrates blood solutes and triggers thirst (McGreevy, 2004). Assateague horses, like other barrier island horses, may dig for water, excavating sand until they reach the water table and waiting for a shallow pool to form. Horses will often choose to dig for water rather than use freshwater ponds, even though these excavations require much more effort. The holes are generally 2.5–4 ft/0.8–1.2 m deep. On Assateague, Kirkpatrick (1994) noted that he never observed horses drinking from pools of fresh rainwater, even though they would contain virtually no salt.

A study of Przewalski's horses showed that their water intake varied from 4 L a day of icy water in winter to 20 L a day in warmer weather (Houpt, 2005). Feral horses in the Namib desert in Africa may drink as much as 30 L a day, but have been known to go as long as 100 hr without water (McGreevy, 2004). Lactating mares with young foals drink almost twice as much water as pregnant mares. A thirsty horse can drink up to 12 L in a single guzzle, as fast as 1 L every 6 seconds (Gill, 1994).

Keiper (1985) writes that the high salt content of their water and the crystallized salt on their food prompts the Assateague horses to drink twice as much water per day as a domestic horse, every 3 hr on average. This is a sharp contrast to the mustangs of the Western deserts who may visit a watering hole every day or two (Houpt, 2005).

Free-roaming horses generally live in small family groups called bands within large, overlapping home ranges. A harem band usually consists of a stallion, several unrelated mares, and their offspring from the past few years. The number of mares depends on the stallion's ability to retain them. Most bands range from four to 12 animals, but Keiper (1985) reports a band as large as 23. Band size is largest when a stallion is between the ages of 9 and 12, his prime years.

As in humans, some behavior patterns are universal throughout the species, and others vary from family to family and situationally. Many studies of horse behavior contradict each other because the horses observed in one project had different habits and preferences than those in another.

The stallion remains with his mares all year and keeps them together and away from other stallions, though indiscretions occur. In the 1990s, Dr. Lisa Ludvico, now an anthropology professor at Duquesne University in Pittsburgh, determined that harem stallions sired only about half the foals in their bands (Frydenborg, 2012)!

The stallion maintains close relations with the members of his band that persist over many breeding seasons and sometimes for the duration of his adult life. Together, stallion and mares socialize their offspring, who generally disperse shortly after puberty. Most horses form tight friendships with only two or three companions, no matter how large the band (Feh, 2005).

Some stallions permit one or two subordinate stallions to remain with the band, and some bands have no stallion at all. Bachelor bands are groups formed by young males dispersed from their natal bands and older stallions displaced from their harem bands, as well as stallions of all ages who have been unable to acquire or maintain a harem.

Ranges overlap at communal sites such as watering holes. Horses claim smaller home ranges when high-quality forage, water, shade, and other desirable features are readily available—if the needs of the horses in the band are met close to home, they do not need to go elsewhere. Because the stallion defends his mares rather than a territory, the band is free to migrate to make use of the best available resources (Boyd & Keiper, 2005). In resource-rich habitats, home ranges are smaller, while in harsh alpine or desert environments, they can span many miles.

Bands tend to be larger when the available forage is abundant. A number of bands that share the same geographic area is a herd. More than an aggregate of horses sharing the same space, the herd appears to be a structured social unit with an interband dominance hierarchy (Boyd & Keiper, 2005).

Barrier island horses graze preferentially on saltmarsh cordgrass, *Spartina alterniflora*. This fibrous and abrasive plant supplies more than half of the usual diet of Assateague ponies; American beachgrass (*Ammophila breviligulata*) provides another 20% (Keiper, 1985). Island horses also consume the American three-square rush (*Scirpus americanus*), the common reed phragmites (*Phragmites australis*), saltmeadow cordgrass (*Spartina patens*), thorny sandburs, thistles, rose hips, and crab apples. In the winter, bayberry (*Myrica cerifera*) and elder twigs and branches become important food sources. Uncomplaining ponies munch greenbrier (*Smilax* spp.), a tough, wiry vine studded with formidable thorns, as well as poison ivy (*Toxicodendron radicans*). Sea lettuce (*Ulva* spp.) tossed onto the tideline is a special treat, rich in protein and micronutrients.

Assateague horses have also adopted an interesting mechanism for reducing the amount of sand ingested with their food, which can contribute to excessive tooth wear and induce sand colic. These resourceful ponies often knock sand from the grass by striking it against bayberry branches. Kirkpatrick (1994) writes that almost all the Assateague horses employ this strategy, but most Western mustangs do not, despite equally gritty food sources.

A horse performs flehmen to investigate odors. Raising the lip exposes specialized vomeronasal organs, located in the hard palate to analyze pheromones and other scents.

Evidently, the animals learn the trick by copying other horses. The author, however, had a Connemara gelding who used a similar technique, banging each tuft of grass against his legs or shaking it in the air to remove the grit, indicating that a clever horse can work these things out for himself.

In July and August, when the air is still and hot, Assateague horses take to the beach, where the differential warming of land and ocean creates local winds called sea breezes. Heated by the sun, the air above the land expands and rises. Cooler air moves in from above the ocean to take its place. In the evening, the process reverses; air above the warmer surface water rises and expands, drawing air seaward from the cooler land surface.

Even when the air on the interior of the island is not moving, land and sea breezes can blow on the shore and discourage vicious insects from biting. Eventually the horses become hungry enough to venture inland to graze—to be met by an onslaught of insects that drives them back to the oceanfront. They move along briskly, distress showing plainly in their eyes as flies slice at their flanks and dozens of engorged mosquitoes pattern their underbellies like tiny red beads.

Some species of bird make a good living following the pony bands and feasting on the multitude of insects that accompany them. Brown-headed cowbirds feed at their feet. Red-winged blackbirds perch on the ponies and pluck swollen mosquitoes from their coats or pull flies from between strands of matted mane. Cattle egrets tend to feed on insects stirred up by the movements of the ponies, but they do pick a significant number of ticks, flies, and lice off the ponies themselves.

Although artificial insecticides are generally not used in national parks to control pests, Assateague State Park has sprayed for insects in the day-use area. Kirkpatrick (1994) writes that bands of ponies have been known to migrate 6–7 mi/9.7–11.3 km to gain a respite from insect attack. In 2000, budget constraints precluded spraying. Resurgent insect populations may have been why an unusually large group of horses mingled with tourists on the beach that summer. Avoiding insects takes valuable time away from feeding, but the ponies have little choice.

Biting insects are most strongly attracted to stallions and are less interested in foals. Flies—which can be larger than a quarter—feed by slicing the skin and lapping the flow of blood. A horse's skin is exquisitely sensitive to touch, and he can feel an insect land. To dislodge the intruder, he twitches a specialized "fly-shaker" muscle across his shoulders, the *panniculus carnosus*. If the fly remains, he snaps at it with his teeth, tosses his mane, stamps his feet, or swishes his tail.

Bloodshed is the price of being a harem stallion. Besides being targets of innumerable bloodthirsty insects, stallions are frequently wounded in physical clashes with one another.

Mature stallions are generally intolerant of unrelated male horses. They do not lay claim to territories, but if a rival dares to invade the personal space of his band, he is obliged to defend it. On the Granite Range of Nevada, 96% of the stallions carried scars from bite wounds. Swiss researcher Claudia Feh (1999) counted battle scars on 20-year-old Camargue stallions—one carried 35, another had 146. Sometimes stallions lose their harems or even die as a result of wounds sustained in altercations with other stallions.

Although stallions are capable of vicious combat, they come to blows only as a last resort. Even fairly superficial wounds can cause lameness, infection, or death, and stallions avoid physical violence whenever possible. Bloodshed is reserved for truly important issues such as the right to mate with a mare in heat. Most of the rivalry between stallions is resolved through ritual displays.

A confrontation between stallions usually begins with the two participants standing well apart, watching each other intently, head raised, ears swiveled forward. Sometimes both stallions will defecate while watching each other at a distance, then move on. This seems to indicate that each accepts that the other owns the mares that accompany him (Feh, 2005).

"Stud piles" serve as a record of which stallions have recently passed. These layered manure piles are often seen along well-traveled pony trails. Stallions stop and smell them thoroughly, curl their upper lips back in a flehmen response for further olfactory investigation, pivot around or step over the piles, depositing their own manure on top, and sniff again. Flehmen is an odd-looking posture in which he stretches his neck up to full length, curls back his upper lip, and inhales deeply. This stance allows the inhaled air to reach the vomeronasal (Jacobson's) organs, specialized olfactory devices within the hard palate that can decipher complex olfactory messages. A horse can identify the scents of the other

On Cumberland Island, Georgia, a stallion (right) harasses an adolescent colt, chasing him away from the watering spring, charging him, biting him, and punishing him when he investigates mares in estrus. The frightened colt responds with submissive gestures. Soon, the colt will leave the band, most likely to join a gang of bachelors in similar circumstances.

horses that have used the pile and apparently recognize their rank in the social hierarchy. Keiper (1985) writes that stud piles on Assateague can grow more than 6 ft/1.8 m) long and 2 ft/0.6 m) high. If a bachelor herd encounters a stud pile, members will often defecate on the pile in order of rank, from lowest to highest.

If the conflict is not resolved at a distance, the stallions approach each other, posturing to determine which is dominant. After much squealing and striking the air with a fore hoof, one of the two will usually defecate, a sign of subordination. The dominant stallion will then defecate on top of his rival's manure, demonstrating higher rank by allowing his scent to prevail. Each horse sniffs at the manure pile. If the subordinate agrees that the other horse is dominant, they both retreat. Often, this alone somehow resolves the conflict.

If the issue remains unresolved, each stallion attempts to intimidate the other by appearing powerful and assertive. They approach each other with necks arched, tails high, and light, airy steps, then jog shoulder to shoulder in an exaggerated trot known as a parallel prance. In these encounters, they appear to be sizing each other up, and if one is clearly weaker or less confident, the encounter ends as the subordinate stallion withdraws.

If the two stallions still feel equally matched, the contest intensifies. The two rivals sniff at each other, especially at the muzzle and genitals and under the tail. Loud squeals and screams punctuate these investigations. The horse with the loudest, longest squeal often wins the conflict at this point (Rubenstein & Hack, 1992).

If not, they begin to shove each other with their weight and bite at each other, especially at the legs. The fights can quickly turn ferocious. Pulling back, they rear and strike with the forelegs—this is often a display of power rather than a means of inflicting wounds. By rearing and lunging, a stallion can knock the opponent off balance. If the conflict escalates, back-kicking and biting can inflict serious injury. When the battle is resolved, each stallion rejoins his mares and drives them away from the rival.

Stallion conflicts can last from a few minutes to the better part of an hour, the action taking place in rounds, or battles, separated by retreats to the band to graze. Eventually one of the stallions will give up and take his band elsewhere.

When a stallion grows older and slower, a younger rival usually displaces him. Sometimes the exiled stud will join a group of bachelors, but many old patriarchs become loners. A horse's drive for companionship is powerful, so this imposed isolation must be torture. Stress makes the outcast susceptible to disease and shortens his lifespan.

Experienced stallions can often bluff their way through battles with younger, stronger rivals, retaining control of a herd into old age. Keiper (1985) writes of a 21-year-old stud named Voodoo who kept a harem of seven mares and their foals despite his poor physical condition. Postmortem examination revealed that his bottom incisors had been long since worn away so that the root spaces in his gums had actually healed over.

Courtship among horses is also ritualized. Each of the pair follows a set sequence of behaviors and responses. A mare not fully in heat is unlikely to tolerate a stallion's sexual advances, so he is very careful to test her receptivity thoroughly before attempting to mate. If the stallion is too hasty and crowds a mare, she is likely to kick violently.

Ovulation is seasonal in the mare and influenced largely by length of daylight. In the northern hemisphere, the breeding season is from March through August, peaking in May and June. As spring and summer approach, days lengthen and the mare begins her heat cycles. Free-roaming horses of any given population tend to develop a breeding pattern that begins and ends more abruptly than that of domestic mares. The goal is to foal at a time when the foal is more likely to survive—after the harshness of winter subsides, but early enough in the season that the foal is well developed and fairly independent by the next cold season.

Out-of-season births have increased on Assateague Island NS from 12% in 1984 to 26% in 2001. Despite harsh weather on the island, out-of-season foals are just as likely to survive as foals born in spring (Kirkpatrick & Turner, 2003).

Domestic mares are more likely to ovulate into the fall and winter because of the artificial lighting in barns. In fact, a domestic mare can be persuaded to ovulate out of season by leaving the barn lights on for 16–24 hr a day and to cease her heat periods when light exposure is less than 9 hr daily (Kirkpatrick & Turner, 2003).

Thoroughbred racehorse breeders manipulate heat cycles in this way to gain a competitive edge. Thoroughbreds race against horses with the same birth year, so it follows that horses born early in the year tend to be larger, better developed, and faster than horses born in late summer or fall. It is desirable for a Thoroughbred to be born in late winter or early spring, and breeders routinely bring mares into heat through artificial lighting to achieve an earlier foaling date.

Just before a wild mare comes into estrus, the stallion usually attends her ardently, perhaps keeping her away from the rest of the band or at least other males. He monitors her every move and rushes to smell and scent-mark her manure and urine. He approaches her with a posture that conveys sexual excitement: arched neck, high tail, and flared nostrils. He nickers winsomely. She encourages his attentions by raising her tail invitingly and spreading her hind legs. A mare in heat rhythmically winks her vulva and flashes her clitoris while excreting small amounts of urine. The stallion will touch his muzzle to the puddle and perform flehmen to detect hormones in her urine.

The stallion "teases" her to determine readiness, sniffing, nuzzling, and licking her muzzle and nickering beguilingly. He licks and nibbles her shoulder, elbow, belly, and udder, then hindquarters and genitals. If she is ready for mating, the mare assumes a sawhorse-like

Two stallions from the Virginia part of Assateague attempt to intimidate each other during the Chinco-teague Pony Penning. Both are well-established dominant males, and neither wants to engage in a serious battle, which would probably result in injury to both.

While most encounters between stallions involve posturing and bluffing, vicious battles do occur. This Pryor Mountain bachelor, about age 5, has been trying to acquire mares and has been repeatedly trounced by alpha males. His coat should be glossy black; all the light marks are healing wounds, mostly from bites.

stance for mounting with the tail deflected to expose her genitals. The actual mating takes less than a minute. After copulation, the stallion often loses interest and resumes grazing, but the stallion and mare remain close and mate frequently over the next few days.

Mares are receptive to mating for about six of every 21 days during the breeding season, but this varies from mare to mare and with the same mare from season to season. Mares can be fickle, soliciting the stallion's advances one minute and rejecting them the next. A mare will show signs of estrus before she is fully in heat, and if approached at this time, she will reject the stallion's advances with squealing, striking, kicking, and tail lashing. She is most receptive near the time of ovulation and goes out of estrus 24–48 hr after she ovulates.

The sexual instinct is present even among very young colts. Male foals test their mother's urine in the flehmen posture and ritualistically urinate over her puddle. Male foals will attempt to mount other members of the herd, climbing aboard sideways or from the front, bouncing on other males as well as fillies, adult mares, their mothers, or their older sisters. This play-acting eventually evolves into adult sexual behavior.

In contrast, fillies do not display sexual behavior until their first heat cycles. Mature Assateague mares go into heat from about March through September if they are not pregnant. They come back into heat about 7–10 days after bearing a foal. If reproduction is left entirely to nature, an average, a mare produces her first foal by age 3, and a stallion fathers foals after the age of 5 although stallions sire most of their progeny after the age of 10.

The herd accepts and indulges male foals for their first two or three years. They can get away with outrageous behavior and social blunders and most members of the herd regard them with overt affection. Colts enjoy playing with the harem stallion and may seek him out for a sparring match or mutual grooming. They impishly chew the legs and tails of adult horses, gallop disruptively through the band in a gang, break up mutual grooming sessions, and prevent adults from napping. They spend their days romping, wrestling, grazing, nursing, and sleeping. Life is good.

The colt begins to feel the effects of testosterone somewhere around his second or third spring. Mares suddenly seem very interesting. Colts are physically able to impregnate mares by the age of 2–3, but are not behaviorally mature until the age of 5–6 (Boyd & Keiper, 2005; Keiper, 1985). Keiper (1985) writes that 57% of Assateague horses leave their natal band between the 12th and 24th month. Studies of the horses of the Granite Range in Nevada showed that colts dispersed when they were just over 2 years old, with an age range of 11–52 months (Berger & Cunningham, 1987). This dispersal is usually voluntary; Rubenstein wrote that on Shackleford Banks, mares and stallions rarely drive off juvenile males (Rubenstein, 1982).

Only stallions in good condition rise to the top of the social hierarchy and acquire bands of their own (Rubenstein, 1982). Because a harem band is dominant and has access to better forage than would a bachelor band, a colt who remains with his natal band until he is 3 or 4 is often larger, stronger, and more confident. This gives him a competitive advantage over colts dispersed at a young age. Adolescent males over the age of 2 often leave and return to the natal band at will until they eventually abandon the band permanently (McDonnell, 2005). Some males remain in their natal bands as subordinate stallions and inherit the mares if the harem stallion dies or is displaced.

Sometimes, stallions (or mares) will begin to harass the colt deliberately. The first approaches are not overtly aggressive. The stallion begins to punish the colt for minor

Although an average mare bears only one offspring every other year, in the absence of predators, confined herds such as those on barrier islands can increase to the point of unbalancing the environment. Note the curly coat on this foal, a trait that occasionally occurs in both Assateague herds.

infractions, biting at the legs and neck in a less-than-friendly way, like a manager motivating an employee to quit to avoid firing him. After holding an established, secure spot in the social ranking for years, suddenly he is not wanted. Most of the time, the youngster comes to understand his new status and sets out on his own.

At first, displaced colts move to the boundary between the home range of their natal band and that of a neighboring band. Neighboring stallions attack and bully the young trespassers, causing them to move on to new ranges—where the stallions of other bands chase them away.

Horses are gregarious animals and are uncomfortable without others of their own kind around. Evicted colts from the same natal band congregate with those from other groups to form bachelor bands, usually of two to four individuals but in some populations as many as 16. Bachelor bands are transitional groups that serve as surrogate family for their adolescent years.

Bachelor bands do more than provide adolescent colts with companionship. Through the frequent mock battles that make up much of their playtime, they learn the techniques necessary to defend a harem, win contests against other stallions, and establish their rank in the dominance hierarchy. They shoulder and shove one another, mutually groom, and stand with their bodies touching, taking comfort from the physical contact. Occasionally they sneak into a harem band and mate with mares in estrus.

Between the ages of four and seven, the bachelor stud reaches full maturity. If he is sufficiently assertive, he begins to steal mares from other stallions to form the nucleus of his own band or challenges harem stallions for herd ownership. Young stallions will often break from a bachelor band and spend some time alone before acquiring a harem (McGreevy, 2004).

The dispersal of adolescents serves to reduce inbreeding. If young horses encounter their sires more than 18 months after leaving the band, they are treated as strangers—colts are regarded as rivals and fillies courted as potential mates (Boyd & Keiper, 2005). The incest rate varies from herd to herd. A study of Granite Range horses noted that about 4% of copulations were between father and daughter, though these matings are less common in other populations (Berger & Cunningham, 1987).

Determining paternity can be challenging. Equine geneticist Dr. E. Gus Cothran of Texas A&M University attempted to establish parentage of Pryor Mountain foals by observing matings, then confirming parentage with DNA testing, and found that his error rate was 43% (U.S. Bureau of Land Management, 2001). Paternity testing of 55 Assateague horses showed 36 to be sired by harem stallions, seven by subordinate males in the harem stallion's band, and nine by stallions outside the band (Eggert et al., 2010).

Adolescent fillies leave their natal bands of their own accord, usually during a heat cycle when sexually attracted to an unfamiliar stallion. Research on the feral fillies of the Camargue region of France revealed no sign of weakening social bonds with herdmates before leaving. There was also no evidence of sexual competition between the fillies and the mature mares. In fact, the fillies refused sexual advances from the dominant stallion (Monard, Duncan, & Boy, 1996); and if the stallion showed persistent interest, the filly's mother would often intervene and drive him away with back-kicks (Feh, 2005).

Although some stallions retain daughters in their harems as mates, observations of wild horses in Nevada showed that stallions typically do not court adolescent fillies within their own band, even if the fillies are not their offspring. One study describes several instances of adolescent fillies who copulated with unfamiliar stallions while their fathers stood nearby, unconcerned (Berger & Cunningham, 1987). More commonly, the stallion will drive away the filly and her consort. When a stallion is familiar with a pubescent mare, he is not usually interested in mating with her; but if she leaves the band and returns as an adult, the stallion appears not to recognize her and courts her as fervently as he would an unrelated mare (Berger & Cunningham, 1987).

The dispersing filly will usually join another harem or keep company with a bachelor group. She has a choice about which band to join, inviting herself by wandering close while in heat and encouraging the interest of the harem stallion. Sometimes the stallion drives her away, sometimes he investigates the romantic possibilities. Fillies prefer a dominant stallion that can protect them from the harassment of others who wish to copulate. Often the mares in the new harem do their best to drive a new filly away (Rubenstein, 1982). If the stallion welcomes the new addition, he will support her when the mares try to exclude her.

Sometimes a mixed group of colts and fillies from the same band will stay together. These adolescent bands are temporary, and their composition changes. Young stallions can acquire mares by claiming a newly dispersed adolescent filly, by challenging and besting a harem stallion, or by staying on the outskirts of a band until they can conquer the stallion or abscond with a number of his mares. As the colts mature, they may depart one by one to form their own bands by stealing mares from harem stallions. Eventually one bachelor is left with the adolescent fillies, allowing him to develop a band with them as his foundation mares.

Typically, young stallions are extremely attentive to newly acquired mares, nuzzling them, grooming them, and following them everywhere. They become the focus of the stallion's whole existence, often more important than even grazing.

Young mares are likely to be happy in their new herds, but older mares often have strong bonds with one another. When stolen from a harem, they may constantly attempt to escape and rejoin their mare-friends. As other mares enter the harem, the stallion will copulate with and defend all his mates, but will typically forge powerful attachments with only one or two mares.

Genetic diversity helps a population survive environmental change and disease. It also reduces the incidence of genetic abnormalities that occur transmitted by recessive genes. When a population has many unrelated individuals, a wide variety of genes combine to create each generation, and dominant genes for healthy traits balance the disease-producing recessive ones. Dominant genes for healthy traits tend to inhibit expression of recessive genes. If horses in a herd are genetically similar, a challenge such as a drought or epidemic that kills one horse is more likely to destroy the whole herd. Genetic diversity improves the chances that a population will survive by increasing the number of different responses to stressors such as drought or epidemic. Maintaining genetic diversity is a serious challenge in small, isolated populations such as the Assateague horses.

DNA analysis shows that there are 11 maternal lineages in the Maryland herd (Eggert et al., 2010). In 2006, the population retained 95.8% of the genetic diversity of the foundation horses. Twenty-one horses out of a herd of 144 showed signs of inbreeding. Fifty years from now, the herd is expected to possess 86% of the diversity of its forebears.

If the Maryland herd can be considered a unique population, there is a balance to strike: inbreeding makes the herd more uniform but decreases disease resistance and increases birth defects, many of them lethal. As Cothran points out, "if you pass a certain point with inbreeding extinction becomes likely" (personal communication, September 10, 2014). On the other hand, outcrossing—introducing a horse from another strain or breed—reduces susceptibility to disease and birth defects, but increases variability among herd members in color, conformation, size, and other characteristics. Inbreeding improves conformation at the expense of health and resilience. Outcrossing improves the health of the population at the expense of uniformity. Fortunately, variability can be restored by adding an outside horse or two per generation.

Some horse breeds, such as the Lipizzan and the Friesian, have a small gene pool with no influx from outside breeds. Without outcrossing, over time, the risk of genetic disorders will increase. The American Paint Horse Association, on the other hand, encourages gene flow by accepting horses that have a Thoroughbred or Quarter Horse parent, as long as the other parent is a registered Paint. As a result, any Lipizzan or Friesian looks similar to any other, but Paints can vary widely in body shape and coloring.

How many horses are required to ensure optimal genetic diversity while keeping the herd true to its original character? How many horses can the Maryland section of Assateague support without degrading the environment? Keiper and others researched the horses and their habitat extensively through the 1970s and 1980s to determine the maximum number of ponies that the island could comfortably support (Keiper & Hunter, 1982; Zervanos & Keiper, 1979a, 1979b, 1979c). They suggested that a herd of 130–150 horses would maintain a healthy balance with available forage and native wildlife. When they collected their initial data in 1979, only 60 horses lived on the Maryland section of the island. By 1994, they had multiplied to 165. In December 2000, the herd census was 175.

At first glance, feral horses do not appear to be fast breeders. Mares can foal every year, but are less likely to produce one foal while nursing another; lactating mares tend to foal every other year. In humans, lactation inhibits ovulation and thereby conception, but mares often conceive at postpartum estrus 7–10 days after foaling. For the next 11 months, a pregnant lactating mare must nourish her unborn offspring while producing large amounts of milk. If she is stressed by disease, poor nutrition, or a cold, wet winter, she may abort the developing fetus. "She has invested less in that fetus than she has in the foal, so she aborts the fetus," says Keiper (personal communication, May 19, 2011).

In many wild horse herds, foal mortality typically is about 20–25% in the first year of life (Kirkpatrick, 1994). On Assateague, foal mortality is so much lower than in the West and on many of the other barrier islands, that it might predispose this herd to to rapid increases in population out of proportion to the available forage.

The U.S. Bureau of Land Management maintains that herds of Western mustangs "grow at an average rate of 20 percent a year and can double in size every four years" (USBLM, 2012, October 2). Free-roaming horse populations *can* grow by 20% annually, but these rapid increases are very short-term, and they usually occur in herds recovering from a period of high mortality or in those taking advantage of improved conditions, such as the end of a drought. Between 1974 and 1994, however, the Assateague population grew by only 7% annually, even after nearly 4 dozen horses were transferred to the Virginia herd (Kirkpatrick, 1994; Zimmerman, Sturm, Ballou, & Traylor-Holzer, 2006). If the population had doubled every 4 years from 1965 to 1994, the original 21 ponies would have produced a herd of more than 3,000.

Wild horse populations fluctuate in response to births, deaths, and dispersal and typically increase when resources are abundant and competition is low. Assateague horses are also long-lived and have no predators. Disease, injury, age, and adverse environmental conditions claim some lives every year; but with annual losses of only about 5%, the death rate does not come close to balancing the birth rate.

Computer modeling predicted that, left to its own devices, the Assateague herd could swell to as many as 280 horses before starvation would begin to kill them off and limit fertility. Once the population reached 100 horses in the mid-1980s, the Park Service was convinced that the horses were having a detrimental impact on Assateague, and if they were allowed to reach their population maximum, the environment would suffer catastrophic damage.

Numerous studies convinced the Park Service that the intensive grazing and browsing of the ponies reduced plant diversity and cover, altered the balance of vegetation on the island, and threatened rare plant species (Zimmerman et al., 2006). Ponies, they said, overgrazed the marsh, disrupting habitat for birds and reducing the buildup of detritus—material broken down by bacteria and fungi and eventually consumed by minute marine organisms (Levin, Ellis, Petrik, & Hay, 2002).

When species evolve together, they produce a balanced, healthy ecosystem. Egrets and snow geese have been in a relationship with Assateague for thousands of years, as have the saltmarsh cordgrass, goldenrod (*Solidago sempervirens*), and mosquitoes. The Park Service points out that the association horses have with the Assateague ecosystem has been in place

Nature prepares foals for the climatic conditions into which they are born. This Assateague filly was born in July with a short, soft coat to help her shed heat. The filly below was born on Assateague in March, bearing a dense, insulated coat to protect her from winter winds and ice storms.

Since the Park service began using immunocontraceptive vaccines to limit the fertility of free-roamng horses on Assateague Island National Seashore, only a handful of foals are born annually. Foal mortality, however, is lower, and horses have longer lifespans, especially mares.

for no more than 400–500 years. As exotics, they are more likely to disrupt the balance by overgrazing, disturbing soil, trampling vegetation, and displacing native wildlife, especially when their population expands beyond the carrying capacity of the island.

It can be difficult for nonspecialists to evaluate plausible-sounding information from experts. While conclusions drawn from peer-reviewed research are usually reliable, hard facts are sometimes woven together with speculation, opinion, or propaganda, even in official reports and studies.

For instance, Park Service officials claimed that the expanding pony population was a direct threat to the endangered seabeach amaranth (*Amaranthus pumilus*) (Zimmerman et al., 2006). This is an annual plant that prefers to grow on wide, naturally functioning barrier island beaches as a fugitive species, thriving or not in response to the changing island landscape. Its preferred habitats are the upper beach, sparsely vegetated overwash fans, and interdune areas. Plant counts fluctuate dramatically from one year to the next as natural forces destroy and re-create habitat (Weakley, Bucher, & Murdock, 1996). The Park service began active conservation efforts when the species became rare on Atlantic barrier islands, contracting to only about one third of its historic range (Center for Biological Diversity, n.d.) Amaranth was first discovered at Assateague Island National Seashore in 1967, but it subsequently disappeared, probably because of Park Service dune construction and stabilization projects in 1973 (Center for Biological Diversity, n.d.). At that time, the horse population totaled fewer than 50 animals. In 1998, for the first time in 30 years, two seabeach amaranth plants reestablished themselves on Assateague even though the herd was near its peak census at 171 horses. The Park Service removed one plant to a greenhouse, where it reproduced bountifully, while the wild plant was destroyed by Hurricane Bonnie.

Park managers returned the propagated amaranth to the wild, placed cages around individual plants to protect against foraging horses and deer, and marked them with signs in the over-sand-vehicle zone. By 2006, efforts at restoring the species brought the total wild plants to more than 1,500 despite a horse census of 158 in 2005 (Hayward, 2007; Zimmerman et al., 2006). Still the Park Service holds that horses are the greatest threat to seabeach amaranth, and when the horse population tops 100, the plant declines sharply (Zimmerman et al., 2006).

Elsewhere, free-roaming horses do not seem to threaten the plant. On the Virginia end of Assateague, where horse range is limited to fenced parcels, the plant has "fluctuated without apparent trend" (Center for Biological Diversity, n.d.). Shackleford Banks, North Carolina, is home to both free-roaming horses and seabeach amaranth. The Park Service sows seed and counts plants every year, and some years the plants are more abundant than others, depending on seed dispersal and environmental conditions. Dr. Sue Stuska, wildlife biologist at Cape Lookout NS, says that she has never seen any evidence that the horses disturb the seabeach amaranth on Shackleford Banks (personal communication, May 26, 2010).

Artificial dunes and beach erosion, however, destroy habitat for seabeach amaranth (Weakley et al., 1996), and dune-building projects have repeatedly reshaped Assateague. When artificial dunes block overwash, other plants colonize the sparsely vegetated areas and ultimately outcompete amaranth. The plant is entirely absent from the stretch of North Carolina coast where the Park Service, Civilian Conservation Corps, and Works Projects Administration built a continuous barrier dune from the 1930s to 1950s; yet it thrives on the undeveloped, natural beaches of Core Banks (Weakley et al., 1996).

The piping plover (*Charadrius melodus*) is another threatened species that thrives on undeveloped, unrestrained barrier islands. Piping plovers are sand-colored on top and white below and are similar in appearance to their larger cousin the killdeer, with a dapper black necklace and a distinctive black stripe that runs across the head from eye to eye. The U.S. Shorebird Conservation Plan listed piping plovers as threatened in 1986, and highly at risk on a global scale (Cohen, Erwin, French, Marion, & Meyers, 2010). Habitat management for seabeach amaranth benefits the piping plovers and vice versa. In recent years this sparrow-sized shorebird has increased its numbers on Assateague (Weakley et al., 1996).

Piping plovers nest in small depressions on sparsely vegetated stretches of sand, gravel, or cobble. They lay a clutch of about four eggs that resemble rocks, well camouflaged in the bare scrape of a nest. For 25–29 days, parents take turns incubating the eggs; then they share brood-raising responsibilities (Cohen et al., 2010). Chicks are precocial—they run after their parents and forage for themselves within hours of birth, feeding on invertebrates and insects in the sand and sometimes covering more than a half mile (0.8 km) in their first day of life (Coastal Waterbird Program, 2008). During the first week after hatching, they stay warm by brooding under their parents' wings. Within 23–35+ days, the chicks can fly.

Piping plovers benefit from storm damage. They forage on sand and mud flats. They nest on overwash fans and interdune areas disturbed by storm surge. The severe storms of the 1990s wreaked havoc with many aspects of the island habitat, but they were a boon to the piping plovers, which enjoyed great productivity and increased when the storms sculpted Assateague more to their liking.

Piping plovers face many threats to reproduction. Birds, mammals, and crabs prey on eggs and chicks. Overwash may sweep away some nests, while deer, horses, hikers, and off-road vehicles can crush them (Cohen et al., 2010). Beach renourishment programs and artificial-dune construction temporarily increase the size of piping plover nesting areas, but over the long term, these interventions cause erosion and habitat loss (Cohen et al., 2010).

It would appear that the presence of island horses has little effect on the piping plover population. Breeding pairs on the Maryland portion of Assateague have increased from 14 in 1990 (horse census about 130) to 66 in 2006 (horse census greater than 140). The majority of these birds nest on the northern end of Assateague where storm overwash has created optimal breeding habitat (Hayward, 2007).

In the interests of a healthier ecosystem, the Park Service officially manages the flora and fauna of our national parks to increase biological diversity of habitat, not carrying capacity. Clearly, horse grazing changes the ecosystem of the marsh—but these changes are not necessarily undesirable. In fact, credible studies show that light to moderate grazing *increases* total diversity of species.

On Shackleford Banks, N.C., one study found horse grazing detrimental to nesting gulls and terns, but strongly beneficial to birds that prey on small invertebrates such as xanthid crabs (Levin et al., 2002). Ungrazed marshes had a higher overall bird count, but grazed marshes showed greatly increased avian diversity (Levin et al.).

Horses were a prominent native species on this continent for tens of millions of years, and it is quite possible that human activity set the stage for their extinction roughly 8,000–10,000 years ago. There is strong paleontological evidence that before humans arrived, wherever horses had access to salt marshes, they foraged on the same vegetation that they consume today (Sanders, 2002). As a result, these ancient marshes may have functioned more like modern marshes grazed by horses than like those that are not (Levin et al., 2002). Science has not yet found evidence that ancient horses foraged on barrier islands, probably because the Pleistocene coastline is now the continental shelf. Modern horses, however, will voluntarily venture to islands if they are an easy swim from their home range, and it is likely that prehistoric horses did the same. Free-roaming horses on Shackleford Banks routinely access smaller islands in Core Sound, and ponies sometimes swim from Assateague to forage on Chincoteague of their own accord.

It may be that if Paleo-Indians reduced or eliminated the native herds of horses, they fundamentally changed the nature of the salt-marsh ecosystem into that found by early European explorers. Which should be considered natural—the ungrazed marsh that followed the arrival of humans thousands of years ago or the grazed marsh that existed for millions of years before that? Which natural state should set the standard for the Park Service?

On the other hand, because there is no evidence that horses are native to *barrier island* environments, one could argue that they did not evolve with those unique ecosystems, even though estuarine wetlands are native habitat. And whether or not they can be considered native wildlife in island salt marshes, when the number of horses exceeds the threshold the habitat can comfortably support, the environment will suffer damage.

Scholarly journals and popular media alike have published countless articles concerning wild horses on public lands. Many have accused these animals, with and without reason, of disrupting ecosystems and degrading habitat. Both sides of the argument are often heavily biased, emotionally charged, and distorted by hearsay.

The Park Service maintains that wild horses interfere with the formation and stability of sand dunes by eating dune-building plants such as American beachgrass (*Ammophila breviligulata*). Some research indicates that when horses degrade dunes, overwash increases and island migration accelerates, upsetting the balance of species living on the island (Hayward, 2007). In recent years, horses have convened in larger numbers on the sensitive northern end of Assateague, and the Park Service has become increasingly concerned that they are causing damage to a vulnerable ecosystem (Zimmerman et al., 2006).

To determine the carrying capacity of the island, researchers surveyed the quantity of vegetation on Assateague in 1978–1979 by evaluating the types and amount of plant life growing in a series of east-west transects (Keiper, 1985). Results indicated that the ponies consumed about 1% of available foliage (Keiper, 1985). Another study quantified pony forage consumption by calculating the amount of forage taken in one bite, the number of bites taken over one hour, and multiplying this by the number of hours of daily grazing (Keiper, 1985). The researchers used this figure to calculate forage consumption per pony, per day and per year.

These scientists calculated the carrying capacity of Assateague Island NS at 1,500 ponies (Keiper, 1985). In other words, if ponies were allowed to eat every wisp of forage on Assateague without competition from any other herbivores, their numbers could rise to about 1,500 before they ran out of food. At this population density, the island ecosystem would become disastrously deranged. Most species of edible plants would disappear from the island, and non-edible vegetation would fill the void. Populations of birds, fish, invertebrates, other mammals, and even the microbial community would also suffer catastrophic imbalance. The ponies themselves would become thin, malnourished, subfertile, and parasite-ridden before their numbers began to fall. Keiper suggested that the pony population

The natural shorefront of Assateague consists of low, rolling dunes that readily overwash. Whenever people cease to bulldoze artificial dunes into place, this natural process resumes. Wild horses at healthy population densities have little impact on natural dunescapes, but can cause damage to artificial dunes. These fence posts once held wire that kept the ponies off the dunes.

could remain in harmony with its environment if the herd were maintained at no greater than 1/10 of the carrying capacity of the island, or 150 ponies (Keiper, 1985).

In 1977–1978 researchers created exclosures and weighed grazed and ungrazed vegetation samples. Results indicated that at the time ponies had little effect on saltmarsh plants, but significantly reduced the amount of vegetation growing on dunes (Keiper, 1985).

Another Assateague study paired plots of land bearing artificial dunes, grazed and horse-excluded, which were selected to be as similar as possible in contour and vegetative cover (De Stoppelaire, Brock, Lea, Duffy, & Krabill, 2001). A photograph taken 7 years later shows a vegetated artificial dune surrounded by horse-exclusion fencing atop a stark, bare overwash flat. The protected area held the only surviving dune in the vicinity, the rest had been grazed bare and washed away in storms (De Stoppelaire et al., 2001). The evidence seems to speak for itself, until we consider that artificial dunes are a landscape feature that is as foreign to the original island as are paved roads and bathhouses.

Contrary to public perception, a long, high, dune barricade paralleling the shorefront is *not* a natural feature of a barrier island. In nature, some areas of island beach receive surplus sand and build higher dunes while others develop low dunes and overwash flats. Influences such as the direction of the prevailing winds and mechanisms of sand supply and transport determine the locations and heights of dunes (Pilkey et al., 2004; Zimmerman et al., 2006). On most of Assateague, low dunes and flat overwash passes are the norm. And on most of Assateague, the Park Service is allowing the natural profile of the island to reassert itself. One could argue that a healthy population of horses poses minimal threat to this more natural landscape.

The high dunes on Currituck Banks, North Carolina, remain despite hundreds of years of grazing. Wild horses roamed the Outer Banks by the thousands as recently as the 1920s, yet the dunes remain intact. Note the irregular contours and the channels where overwash flows through during storms. Natural dunes function not as a wall to block waves, but as a cushion to absorb their energy.

Nobody knows for sure how barrier islands were created, but the prevailing hypothesis is that they originated on the mainland during a sea-level nadir. When sea level rose, high dunes along the coast apparently became separated from the mainland, forming the nucleus of the chain, which became more substantial over time. As sea level continued to rise, waves, storms, wind, and tides redistributed shorefront sediments and pushed the barrier islands toward the receding mainland by eroding the shore, overwashing a blanket of sand, and building the marsh on the opposite side.

Sea level has repeatedly retreated during the ice ages, when the glaciers were at their peak, and advanced during interglacial periods, when the glaciers melted. Since the last Ice Age, about 18,000 years ago, sea level has been continually rising, at first rapidly, then more slowly beginning about 3,000 years ago—and recently, rapidly again (Allen et al., 2010).

Within this long-term trend there are short-term variations. The tideline fluctuates sporadically in response to strong winds and daily in response to the pull of the moon. Ocean water directly opposite the moon collects in a mound, and on the opposite side of the globe, a similar bulge forms because of the reduced lunar gravity and the rotational force of the planet (Pilkey et al., 2004). The two bulges are high tides, and the troughs between them are low tides. The times change from one day to the next because the moon's orbit is 50 min longer than our 24-hr day. The highest tides, "spring tides," correspond with a full or new moon; the lowest tides occur when the moon is in its first and last quarters. On Assateague, the mean difference between the high and low lunar tides is about 5 ft/1.5 m.

Overwash occurs normally when storms and lunar tides push water up onto the beach and through gaps in the dunes. Sea creatures caught in the flow are doomed to certain death

Just north of Hatteras, N.C., overwash breaches artificial dunes with most major storms, blocking the roads with sand until bulldozers can clear it. In 2003, Hurricane Isabel, a Category 2, created two new inlets on Hatteras Island, including one 500 yards (457 m) wide at this spot. The U.S. Army Corps of Engineers and the N.C. Department of Transportation filled the inlet and rebuilt the highway. Hurricane Irene, a Category 1, also created two new inlets in 2011, necessitating restoration of the island and costing $12 million in highway repair.

as the water retreats or sinks into the sand. Plant life intolerant of salt establishes itself safely away from the ocean, but overwash often kills it. If overwash reaches clear across the island, it dumps a load of sand in the marsh, killing vegetation.

This destruction is temporary, and grasses quickly root in or push through the new sediments. In this way, the western shore of Assateague gains 0.53 ft/16 cm a year. When overwash is excessive, vegetation decreases and landward migration accelerates (De Stoppelaire, Gillespie, Brock, & Tobin, 2004).

Over the long term, the island naturally rolls over itself and migrates slowly to the west. Stumps of cedar forests, clumps of peat, and Indian artifacts are often seen on the beach after a storm; all these were once found on the bayside.

Storms carve inlets through the island, and ocean currents refill them with sand. Eleven navigable inlets and numerous smaller ones have been noted along the length of Assateague just in the short time since Europeans first settled the area. Communities of flora and fauna have evolved to cope with the relentless challenges of the barrier island environment.

In order to build "permanent" structures on low, unstable, overwash-prone islands like Assateague, developers erected artificial foredunes—substantial sand bulwarks—to arrest island movement and protect roads and buildings from flooding and other damage. The state of Maryland, working with the Army Corps of Engineers and the Park Service constructed artificial dunes along the shoreline of Assateague during the 1950s and 1960s. The artificial dunes were effective at preventing the natural process of overwash for decades until powerful storms in 1991, 1992, and 1998 finally overwhelmed them.

Contractors manufacture, repair, and replenish artificial dunes through a three-step process. First, workers obtain large quantities of sand though offshore dredging and other sources, then pile it into dunes with bulldozers. Sometimes bulldozer operations scrape

A bulldozed wall of sand stands between the campground and the beach at Assateague State Park.

sand from the beach. Workers plant the resulting sand wall with rows of American beach-grass and panic grass, which grow deep roots and tall leaves that catch and stabilize blowing sand. Fences exclude the horses, deer, and people that would damage the grass and weaken the dune. The fences also catch and hold sand, augmenting dune growth.

The wall-like artificial dunes differ in shape and function from dunes naturally shaped by wind, sea, and sand. Natural island dune systems are a labyrinth of sand hills, hollows, and channels called *overwash passes*. The roots of naturally colonizing plants support and fortify the dune. Electrical bonds called van der Waals forces increase cohesion by bonding the uniform-sized grains of sand with the water between them. Natural dune systems are resilient and generally remain intact through centuries of wind, storm, and grazing pressures, even though they may migrate significant distances. In natural circumstances, dunes, beach, and offshore sand supply sustain a primordial rhythm.

On encountering a natural dune line, high water meets resistance, but can continue to overwash portions of the island (De Stoppelaire et al., 2004). Sand is dropped in the overwash passes and builds the island higher. When the island is pummeled by storm waves, a small bluff or scarp forms, which partially deflects further waves (Pilkey et al., 2004).

Natural dunes typically change shape as they age. Flats are recently overwashed areas without foredunes where new vegetation is just beginning to take hold. As plants claim the flats, they catch sand, and the dune grows into a rounded knoll, or "embryo" dune. At the ridge stage, vegetation becomes well established. The deep and branching root systems create a dynamic scaffolding that holds the sand in place, allowing the dune to reach greater heights. Buttes are the oldest of the dunes and are probably the remnants of old ridges that have been sculpted and steepened by erosion and storms (Seliskar, 2003).

Overwash is essential to keeping the island in balance with natural forces. Barrier islands in harmony with sea and storm migrate westward in sync with the rising sea level.

A band of Assateague horses rests on the dunes. Note the difference in stability evident between the natural dune, under the horse at the far right, and the smooth, easily eroded bulldozed dune to the left. In a storm, the natural dune is likelier to hold fast, and the artificial dune will probably erode.

To build and brace artificial dunes is to ignore barrier island dynamics. Yet when people decided to build roads and buildings on barrier islands, the natural progression of beach migration and dune attrition became a "problem," and the response was to attempt to arrest the natural process and eliminate livestock that could hinder stabilization (Pilkey et al., 2004).

When people first attempted to immobilize dynamic islands, unnatural sand movement patterns developed. Rather than stopping erosion on Assateague, man-made dunes actually accelerated it (Pilkey et al., 2004). Artificial sand walls created a barricade that stood fast against storm waves, preventing overwash and causing erosion as the deflected water ran off the beach. Without overwash, the beach became narrow and sand-deprived. The natural balance of vegetation was skewed, and sand was unable to cross the island to continue the natural migration process (Hayward, 2007).

The foredunes are but one small feature of the shoreface and do very little to hold the shoreline in place (Pilkey et al., 2004). With or without the incursion of horses, bulldozed sand dikes are inherently weaker than natural dunes. They are composed of grains that differ greatly in size and shape, so van der Waals forces are less able to stabilize the sand. Bulldozing sand is detrimental to beaches. It kills mole crabs, coquina clams, and the beach microbiota, disrupting the food chain. Often the blades scrape deeper than the legal 1-ft/30-cm limit (Pilkey et al., 2004).

Studies that evaluate the effect of horses on dunes do not distinguish between natural and artificial systems. Horse grazing accelerates the disintegration of *artificial* dunes. Put another way, horse-grazed artificial dunes revert more quickly to their natural flat state.

Naturally forming dunes, such as those near the Life of the Dunes trail, endure even when grazed by horses. And in North Carolina, natural dunes tower more than 35 ft/11 m at the west end of Shackleford Banks, where horses have roamed free for centuries. Horses are not allowed on the beach area or hook at Toms Cove, yet that section of Assateague is rapidly overwashing and migrating southwest. On Pea, Hatteras, and Ocracoke islands, N.C., where free-roaming horses have been absent for decades, barren artificial dunes persistently crawl across the highway with every storm and resist rebuilding efforts with a gap-toothed sneer. Great tongues of seawater flow through the breaches and rip inlets with the force of their outflow. Over the long term, artificial dunes cannot halt island migration or beach erosion—with or without horses.

The Park Service and the Fish and Wildlife Service are now trying to work with nature on Assateague rather than restricting its action. On the Maryland end, the Park Service has allowed the high artificial dune to break down, permitting nature to take its course. On the south end of Assateague Island in Virginia, natural island migration has claimed much of the beach parking area. "Land managers must spend taxpayers' money appropriately and shouldn't waste it fighting change," said Lou Hinds (personal communication, May 21, 2010). "Beach re-nourishment is not sustainable, and therefore is not responsible. Big storms will wash all that sand back into the ocean."

While the Park Service has abandoned artificial dunes, Assateague SP has gone on building them. According to the *Assateague State Park Land Unit Plan* (Maryland Department of Natural Resources, 2005, p. 24):

> Perhaps the most important capital improvement on Assateague Island is the replenished foredune that stretches the length of the State Park shoreline and beyond, which, among its many benefits, serves to protect the capital improvements behind it from flooding and other damage . . . . The barrier dunes are perennially subjected to climatic and geologic forces and are in periodic need of upkeep and repair.

In August 2001 the state park began a major dune replenishment project, pumping sand from the ocean to create 14-ft/4.3-m dunes fenced at the bottom to exclude horses and people. In May 2003, an emergency dune-restoration project transported and distributed 13,200 cubic yards/10,100 m³ of sand to the park's dune system at a total cost of about $256,515. After that, the agency planted beachgrass and installed a fence at a cost of $139,404. As of 2005, annual maintenance of the artificial dune involved the addition of 95,000 cubic yards/72,600 m³ of sand at a cost of $717,655 (Maryland DNR, 2005).

Horses are held responsible for damaging Assateague's dunes; but in truth, much of the so-called dune damage is simply evidence of a barrier island's true nature. From another perspective, the artificial dunes themselves damage the island by taking it farther from a natural state.

Horses are not the only herbivores to cause environmental damage. Two species of deer inhabit the island, the native white-tailed deer (*Odocoileus virginianus*) and the exotic sika \'sē-kə\ (*Cervus nippon*), actually a small Asian elk. In the early 1900s, Clemment Henry of Cambridge, Md., kept a small herd of sika deer, which he released nearby on James Island, in Chesapeake Bay, in 1916. Another free-roaming herd was released elsewhere in Dorchester County, Md. Later, Dr. Charles Law of Berlin, Md., purchased seven of these deer, many of which apparently had been obtained by Boy Scouts for a petting zoo at Ocean City (Lowney, Schoenfeld, Haglan, & Witmer, 2005). Seven sika deer were introduced to Assateague in 1923 (Diefenbach & Christensen, 2009).

Exotic sika deer ruminate peacefully by the Bayside Drive. Some wear radio collars fitted by graduate student Sonja Christensen to study habitat use (Diefenbach & Christensen, 2009).

Sika is the Japanese word for deer, though pronounced \ˈshē-kə \ in that language. *Sika deer*, then, translates to "deer deer" (McCullough, Takatsuki, & Kaji, 2009). Smaller than white-tailed-deer, fawns and adults alike are patterned with white spots that provide camouflage in the dappled light of their forest habitats. Gracefully proportioned and lovely to behold, they are native to Japan, Korea, Siberia, China, Vietnam, and Taiwan. They are widespread as a feral species, with free-roaming populations in New Zealand, South Africa, Morocco, Australia, New Guinea, Great Britain, Ireland, Denmark, France, Austria, Switzerland, Poland, the Czech Republic, Russia, and parts of North America such as Texas. In 2006, the Assateague census counted roughly 400 sikas and 100 native white-tailed deer. Sikas can cause great damage to forest habitat. McCullough et al. (2009, p. 3), write, "no other deer can match the sika in its ability to strip the vegetation bare and expose soils to massive erosion—thereby creating a wasteland . . . [the species has an] extraordinary capacity to negatively impact its own habitats." Unlike white-tailed deer, sikas are unlikely to confine themselves to home ranges. They move freely around the island and may also migrate seasonally, sometimes traveling up to 12 mi/19 km in a single 24-hr period (Diefenbach & Christensen, 2009).

The white-tailed-deer roaming the island are larger than sikas, but somewhat smaller in stature than their cousins on the mainland. In spring and summer, deer of both species are often bold, especially at dawn and dusk when they are the most active. During the hunting season, they become shy and sequester themselves in heavy brush.

Horses and deer have different dietary predilections. Deer tend to browse shrubs; horses do this infrequently. Sika and white-tailed deer tend to focus on devouring their favorite foods; horses consume a broader sampling of whatever is growing (Sturm, 2007). Sikas eat bark from trees circumferentially, killing them; horses do not.

Assateague Island NS controls the sika and white-tailed deer populations through archery, shotgun, and muzzle-loader hunting seasons. Assateague is one of two public areas east of the Mississippi River where sikas can be hunted (Hayward, 2007). Hunters remove about 130 sikas from the herd each year, keeping the population stable.

In January 2011, the sika hunt made national news when a hunter shot and killed a 28-year-old mare, which was found by another hunter some time later. It is assumed that the killing was accidental, but one wonders how an alert hunter could mistake a mature pony for a dwarf deer. The Park Service offered a $1,000 reward, and more than a year later, Justin B. Eason, 26, and his father, John A. Eason, 51, were convicted of the crime. The younger Eason, who shot the horse and left her to die, was ordered to pay $3,000 in fines and $2,000 in restitution for the horse and put on supervised probation for 18 months. The elder Eason was fined $1,000 and faced a year of probation for providing a false report, and both men were forbidden to hunt on federal lands for five years (Soper, 2012).

In the 1970s and early 1980s, the herd of 50–100 horses appeared to cause little damage to the Assateague ecosystem. Horse bands spread themselves widely over the island without overlapping of home ranges. As the herd passed 100 horses in the mid-1980s, horse grazing began to alter the tidal marsh environment. Growing bands of horses began to compete for resources where their ranges overlapped.

By 2000, the pony herd strained not only the island resources, but also the patience of Park officials. A larger herd led to more interaction between people and ponies. In 1975 only three young bachelors frequented the campground areas. Eight years later, 40 horses preferred to forage near the campsites. With more contact, more people were getting kicked and bitten, more private possessions were being damaged, and more horses were getting injured or killed on the road. Habituated to human doings, the ponies became bold and raided tents, screen houses, and garbage cans in search of food, scattering litter in the process.

The maintenance of a "desirable feral species" is allowed in certain circumstances under Park Service management policies as long it does not displace native wildlife. The horses of Assateague have cultural and economic value in that they are historically significant, people enjoy seeing them, and many visitors choose the park primarily to share their campsites with ponies.

One stated Park Service goal is to preserve a healthy herd of free-roaming horses that are subject to natural processes. Another goal is to protect key threatened species and island ecosystems, all of which are in jeopardy if habitat is overgrazed by voracious horses. Clearly it is important for the Park Service to control the equine population on the Maryland end of Assateague, but a balance must be struck between maintaining feral horses and protecting native and endangered species. How should one limit reproduction in a wild herd of horses?

The Chincoteague Volunteer Fire Company and the Fish and Wildlife Service limit equine population growth on the Chincoteague NWR by gathering the herd annually and selling most of the foals at auction. This roundup raises money for the Fire Company and keeps the herd population in balance. The Park Service was not interested in adopting a similar program for the Maryland herd, fearing that it would cause injury to horses and humans and increase the workload of an already overtasked staff.

By the early 1970s the U.S. Bureau of Land Management faced a similar problem in the herd management areas of the Western states and placed it in the hands of biologists Jay

How many horses are enough? How many are too many? The current population target is 80–100 animals on the Maryland end of Assateague. The Park Service allows every mare to have a single foal, then contracepts her for the rest of her life. Occasionally mares do not respond to contraception and continue to foal regularly.

Kirkpatrick and John Turner. Their solution was to develop a method of birth control for use on free-roaming horses. In 1988, with help from Irwin Liu of the University of California at Davis, they created a contraceptive that tricks a mare's immune system into attacking sperm as they try to penetrate the zona pellucida, the transparent, noncellular protein layer that surrounds all mammalian egg cells. Using the zona pellucida of pigs, they created an inexpensive, evidently harmless, and mostly reversible injection that essentially vaccinates wild mares against pregnancy without capture or restraint.

Since 1994, Assateague Island NS officials have maintained contraception through a yearly PZP booster, easily delivered by dart gun. At first, the seashore used the PZP vaccine to keep the herd's birth rate close to zero until it could decline from 166 horses to the target of 150. Foaling in 1995 in increased the population to 173, and decline proved much slower than expected. Even the scientists were surprised to find that contracepted mares live significantly longer than those allowed to reproduce at will. Mean age at death for untreated mares at the beginning of the program was just 6.5 years. By 2007, MAD for mares contracepted 3 or more consecutive years increased to 20 (Kirkpatrick & Turner, 2007). By 2013, the figure reached an astonishing 26+ years, comparable with the lifespan of a domestic horse (J. Kirkpatrick, personal communication, May 29, 2014). Treated mares not only lived longer, but also had a higher quality of life—their body-condition scores improved (Kirkpatrick & Turner, 2002)—and mortality for foals and other horses also dropped (J. Kirkpatrick, personal communication, May 29, 2014). Consequently, although the Assateague herd is no longer growing, it has taken a long time to shrink.

At first, the plan was to vaccinate all adolescent mares, then allow them to produce three foals each before halting their reproduction permanently (Eggert et al., 2010). In 1998, the plan was revised to allow each mare to produce only two offspring, and in 2000, each mare was allowed to produce a single foal. Currently, fillies are vaccinated at puberty and contracepted for 3 years, after which vaccine is withheld until they foal. Once they have foaled, the vaccinations resume and they are contracepted indefinitely.

The equine population of Assateague has been extensively studied over recent decades, and the Park Service keeps records on the lineage and habits of each horse. This information allows biologists to determine which horses to dart in order to prevent specific family lines from being lost from the gene pool.

Keiper's research suggested that minimal environmental damage would occur to the north end of Assateague if the pony population were kept below 150 (Keiper, 1985). Immunocontraception limited the growth of the herd, but by 2006 it became apparent to the Park Service that the 144 horses on the Maryland end were still disrupting the island ecosystem. From 1994 to 2000, horses grazed upon saltmarsh cordgrass (*Spartina alterniflora*) and avoided saltgrass (*Distichlis spicata*), giving the latter a competitive advantage and changing the balance of each species in the marsh. Moreover, saltgrass cannot tolerate prolonged immersion in brackish water and does not trap and filter sediment as *Spartina* does. The Park Service concluded that the horses were doing irreversible harm to the marshes of Assateague.

The agency recruited the Conservation Breeding Specialist Group to assess the free-roaming horse population and its habitat and evaluate how best to manage the herd (Zimmerman et al., 2006). A diverse team of participants convened to discuss management objectives, past management plans, and the results of computerized population modeling. Its stated goal was to balance the desire for a large, healthy herd of horses with a healthy island ecosystem that favors endangered species.

An important consideration in any wildlife management project is effective population size, the number of individuals in a population that contribute offspring to the next generation. A healthy horse herd requires harems containing animals of reproductive age with enough genetic diversity to minimize inbreeding. A large population provides a diverse gene pool, but causes heavy damage to the ecosystem, which eventually harms the health of the horses themselves. A very small population would have more environmental resources at its disposal, but inbreeding would eventually impair its health and render it vulnerable to disease and birth defects. Clearly, the optimal number of horses lies somewhere between the two extremes.

Effective population size is influenced by the fertility of individual animals, longevity, breeding cycles, sex ratio, mating strategy, and historic population size. Effective population size and census population size are not synonymous. A herd of 1,000 mares (none of which is pregnant) would have an effective population size of 0 because it is unable to pass its genes to a new generation without a stallion.

The definition of effective population differs with context and can have different mathematical and biological meanings, which are often confused or misunderstood. "Effective population size is a very complex term," says equine geneticist Gus Cothran (personal communication, April 20, 2011). "I have a book on my shelf with an equation for calculating it, and it takes about a page and a half. In a real population, the effective size is generally

Wild ponies are a unique part of the character of Assateague Island National Seashore. People from all over the world visit the park to enjoy the sight of horses against the backdrop of relatively pristine beach.

one quarter to one third of the census size. So that means that 150–200 individuals are required to have an effective size of 50."

Franklin (1980) proposed the 50/500 rule, derived from populations of fruit flies. According to this rule, populations with an effective population size less than 50 are at risk of inbreeding depression and extinction in the short term, and populations with an effective size of less than 500 risk extinction over the long term (Franklin, 1980). The minimum population of 50 animals corresponds to rate of loss of genetic variability at 1% per generation, roughly half the maximum rate accepted by domestic animal breeders. Most conservation administrators recognize this "rule" as an overly generalized hypothesis that cannot guide management for species as different as snow leopards and salamanders.

The minimum viable population is an estimate of the smallest size at which a population can exist without facing extinction from natural disasters or loss of genetic variability. Conservationists usually consider MVP the number of individuals necessary to ensure 90–95% probability of survival of the group for 100 to 1,000 years.

This number, however, might not be sufficient if environmental catastrophe should wipe out a large number of animals. Drought, a brutal winter, or fire can quickly deplete a herd, as can disease. In the Assateague herd, between 1989 and 1993, at least 15 horses died of mosquito-borne eastern equine encephalitis, and 12 drowned in a single storm.

The concept of the MVP is useful, but implementation of any management plan involves reconciling scientific abstractions with environmental, political, financial, and logistical realities (Traill, Brook, Frankham, & Bradshaw, 2010). Federal agencies, as a rule, consistently revise minimum numbers down. Regarding the Assateague herd, Zimmerman et al.(2006, p. 51) reported, "The PHVA participants recommended a short-term target population size of 80–100 horses, perhaps managing toward the lower end of this range." Conversely, horse advocates typically push for the largest sustainable population, arguing that lost genes cannot be restored, and outcrossing to increase variability may erode unique characteristics of the herd. Usually the target population represents a compromise between the two factions. Says Cothran "In general, management agencies such as the BLM or the NPS . . . want to keep the lowest numbers they can, and so they go for the compromise numbers. For example, on the Shackleford Banks, we figured that about 120 animals would be acceptable, at least for a long time" (personal communication, April 20, 2011).

To better analyze population dynamics, scientists create computer models that transform complex biological systems into simplified representations in hope of making accurate predictions. These models depict key features of the natural processes, but are easier to evaluate.

To explore the horse-ecosystem balance in the hypothetical realm, the Park Service used Vortex Population Viability Analysis software to simulate demographics, population trends, and environmental events. Data for each of the 56 stallions and 89 mares of Assateague gleaned from decades of research were fed into the software.

Vortex takes the horses through a "what if" scenario, inventing normal life-cycle events such as birth, death, and catastrophic events such as storms and disease. Every time the program is run, the result is different—horses have more offspring or fewer, horses drown in storms or do not, individuals are contracepted or are not. After the program had created 500 scenarios, participants gained insight into the probable outcomes of different management practices (Zimmerman et al., 2006).

In creating these simulations, however, the Park Service assumed that the horses' reproduction is density-independent (Zimmerman et al., 2006). But because horses are a K-selected species, equine herd growth is density-*dependent*. Since the creation of the national seashore, the herd has never been large enough to limit its own expansion.

Among zoo animals, when gene diversity falls below 90%, litter sizes become smaller, birth weights decrease, and young are less likely to survive. If the number of horses on Assateague exceeds 80 individuals, genetic diversity should remain above 90%, in the absence of natural disaster, and a visitor would probably see horses in the park on any given day (Zimmerman et al., 2006). The Park maintains that a herd of 60 horses can remain more than 90% diverse if mares that carry the rarest bloodlines are allowed to bear most of the foals while the rest are strictly contracepted. A herd of 60 can also remain optimally diverse if two outside mares are added every 10 years.

Cothran disagrees. "Although 60 horses could retain 90% of the diversity, that is not a sustainable number in the long term," he said. An effective population of 50 is considered by conservation genetics people to be the lowest that could be sustainable, but effective size is ⅓ to ⅕ census size, so we really are talking about a lot more than 60" (personal communication, September 10, 2014).

In the future, the Park Service may need to add horses to the Maryland herd if genetic variability declines or if disease or disaster threatens the population. The agency could choose these additions from other barrier island populations, such as Shackleford Banks, Currituck Banks, and Cumberland Island. The horses from these herds are genetically similar to the Assateague horses, yet different enough to revitalize the gene pool. They are also well adapted to life on a barrier island and so are likely to survive (*Environmental*

Assessment, 2009). Although the Chincoteague ponies originate from the same ancestral population, Cothran does not consider them good prospects for outcrossing. "They are so mixed genetically, I would not pick them unless the only objective was variability," he said (personal communication. September 10, 2014).

"Genetic diversity is currently good," says Allison Turner, biological science technician at Assateague NS (personal communication, February 17, 2011). "Barring a catastrophic event, additions should not be necessary in our lifetimes. Maintaining a genetically healthy viable population was a key consideration in determining the population size goal."

The Conservation Breeding Specialist Group final report in 2006 suggested a target population of 80–100 horses, maintaining the herd at the lower end of the range but adjusting the number upwards as needed (*Environmental Assessment*, 2009; Zimmerman et al., 2006). The Park service adopted their recommendations. Computer modeling suggested that under the 2006 management plan, the horse population would shrink to about 100 individuals by 2012, 80 by 2014, and 50 by 2016. By adjusting the rate of contraception, the Park Service might maintain the herd at a desirable level within this range or decreased further if necessary. A recent genetic study lowered the MVP threshold even further, concluding that the herd was genetically adequate with only 47 horses (Eggert et al., 2010). Others argue that a herd of this size would eventually suffer the effects of inbreeding and would be more vulnerable to extinction through disease or disaster. Cothran comments, "I strongly disagree with the number of 47 being adequate. In 2–3 generations a serious decline in variation would be almost certain" (personal communication, September 10, 2014).

The Park Service hopes to offset the drawbacks of a smaller herd size by offering the visitor improved information and guidance on where to find the horses within the developed parts of the national seashore and Assateague State Park. It also intends to build an elevated observation platform to help visitors spot horses from a distance. The Park Service also plans to solicit community input through public meetings and online forums before making future management decisions.

Has maintaining fewer horses made a difference? Significant reduction in horse numbers is very recent, and research has yet to discover whether reduced grazing pressures have caused environmental improvements. Unofficially, however, park personnel point out benefits. "Historic data shows that beachgrass (*Ammophila spp.*) was the second favorite forage for horses, and horse grazing was significantly impacting dune formation and beachgrass proliferation," says Turner (personal communication, February 17, 2011). "We have noted reduction in those impacts recently."

Assateague and its horses are a "natural laboratory" and a valuable research resource. Many procedures used in the study of wildlife were perfected on the horses of Assateague Island, including remote pregnancy testing, fetal health evaluation, remote evaluation of endocrine function, immunocontraception, and fecal DNA analysis.

The Assateague horses are the best known and most loved of the wild herds of the Eastern Seaboard, and the closest to major cities and other dense populations. Over 2 million people visit Assateague's three parks every year, most of them remaining in the developed areas. Many come just to see the ponies. Attendance is heaviest from Memorial Day through Labor Day, peaking in July and August and reaching a nadir in January and February.

Visitors to the Maryland part of Assateague can readily see horses at roadside pulloffs, nature trails, and campgrounds, or they can hike out to find the more reclusive bands that keep to the wilderness areas. Despite heavy visitation, it is still easy to find empty beach and seclusion in nature if one is willing to walk a short distance. Free-roaming horses go about their business unmindful of human activity, and to watch them moving across the dunes in the heat of the day is to abandon oneself to a sense of affinity with the natural world. With thoughtful foresight, a balance can be achieved between horses, humans, and environment so that Assateague Island NS might be preserved in its natural state for the enjoyment of generations to come.

References

Allen, A., Gill, S., Marcy, D., Honeycutt, M., Mills, J., Erickson, M., . . . Smith, D. (2010, September). *Technical considerations for use of geospatial data in sea level change mapping and assessment.* Silver Spring, MD: National Ocean Service.

Assateague Island National Seashore. (2006). *Feral horse management at Assateague Island National Seashore.* Retrieved from http://www.nps.gov/asis/upload/feralhorsemanag.pdf

Associated Press. (1962, May 7). Former S&L head gets prison term. *Salisbury Times* (Salisbury, MD), *39*(122), p. 4.

Berger, J., & Cunningham, C. (1987). Influence of familiarity on frequency of inbreeding in wild horses. *Evolution, 41*(1), 229–231.

Boyd, L., & Keiper, R. (2005). Behavioural ecology of feral horses. In D.S. Mills & S.M. McDonnell (Eds.), *The domestic horse: The origins, development and management of its behaviour* (pp. 55–82). Cambridge, United Kingdom: Cambridge University Press.

Buchanan, M. (2003, December). *Horse manure management: A guide for Bay area horse keepers.* [Petaluma, CA: Council of Bay Area Resource Conservation Districts]. Retrieved from http://www.acrcd.org/Portals/0/Equine Fact

Center for Biological Diversity. (n.d.). *Seabeach amaranth.* Retrieved from https://www.biologicaldiversity.org/campaigns/esa_works/profile_pages/SeabeachAmaranth.html

Clean Water Act (Federal Water Pollution Control Act Amendments), 33 U.S.C. §§1251–1387 (1972).

Coastal Waterbird Program. (2008, September 28). *Are there piping plovers nesting on your beach?* Lincoln, MA: Mass Audubon. Retrieved from http://www.massaudubon.org/PDF/cwp/piping_plover_landowners.pdf

Cohen, J.B., Erwin, R.M., French, J.B., Jr., Marion, J.L., & Meyers, J.M. (2010). *A review and synthesis of the scientific information related to the biology and management of species of special concern at Cape Hatteras National Seashore, North Carolina* (U.S. Geological Survey Open-File Report 2009–1262). Retrieved from http://pubs.usgs.gov/of/2009/1262/

De Stoppelaire, G.H., Brock, J., Lea, C., Duffy, M., & Krabill, W. (2001, July). *USGS, NPS, and NASA investigate horse-grazing impacts on Assateague Island dunes using airborne lidar surveys* (U.S. Geological Survey, Open-File Report 01-382). Retrieved from http://pubs.usgs.gov/of/2001/of01-382/

De Stoppelaire, G.H., Gillespie, T.W., Brock, J.C., & Tobin, G.A. (2004). Use of remote sensing techniques to determine the effects of grazing on vegetation cover and dune elevation at Assateague Island National Seashore: Impact of horses. *Environmental Management, 34*(5), 642–649. doi: 10.1007/s00267-004-0009-x

Diefenbach, D.R., & Christensen, S.A. (2009, August). *Movement and habitat use of sika and white-tailed deer on Assateague Island National Seashore, Maryland* (Technical Report NPS/NER/NRTR—2009/140). Philadelphia, PA: National Park Service, Northeast Region.

Eggert, L., Powell, D., Ballou, J., Malo, A., Turner, A., Kumer, J., . . . Maldonado, J.E. (2010). Pedigrees and the study of the wild horse population of Assateague Island National Seashore. *Journal of Wildlife Management, 74*(5), 963–973. doi: 10.2193/2009-231

Environmental assessment of alternatives for managing the feral horses of Assateague Island National Seashore: Finding of no significant impact. (2009). Berlin, MD: U.S. National Park Service, Assateague Island National Seashore.

Feh, C. (1999). Alliances and reproductive success in Camargue stallions. *Animal Behaviour, 57*(3), 705–713. doi: 10.1006/anbe.1998.1009

Feh, C. (2005). Relationships and communication in socially natural horse herds. In D.S. Mills & S.M. McDonnell (Eds.), *The domestic horse: The origins, development and management of its behaviour* (pp. 83–109). Cambridge, United Kingdom: Cambridge University Press.

Franklin, I.R. (1980). Evolutionary change in small populations. In M.E. Soule & B.A. Wilcox (Eds.), *Conservation biology: An evolutionary-ecological perspective* (pp. 135–140). Sunderland, MA: Sinauer Associates.

Frydenborg, K. (2012). *The wild horse scientists*. Boston, MA: Houghton Mifflin Harcourt.

Gill, E. (1994). *Ponies in the wild.* London, United Kingdom: Whittet Books.

Goldschmidt, P.G., & Dahl, A.W. (1976, April). Estimating population in seasonal resort communities. *Growth and Change, 7*(2), 44–48. doi: 10.1111/j.1468-2257.1976.tb00305.x

Hayward, L. (2007). *State of the parks: Assateague Island National Seashore, a resource assessment.* Washington: National Parks Conservation Association.

Houpt, K. (2005). Maintenance behaviours. In D.S. Mills & S.M. McDonnell (Eds.), *The domestic horse: The origins, development and management of its behaviour* (pp. 94–109). Cambridge, United Kingdom: Cambridge University Press.

Houston, D., & Schreiner, E. (1995). Alien species in national parks: Drawing lines in space and time. *Conservation Biology, 9*(1), 204–209. doi: 10.1046/j.1523-1739.1995.09010204.x

Ingle, M.C. (2005). *The development and testing of a procedure for monitoring visitor-horse interactions at Assateague Island National Seashore* (Unpublished master's thesis). North Carolina State University, Raleigh.

Keiper, R. (1985). *The Assateague ponies.* Atglen, PA: Schiffer Publishing.

Keiper, R.R., & Hunter, N.B. (1982). *Population characteristics, habitat utilization, and feeding habits of the feral ponies, sika deer, and white-tailed deer within Assateague Island National Seashore.* Research/Resources Management Report, MAR-4. Philadelphia, PA: National Park Service, Mid-Atlantic Region.

Kirkpatrick, J.F., & Turner, A. (2002). Reversibility of action and safety during pregnancy of immunizing against porcine zona pellucida in wild mares (*Equus caballus*). *Reproduction* (Suppl. 60), 197–202.

Kirkpatrick, J.F., & Turner, A. (2007). Immunocontraception and increased longevity in equids. *Zoo Biology, 25,* 237–244. doi: 10.1002/zoo.20109

Kirkpatrick, J. (1994). *Into the wind: Wild horses of North America*. Minocqua, WI: Northword Press.

Kirkpatrick, J.F., & Turner, A. (2003). Absence of effects from immunocontraception on seasonal birth patterns and foal survival among barrier island wild horses. *Journal of Applied Animal Welfare Science, 6*(4), 301–308.

Lambert, M.S., Ozbay, G., & Richards, G.P. (2009). Seawater and shellfish (*Geukensia demissa*) quality along the western coast of Assateague Island National Seashore, Maryland: An area impacted by feral horses and agricultural runoff. *Archives of Environmental Contamination and Toxicology, 57*(2), 405–415. doi: 10.1007/s00244-008-9277-4

Levin, P., Ellis, J., Petrik, R., & Hay, M. (2002). Indirect effects of feral horses on estuarine communities. *Conservation Biology, 16*(5), 1364–1371. doi: 10.1046/j.1523-1739.2002.01167.x

Lowney, M., Schoenfeld, P., Haglan, W., & Witmer, G. (2005). Overview of impacts of feral and introduced ungulates on the environment in the eastern United States and Caribbean. In D.L. Nolte & K.A. Fagerstone (Eds.), *Proceedings of the 11th Wildlife Damage Management Conference* (pp. 64–81).

Mackintosh, B. (2003, October 27). *Assateague Island National Seashore: An administrative history*. Washington, DC: National Park Service, History Division. Retrieved from http://www.nps.gov/asis/parkmgmt/upload/asisadminhistory.pdf (Original work published 1982)

Maryland Department of Natural Resources, Resource Planning. (2005, October). *Assateague State Park land unit plan*. Retrieved from http://www.dnr.state.md.us/irc/docs/00011180.pdf

The Maryland Wetlands Act of 1970, Md. Env. Code Ann. §§ 16-101–16-503 (1970).

McCullough, D.R., Takatsuki, S., & Kaji, K. (Eds.). (2009). *Sika deer: Biology and management of native and introduced populations*. Tokyo, Japan, & New York, NY: Springer.

McDonnell, S.M. (2005). Sexual behaviour. In D.S. Mills & S.M. McDonnell (Eds.), *The domestic horse: The origins, development and management of its behaviour* (pp. 110–125). Cambridge, United Kingdom: Cambridge University Press.

McGreevy, P. (2004). *Equine behavior: A guide for veterinarians and equine scientists*. London: W.B. Saunders.

Monard, A.M., Duncan, P., & Boy, V. (1996). The proximate mechanisms of natal dispersal in female horses. *Behaviour, 133*(13/14), 1095–1124.

Naked rodeo on Assateague. (2007, August 24). *The Dispatch* (Ocean City, MD). Retrieved from http://www.mdcoastdispatch.com/articles/2007/08/24/Cops-and-Courts/Naked-Rodeo-On-Assateague

National Environmental Policy Act of 1969, 42 U.S.C. § 4321 et seq. (1970).

Ocean City Planning and Zoning Commission. (2006, August). *Town of Ocean City Maryland comprehensive plan*. Ocean City, MD: Author.

Pendleton, E.A., Thieler, E.R., & Williams, S.J. (2004). *Coastal vulnerability assessment of Cumberland Island National Seashore (CUIS) to sea-level rise* (U.S. Geological Survey Open-File Report 2004-1196, Electronic Book). Retrieved from http://pubs.usgs.gov/of/2004/1196/ images/pdf/CUIS.pdf

Pilkey, O., Rice, T., & Neal, W. (2004). *How to read a North Carolina beach*. Chapel Hill: University of North Carolina Press.

Russo, B. (2009, October 30). Major sewer pipe project to disrupt OC highway. *The Dispatch* (Ocean City, MD). Retrieved from http://www.mdcoastdispatch.com/articles/ 2012/09/28/Top-Stories/Major-Sewer-Pipe-Project-To-Disrupt-OC-Highway

Rubenstein, D. (1982). Reproductive value and behavioral strategies: Coming of age in monkeys and horses. In P.P.G. Bateson & P.H. Klopfer (Eds.), *Perspectives in Ethology* (Vol. 5, Ontogeny, pp. 469–487). Princeton, NJ: Princeton University Press.

Rubenstein, D.I., & Hack, M. (1992). Horse signals: The sounds and scents of fury. *Evolutionary Ecology, 6*, 254–260.

Sanders, A.E. (2002). Additions to the Pleistocene mammal faunas of South Carolina, North Carolina, and Georgia. *Transactions of the American Philosophical Society, 92*, Part 5.

Seliskar, D.M. (2003). The response of *Ammophila breviligulata* and *Spartina patens* (Poaceae) to grazing by feral horses on a dynamic mid-Atlantic barrier island. *American Journal of Botany, 90*(7), 1038–1044.

Soper, S.J. (2012, February 24). Assateague horse shooter sentenced. *The Dispatch* (Ocean City, MD). Retrieved from http://www.mdcoastdispatch.com/articles/2012/02/24/ Top-Stories/Assateague-Horse-Shooter-Sentenced

Sturm, M. (2007, May). *Assessment of the effects of feral horses, sika deer, and white-tailed deer on Assateague Island's forest and shrub habitats: Final report.* Berlin, MD: Assateague Island National Seashore. Retrieved from http://www.nps.gov/nero/science/FINAL/ ASIS_horsedeer/ASISHorseDeerVegFinal_May07.pdf

Traill, L., Brook, B., Frankham, R., & Bradshaw, C. (2010). Pragmatic population viability targets in a rapidly changing world. *Biological Conservation, 143*(1), 28–34. doi: 10.1016/j.biocon.2009.09.001

U.S. Bureau of Land Management, Billings Field Office. (2001). *Environmental Assessment and Gather Plan, Pryor Mountain Wild Horse Range, FY2001 Wild Horse Gather and Selective Removal* (EA #MT-010-1-44). Billings, MT: Bureau of Land Management, Billings Field Office.

U.S. Bureau of Land Management, Elko District, Wells Field Office. (n.d.). *Proposed northeast Nevada wild horse eco-sanctuary.* Retrieved from http://www.blm.gov/pgdata/etc/ medialib/blm/nv/field_offices/elko_field_office/information/nepa/eiss/archives/ nenvwh_ecosanctuary.Par.21472.File.dat/EcoSanctuaryScopingBrief.pdf

Weakley, A., Bucher, M., & Murdock, N. (1996). *Recovery plan for seabeach amaranth* (Amaranthus pumilus*) Rafinesque.* Atlanta, GA: U.S. Fish and Wildlife Service, Southeast Region.

Zervanos, S.M., & Keiper, R.R. (1979a). *Ecological impact and carrying capacity of feral ponies on Assateague Island National Seashore.* Final contract report to U.S. National Park Service, Mid-Atlantic Region, Philadelphia.

Zervanos, S.M., & Keiper, R.R. (1979b). Factors influencing home range, movement patterns, and habitat utilization in Assateague Island feral ponies. *Proceedings of Ecology and Behavior of Feral Equids Symposium* (pp. 3–14). Laramie: University of Wyoming.

Zervanos, S.M., & Keiper, R.R. (1979c). *Winter activity patterns and carrying capacities of Assateague Island feral ponies.* Report to U.S. National Park Service, Denver, CO.

Zimmerman, C., Sturm, M., Ballou, J., & Traylor-Holzer, K. (Eds.). (2006). *Horses of Assateague Island population and habitat viability assessment workshop: Final report.* Apple Valley, MN: IUCN/SSC Conservation Breeding Specialist Group.

Chapter 4

Chincoteague–Assateague, Virginia

The Wild, Wild East

Chincoteague
and Vicinity
VIRGINIA

Legend:
- ▇ Highway
- ▬ Road
- ▮▮▮ ORV route
- ▪▪▪ Trail
- ☐ Beach
- ☐ Marsh
- ☐ Maritime forest
- ☐ Private

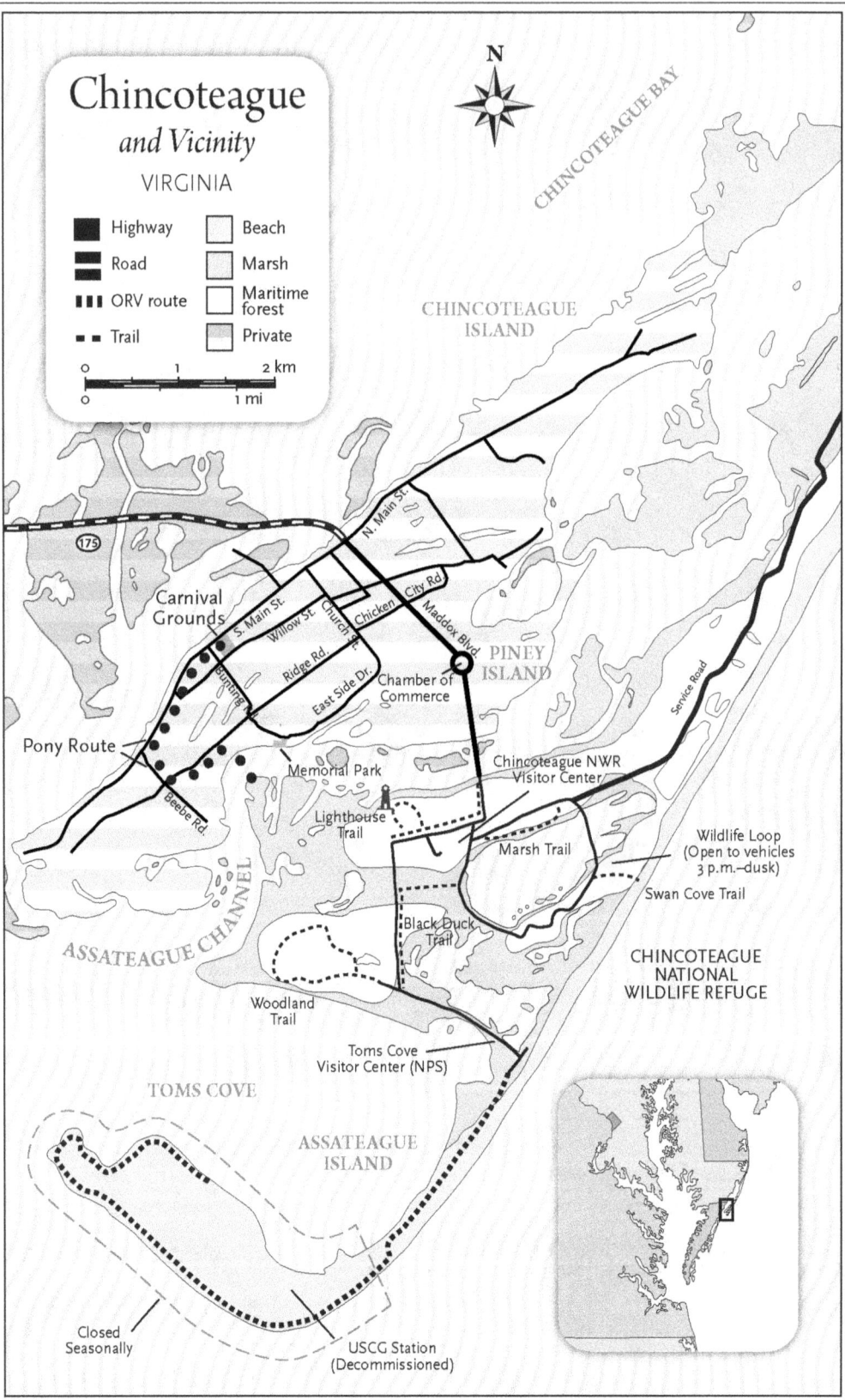

0 ─── 1 ─── 2 km
0 ─── 1 mi

N

CHINCOTEAGUE BAY

CHINCOTEAGUE ISLAND

175

N. Main St.

Carnival Grounds

S. Main St.

Willow St.

Church St.

Chicken City Rd.

Ridge Rd.

Maddox Blvd.

PINEY ISLAND

Bunting

East Side Dr.

Chamber of Commerce

Pony Route

Service Road

Beebe Rd.

Memorial Park

Chincoteague NWR Visitor Center

Lighthouse Trail

Marsh Trail

Wildlife Loop (Open to vehicles 3 p.m.–dusk)

Swan Cove Trail

ASSATEAGUE CHANNEL

Black Duck Trail

CHINCOTEAGUE NATIONAL WILDLIFE REFUGE

Woodland Trail

Toms Cove Visitor Center (NPS)

TOMS COVE

ASSATEAGUE ISLAND

Closed Seasonally

USCG Station (Decommissioned)

The spectacle was mesmerizing. At the urging of the cowboys, a multitude of wild horses charged into the stillness of Assateague Channel and swam, heads dotting the slate-blue surface, lips curled and nostrils narrowed. Mares nickered encouragement to wild-eyed colts as they plunged into water made turbulent by hundreds of flailing hooves. Saltwater Cowboys shouted and cracked bullwhips for emphasis, but the ponies did not need much urging.

Swimming comes naturally to them. In the wilds of Assateague, they often ford tidal creeks and seek refuge from insects by immersing themselves in the waters of the bay. Sometimes ponies make the swim to Chincoteague of their own accord, singly or in groups, forcing the cowboys to conduct an impromptu roundup to restore them to their Assateague range.

As their hooves found purchase in the sticky marsh mud, they emerged onto Chincoteague. A prancing, snorting throng of sleek, dripping horses rose out of the bay—palominos, chestnuts, buckskins, and pintos—like mythical beasts rising from enchanted waters in a fairy tale. It was enough to make a photographer put down her camera and gawk in slack-jawed astonishment.

Chincoteague, a small island community on the Eastern Shore of Virginia, historically made its living from the sea. Chincoteague's delectably salty oysters put the town on the map, but town was thrust forever into the limelight in 1947, when Marguerite Henry's book *Misty of Chincoteague* brought the tradition of Pony Penning to the attention of children all over the world. Today the promise of seeing wild ponies (and eating tasty seafood) brings thousands of annual visitors to this unique little island.

Although there are festivals and activities in Chincoteague all year long, the most famous attraction is the herd of free-roaming ponies that lives across the channel on Assateague Island in the Chincoteague National Wildlife Refuge. Traditionally, on the last Wednesday and Thursday of July, locals round up the ponies on their Assateague home ground, swim them across the channel to Chincoteague, and sell spring foals at auction. Every year Pony Penning draws crowds in the tens of thousands and attracts international media coverage.

The periodic penning of Assateague Ponies has gone on for hundreds of years, and the Chincoteague Volunteer Fire Company has maintained the tradition since 1924. The fire company has held the Firemen's Carnival, Pony Penning, and sale annually, except for a lapse in 1943 and 1944 due to World War II. The proceeds of the auction go toward maintaining the herd and covering operating expenses for the fire company, the legal owner of the ponies on the Virginia end of Assateague. "The men and women of the fire company work long, long hours to do what we do," says Denise Bowden, executive secretary and spokesperson for the firefighters (personal communication, February 5, 2011).

Every July, almost 200 wild ponies enter the waters of Assateague Channel for the annual swim. Heads and backs above the water, the ponies move to shore in a cohesive mass.

Pony Penning is the culmination of the month-long Firemen's Carnival. Each year, approximately 30–40 Saltwater Cowboys participate in the roundup, swim, and parade. About half are members of the fire company; the rest take their horses on an annual trek to the island to participate. The Cowboys typically work their way up through the ranks, helping with menial tasks at first, and only sliding into the coveted cowboy role if a rider dies or retires (Kobell, 2006). "If a Cowboy or fireman has a son or male relative who wants to get involved, they will take him on," says Pam Emge, co-author of *Chincoteague Ponies: Untold Tails* (Szymanski, 2012). "The Fire Company says that women can become Saltwater Cowboys, but you never see a woman rider. They are all men, every one"(personal communication, September 7, 2014).

On the weekend before the drive, the Cowboys take their own horses to Assateague and gather the wild ponies that live on the Chincoteague NWR. They confine the 50-odd adult horses and their foals from the southern end of the refuge in a corral on Beach Road. To the north, a second pen near Swan Cove holds the larger part of the herd, about 100 adult ponies and their foals. The ponies are intelligent and have indelible memories—they know the drill.

The joining of the two groups, an event known as the Beach Walk, takes place on the last Monday in July. Crowds start to form before daybreak as spectators claim prime spots along the beach. The Fish and Wildlife Service scouts the beach for piping plover nests before the pony drive, and sometimes requires the cowboys to use an alternative route to avoid disturbing broods (Grey, 2014). It seems there is always a light mist rolling off the ocean as the restless crowd waits, sometimes for hours. Then, emerging like a mirage from the shimmering morning mist, the herd materializes. About 150 ponies, flanked by slicker-clad Saltwater Cowboys, move slowly down the shoreline through

As their hooves find the bottom, they rise from the water like mythical sea creatures.

the lap of breaking waves. Occasionally a cowboy must reclaim an excited nervous foal which darts across the beach or into the surf. Making a sharp right-hand turn toward the crowd, the ponies surge by the spectators and down Beach Road to join the southern herd at the corrals.

Within the corrals, excited horses collect into groups, break into a run, scatter, and collect again. Foals whinny frantically for their dams. Stallions suddenly thrown together battle to establish dominance and reassemble their scattered mares. Equine family groups cluster around hay piles, squealing and arguing with others as they jockey for the best feeding spots. Occasionally, a pregnant mare gives birth. All the while, people cluster along the fence, fascinated by equine dramas witnessed at close range. Prospective pony owners eye the foals critically, choosing favorites for Thursday's bidding.

The fire company extends a hose from a fire truck and fills the troughs with cool water. All day Monday and Tuesday, people flock to the pony pens to watch the milling animals with insatiable interest. A veterinarian evaluates the health of each animal and looks for illness and injuries. Heavily pregnant mares, old or weak ponies, and newborn foals are exempt from the swim and are taken to the carnival grounds by trailer.

People flock from all over the world to witness the Pony Swim. Countless others watch the event live on television programs such as *Good Morning America*. On the last Wednesday of July, Saltwater Cowboys herd the ponies from the pens to the edge of the channel that separates Assateague from Chincoteague. There they await slack tide, the interval between high and low tides when the current in the channel is at its weakest. Well in advance of the event, boats form a corridor through which the ponies will swim. The wait for slack tide seems interminable.

The crowd suddenly snaps to attention as the Coast Guard fires a red starburst rocket to signal the start of the main event. As thousands watch, cowboys herd the horses into the channel. Initially they hesitate, and then a seasoned pony who has been though previous swims splashes into the water and heads for the distant shore, followed by another and another. The crossing takes only about 7–10 min, whereupon the ponies emerge from the channel waters onto Chincoteague soil.

Suddenly there are ponies rising from the water *en masse*, dripping brackish water and looking very pleased with themselves. The first foal to reach Chincoteague is dubbed King or Queen Neptune and is traditionally raffled off that night—in 2009, the fire company sold more than 6,000 tickets at a dollar each (Fried, 2010). The seasoned adults and bewildered youngsters move onto a small meadow of lush green grass to rest and recover.

Then the cowboys press them into motion again. The large, lively herd of ponies parades up Main Street to another set of holding pens at the carnival grounds, where they will stay until Friday morning. Moving shoulder to shoulder, the animals snort and prance past wide-eyed spectators at the curbside. Occasionally, the cowboys must guide a rogue pony back to the herd after it breaks away from the group and runs across a lawn or driveway.

Thursday is auction day. The best seats are staked out well in advance, but any persistent bidder can get close enough to the action to participate. It is usually very hot, with abundant flies and mosquitoes and frequently a passing shower. Anyone may bid, and it is possible to place a bid unintentionally by waving to a friend or swatting a fly.

Beginning early in the morning, staff members separate foals from their mothers and begin the process of presenting each youngster to the admiring crowd for bidding. Volunteers lock hands around previously unhandled foals, grabbing the tail and under the belly, and attempt to guide them around the arena while the weanlings leap, pitch, and try to escape. Some foals recognize the futility of rearing and kicking, so they sit or lie down and refuse to move.

By midday, nearly every foal will have a new owner. The veterinarian is available to draw a Coggins, test, vaccinations, de-worming, and issue a health certificate before the foal leaves the island. Most are loaded into trailers by 5:00 p.m. Friday and trundled to faraway new homes. Foals younger than one month old are usually allowed to remain with their mothers after the sale, to be claimed in the fall by their buyers. Some youngsters are sold as "turn-backs" or "buy-backs" and are returned to the herd.

Children venture to the island hoping to place a winning bid, but not everybody gets a pony. Because about 1,000 bidders compete for about 60–70 foals, most would-be pony owners return home disappointed (Fried, 2010). And Chincoteague foals sell for prices that would challenge the wallet of the most hardworking child.

Horses tend to bring out the best in people, and it may be true that the folklore and legends surrounding the Chincoteague Ponies make that equine magic all just a little more special. In 2004, the Feather Fund was founded to help children of limited means purchase a Chincoteague Pony.

The story is poignant. In the summer of 1995, Carollynn Suplee, feeling grateful to have survived her recent brain surgery, attended the auction with the hope of buying a turn-back pony. By the time she arrived, the turn-backs had all been sold. Deeply spiritual and feeling the need to give something back after surviving her illness, she donated money that allowed two horse-loving sisters to bid on and win Sea Feather, a pony that they otherwise could not afford.

She returned to the island annually, prayed for direction, and each year helped a child purchase a foal. She became known as the "Pony Fairy" until her death in 2003. The following year the Feather Fund was established as a nonprofit in her memory. Children age 10 to 14 may apply for a pony and demonstrate that they have saved a portion of the money on their own, have experience with ponies, and are able to maintain a foal. The Feather Fund

At the joining of the herds, the world-famous Saltwater Cowboys drive the north herd down the beach to the corrals on Beach Road, where the south herd awaits. Spectators lined up along the beach get an up-close view of lively ponies against the backdrop of sunrise over the ocean.

committee chooses a winner from letters, photographs, and videos, and then helps the child select and bid for a foal. Sometimes two or three foals are awarded to lucky children.

After the auction, unsuccessful bidders may acquire ponies though alternative means. Foals born after Pony Penning and into the fall are sold during the October roundup, and several breeders on Chincoteague Island offer foals and trained horses for sale each year. The Chincoteague Pony Rescue is a nonprofit organization in Ridgely, Maryland, that rehabilitates abandoned, neglected, abandoned and abused Chincoteague Ponies and places them in loving homes.

Desirable coat color inflates the prices more than anything else—chestnuts typically fetch lower prices than pintos with flashy markings. One can usually determine whether a foal will be buckskin, palomino, or chestnut by the color of the juvenile coat, but there is always the possibility that a foal will mature with a different color than expected. Foals usually go through several coat-color changes before arriving at their adult coloration, and even then many horses change color seasonally. R. Owen Hooks (2006) estimates that over the last 50–60 years, about 40% of the Chincoteague herd has been pinto; 30% chestnut; 20% bay; 8% buckskin, dun, or grulla; and 2% other colors such as palomino, gray, cream, roan, or black. Rare colors usually fetch higher prices, except dun, which tends to sell for less than average.

Prior to the 1920s, the horses were solid-colored. E.L. Vallandigham (1893, p. 28) wrote, "The ponies are from 11 to 14 hands high, and weigh from 650 to 750 pounds. There are many bays, sorrels and blacks, a few grays and occasionally a roan."

In 1910, the Phoenix *Arizona Republican* described the herd thus:

The ponies are very irregular in size, for they are often found as tall as the bronco and the mustang of the west, and again as small as the Shetland pony. They are as a rule weedy and inclined to be leggy, with rought [sic], uneven, sunburned coats. Light bays and sorrels predominate. A fair number of brown bays will be found, while black, white and [d]un colored ponies are exceedingly rare. ("The Wild Ponies of a Virginia Island," 1910, p. 9)

Marguerite Henry's best-selling *Misty of Chincoteague* (1947) remains very popular and is responsible for a good amount of the tourist traffic to Chincoteague each year. In 1961 the story became the successful movie *Misty*.

Misty was a real pony, born on the Beebe ranch—not on Assateague, as in the book. Henry fell in love with the week-old Misty while visiting Chincoteague and bought her from Clarence Beebe for $150. Paul and Maureen Beebe, who became characters by the same names in the book, halter-broke and gentled the pony during her stay on the Beebe ranch. When Misty was weaned, Henry had her shipped out to her home in Illinois to provide inspiration while she wrote her famous story. Although *Misty of Chincoteague* is not strictly factual, the setting is true to life and gives a fairly accurate portrayal of Pony Penning in the 1940s. Misty died in 1972 at the age of 26.

More than 60 years after the publication of *Misty of Chincoteague*, the story is as popular as ever. Visitors can see Misty and her daughter Stormy at the Museum of Chincoteague Island—stuffed! There is a bronze statue of Misty as a foal prominently displayed on Main Street, and Misty's hoofprints still grace the concrete walk in front of the Island Roxy Theater, where her movie premiered. The Beebe ranch still stands.

Each year charitable individuals purchase several foals and donate them to the fire company for re-release on Assateague as breeding stock. This custom began after the Ash Wednesday storm in March 1962, which drowned many ponies. A 1964 article in the Norfolk *Virginian-Pilot* mentioned "Yankee" stock added to the "storm-wasted" Assateague herd so that the pony penning could continue (Grey, 2014).

Pregnant Misty weathered the storm safely in the Beebes' kitchen and soon after delivered her third and final foal, aptly named Stormy, at a veterinary clinic in Pocomoke City., Md. In 1963 Stormy became the heroine of her own book: *Stormy, Misty's Foal* (1963/2007). This book was also a fictionalized account—three of the protagonists (Grandpa Clarence, Grandma Idy, and Paul) were dead by then.

The Ash Wednesday storm devastated the wild herd. Twentieth Century Fox made the movie *Misty* available for special showing along the East Coast with the proceeds used to purchase ponies previously sold at auction to replenish the herd. This tradition was eventually formalized into the Buy-Back program in the 1990s.

When the foals are taken from their mothers on auction day, they are initially frantic to reunite, and they whinny incessantly to one another. After witnessing the affection evident between the mare and her offspring, some people are upset by the forced separation. They say it is heartless to take a foal from its mother at such a young age. Indeed, in the wild, horses often nurse for a year, or even two. Domestic horses are typically weaned between 4 and 6 months of age.

Additionally, at the time of the auction, these mares are often already pregnant with next year's foal. Mares who nurse one foal while gestating the next have higher rates of spontaneous abortion. Weaning the foals in July increases the likelihood that they will deliver

Marguerite Henry's book *Misty of Chincoteague* has been loved by children since its publication in 1947. A bronze statue of Misty stands prominently on Main Street.

Although many of the events in the Misty books are fictitious, Misty was a real pony, and her hoofprints remain in the sidewalk in front of the Island Roxy theater. Misty and her foal Stormy can be seen at the Museum of Chincoteague Island—stuffed!

healthy foals the following spring (R. Keiper, personal communication, May 19, 2011). On the other hand, mares burdened with nearly continuous pregnancy and lactation tend to have shorter lives and poorer health than mares who have few pregnancies (Kirkpatrick &. Turner, 2007).

Although early weaning is associated with increased mortality in foals living wild on barrier islands, it appears to have no adverse effect on the health of foals bought at auction. Research on orphaned foals, most of them weaned at birth, has shown that early weaning presents no long term disadvantage if the foals are properly fed and socialized. In the first weeks of early weaning, foals grow more slowly than their suckling counterparts, but quickly recover lost ground (Anderson, 1995). There is no difference in size or behavior between foals weaned early, including those weaned at birth, and those allowed to suckle their dams for 4–6 months (Tateo, Maggiolino, Padalino, & Centoducati, 2013). In fact, many of the mainland-raised ponies grow far taller than their counterparts remaining on the island. Like a puppy who cries her first night away from her mother, the foals call and search at first, but quickly settle into the new routine. The youngest foals stay with their mothers until the October roundup.

On Friday, most of the remaining horses are trailered back to the North portion of the refuge, and the Cowboys herd the rest to the water's edge for a short swim to swim to the south end. Generally, by the time the herd steps out of the water on Assateague, the mares have stopped looking for their foals, although there are anecdotal accounts of mares swimming back to Chincoteague in an attempt to retrieve them (Spies, 1977).

In 1971, the Humane Society of the United States and the American Horse Protection Association protested the Chincoteague Pony Penning, claiming that it was blatantly abusive to the animals. In its November 1971 newsletter (quoted in Spies, 1977, p. 20), the organization said, "Three HSUS investigators observed the annual round-up and auction of the wild ponies at Chincoteague, Va., last summer and concluded it was the cruelest activity they had ever witnessed." The agency claimed that newborn, sick, and dehydrated horses were being forced to make the swim. They alleged that day-old foals were being taken from their mothers and loaded into the backs of station wagons for long drives to distant states. Often their feet were tied to minimize struggle (Spies, 1977). Buyers purchased nursing foals without realizing that they would need to bottle-feed them every two hours around the clock. People ignorant of proper horse care and not equipped to maintain livestock would get caught up in the excitement and rashly buy a foal because it was cute. It is not easy to raise a foal from undisciplined babyhood to become a reliable mount, even in the best of circumstances, and these unprepared non-equestrians were clearly out of their depth, much to the foals' detriment.

Ronald Rood's 1967 book *Hundred Acre Welcome* (p. 76) offers some corroboration of these claims. "Ruefully we had to admit that not a few of those ponies, bought so hastily and with so many good intentions, would probably end up as mere curiosities, pets—sort of summer romances that hung around through the winter. Worse still, if they went to homes smaller than our farm, they might become definite liabilities, finally to be sold, or given away to any taker." Rood allowed his son to impulsively bid on a young foal. After the transaction was completed, Rood wondered, "But now what do we do? How in the dickens do I get a pony out of here and back to Vermont?" (1967, p. 79). Since Rood had a commitment in Raleigh, N.C., after their stay in Chincoteague, they could not send the colt back to

A young foal resists the efforts of a volunteer fireman to load him into the motor home that stands ready to take him to his new home. July 30, 1970. Photograph courtesy of the *Baltimore Sun*.

Vermont; so a local friend recommended taking the colt with him. "In your station wagon. It will be a cinch!" (Rood, 1967, p. 80). Rood writes, "Others besides ourselves were adopting makeshift means with regard to their ponies." He described a man who put a brown-and-white filly in the back of his pickup truck. She "slipped in her struggle to keep her balance as he drove away" (1967, p. 80). Another woman removed the back seat of her car, shoved it onto the trunk, and loaded her foal into its place. "I wondered how they would make out on their trip back to New Jersey" (1967, p. 81). Rood transported his pony crated in a Volkswagen Microbus, unloading him at rest stops, tidying his stall, and dumping the dirty bedding in trash cans.

Feeding the pony was another matter. After the foal was purchased, a firefighter told Rood, "This one's smaller than most of 'em. . . . So I doubt if he's weaned from his mother very much. You might have to feed him with a bottle for a couple of weeks" (1967, p. 86). The foal rejected the bottle, but eventually accepted soaked grain. The colt found a new home on Rood's farm in Vermont as a family pet and a tourist attraction.

In the 1970s, the fire company enlarged the corrals by 30 ft/9 m. Old, sick, newborn, and pregnant mares at term were exempted from the swim. It discontinued hot branding and later implemented nondestructive freeze-branding. It sold very young foals only to those willing to take on the responsibility of bottle-feeding and supplied samples of milk replacer. The fire company also ensured that new owners transported their foals in approved conveyances. A 1973 follow-up report (in Spies, 1977, p. 20) stated, "HSUS attended the annual wild pony swim and auction in Chincoteague, Va., in July and concluded that the event has improved to the point of being completely humane."

Yet in 1992 and 1995, The Humane Society of the United States again protested the event. This time they disapproved of the "wild pony rides," traditionally held on Thursday after the auction (Willis, 1995). Unbroken wild ponies, many of them lactating mares that had just lost foals to the auction, were ridden bareback rodeo-style, bucking, wheeling, and frantic beneath courageous contestants. This practice was halted in 1996.

The fire company says that the accounts of cruelty were exaggerated or false and assert that they always try to act in the best interest of the ponies. Humane organizations made a fuss over the use of bullwhips in rounding up the horses; but on closer investigation, they found that the cowboys used whips as noisemakers, not to strike the ponies.

Sometimes natural disaster befalls the penned ponies. The fire company reports that in 2012, a lightning strike in the corral tragically killed the pregnant 2008 mare Dream Dancer, unique for her curly black-and-white coat. Similarly, in 1981 the fire company reported two mares and a colt dead from a lightning strike in the Assateague corral ("Move' Em Out," 1981, p. 4).

At every Pony Penning, the fire company, which legally owns all the ponies, inspects the new crop of foals and determines whether any exceptional colts and fillies should be returned to the herd as breeding stock. In the past, unsold foals were also returned to Assateague—but in recent years, every foal has sold easily. Each year, the fire company generously donates the purchase price of a buy-back foal to a charity, such as Ronald McDonald House or the Hospice of the Eastern Shore.

As tax-deductible contributions to the fire company, buy-backs typically fetch much higher prices than the ponies sold to leave the island, and bidding is fierce. The winning bidder chooses a name for the pony, poses for a photograph, and then signs the pony over

A mare named 15 Friends of Freckles shepherds her youngster out of harm's way as two stallions, Wild Bill (left) and the mare's sire, the half-Arabian North Star (right) move into battle.

to the fire company. For the life of the buy-back ponies and their descendants, purchasers feel a personal connection to Assateague and enjoy knowing that "their" horses remain wild on the island.

The "Buy-Back Babes," a group of women from all over the United States—Florida, Illinois, Pennsylvania, California—have collectively purchased a number of ponies for re-release on the island. "We got our name, the Buy-Back Babes, because our husbands will not let us buy one and bring it home, thus we have to buy the buy-backs!" says Jean Bonde, who has been a Buy-Back Babe since the group came together in 2002 (personal communication, July 23, 2010). Anyone can become a Buy-Back Babe by approaching a member of the group at the pony auction and contributing money toward the purchase of a turn-back foal. "Whoever donates money to buy a pony is a Buy-Back Babe," says Bonde (personal communication, July 23, 2010). "There are approximately 36 Buy-Back Babes and about 32 others interested in ponies who want to receive e-mails talking about pony happenings." Some Buy-Back Babes purchase other turn-back foals independently of the group. They also assist the Feather Fund to help children buy ponies. Once, when a woman from Connecticut bought a foal but could not afford to ship it home, the Buy-Back Babes purchased the foal from her, a chestnut named Gideon, and donated it to the Feather Fund to be given to a child.

Between Pony Pennings, the Buy-Back Babes make frequent visits to the Chincoteague NWR to observe the horses. They take photographs and share them via e-mail, discussing Mystery's new pinto foal to Spirit and how 4-year-old Prince was beginning to challenge the harem stallions for dominance. They know each animal intimately by name or nickname, habits, preferred ranges, and sometimes lineage. "We are like little girls, we get so excited when we do see ones we recognize," says Bonde (personal communication, July 23, 2010).

In 2007, the Buy-Back Babes paid a record $17,500 for Tornado's Prince of Tides, a palomino-and-white colt colored much like the legendary Misty. He was doubly prized because few stallions are returned to Assateague. "Usually the cowboys pick out the ones they want to keep, but in several instances we had picked out a few before the auction, went to the cowboys and said we wanted to bid on one particular one," says Bonde (personal communication, July 23, 2010). "These are the ones that cost us the most money because we said we would take the bids all the way to the top."

Prince reigned as the record-setting pony for 7 years. Then in 2014, Catherine Miller of Peru, Illinois, bid $21,000 for a black and white Buy-Back filly out of Leah's Bayside Angel by the stallion Sockett To Me. That year, the average price per pony was $2,772, and the fire company earned $149,700 from the pony sale.

The winter of 2009–2010 was brutal, with powerful storms and significant flooding. Four buy-back foals went missing and were presumed dead, but no bodies were found. The other three buy-backs from that year were in good health. One of the foals lost was a very special filly named Suzy Q. She was named for Buy-Back Babe Suzanne Craig, who succumbed to pancreatic cancer in 2009. Another of the missing foals was the offspring of E.T., a buy-back purchased solo by Bonde in 2002 (Boswell & Mason, 2010).

At the Maryland end, horses also died during the winter of 2009–2010, but these were all elderly. "We lost 6 horses during the winter," says Allison Turner, biological science technician at Assateague Island National Seashore (personal communication, February 17, 2011). "Three were 30 years old, and the others were in their mid-20s. It was normal seasonal mortality."

Saddened over the loss of the foals, the pony committee agreed to give the Babes a filly to replace Suzy Q. The women selected Gidget's Beach Baby, a pinto filly out of Gidget, their 2002 buy-back mare. To reduce future losses, the pony committee agreed to shelter weanling buy-back foals at the carnival grounds for their first winter, a trade-off between "wildness" and an increased chance for survival. The pony committee monitors them and ensures they have plenty of hay, grass, and water, and visitors can peer at them through the chain-link fence.

Monitoring the wild ponies is labor intensive for the fire company. Denise Bowden explains, "The Chincoteague Volunteer Fire Company's number one priority is to save lives and protect property—after that it's all about the ponies and nothing else" (personal communication, February 5, 2011).

The brawny sorrel stallion Surfer Dude apparently believes that he and his mares should have the run of the entire refuge. His band regularly escapes from their fenced rangeland at the end of Woodland Trail. They swim out into the bay and around the fence to graze on what they must see as the greener grass. "Beautiful Surfer Dude. I think more people know him than any other horse," says Bonde (personal communication, July 23, 2010). "He has two lead mares to guide him along, so they basically go where they want to go and he goes along with it." Every time they make a foray into forbidden territory, the refuge staff calls the fire company. Volunteers leave their day jobs to round up the escapees and replace them in their enclosure. In 2010, they escaped so persistently that the fire company confined them in the Beach Road corral for the duration.

Managing a wild horse herd presents numerous challenges not typically faced by volunteer fire departments. "We regard these beautiful animals as a gift from God," says Denise Bowden. "As much as they are wild animals—and believe me they are wild—they are also gentle creatures that sometimes need a little extra care when the elements get rough" (personal communication, February 5, 2011).

Dripping wet and looking self-satisfied, the ponies emerge from the water to graze on the lush marsh grass. In 2011 Kimball's Rainbow Delight (right), a daughter of North Star, was first to emerge, followed closely by Island Breeze.

Each April, as spring starts to take hold, the ponies are gathered and a veterinarian assesses the health of each pony, administers vaccinations for Eastern and Western encephalitis, tetanus, West Nile virus, and rabies, draws blood for necessary tests such as Coggins, and checks mares for pregnancy (Grey, 2014). At the July roundup, the veterinarian performs another health check, and separates the ponies too young, old or debilitated to participate in the swim. He estimates the age of the foals before they enter the auction ring, and signs health certificates for the foals that have been sold. In October, the ponies are gathered for a health assessment before winter and to be wormed with Eqvalan®, a drench dewormer. Foals that were too young for weaning in July or that were born after Pony Penning are removed from the herd and sold. In 2014, Dr. Charles Cameron, DVM, of the Eastern Shore Animal Hospital was recognized for serving the wild pony herd for 25 years.

Even though the fire company is the legal owner of all horses on the refuge, people who have purchased buy-back foals sometimes become heavily invested in their welfare and complain to the pony committee if they show any signs of compromised health. The fire company walks a very thin line, committed to increasing the health and survival of the ponies while keeping them as wild as possible. "The Chincoteague Volunteer Fire Company is doing the best we can with what we have," said Roe Terry, former public relations officer for the fire company (personal communication, July 27, 2010).

The public generally prefers that the ponies live wild and free with no interference, but also prefers that they always have adequate food, water, and health care. These two states are often mutually exclusive.

Over the centuries, the fittest—and sometimes the luckiest—ponies have survived on Assateague to contribute their genes to the next generation. On the Maryland end of Assateague, the ponies are managed as wild animals and live with minimal interference by humans. Disease is seen by the National Park Service as a normal occurrence for wild horses. The horses at the north end of the island may be weakened by a harsh winter or a dry summer, or they may stand strong until conditions improve.

Every organization charged with the responsibility of managing wild horses needs to decide how much human intervention is appropriate. Management options run the gamut from keeping hands off, as with the wild horses of Nevada, to providing food, shelter, reproduction assistance, and health care, as with the herd on Ocracoke, N.C.

Though their ancestors ran free on the island for hundreds of years, Ocracoke Banker horses have been stabled in barns, corrals, and pastures since 1959. Not one individual alive today has experienced any more freedom than the average stabled horse. In fact, when Hurricane Isabel flattened the Ocracoke pony pen in September 2003, the horses stood out in the sand-covered road and waited patiently for people to come with hay and grain. Park Service ranger Laura Michaels jokes, "They were looking at their pony watches, saying 'It's been a day, where's my food?'" (personal communication, May 22, 2010).

Western ranches such as the famed Montana Horses allowed their several hundred horses to run free in the hills from autumn to mid-spring with minimal human intrusion. The horses lived as wild until the April roundup, when they were gathered and herded 30 mi/48 km to the ranch. From there, they were leased to work at guest ranches, camps, and trail-riding stables for the summer. These ranch horses experienced a much greater degree of freedom most days of the year than the Ocracoke horses have at any point in their lives.

At the "total dependence" end of the continuum, horses cannot be considered wild. Because humans are entirely responsible for the welfare of the Ocracoke horses, they are halter-broken for safe handling and maintained at a healthy weight, with proper hoof and veterinary care. Breedings are planned for optimal genetic diversity while preserving historical Banker bloodlines. When Hurricane Isabel blew down the barn, a stronger, safer one was built. They are fed hay and grain and watered in troughs, and they enjoy grooming, scratching, and pats from their handlers. At the price of freedom, they live longer, are better nourished, and are healthier overall than their free-roaming ancestors. Rather than survival of the fittest, this herd persists through survival of the favored.

At the other end of the continuum, the Cumberland Island, Georgia, herd roams free without any interference from people. These horses live as they wish from birth to death, choose their own mates, live with parasites, find their own water sources, eat whatever the island provides, and generally run the course of any illnesses without intervention from people. They may have more health issues and often have shorter lifespans than domesticated horses. Valuable genes may be lost because a particular horse never finds a mate. But they are free.

Autonomy versus greater safety and health is a trade-off for people, too. Helmet laws, smoking restrictions, emissions testing, mandated seat belts and infant car seats set limits on our behavior, for good or ill. People want safer environments and good health, but chafe when laws restrict personal freedoms.

"The Buy Back Babes, fire department and the refuge struggle with where the line should be drawn," said Lou Hinds, former Chincoteague NWR manager (personal communication,

On the day of the auction, Chincoteague Volunteer Fire Company staff parades the foals in front of the crowd and sells them to the highest bidders. Most youngsters resist handling, rearing and leaping in an effort to break free or simply lying on the ground and refusing to budge. This colt's dished face, broad forehead, large eyes, and small muzzle speak of Arabian lineage.

May 21, 2010). "If they are truly a wild population of ponies living their lives on a barrier beach island, they should be living at the mercy of Mother Nature. If they are dependent on people taking care of all their needs, they become just another managed herd of horses."

Horses cannot tell us whether they prefer an indulged existence in domestication or unfettered autonomy with a more precarious health status. The entities that decide the fate of the horses must place each herd somewhere on the wildness continuum, and no matter what they decide there will be compromises and trade-offs. Every intervention intended to improve the health and wellbeing of the horses takes them farther from the wild end of the gamut. Keeping horses wild means minimal intervention, but failure to provide adequate food or water is usually seen as neglect and cruelty. There are no indisputably correct answers and few easy decisions.

The Chincoteague firefighters are volunteers who balance their responsibilities to the community and to the horses with day jobs and families. They maintain the fences on the

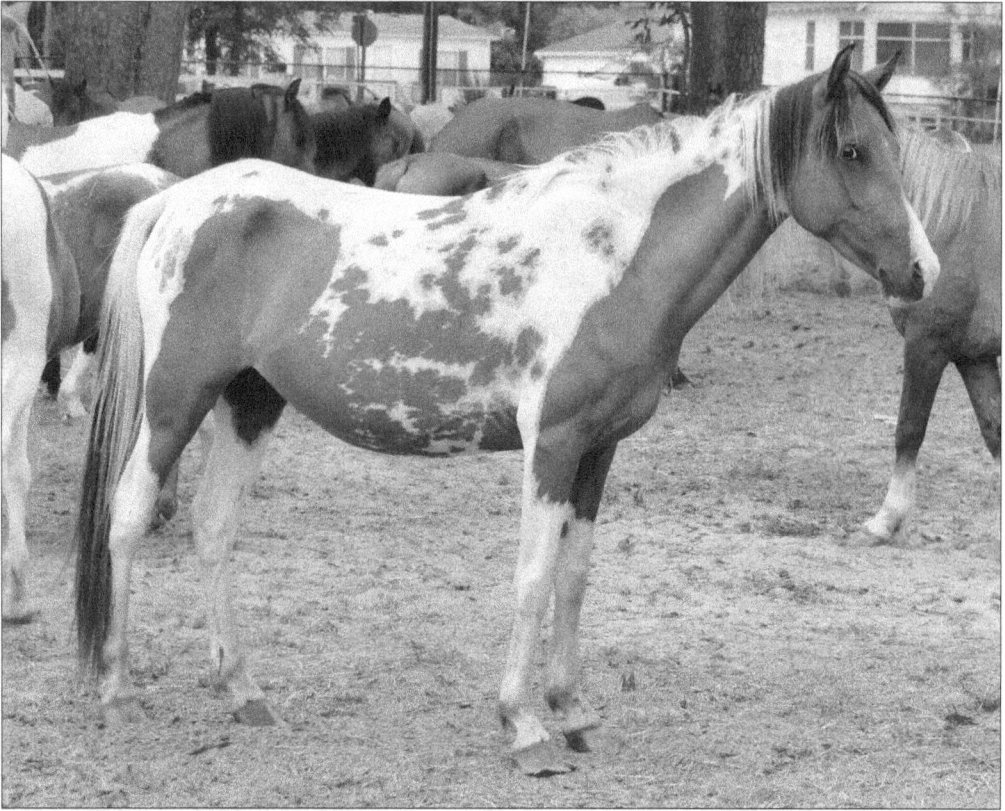

15 Friends of Freckles, a loudly-marked pinto mare, listens to her foal calling to her from the auction pen. Chincoteague Ponies tend to have good conformation and excel in disciplines from hunter-jumper to endurance and Western pleasure.

refuge and run the carnival and the events of Pony Penning week. When there is a fire, medical emergency, or a problem with the ponies—for example, when Surfer Dude's band escapes from the south enclosure—they respond with alacrity. Many, like Roe Terry, have belonged to the organization for much of their lives.

Assateague Island is fortunate to have escaped the clutches of civilization, although man has left his mark in many places. Archaeological sites on and around the island include the remains of eight shipwrecks, two of them Spanish, U.S. Lifesaving Service stations built beginning in the 1870s, a presidential yacht that sank in 1891, several abandoned villages, and 19th-century fish and salt factories (Hayward, 2007).

On the Chincoteague NWR, the buildings of the U.S. Fish and Wildlife Service and the Park Service cluster along Beach Road, leaving most of the refuge undeveloped and for the most part inaccessible to the public. At least five federally listed threatened or endangered species breed on Assateague, and protected marine animals regularly visit Assateague's offshore waters (Hayward, 2007). A number of paths and roads, including a scenic loop around an artificial pond, give the visitor an overview of the refuge's features and habitats.

The Chincoteague NWR allows up to 150 adult ponies to live on its property in two barbed wire enclosures through a special-use agreement with the Chincoteague Volunteer Fire Company, which owns and manages the herd. The actual number of ponies is often below this maximum threshold: on August 31, 2012, there were 134 ponies on the refuge; 22 stallions and 112 mares (Grey, 2014). Each November, the fire company pays the refuge $1,500 for an annual grazing permit. Although the ponies are allowed on the refuge, they are not necessarily welcome. Hinds commented, "I've been told that the Fish and Wildlife Service has made it very clear to the firemen that we prefer not to have the ponies on the refuge" (personal communication, May 21, 2010).

According to Keiper & Houpt (1984), approximately 74% of the Virginia mares foaled in any given year, typically April through June, and about 80% of the foals were removed at Pony Penning. In contrast, the foaling rate for the Maryland herd was 57% in the days before immunocontraception.

While conducting a study of compensatory reproduction in feral horses in October 1989, Turner and Kirkpatrick collected urine or fecal samples from 40 Maryland mares (10 of which were lactating) and 48 Chincoteague mares (2 of which were lactating). The team tested urine for creatinine, estrone conjugates, and progesterone metabolites and tested feces for total estrogen and progesterone metabolites. These methods have proven 100% accurate in detecting pregnancy in domestic horses. The team found no difference in abortion rate between the two herds, even though the pregnancy rate was nearly twice as high in the Chincoteague herd, and concluded that "the differential foaling rates between the 2 herds is determined by October pregnancy rates, and not by fetal loss after approximately 90–150 days postconception" (Kirkpatrick & Turner, 1991, p. 650).

The paper went on to say that other researchers had shown the critical period for pregnancy loss in domestic mares to be days 25–31, and that the 1989 research showed a fetal loss rate at 90 days similar to that reported in domestic mares after day 45 postconception. They attributed the differential in foaling rates to lactational anestrus, the suppression of ovulation in lactating mares, and speculated that when foals are weaned early, the mare is likely to come into heat and become pregnant soon after.

Equine gestation averages 340 days, usually ranging from 327 to 357 days, but pregnancies as long as 399 days have been documented (Giffin & Darling, 2007). This variation seems to have evolved to improve the survival of each foal by synchronizing reproduction with the seasons (Giffin & Darling, 2007). Mares typically come into heat a week or so after foaling. If no pregnancy results, she will return to estrus at about 30 days postpartum, and every 3 weeks thereafter until the end of the breeding season. A mare that conceives during her foal heat will typically maintain a foaling interval of about one year (McCue, 2009). Pregnancy rates are 10–20% lower when a mare is mated during her foal heat (Giffin & Darling, 2007).

Chincoteague foals are typically born in April, May, and June each year. If weaning at Pony Penning brought mares into estrus, then the following year's foals would be conceived in August and born in predominantly in July. A mare bred on August 15th would deliver a foal between July 12 and July 27 the following year. "Mares come into post-partum estrus 7–10 days after foaling, and in many cases they are successfully bred during that post-partum estrus, so some mares only have one estrus each year," says Dr. Ronald Keiper, a zoologist who has studied free-roaming barrier island horses for 40 years (personal communication, May 19, 2011). Keiper explains,

> Nursing a foal does not prevent that post-partum estrus. Mares become pregnant quickly and then for the next 11 months, they must "feed" their developing fetuses

and provide milk for lactation. That strains the health of the mare, so if she is stressed by disease or a cold, wet winter with poor-quality winter food, she may abort the developing fetus. She has invested less in that fetus than she has in the foal, so she sacrifices the fetus (personal communication, May 19, 2011).

Lucas, Raeside, and Betteridge (1991) examined estrogen levels in the feces of 154 unmanaged free-roaming mares on Sable Island, Nova Scotia, over 4 years and found that fetal loss after day 120 was 26.0% overall (yearly variation 9.6–37.3%), and yearlings abort about 70% of conceptions. Of mares greater than 2 years of age that did not foal in the spring, about half had aborted foals after 120 days' gestation. Kirkpatrick and Turner (1991) noted a comparable rate of late pregnancy loss in both the Maryland and Virginia herds and concluded that mares of both herds pregnant in October have a fetal loss rate similar to that reported for domestic mares after 45 days of conception. Lactation imposes great energy demands on the mare, particularly in the first 12 weeks. A domestic mare produces an average of 3 gallons/11.4 L of milk a day during the first 5 months of lactation, the equivalent of 3% of her body weight. On the Maryland end of Assateague in the late 1980s, lactating mares often lost weight, had low body condition scores by the end of the summer, and appeared "unable to ingest enough food for the demands of peak lactation" (Rudman & Keiper, 1991, p. 456). Malnutrition is a known cause of abortion in mares. It appears that nutritionally challenged lactating mares are more likely to miscarry foals before October.

In other words, weaning a foal during Pony Penning does not make a mare more likely to get pregnant. But if she is already pregnant from an early postpartum estrus, weaning in July makes her more likely to carry her current pregnancy to term and deliver a healthy foal the following spring. Veterinary care and supplemental feeding of the Chincoteague NWR herd may also decrease the rate of pregnancy loss.

Two decades of managed reproduction have changed the Maryland herd so that researchers are unlikely to duplicate Kirkpatrick and Turner's findings from the late 1980s. Whereas Virginia mares continue to foal predictably in mid–late spring, out-of-season births rose to 26% among their Maryland kin by 2001, under the influence of immunocontraceptives (Kirkpatrick & Turner, 2003).

The majority of Chincoteague Ponies are maintained on a parcel within the refuge that runs from north of the wildlife loop to the Virginia-Maryland line. About 805 acres are occupied by the freshwater impoundments of South Wash Flats, Old Fields, Ragged Point, and a portion of North Wash Flats (Grey, 2014). During the summer, the ponies graze on 2,695 acres/1,091 ha of lush forage. During severe storms, gates are opened to give the ponies access to the White Hills, a maritime forest growing atop an old ridge of dune, the highest ground in the refuge. An additional 704 acres/285 ha of the north range is managed during the summer as piping plover habitat to mitigate for the nesting habitat used as a parking area for the public recreational beach. In late fall the north refuge ponies use this parcel as winter habitat.

The 547-acre/221-ha section at the south end of the island supports another group of adult horses and their foals. This parcel includes Black Duck Creek and all of Black Duck Marsh and comprises 70% marsh and grasslands and 30% maritime forest. A number of natural freshwater pools usually provide drinking water, but in drought conditions, they use troughs filled by the fire company or drink brackish water (Grey, 2014).

A mare with hypocalcemia rolls in pain while her hungry young foal urges her to rise. Intravenous infuson of calcium resolved the problem for this mare. The foal was weaned to a bottle and sold at auction to a buyer willing to become his surrogate mother by feeding him every 2 hours around the clock.

Historically, about 100 horses were maintained on the north range and 50 on the south. Rising sea level, however, has rendered the south parcel increasingly boggy, and in August 2012, 21 horses lived on the south compartment, with 113 on the north (Grey, 2014).

Horses are not the only potential threat to piping plovers (*Charadrius melodus*), seabeach amaranth, (*Amaranthus pumilus*) and other endangered species. On Assateague, over-sand vehicles such as pickup trucks are allowed to drive on 16 mi/26 km of open beach (12 mi/19 km in Maryland, 4 mi/6 km in Virginia). Beach driving is a popular activity, and as many as 145 vehicles are allowed access to the Maryland over-sand route at any one time. This heavy traffic disrupts the beach surface, leads to increased sand movement, impedes the formation of new dunes, and endangers shorebird habitat (Hayward, 2007).

Whereas the Park Service works to help Assateague return to a more nearly natural state, the Fish and Wildlife Service actively manages its natural areas to make them more hospitable to desired wildlife species. Fourteen artificial freshwater ponds cover more than 2,623 acres/1,061 ha of the refuge. These impoundments create freshwater wetlands for waterfowl in winter and wading birds in summer as well as reptiles, amphibians, and other wildlife. Water is deeper in the spring and dries up over the summer, concentrating the populations of fish and eels; so when wading birds have young in the nest, food is easier to obtain. Alternately, the impoundments can be drained to provide nesting habitat for the endangered piping plover.

The ponds were dug 20–30 years ago to provide habitat for waterfowl. At that time the ocean was farther away, and the dunes were higher. Now only a series of small dunes separates the ponds from the ocean, and with Hurricane Sandy in 2012, water ripped through the beach, flooded the ponds, tore apart the pavement on Beach Road, and opened Avocet

A female yellow-rumped warbler waits helplessly in a mist net for an ornithologist to free her. Before she is allowed to fly away, she will be weighed, measured, and banded. Thousands of migrating songbirds pass through the refuge each fall, and biologists collect important data on them.

Inlet from Swan Cove to the sea. When the inlet spontaneously closed a short time later, only a narrow ribbon of beach divided the pond and the ocean.

The Chincoteague NWR tries to conserve and protect wildlife, giving priority to threatened and endangered species, and it is a breeding ground and sanctuary for songbirds and migratory waterfowl. The diversity of habitats on the refuge—beach, dunes, salt marshes, freshwater wetlands, maritime forests, estuaries, and ocean—appeals to a wide range of wildlife.

These habitats make Assateague a desirable stopover for birds migrating along the Atlantic Flyway. Spring and fall see a procession of species from songbirds to raptors and abundant waterfowl, some resident species and some passing through. The National Audubon Society designated the Virginia end of Assateague a Global Important Bird Area (National Audubon Society, 2007). The seashore is a primary feeding area for shorebirds, and the birds in turn are food for migrating peregrine falcons.

Four species of sea turtle swim the offshore waters, and loggerhead (*Caretta caretta*) and green sea turtles (*Chelonia mydas*) occasionally nest on the sandy beaches. The ocean off Assateague is also home to six species of baleen whales, five of which are endangered; 16 species of toothed whales and dolphins; four species of seals: and occasionally the endangered West Indian manatee (*Trichechus manatus*).

Right whales winter off Assateague and are occasionally spotted from the beach. The Atlantic northern right whale (*Eubalaena glacialis*) is highly endangered, with an estimated North Atlantic population of only 200 (Hayward, 2007).

The endangered Delmarva fox squirrel (*Sciurus niger cinereus*) was introduced to Chincoteague NWR in an attempt to increase its population. Once common from New Jersey and Pennsylvania to the Delmarva Peninsula, by the 1920s it was extinct in all states except Maryland. Between 1968 and 1971, 30 of these squirrels were released on the refuge to form a successful breeding colony.

Species considered detrimental to the mission of the Fish and Wildlife Service are controlled or eliminated, including native raccoons and introduced red foxes and mute swans. The three key unwanted species that compete at the table with the island's wanted wildlife are exotic sika deer, non-migratory nuisance Canada geese, and reintroduced Chincoteague Ponies.(L. Hinds, personal communication, May 21, 2010). The refuge keeps these species under tight control—the ponies through annual removal, the deer through hunts, and the geese through egg addling.

To "addle" an egg is to remove it from the nest, kill the embryo without changing the appearance or texture of the egg, then return it to the nest so that the mother will continue to incubate it. If the mother believes the egg will not hatch, she will begin laying again. A common method endorsed by the Humane Society of the United States is to suffocate the embryo by painting the shell with mineral oil.

Shrub thickets provide nesting habitat for songbirds and a windbreak for wildlife. The White Hills stabilize the island and provide habitat and high ground for raccoons, ponies, deer, and other animals.

Where Assateague Island is widest, high ground develops and habitats diversify maximally. Where the island is narrow, most notably at the north end, overwash and salt spray reduce diversity of plant communities, but provide optimal habitat for seabeach amaranth and piping plovers. Assateague widens from north to south, and therefore the Chincoteague NWR supports a greater variety of habitats (Hayward, 2007)

According to Hinds (personal communication, May 21, 2010), there is good reason to believe that Assateague will not be one long island over the next 100 years, but an archipelago, divided by a series of new inlets. This is a reasonable prediction because geological processes have torn Assateague asunder, then made it whole again several times just since European contact. Hinds said,

> It would not be responsible or sustainable for the federal government to attempt to prevent the division of the island, or to repair it if it should happen ... we at the refuge are following the lead of the Park Service; we are just letting Mother Nature to do her thing. We are not building dunes or replacing dunes. We are allowing the island to migrate to wherever it wants to go.

The nor'easter of November 2009, which included the remains of Hurricane Ida, was a lesson in the futility of trying to restrain the forces of nature. Locals say that Assateague had not seen a storm of that magnitude since the early 1990s. It washed away the parking lot at Toms Cove, removed artificial dunes, and kicked up about 40 tons/36 metric tons of debris onto the shore.

Every time a storm wreaks chaos with the parking area, the Park Service restores it. Between 2002 and 2012 the parking lot was replaced four times, each episode draining between $200,000 and $700,000 of taxpayer money (*Proposed Comprehensive Conservation Plan*, 2012a). Even small storms cause considerable damage, while significant hurricanes such as Irene in August 2011 narrow the beach and render the parking area unusable.

The 2009 nor'easter defiantly disarticulated an artificial reef built in the 1970s and spat nearly 2,000 tires onto the beach to become a cleanup headache for the Park Service and the refuge. When offshore sediment deposition and loss of shell bottom decreased the diversity and abundance of fish, in the 1950s private organizations developed tire reefs to provide habitat for marine life. Th¬e state of Virginia took over reef management in the 1970s, and it makes these structures not only from tires, but also from trees, defunct automobiles, brush, rock, concrete, steel, cable, retired New York City subway cars, and old ships. For the most part the reefs hold fast, but between storm action and the natural movement of barrier islands and offshore sediments, disrupted reefs can wreak havoc with Assateague beaches.

The Chincoteague Refuge is one of the top five most visited National Wildlife Refuges in America, with peak visitation between Memorial Day and Labor Day. The town of Chincoteague and Accomack County's major tourist attractions are the Assateague Island recreational beach, the ponies, and the Refuge (Grey, 2014). The beach lot, now unpaved, has continued to provide parking space for 961 vehicles. Plenty of beach remained for recreational activities such as kite flying, Frisbee tossing, swimming, sunbathing, strolling, and shelling. But for how long?

The National Wildlife Refuge System Administration Act (1966) mandates that each refuge must develop a comprehensive conservation plan (CCP) to guide long-term management of resources, conservation efforts, and public uses and uphold the mission of the refuge system. This plan should be developed with input from the local community and the public at large (*Proposed Comprehensive Conservation Plan*, 2012a). Chincoteague NWR implemented its last such plan in the early 1990s.

In December 2011, the Fish and Wildlife Service and Park Service met with representatives from the town of Chincoteague, Accomack County, the National Aeronautics and Space Administration's Wallops Island facility, the state of Virginia, the Accomack-Northampton Planning District Commission, and the Volpe National Transportation Systems Center. Much of the discussion centered on the recreational beach and its parking facilities, off-site beach parking, and an alternative public transportation system. The group fine-tuned the three proposed alternatives.

Alternative A, a "status quo" option had the Fish and Wildlife Service managing the refuge essentially as it has been run since 1992, with beach access and parking unchanged until the ocean swept it away (USFWS, 2012, p. 1). Alternative B offered compromises such as relocating the recreational beach 1.5 mi/2.4 km to the north on a more stable part of the island near Snow Goose Pool. A and B both includes maintaining the current 961 beach parking spaces. Alternative C prioritized habitat and wildlife management over public use and access, reduced the number of number of parking spaces to 480, and relocated the recreational beach and parking 1.5 mi/2.4 km north of the current beach.

The refuge also considered implementing a voluntary shuttle system to move visitors from off-site locations to the beach and refuge, allowing them to access the beach when the parking lot is full or unusable due to storm damage (USFWS, 2012). The Maddox family of Chincoteague evidently hoped to sell the refuge about 200 acres of land, including the Maddox Family Campground, for off-site beach parking and a staging area for a shuttle system. The price tag for land acquisition alone in 2012 was $7.5 million (*Proposed Comprehensive Conservation Plan*, 2012a).

Wearing thick winter coats, these foals spent their first winter well fed and protected from the elements at the Chincoteague fairground. Though this confinement temporarily removed them from freedom, it increased the likelihood of their survival.

Chincoteague Mayor John Tarr and other local elected officials and business owners have expressed fear that restriction of beach access will cause the economic collapse of the town (*Proposed Comprehensive Conservation Plan*, 2012b, 2012c). Chincoteague is a gateway community that feeds, lodges, and entertains 1.5 million visitors to the refuge each year.

Visitation to the refuge climbs annually, but the number of resident watermen and their families is dropping. The population of Chincoteague increased from 3,572 to 4,317 between 1990 and 2000, but dropped to 2,941 (-32%) by 2010 (Bonetti, 2013). Much of the town is empty in the off season—there are almost three times as many housing units as households, 40% of which are made up of people 65 or older. Many of those older residents are retirees from elsewhere. A lot of natives have evidently left the island or turned their former residences into rental properties. Tourism is the economic mainstay of Chincoteague now.

In 2010, the town of Chincoteague conducted a beach access survey of more than 11,500 visitors (Town of Chincoteague, 2010). Its findings confirmed the fears expressed by some critics. Eighty-two percent of respondents reported that they visited Chincoteague primarily to go to the Assateague beach and that they would not return if the only beach access were by shuttle from an off-island parking area. Many local business owners feared that the Fish and Wildlife Service will develop a parking lot on Chincoteague, then fail to maintain onsite beach parking because of environmental and budget concerns (*Proposed Comprehensive Conservation Plan*, 2012b, 2012c).

Moreover, Assateague Island offers the only public beach accessible by automobile in the roughly 100 mi/160 km of coastline between Ocean City, Md., and Virginia Beach, Va. Virtually all of this shorefront is managed by the Fish and Wildlife Service as habitat for waterfowl and other wildlife (*Help preserve access*, n.d.). Other national parks limit vehicles to public buses. Before Zion National Park in Utah implemented a system of 30 propane-powered shuttle buses, 5,000 private vehicles drove in the park daily. Public transportation has eliminated more than 13,000 tons of greenhouse gas emissions annually. At Acadia National Park in Maine, a propane-powered shuttle bus has eliminated more than 800,000 vehicle trips since 1999 (Rennicke, 2007).

"We wish to maintain the existing [beach] access and the existing [management] plan as much as possible," said Denise Bowden, whose family has lived on the island for generations (personal communication, February 5, 2011). "We realize the importance of Assateague Island as not only a wildlife refuge but also as a major source of income for Chincoteague Island, and the Eastern Shore as well."

The CCP includes a vision of how the pony herd will be managed 25, 50, 75, and 100 years from now in relation to sea-level rise, wildlife management, and other concerns (Grey, 2014). This plan is a cooperative effort between the refuge and the fire company, making decisions that will influence future generations of firefighters, park managers, and ponies.

Chincoteaguers live in intimacy with nature's rhythms: the seasons, the tides, the storms and calms, the sowing and growing, the reaping and resting. The firemen have always held to traditional practices, managing the pony herd by drawing from the wisdom of their ancestors to solve problems. They pen ponies in the spring, summer, and fall. In July, they sell the spring foals at auction. The seasons flow from year to year in a comfortable sameness.

Even though the fire company will work with the Fish and Wildlife Service to develop a strategy for pony management, the CCP will impose restrictions on these traditional practices. Local people are concerned about sustaining their traditional way of life and the culture of Chincoteague Island. Unfortunately, traditional lifestyles are often at odds with an evolving world.

Evidence for global climate change is all around us. The National Academy of Sciences reports that over the past century, the earth's surface temperature has risen by 1.8F°/1C°). The most credible sources predict a global temperature rise 3.2–7.2 F°/1.8– 4.0C° by 2100 (Doney et al., 2012; National Oceanic and Atmospheric Administration, 2012, August 21). The Intergovernmental Panel on Climate Change *Special Report on Emission Scenarios* estimates a possible range of increase of 2.0–11.5 F°/1.1–6.4C°, depending on the amount of greenhouse gas emissions (U.S. Environmental Protection Agency, 2013). NOAA reported that in August 2012, the global temperature exceeded the 20th century average for the 36th consecutive August and 330th consecutive month (NOAA, 2012, September). Most of this increase has occurred over the past two decades. Montana's Glacier National Park has seen 73% of the area once covered with glacial ice turn to naked rock. By 2030, all of the glaciers in the park may be gone (Rennicke, 2007).

Cahill et al. (2012) argue that rising temperatures can cause decline and extinction long before heat itself becomes directly lethal to a species. Among the indirect effects of warming are not only rising sea level, but also changes in salinity, precipitation patterns, vegetation, and fire frequency; harm to beneficial species; benefits to harmful species; and "temporal mismatch between interacting species" (p. 2). As an example of the last-named effect, shorter winters may awaken hibernating animals while plants in their diet that respond to light, not temperature, are still dormant.

When asked what he thought was the most important issue facing the refuge, Hinds answered without hesitation, "Sea level rise." He continued,

Chincoteague Ponies receive veterinary inspections in April, July, and October and are treated for illness and injury when necessary. The veterinarian places a sticker on the back of each horse upon vaccination. The buckskin mare escorting her foal is Poco Latte, the smallest adult pony in the herd.

For the last 10,000 years, give or take, sea level has been rising. We have enjoyed a very stable period where the rise has been slow for the last 3 or 4 thousand years, but it *is* rising. If the scientists are correct, that rate of sea level rise will increase dramatically over the next 100 years. And that's what we are planning for right now. (Personal communication, May 21, 2010)

Over the past 3 million years, world average sea level has fluctuated by more than 300 ft/91 m (Pilkey, Rice, & Neal, 2004). Over the past 400,000 years, sea level has fallen when glaciers advanced and risen when glaciers melted. About 20,000 years ago, the coast was approximately 62 mi/100 km east of its current location because much of the world's water was locked in the glaciers—the sea-level increase we see today is part of a trend that started at about that time (Maryland Department of Natural Resources, 2005). Sea level rose rapidly until about 6,000 years ago, submerging an older line of islands and pushing sediment deposits up to form the present-day Eastern Seaboard barrier islands (Graham, 2007). About 3,000 years ago, coastal features struck a balance with the ocean: as the sea level rises and the coastline continues to retreat westward; the barrier islands migrate landward; and the islands, wetlands, lagoons, and other features maintain roughly the same relationships to one another in the absence of interference from humans (Graham, 2007).

Assateague Island is no more than 46 ft/14 m above sea level at its highest point, and it is extremely vulnerable to changes in sea level. Since Assateague came into existence, oceans have risen about 33 ft/10 m around the world. Global sea level has risen about 7.1 in/18 cm in the last 100 years (Pendleton, Williams, & Thieler, 2004). In response, the island has gradually migrated up the continental slope through overwash and accretion.

Park Service staff collaborated with the U.S. Geological Survey in 2004 to estimate the sea-level rise and its effects on Assateague. Climate models predict a further rise of 18.9 in/48 cm by 2100, which is more than double the rate of rise for the 20th century (Pendleton, Williams, & Thieler, 2004). According to the 2007 National Parks Conservation Association assessment, sea level is rising 0.124 in/3.15 mm annually, but may increase by a factor of 2–5 in the next century, enough to overwhelm the island's ability to adjust (Hayward, 2007). Interestingly, the Atlantic coast from Cape Hatteras to Massachusetts appears to experience sea-level rise 3–4 times higher than the global average, possibly due to oceanic currents, plate tectonics, and gravitational changes influenced by icemelt (Sallenger, Doran, & Howd, 2012).

Barrier islands respond to accelerated sea-level rise by becoming narrower. Thinning allows for frequent cross-island overwash that permits these islands to migrate quickly and efficiently in step with sea-level rise (Pilkey et al., 2004). With climate change comes increasingly violent storms, which accelerate overwash, erosion, inlet formation, and other processes. Farther south, the narrower parts of North Carolina's Outer Banks could disappear entirely (N.C. Office of Conservation, Planning, and Community Affairs, 2010; Pippin, 2005).

Rising sea level affects all aspects of the Chincoteague NWR—loss of wildlife habitat, shoreline erosion, inundation of wetlands, saltwater intrusion into estuaries and freshwater aquifers, and damage to cultural and historic resources and infrastructure (Pendleton et al., 2004). As climate change raises the water temperature, algae blooms may increase in estuarine waters, decreasing water quality and stressing aquatic grasses and fish communities (Hayward, 2007).

Hinds said (personal communication, May 21, 2010),

> At first, people will want to fight Mother Nature, putting up barriers against sea level rise, because they will want to protect their homes, their business, or their way of life. They don't want to change the way they have always lived, but change is inevitable and nature will have her way in the end.

Using a computerized Sea Level Affecting Marsh Model (SLAMM) analysis, Delissa Padilla Nieves (2009) showed that if sea level rose 1 m by the year 2100, 57% of the salt marsh would be swallowed by the sea, including most of the grazing area within the southern compartment. Grazing appears to accelerate this loss by reducing or eliminating the accumulation of detritus necessary for salt marsh root systems to keep pace with rising sea levels.

Grazed marshes have a lower profile and may be more prone to flooding as sea level rises (Taggart, 2008). A study involving feral horse grazing in the Rachel Carson National Estuarine Research Reserve in North Carolina observed that grazed marshes showed sparser growth of *Spartina* grasses, with decreased cover, blade length, and seed production, and a decrease in sediment buildup. Chincoteague NWR is using a projection of 3.3 ft/1 m of sea-level rise over the next 100 years. That figure has dramatic implications for wildlife management and visitor services. Hinds said that within the next 50 years, the south corral area will be too wet to support ponies, and large portions of the present marsh will be open water.

In fact, after a series of nor'easters slammed the East Coast in January, 2011, a number of ponies were moved from the flooded pasture to the north end of the refuge. The north end is higher, contains more acreage, and has more varied terrain. Although that area may also be

Egrets rest on the backs of ponies in the marshland of the south pasture. This lowland has grown progressively wetter over the years.

reduced by sea-level rise, it will probably remain adequate for pony habitat. If rising sea level significantly decreases the size of the island, Hinds said that the number of Chincoteague Ponies permitted in the refuge will need to decrease correspondingly to remain in balance with the ecosystem (personal communication, May 21, 2010). The Fish and Wildlife Service recommends, but does not demand, that the fire company maintain the herd at 134 adult ponies or fewer until the year 2023, when stakeholders will consider updated climate-change data (Grey, 2014).

Currently, the refuge uses intentional disturbances within the impoundments to enhance the habitat for migrating birds. Wintering waterfowl benefit from the cutting of emergent vegetation—the plants that grow in water but extend above the surface. Resource managers sometimes mow, disc with tractors, or set fires to reduce vegetation and allow sunlight to reach the ground.

In the future, refuge managers may enlist the help of willing conspirators—the Chincoteague Ponies—to create such disturbances (L. Hinds, personal communication, May 21, 2010). Utilizing Chincoteague Ponies in the management of impoundments would reduce the use of staff, tractors, and fuel (Grey, 2014). Grazing ponies could clear away undergrowth and punch in seed with their hooves. Their waste, broken down by invertebrates and microorganisms, could help sustain the small creatures on which shorebirds feed.

Research has validated the benefits of using horse grazing to manage estuarine habitats. In managing habitat for waterfowl, grazing encourages a diverse assortment of plants and reduces woody vegetation in moist soil areas (Vavra, 2005). Planned rest periods between intervals of grazing allow tall grass cover to reestablish itself for waterfowl nesting (Vavra, 2005). Land that has been grazed develops a patchy herb layer that is desirable habitat for some birds. Despite its inflexible stance against grazing, elsewhere in the country, the Fish and Wildlife Service has advocated grazing to create desirable habitat for Attwater's prairie chickens (Vavra, 2005). And in the Virginia mountains, the U.S. Forest Service introduced Shetland ponies in 1974 to reduce overgrowth on Mt. Rogers and the Grayson Highlands.

There is good evidence that grazing by horses on barrier islands increases the population of small crabs, which in turn attract diverse species of waterfowl (Levin, Ellis, Petrik, & Hay, 2002). Wigeon, gadwall, and pochard feed on submerged plants that grow better when the marsh is grazed. Horse grazing in the Camargue of France has proved useful for management of marshes for waterfowl by opening up the emergent vegetation, especially where the water level is controlled by the manager (Duncan & D'Herbes, 1982).

The invasive *Phragmites* reed is nutritious and palatable to horses, and grazing pressures weaken its unwanted foothold on estuarine margins. When *Phragmites* shoots are bitten off below water level, the plant often starts to decompose and may die. Under heavy grazing *Phragmites* declines quickly in height and density (Duncan & D'Herbes, 1982). *Phragmites* grows tall and dense in ungrazed exclosures, but it is low and sparse in heavily grazed areas. Herbicides and regular grazing may be the only things that can kill *Phragmites*, and the latter is much less destructive.

On the Chincoteague Refuge, the ponies are abundant and not at all shy of visitors, but they are confined behind barbed wire, which keeps them off the pavement and away from people. Visitors can view ponies with binoculars from the observation platform on the Woodland Trail or across the roadside fences, but the animals are often distant.

The federal agencies charged with managing feral horses on Assateague have philosophical differences regarding interaction between the American public and wild equids. The Fish and Wildlife Service claims it puts distance between people and ponies on the Chincoteague refuge "to keep the wildness in the animals" (L. Hinds, personal communication, May 21, 2010), an unusual stance for an agency that does not classify them as wildlife. More likely, the separation is imposed to prevent injuries and lawsuits—and perhaps to prevent the public from forming stronger attachments to the ponies and advocating for their retention.

The Park Service allows horses to remain in certain parks, such as Cape Hatteras NS, as a cultural resource, and manages them as wildlife on Assateague Island, Cape Lookout, and Cumberland Island National Seashores, in Theodore Roosevelt National Park, and at other sites even though it officially classifies them as feral livestock. While prohibiting direct contact, Assateague National Seashore and Assateague State Park on the Maryland end of the island allow ponies to roam freely into areas where people congregate.

Visitors can see the northern part of the Chincoteague NWR by walking the 8-mi/12.8-km service road or by taking a guided bus tour operated during the summer months by the Chincoteague Natural History Association, a nonprofit organization that works with the Fish and Wildlife Service in providing educational opportunities. A number of Chincoteague watermen offer scenic boat cruises to watch ponies and other wildlife on Assateague at close range.

People tapped Assateague's resources long before any records were kept. Paleo-Indians probably foraged along the East Coast and its islands for thousands of years, but we have little evidence of their presence, nor are we likely to find much useful information about them. Sea level has risen considerably since these early inhabitants took up residence. The land where they lived was swallowed by the ocean, by westward-moving estuaries, or by the migrating barrier chain, and in some cases may be 200 mi/322 km out to sea on

After the annual July Pony Penning, the south-end horses swim back to Assateague and regroup on their home ranges.

the continental shelf. Archaic tools are sometimes found in the sand dredged from off-shore bars and piped onto Ocean City beaches, but out of context they reveal few insights (Langley & Jordan, 2007).

Before Europeans arrived, various native tribes of the Algonquian linguistic family used the island seasonally for hunting and fishing. The Chincoteague Indians lived not on Chincoteague, but on the mainland, alongside a number of native tribes who named themselves after the creeks on which they lived, or vice versa. Popular tradition and nearly every guidebook describing the island hold that *Chincoteague* translates to "beautiful land across the water." Kirk Mariner writes (2003, p.4) that it actually means "large stream or inlet." Similarly, Dunbar (1958) notes a possible etymological connection between *Chincoteague* and *Ginguite Creek* on the northern Outer Banks. The 1651/1667 Farrer map, below, shows "Cingoto" as a specific island.

The Chincoteague Indians lived in settlements on the mainland and migrated seasonally to barrier islands to make use of food sources. Their villages were located where they were sheltered from storms, above the reach of the tides, and near freshwater. These natives visited Chincoteague and Assateague islands to forage for shellfish, fish, and game and to collect shells used to make beads, which they used as currency.

In the early 1500s, Spain explored the East Coast as far north as South Carolina while England had explored at least as far south as Nova Scotia (Mariner, 2003). The first recorded European to set foot on Maryland or Virginia was an Italian named Giovanni da Verrazzano, who sailed for the King Francis I of France in 1524 aboard a ship named *La Dauphine*. Some historians believe that he may have landed on or near Assateague Island. Others place his landing at about 10 mi/16 km north of Cape Charles. He named his find "Arcadia," which he described as "beautiful and full of the largest forests" (Covington, 1915, p. 205).

His was not a welcome landing. Verrazzano wrote in his letter to the king that the first person his party encountered was a man, "handsome, nude, with hair fastened back in a knot, of olive color," who extended a burning stick to the party "as if to offer us fire" in what was probably a friendly gesture (Covington, 1915, p. 207). Verrazzano's crew responded with fire of its own. "We made fire with powder and flint-and-steel, and he trembled all over with terror, and we fired a shot. He stopped as if astonished and prayed, worshipping like a monk, lifting his finger toward the sky, and pointing to the ship and sea, he appeared to bless us" (Covington, 1915, p. 207).

European explorers often abducted native children so that they might learn the language of their captors, convert to Christianity, and act as interpreters and mediators in converting their people (Langley & Jordan, 2007). English colonists also left their own boys in Indian villages to learn the native language (Langley, Van Driessche, & Charles, 2009). Verrazzano describes successfully kidnapping an 8-year-old native boy to take back to France. He wrote of his unsuccessful attempt to abduct a teenage girl "of much beauty and tall of stature, but it was not possible on account of the very great cries which she uttered for us to conduct her to the sea" (Wise, 1911, p. 6). The Spanish abducted children as well, taking a boy from the Potomac River area and another from the Eastern Shore in 1588 (Langley et al., 2009).

Verrazzano describes the natives as lighter-skinned than those he had previously seen. They clothed themselves in Spanish moss (which can still be found hanging from cypress trees along the Pocomoke River), hunted, fished from dugout canoes, and ate wild peas. Their arrows were made from reeds with tips fashioned from bone.

There were about 2,000 Native Americans on the Eastern Shore in 1608, including the Pocomoke, Annamessex, Manokin, Nassawattex, Acquintica, Assateague, Chincoteague, and Kickotank (Langley & Jordan, 2007). These tribes are represented in the historical record as "timid, harmless, kind-hearted people" (Upshur, 1901, p. 91). It is unlikely that any of these tribes lived on the barrier islands year-'round because of insufficient game, poor soil, and horrendous storms (Langley & Jordan, 2007).

In September 1649, Henry Norwood and 329 other colonists left England bound for Jamestown aboard the *Virginia Merchant*. After three wretched months at sea, Norwood and a dozen of the sickest men and women went ashore on a Delmarva barrier island to find drinking water. They were abandoned on the isolated beach in the dead of winter when their ship left without them! Historians dispute the location where Norwood was stranded, but evidence suggests that it was Assateague or Assawoman Island (Mariner, 2003).

They had little ammunition, but managed to feed themselves on oysters and game birds. When Norwood sent his cousin to look for friendly Indians, he realized that he and his party were on a barrier island far from the mainland villages. One by one, members of the group died of cold, exposure, and hunger, and the rest survived by turning to cannibalism (Langley & Jordan, 2007). Certainly if horses or other livestock had been on the island when Norwood arrived, his party would have eaten them rather than one another, or at least he would have mentioned them. Wild game was also scarce and elusive—Verrazzano had commented that "the animals in these regions are wilder than in Europe from being continually molested by the hunters," and deer were uncommon (Langley et al., 2009). Norwood was about to try swimming to the mainland to seek help from the local tribes when on the ninth or 10th day, they were rescued by Kickotanks.

The Indians showed them great hospitality and dispatched a messenger who brought back Jenkin Price, a fur trader, to accompany them to the nearest English settlement. They traveled on foot for a day and lodged with the Chincoteague tribe overnight in their home

This map, drawn in 1651 by John Farrer and updated by his daughter Virginia, depicts Chincoteague ("Cingoto Ile") but not Assateague, places the Pacific Ocean just to the west of the Appalachians, and suggests a Northwest Passage through the Hudson River (Farrer & Farrer, 1667). Courtesy of the Library of Congress, Geography and Map Division.

on the mainland, and Norwood woke in the morning to find the chief's daughter sharing his bed, a gesture of hospitality from his hosts. During their stay with the natives, Norwood and company feasted on oysters, deer, duck, goose, curlew, and swan—a veritable bounty of game and seafood obtained from the same environs where Norwood's party nearly starved. "The shore swarmed with fowl," he wrote, and the Indians told him that wolves "'did greatly abound in that island'" (Covington, 1915, p. 213).

When Europeans began to settle the Choptank River watershed on the Eastern Shore of Maryland, only about 2% of the region was cleared by the local Native people; 92%–94% of the landscape was forest and about 6% wetlands (Benitez & Fisher, 2004). By setting periodic forest fires, the Natives maintained open woodland with minimal underbrush to obstruct their passage. Native people typically lived in rectangular homes made of perishable local materials with the capacity to house about 20 individuals. When the group relocated seasonally to access food sources or moved to till new ground, their belongings went with them, leaving little enduring to mark where the village once stood.

Initially, interactions between settlers and the Native people were amicable, centering on trading for corn in Virginia and for furs in Maryland. The Eastern Shore tribes did not participate in the Jamestown massacre of March 1622, so relations between settlers and natives remained guarded but friendly.

Settlers obtained Indian lands through purchase, trade, or patenting apparently abandoned lands, ignoring laws passed in 1652, 1654, and 1658 to regulate and limit acquisition of land by Europeans (Langley & Jordan, 2007). By the 1650s, natives were losing rights to residential lands as well as to hunting grounds and areas used for agriculture.

As Europeans grabbed more territory, relations with the natives soured. When rumors circulated that the natives were plotting to poison the wells of the settlers, Edmund Scarborough, infamous for his fanatical hatred of all Indians, repeatedly incited conflict against them, a penchant that led to his prosecution in Jamestown (Langley & Jordan, 2007). In 1651, he led a brutal attack on the Pocomoke tribe, which served to increase resentment toward whites and strengthen cohesion among tribes.

In 1659, backed by the governor of Virginia, but not that of Maryland, Scarborough descended on the peaceful Assateague tribe with 300 men and 60 horses. The Assateagues could not have been numerous, because they were described as "harder to find than to conquer" (Langley & Jordan, 2007, p. 19). His intention was to ensure that the tribe "may neither plant corne, hunt, or fish, soe make him poore and famish him" (Mariner, 2003, p. 10). Later called the Seaside War of 1659, this encounter weakened the tribe, and thereafter it shrank in population and in territory. In 1660, the Indians of Accomack complained that they had lost so much land that they could not survive.

The Delmarva Indians may already have been in decline by the time permanent settlers arrived from Europe. In the summer of 1608, John Smith reported that the Accomack had recently suffered a "strange mortalitie" (1624, p. 55). Some researchers have suggested that they succumbed to a European disease, perhaps smallpox or influenza, from earlier contacts (Mires, 1994).

Conflict, disease, and flight removed many Indians from the landscape. Assimilation, poor documentation, lost or destroyed papers, and the advantages of passing for white removed many of the survivors from the historical record. The final blow to many Virginia Indians may have been the Racial Integrity Act of 1924, a result of the eugenics movement, which reduced the list of officially recognized racial identities to two: white and colored. The Act mandated that officials record the race of every person at birth using the "one drop" rule—anyone with African or Native American ancestry, no matter how remote, was classed as colored, and as such was excluded from white schools and forbidden to marry a white. The First Families of Virginia, a lineage society comprising descendants of the original Virginia colonists, including Pocahontas, protested. The "Pocahontas exemption" granted white status to anyone less than ⅛ Indian.

The boundary between Maryland and Virginia was established in 1668, and within a few decades European settlers had claimed "every foot of the territory from the Western shore of Smiths Island, in the Chesapeake bay, to the Eastern shore of Assateague island, on the Atlantic Ocean, on the whole course of the divisional line run by Calvert and Scarbrough" (Virginia Commission on Boundary Lines, 1873, p. 133).

In 1667, a sailor with smallpox wandered into an Indian village and succeeded in spreading the disease all along the Eastern Shore, devastating the native population. The native tribes believed this sailor had been sent among them by the whites to kill them. Whether or not this was true, the effect was the same. (There is a satisfying irony in reports that Scarborough himself also died of smallpox.)

Wise (1911, p. 66) wrote,

> The peaceful Indians of the Eastern Shore, among whom the first colonists of the peninsula had settled, had greatly diminished by the end of the 17th century, and

Following the July Pony Swim, Saltwater Cowboys drive the ponies though Chincoteague to the fairground. Herd instinct takes over, and the ponies follow one another in a sinuous mass.

the dying out of the Savages was followed by the arrival of Negroes in large numbers, of whom up to that time there had been but few.

He also describes the first horse on Virginia's Eastern Shore as

one conveyed to Colonel Argoll Yeardley by George Ludlow of the Western Shore
by a bill of sale dated January 30th 1642. None of the many inventories on record
prior to that date includes horses. . . . In 1645 Stephen Charlton also owned a horse
and in November of that year a consignment of horses arrived from New England,
many of the animals having died on the passage south. The custom of branding
stock was begun at this time. (1911, p. 307)

Wise considered it unlikely that the wild horse herds of Assateague had arrived by shipwreck ahead of the colonists, commenting that if the horses had been on these islands, early
settlers would have made use of them. He writes (1911, p. 308),

When Chincoteague Island was first prospected and granted to one of the colonists
in 1670 by James II no mention of horses occurs. Again while Colonel Norwood,
who was shipwrecked on the nearby coast and spent some time in the neighborhood as the guest of the hospitable Kickotanke chieftain, mentions the presence
of large numbers of hogs in the marshes near Gingo Teague, he does not mention
horses. Colonel Norwood passed right by the island in 1649 and would certainly
have mentioned the wild horses had they been there at that time. It has also been
said that the wild ponies which rove in great herds over the Accomac island owe
their origin to horses left there by pirates in the early days but this too is doubtful . . .

the number of horses in the colony in 1631 was very small and prior to 1649 references in the records of Virginia to horses are exceedingly rare.

Horses were scarce in the earliest Virginia settlements. John Smith wrote that in the fall of 1609 nearly 500 colonists at Jamestown had "fiue or sixe hundred Swine; as many Hennes and Chickens; some Goats; some sheepe," but only "six Mares and a Horse" (1624, p. 93). Although recent scholarship suggests that salty drinking water (made saltier by prolonged drought) and poor sanitation may have killed many people (Cohen, 2011), hunger was severe and drove a few to cannibalism. After the winter of 1610, the infamous "starving time," just 60 colonists remained alive. Livestock fared worse. "As for our Hogs, Hens, Goats, Sheepe, Horse, or what liued," Smith wrote, "our commanders, officers & Saluages daily consumed them . . . till all was deuoured" (p. 105).

The suffering and desperation of the settlers were evident in George Percy's account:

> Haveinge fedd [upon] horses and other beastes as longe as they Lasted, we weare gladd to make shifte w[i]th vermin as doggs Catts Ratts and myce . . . to eate Bootes shoes or any other leather. . . . notheinge was Spared to mainteyne Lyfe and to doe those things w[hi]ch seame incredible, as to digge upp deade corpes outt of graves and to eate them. . . . one of our Colline murdered his wyfe Ripped the Childe outt of her woambe and threwe itt into the River and after Chopped the Mother in pieces and sallted her for his foode. (Percy, 1625, p. 7)

The meat from butchered livestock meant survival for individual colonists, but to survive as a colony, Jamestown needed breeding stock. The horses were replaced in the spring of 1611, when Sir Thomas Dale brought 17 stallions and mares (Anderson, 2002). In 1612, Virginia's deputy governor, Sir Thomas Gates, made it a capital crime to kill "any Bull, Cow, Calfe, Mare, Horse, Colt, Goate, Swine, Cocke, Henne, Chicken, Dogge, Turkie, or any tame Cattel, or Poultry, of what condition soever" without permission (Anderson, 2002, p. 381).

In October 1613 Captain Samuel Argall (who had kidnapped the famed Pocahontas in April of that year) sailed 500 miles to Port Royal, Nova Scotia, destroyed the French settlement, and absconded with whatever he could use, including a number of horses (Harrison, 1927). Sources differ on how these horses had reached Canada. One says that "In 1604, M. L'Escarbot, a French lawyer, brought several horses to Acadia. From this stock sprung the famous Canada ponies, which, owing to the bitter climate in which they live, do not represent the size of their Norman ancestors, but are still the knottiest, naughtiest, hardiest little creatures of their kind in the world" (Washburn, 1877, p. 379). Others say that King Henry IV of France had supplied them and that they were probably similar to the horses sent to Quebec in 1665 by Louis XIV from his royal stables in Normandy and Brittany. The Quebec horses were to become the foundation stock for the rugged Canadian Horse breed, carrying Spanish, Barb, and Arabian bloodlines infused with the draft blood of the Breton and the French Norman (Lynghaug, 2009).

The horses Argall brought to Jamestown were probably representative of the finest French bloodlines of the time. This importation is another means by which Spanish genes might have reached the horses of colonial Virginia without direct introduction by the Spanish, who tended to take southern routes to the New World. It is unclear, however, whether any of the Canadian horses survived to contribute their genes to the Virginia colonial stock.

Until 1624, the Virginia Company was the principal source for imported livestock to the colonies. Ralph Hamor noted the presence of horses in his *True Discourse on the Present*

Mares and foals snack on marsh grasses after swimming to Chincoteague. Before the early 20th century, there were no spotted horses in the herd. Today, pintos are very popular with visitors and typically fetch the highest prices at auction.

State of Virginia (1615, cited in Wallace, 1897, p. 109), "The colony is already furnished with two hundred neat cattle, infinite hogs in herds all over the woods, some mares, horses and colts, poultry, etc." A census in 1620, though, counted only eleven horses in the colony. The Virginia Company sent out 20 English mares later that year to bolster the breeding herd (Harrison, 1927). Sir Francis Wyatt reported in 1626 that more horses were needed for military purposes and for the manufacture of pitch and tar (Harrison, 1927). The equine population was to grow very slowly—almost 30 years later, there were still only about 200 horses in Virginia.

The colonists of the Chesapeake Bay region initially imported all the livestock typically found on English farms, but over time the barnyard census shifted to reflect local conditions. Sheep were impractical because of predation, so wool and mutton were scarce until the end of the 17th century, when many predators had been eradicated (Anderson, 2002). Historian and novelist Edward Eggleston points out another obstacle to sheep raising:

> The only domestic animal that did not multiply to excess in the wild pastures of America was the sheep, which had for deadly foes the American wolf and the English woolen manufacturer.... There was nothing that English legislation of the time sought more persistently than the development of the English woolen trade.... the importation of a sheep for the improvement of the colonial breed was punishable with the amputation of the right hand. (1884, pp. 445–446)

After a week of stressful roundups, handling, separation from foals, and the proximity of people, ponies release pent-up anxieties when returned to the refuge, galloping and kicking up their heels.

Encouraged by generous bounties, stockmen eliminated predators, but were otherwise inattentive to their free-range flocks. "The negligent methods prevalent in a new country bore more hardly on sheep than on other animals, and it was estimated that about one-third of all the sheep in the northern colonies perished in a single hard winter, a little before the middle of the eighteenth century" (Eggleston, 1884, p. 446).

Following the lead of the Indians, farmers raised tobacco and corn largely with hoes, not plows; so few oxen were needed, and horses were generally not used for plowing.

Between 1637 and 1777, Virginia colonists practiced "open-woodlands husbandry," allowing animals to range freely. The major cash crop was tobacco, and by allowing the livestock to range freely in the woods and fields the settlers reaped maximal animal production with minimal labor, freeing the men to work in labor-intensive tobacco cultivation. Colonists also used the woods as a way to avoid paying taxes on animals, a problem addressed by the Virginia statutes in 1646: "such persons who have concealed the number of their persons tithable, lands, horses, mares & c. shall for every tithable person, lands, & c. pay double the rate that this present Grand Assembly hath assessed" (Hening, 1823, vol. 1, p. 329).

Unfortunately the free-roaming animals did not differentiate between what they were supposed to eat and what was forbidden to them, and they often destroyed crops. In 1643, the Virginia Assembly ruled that colonists had to fence in their crops, not their animals. Three years later, Assembly required crop fencing to be 4.5 ft/1.4 m high (subsequently increased to 5 ft/1.5 m) and "substantiall close downe to the bottome" (Hening, 1823, vol. 1, p. 332). In other words, the barrier was to be tall enough that horses were unlikely to jump it and low enough to thwart rooting pigs. Those who did not enclose their crops would "plant, upon theire owne perill" (Hening, 1823, vol. 1, p. 199).

The landscape of the English colonies functioned as a vast grazing commons for unsupervised livestock, with fenced patches of tobacco and corn surrounded by large open meadows and forest. For the next 250 years, fencing laws in many states followed this model; cultivators were responsible for fencing their crops and had no recourse if free-roaming livestock invaded their fields (Hayter, 1963).

In 1649, there were 200 documented horses in the colony, 20,000 head of cattle, 50 asses, 3,000 sheep, 5,000 goats, "innumerable" swine, and poultry "without number" (Farrer, 1649, p. 3). In the late 1660s, the Assembly first repealed its prohibition of exporting horses, then forbade their importation "to restraine the numerous increase of horses now rather growing burthensome then any way advantagious to the country" (Hening, 1823, vol. 2, pp. 267, 271).

Farrer (1649, p. 3) describes Virginia horses as "of an excellent raise [race]." Thomas Glover (1676, cited in Harrison, 1927, p. 332) pronounced Virginia's horses to be "as good as we have in England." John Clayton, parson of Jamestown, wrote to the Royal Society (1688, cited in Harrison, p. 332),

> There are good store of Horses, though they are very negligent and careless about the breed. It is true there is a law that no Horse shall be kept stoned [uncastrated] under a certain size, but it is not put in Execution. Such as they are there are good stock, and as cheap or cheaper than in England, worth about five pounds apiece. They never shoe them nor stable them in general.

In *Un Français en Virginie* (1687, cited in Harrison, 1927, p. 332), Durand commented,

> I do not believe there are better horses in the world, or worse treated. All the care they take of them at the end of a journey is to unsaddle, feed a little Indian corn and so, all covered with sweat, drive them out into the woods, where they eat what they can find, even though it is freezing.

In the late 1600s colonial cattle and horses alike were denied shelter even in extremes of weather and were expected to forage for themselves in all seasons. Sometimes they survived a harsh winter by eating the bark from trees. Perhaps as a result, English livestock apparently matured to a smaller size when raised free-range in the Chesapeake area. Wild stallions stole mares in heat or made forays into the plantations to breed with them. Colonists complained that random mating "'doth both Lessen & spoyle the whole breed and Streyne of all horses'" (Anderson, 2002, p. 403).

Nobody really knew how many domestic animals roamed Virginia during colonial times. Robert Beverley wrote in the early 1700s, "Hogs swarm like Vermine upon the Earth, and . . . find their own Support in the Woods, without any Care of the Owner" (1705, p. 81). He added (p. 76), "the wild Horses are so swift, that 'tis difficult to catch them; and when they are taken . . . they are so sullen, that they can't be tam'd." To increase horses' stature, in 1686 the Virginia legislature forbade planters to turn into the woods "any ston'd [uncastrated] horse two years old or more and under thirteen and one-half hands" (Harrison, 1927, p. 332).

As the number of English planters increased, so did the number of livestock roaming the country and destroying the unfenced crops of local Indians. Frustrated by shrinking hunting grounds and the depredations of ravenous livestock, the Assateague tribe staged a minor and fruitless raid on white settlements in 1659 (Laing, 1959). Surviving records do not clarify how this event was related to the Seaside War of the same year.

By 1669, there were so many semi-wild horses in Virginia that further importation was prohibited and stallions were to be gelded if caught (Wise, 1911). Planters even shipped surplus animals to Barbados and New England (Anderson, 2002). Laws were passed requiring horse owners to reimburse farmers for damaged crops and fences, but there were many horses—some without owners—and crops were under constant assault.

Through the early 1700s, horse owners sometimes employed various devices to keep horses close to home and out of the fields. A horse lock, a piece of wood fastened between the fore and hind legs of the horse, prevented the animal from galloping or leaping (Peck, 2008). Hobbles were created by tying the horse's neck to a hind fetlock. Some owners used a modified yoke with a hook on the bottom that would catch on the top rail of a fence if a horse tried to jump it (Peck, 2008).

When tobacco prices collapsed in the late 1600s, the settlers compensated by growing wheat and keeping sheep (Peck, 2008). Free-range horses became such a nuisance, young men hunted them with dogs for sport—though it appears the object was capturing, not killing, the horses. A planter could shoot horses found depredating his crops. In 1662, a tax was levied on horses, and owners were required to keep them confined from July 20 until October 20 (Bruce, 1907). As marauding livestock grew more vexing, laws were passed demanding that all livestock be fenced, and anyone who captured unbranded free-roaming livestock gained legal ownership of them.

As early as 1657, Virginia colonists circumvented the maze of disputes, laws, and taxes and avoided the expensive trouble of fencing by turning their stock loose on nearby islands (Dunbar, 1958). In the late 1600s, wealthy landowners of the Eastern Shore acquired many islands on the coast and in Chesapeake Bay for this purpose; Saxis, Watts, Tangier, Smith, Hog, Assateague, Chincoteague, Parramore, Metompkin, and Wallops islands were all put to use grazing hogs, goats, sheep, cattle, and horses by the end of the 17th century (Mariner, 2003). Owners singly or jointly put their livestock on necks and barrier islands to forage without endangering crops and survive at minimal expense (Mariner, 2003).

Captain Daniel Jenifer held the first patents on both Chincoteague Island (1671) and the Virginia part of Assateague Island (1687) (Langley & Jordan, 2007). The Maryland part of Assateague was not patented until the 1700s.

In *Harper's Monthly* novelist, professor, and children's author Maude Radford Warren wrote,

> Some of the islanders vigorously oppose the tradition that Chincoteague was originally settled by convicts, but the evidence tends in that direction. In the old days a planter was allowed fifty acres of land for each settler he introduced. In 1687 Captain Daniel Jenifer brought over a number of convicts, perhaps seven, perhaps thirty-five, and in return Chincoteague and Assoteague [sic] were patented to him. Twice the patent of Chincoteague lapsed, but finally, in 1692, twenty-five hundred acres of the lower half were given to John Robbins and twenty-five hundred of the upper half to Col. William Kendall, and from these two men almost all the people now on the island got their titles (1913, p. 775)

Thomas Welburn and his four employees were the first white settlers on Chincoteague Island, arriving in 1680 to "seat" the island by building a house and planting tobacco, apple trees, and corn (Mariner, 2003). Welburn's tenant, Robert Scott, was probably the first white man to live on the island—for the year required by law to cement Welburn's claim to the island. When Edward Hammond visited Chincoteague in 1681, he found that the crops had gone wild, and the local natives were enjoying the use of both the house and the corn.

Records are vague, but livestock seems to have been present on Assateague since the late 1600s. If the present-day horses of Assateague descend from this population, they can boast more than three centuries of continuous occupancy (Mariner, 2003). By 1684 the Welburn claim was considered abandoned, and it wound up in the possession of William Kendall, an affluent and influential slave-owner. Welburn, believing himself the rightful owner of Chincoteague Island, threatened to shoot Kendall if he dared "seat" there himself. Nonetheless, Kendall went to the island with Major John Robins and Thomas Eyre and erected "a small house . . . about ten foot long like ye roof of a house upon ye ground" (Mariner, 2003, p. 13). Welburn fought the claim in court, but in 1691, the general court awarded half of the island to Kendall and half to Major Robins. The dividing line was near modern-day Church Street. Chincoteague was on its way to becoming a town.

Chincoteague had little arable land, but like most of the barrier islands of the Eastern Shore, it was ideally suited for raising livestock. Horses, cattle, goats, hogs, and sheep roamed freely in the late 1600s. Robins's will, recorded May 28, 1709, bequeathed to his sons his land on "Jingoteague island . . . where my man and woman George and Hannah Blake look after the stock" (Mariner, 2003, p. 18). Another heir received a horse from the livestock grazed on "Gingoteague Island," indicating that horses roamed the island at that time (Mariner, 2003, p. 18). (*Chincoteague* is spelled three different ways in this document alone—spellings of English and Indian names did not become fixed until much later.)

Through the 1700s, the names of the wealthy absentee landowners appear on record, but it was the tenants that lived and worked on the island, and their names were often lost to history. Oral tradition maintains that early settlers often intermarried with Native Americans. Mariner describes simple homes—one-story windowless frame buildings with sand floors and a hole cut in the ceiling to let smoke escape from the primitive fireplace—but some of these dwellings may have been seasonal (Mariner, 2003). Light was provided by fish oil burned in clam shells.

Warren wrote about a later period, the mid-19th century, and offered a different description of local architecture:

> In those early days they had log houses a story and a half in height, boarded outside, plastered inside, and supported on great cedar blocks. Most of the houses had great hearths which would hold logs as large as a man, and fine brick chimneys; the poorer people, however, had "andiron" chimneys made of lime and laths. In 1840 there were about five hundred people living in twenty-six houses. They did not build more, for in those times the young people would "win away" to Delaware and Pennsylvania. (1913, p. 782)

DeVincent Hayes and Bennett (2000, p. 69) published an old photograph of Kendall Jester's sturdy residence, purportedly constructed in 1727. It had front steps and a masonry chimney, and it was substantial enough to form the nucleus of the Lighthouse Inn, which stood till the 1960s.

In 1776, the Virginia Convention ordered livestock removed from the Eastern Shore islands to prevent British ships from raiding them as a food source. At that time, Chincoteague was home to 20 families plus horses, cattle, and about 400 sheep; more livestock ran free on Assateague. The islanders petitioned to keep the livestock on the islands, stating that they had themselves organized a militia and a guard of 30–40 men stationed in the area and would bear the loss themselves if they could not protect their livestock from "'small cruising vessels of the enemy'" (Force, 1846, p. 1563). The Convention rescinded its order.

Dr. Thompson Holmes owned a mainland plantation on Chincoteague Bay and built a home on the island in 1811. Almost 25 years later, as he departed from the island, he wrote in eloquent detail about the local horses and their habitat (1835, p. 418):

> Assateague and Chincoteague islands are flat, sandy and soft, producing abundance of excellent grass, upon which they become very fat during the summer and autumn, notwithstanding the annoyance of flies, with which those islands frequently abound. . . .
>
> Their winter subsistence was supplied abundantly by nature. The tall, dense, and heavy grass of the rich flat lands, affording them green food nearly the whole winter, the tops of which alone were killed by the frosts, mild, as usual, so near the ocean. They never suffered for provender, except in very deep snows, with a crust upon the top, or when high tides were immediately succeeded by intense cold, which covered the marsh pastures with ice, both of which accidents were of rare occurrence, and very transient in their duration.

He added (1835, p. 419),

> They are hardy, rarely affected with the diseases to which the horse is subject, perform a great deal of labor, if proportioned to their strength, require much less grain than common horses, live long, and are, many of them, delightful for the saddle. I have a beautiful island pony, who for fifteen years has been my riding nag in the neighborhood and upon the farm, who has given to my daughters their first lessons in equestrian exercise, and has carried us all many thousands of miles in pleasure and safety, without having once tripped or stumbled; and he is now as elastic in his gait, and juvenile in his appearance, as he was the first day I backed him, and is fatter than any horse I own, though his labor is equal, with less than two-thirds of their grain consumption. His eye still retains its good natured animation, and to one unskilled in the indications of a horse's teeth, he would pass readily for six or seven years old. My regrets at parting with this noble little animal, are those of the friend.

Nearly 200 years ago, this educated observer saw nothing unusual about these horses except their diminutive size and greater stamina. He supports the idea that all Delmarva horses—island and mainland—originated from the same colonial stock and discredits the shipwreck hypothesis:

> In regard to the origin of the race of our insular horses, there is no specific difference between them and those of the main land: the smaller size and superior hardihood of the former are entirely accidental, produced by penury of sustenance through the winter, occasional scarcity of water, continual exposure to the inclemency of the seasons, and the careless practice of permitting promiscuous copulation among them, without regard to quality. (Holmes, 1835, p. 418)

He concluded that little effort would be needed to bring these horses up to their ancestors' stature:

> The largest and finest work-steers of the Eastern Shore, are raised upon these islands, without any expenditure for winter support; a proof that horses of full size, might also be reared there, with judicious attention to the breed, proper selection of

A mare reassures her frightened foal that he is safe, even as her own nervousness shows in her expression. Tomorrow, however, the fire company will abruptly wean him and separate him from his mother forever. While the first day or two apart is stressful for both, all foals must leave their mothers at some time.

stallions, and care to provide water. No other attention is necessary, except to watch the winds and weather about the periods of the equinoxes, when desolating tides are threatened, and to drive the stock upon high grounds, secure against inundation. (Holmes, 1835, p. 419)

Two decades later, Charles Lanman (1856, p. 235) wrote of Assateague,

From time immemorial it has been famous for its luxuriant grass, and from the period of the Revolution down to the year 1800, supplied an immense number of wild horses with food. When the animals were first introduced upon the island has not been ascertained, but it is said that they were most abundant half a century ago. At that period there was a kind of stock company in existence, composed principally of the wealthier planters residing on the main shore. The animals were of the pony breed, but generally beautifully formed and very fleet, of a deep black color, and with remarkably long tails and manes. They lived and multiplied on the island without the least care from the hand of man, and though feeding entirely on the grass of the salt meadows, they were in good condition throughout the year. They were employed by their owners, to a considerable extent, for purposes of agriculture, but the finer specimens were kept or disposed of as pets for the use of ladies and children. The prices which they commanded on the island varied from ten to twenty dollars [about $250–500 today (Friedman, n.d.)], but by the time a handsome animal could reach New York or New Orleans, he was likely to command one hundred and fifty or two hundred dollars [roughly $3,800–5,000].

These ponies were solid-colored—bays, blacks, and sorrels. An account from 1897 says, "They are about thirteen hands high, uniform in shape, and resemble each other except in color, for all colors prevail" (Wallace, p. 111). A comfortable pacing gait (legs moving in lateral pairs rather than the diagonal pairs of the trot) or a four-beat amble was desirable in Colonial days. Horses were chiefly used to carry a rider long distances along unimproved trails. Hugh Jones (1724/1865, p. 49) wrote that the Virginia horse "will pace naturally and pleasantly at a prodigious rate." Like their brethren on the mainland, some of the horses of Chincoteague and Assateague also paced.

A "gaited" horse naturally performs a smooth four-beat amble, running walk, or single-foot gait. Modern horses such as the Paso Fino, Peruvian Paso, Tennessee Walking Horse, Rocky Mountain Horse, and Missouri Fox Trotter are gaited, and this characteristic is common in breeds with old Spanish lineage. Spanish Jennets apparently also carried the genes for the two-beat pace, as demonstrated by modern Standardbreds and the extinct Narragansett Pacers.

In the Chincoteague herd, pacing and ambling were more common in the 18th and early 19th century than today, probably because of a higher proportion of Spanish blood in previous centuries. Strong infusions of Arabian and lesser influences of Morgan ("Nevada ponies imported," 1977, p. 5) and Shetland over the last 90-odd years have refined the build of the ponies and decreased the proportion of horses with a natural amble. Mustangs have also been added to the Chincoteague Refuge herd—some of them gaited and carrying Spanish bloodlines, others with lineage tracing to Thoroughbreds, Arabians, and American saddle and draft breeds.

DNA analysis confirms that the Assateague horses have old Spanish heritage. Blood samples taken in the late 1980s revealed shared characteristics with the Paso Fino breed, which descended from animals brought to the New World by the Spanish (Goodloe, Warren, Cothran, Bratton, & Trembicki, 1991). The genes of the Assateague herd also closely resemble "cold bloods" such as draft horses and ponies, perhaps a lingering genetic contribution from the Shetland Ponies introduced to promote pinto coloration (Goodloe, et al., 1991).

These little horses have long been renowned for their endurance and stamina. Skinner (1843, p. 26) wrote,

> There has been, since long before the American Revolution, on the islands along the sea-board of Maryland and Virginia, a race of very small, compact, hardy horses, usually called beach-horses.... They run wild throughout the year, and are never fed. When the snow sometimes covers the ground for a few days in winter, they dig through it in search of food. They are very diminutive, but many of them are of perfect symmetry and extraordinary powers of action and endurance. The Hon. H. A. W[ise]. of Accomac, has been heard to say that he knew one of these beach-horses, which served as pony and hack for the boys of one family, for several generations;

and another that could trot his 15 miles within the hour, and was yet so small that a tall man might straddle him, and with his toes touch the ground on each side [perhaps an exaggeration]. He spoke of another that he believes could have trotted 30 miles in two hours. As an instance of their innate horror of slavery, he mentions the fact of a herd of them once breaking indignantly from a pen into which they had been trapped, for the purpose of being marked and otherwise cruelly mutilated; and rather than submit to their pursuers, they swam off at once into the wide expanse of the ocean, preferring a watery grave, to a life of ignominious celibacy and subjugation!

Legends of Spanish shipwrecks that brought ponies to Assateague Island tend to be vague and produce confusion. John Wallace (1897), for example, acknowledged that the horses had occupied the island "more than a hundred years" (p. 111), "probably two hundred years" (p. 10), or "from time immemorial" (p. 111). Although he conceded that "The traditions relating to their origin are very hazy and improbable," he added, "the most reasonable one, because it is within the range of possibilities, is that a Spanish ship was wrecked off this part of the coast and the original ponies were on board and swam ashore" (p. 111). As late as 1900, locals called Popes Island and Popes Bay Spanish Point and Spanish Bar, claiming that in the 1500s a Spanish galleon wrecked there (Langley et al., 2009).

Locals have perpetuated the Spanish legend, though it appears that over 130 years ago, they told a version that featured the wreck of an English vessel. In an 1877 article about Chincoteague in *Scribner's Monthly*, Howard Pyle (p. 737) wrote,

When the first settlers came there, early in the eighteenth century, they found the animals already roaming wild about its piney meadows. The tradition received by the Indians of the main-land was that a vessel loaded with horses, sailing to one of the Elizabethan settlements of Virginia, was wrecked upon the southern point of the island where the horses escaped, while the whites were rescued by the then-friendly Indians and carried to the mainland, whence they found their way to some of the early settlements. The horses, left to themselves upon their new territory, became entirely wild, and probably, through hardships endured, degenerated into a peculiar breed of ponies.

It appears that Pyle confused Chincoteague with Assateague. He was also confused about the "Elizabethan settlements," which were established on and around Roanoke Island, N.C. None of the 16th-century English colonists who left a record reached the Outer Banks from the north or explored the Delmarva Peninsula.

Other dubious stories abound. For example, in a Sunday feature for the *Baltimore Sun*, Donald F. Stewart, director of the Baltimore Maritime Museum, described a Spanish ship, the *San Lorenzo*, wrecked off Assateague in 1820 *en route* to Spain with a cargo that included "3,973 gold doubloons; 173,700 silver pieces of eight; 255 bars of gold; 303 bars of silver; plus a statue of the Madonna and a baptistery, also of solid gold" (Stewart, 1977, p. 17). The shipwreck purportedly carried 95 (or 110) ponies, blinded so they would better accept working in a mine, and released some of them onto Assateague. Langley et al. (2009, p. 72) write, "John Amrhein, Jr. (2007, 187) disputes the authenticity of the existence of this vessel and the associated stories (p. 27) and alleges it was a fiction created by Donald Stewart and believes it is a myth perpetrated through repetition. This needs to be considered seriously as a possibility."

The *San Lorenzo* story has many problems. If the ship had been bound *for* the mines of Latin America, it might have carried horses, but not treasure, and it would not have passed near Assateague. Spanish ships took horses to the New World in the early days of exploration and settlement, but by the early to mid-1500s Spain was breeding large numbers of fine horses in the Caribbean and on the mainland, enough to support the conquest of two continents.

Moreover, Spain did not have any miniature pony breeds to send to its colonies. Sponenberg states that several breeds in the *jaca serrana* group of Spanish horses were used for pit ponies, and may have influenced several of the British breeds over centuries, beginning with Roman times (personal communication, June 5, 2014). The Shetland Pony, a native of Scotland, was the preferred breed for "pit ponies," and most of these worked the British mines. Pit ponies were almost exclusively stallions and geldings. "The usage is due solely to the fact that you cannot, in such a limited space as a coal mine, have mares and stallions working alongside one another without trouble and loss of work" (Nova Scotia, House of Assembly, 1892). Once employed in a mine, the ponies usually spent their entire lives there. Chronic lung disease often shortened their lives and those of their human handlers. Pit ponies were not blinded, but a persistent myth holds that working underground causes them to eventually lose their sight.

Although ships bound *to* Spain from South American mines could have carried treasure, they would not have carried a cargo of maimed miniature horses back to Spain, and the trade routes would not have taken ships bound for Europe anywhere near Assateague. A wreck in the early 1800s would not have provided the foundation stock for the Assateague herd; free-roaming horses had been there for well over a century by then. Any horse restrained in the hold of a ship would have been unlikely to survive a shipwreck. Blind horses would have had virtually no chance at all, though Stewart insisted with apparent seriousness, "blind animals have a greater sense of direction than animals with normal vision" (1977, p. 17). Thompson Holmes and other 19th-century observers who wrote extensively about local livestock did not mention the presence of blind horses on Assateague. Finally, Dayna Aldridge of Historic Ships in Baltimore (formerly the Baltimore Maritime Museum) admitted no knowledge of the *San Lorenzo* and explained, "our collections . . . are limited to information related to the four ships and lighthouse in our care, thus it is extremely unlikely that we would have any information" (personal communication, January 25, 2012).

At the time of the supposed wreck of the *San Lorenzo*, horses grazed not only on Assateague, but also on many of the other barrier islands of the Eastern Shore; and they ran wild throughout Delmarva and on the mainland. Although it is possible that at least some horses did arrive on Assateague by way of shipwreck(s), just about all of the livestock grazing on necks and islands along the Atlantic coast was apparently placed there by stockmen, and Assateague is probably no different.

Whether these legends are true or not, ships often succumbed to offshore sandbars, bad weather, illusory inlets, and other hazards. Wrecks kept the U.S. Lifesaving Service, established in the 1870s, busy for decades. From the colonial period through the 1960s, at least 156 ships wrecked within the boundaries of Assateague NS. Another 100 or more, including Spanish ships, may have met their fate there or nearby; and wind and currents may have moved parts of offshore wrecks onshore (Langley et al., 2009). Blackbeard and other pirates are believed to have sailed the waters surrounding Assateague and used the

Ponies return to Assateague in July to resume a wild existence until October, when the Saltwater Cowboys will gather them for inspection and routine care.

island's secluded bays and uncharted inlets as hideouts from the Royal Navy and colonial law enforcement (Hayward, 2007).

Coins from many nations have been found along the beach at the north end of Assateague, including 18th- and 19th-century Spanish silver pieces (Voynick, 1984). Spanish treasure ships or pirate vessels that preyed on them may have run aground and spilled their riches on or near Assateague, but Spanish coins found on the dunes do not necessarily denote even a brief Spanish presence. The British Empire prohibited its colonists from minting their own money, so they often used Spanish coins acquired directly, through illegal trade with Latin America, and indirectly from nearly everywhere else. Spanish dollars (pieces of eight) and other currency were accepted around the world and virtually inescapable. After the Revolutionary War, the newly independent United States based its coinage not on the British pound, but on the Spanish dollar. The first silver dollars that it minted in 1794 were the same size and weight as the latter, and pieces of eight remained legal tender in this country until 1857.

Spanish coins found on Assateague may have come from wrecked Spanish treasure ships, but such vessels riding the westerlies and the Gulf Stream to Spain rarely strayed so far north. Pirates may have stashed the coins, but professional buccaneers were unlikely to hide their loot on a changeable, featureless barrier island. It seems more likely that these coins came from the more mundane wreckage of merchant ships.

The decaying bones of numerous ships have been identified over the years, but new finds keep surfacing. In 2004, an archaeological study found four previously unidentified shipwrecks within the boundary of the national seashore, including the remains of the USS *Despatch*, a 174-ft/53-m wooden-hull steamer that served as the first presidential yacht. Presidents Hayes, Garfield, Cleveland, and Benjamin Harrison used it before it sank in a storm on October 10, 1891. It is probably eligible for the National Register of Historic Places (Hayward, 2007).

Islanders profited from the salvage of shipwrecks, which the residents of Maryland's barrier islands regarded as gifts from the sea or from God (Langley & Jordan, 2007). They collected food, furniture, and other items from the wrecks and kept, bartered, or sold them. Many made a very good living from the numerous wrecks, and villages became efficient at dismantling ships that the ocean had sent to them. This became a serious headache for officials in both Maryland and Virginia, who had difficulty enforcing the law on remote Assateague.

In 1799, the Maryland General Assembly appointed a wreck master to control shipwreck sites. He was authorized to command local constables and captains of vessels in the area to cooperate or be fined. Anyone caught plundering a wrecked ship could be sentenced to death. Soon wreck looting was under control and no longer a desirable way for an islander to earn a living, but opportunities for legitimate salvage remained.

From 1522, Spain provided armed escorts for its treasure ships to protect them from pirates and discourage mutineers. Nearly every year from 1566 through 1790, two fortified convoys left Seville (later Cadiz) for New World ports. The *flota* sailed in spring for the Antilles and Mexico; the *galeones*, in summer for South America and Panama. The fleets met at Havana the following year, headed up the Eastern Seaboard to the approximate latitude of Bermuda or the Carolina capes, then steered northeast to take advantage of the Gulf Stream and the westerlies. By the late 1500s some convoys had 100 or more vessels.

Armed escorts protected treasure fleets from hostile vessels, but they remained vulnerable to navigational hazards and bad weather. Storms sank the whole fleet in 1622 and 1715 and caused major losses in 1554 and 1733.

In 1750, a hurricane devastated seven Spanish treasure ships that sailed recklessly into British colonial waters, still hostile despite the formal end of decades of conflict. These ships carried a variety of treasure and goods, as well as a number of distinguished travelers, including the president of Santo Domingo, the viceroy of Mexico, and the governor of Havana. The cargos are well documented and included European prisoners, sugar, medicinal plants, cotton, vanilla, cacao, tobacco, indigo and other dyes, seedlings, hides, gold, silver, mahogany, copper, and large sums of money, for example, about 400,000 pieces of eight whose silver content alone is worth $10 million or more today (Shomette, 2008).

La Galga (*Greyhound*) probably ran aground on Assateague. Most on board survived, but looters and the churning sea took much of the cargo. *Los Godos*, evidently the only ship carrying livestock, limped into Chesapeake Bay with *San Pedro*, and most of their remaining occupants and cargo found alternate passage to Spain. *Nuestra Señora de Guadalupe* reached Ocracoke, N.C., and set off events that may have inspired Robert Louis Stevenson's Treasure Island (Amrhein, 2007). The other ships wrecked on the Outer Banks or disappeared.

In 1998, a commercial salvage company, Sea Hunt, located two Spanish wrecks believed to be *La Galga* and *Juno* (1802) and recovered more than 100 artifacts, including coins, anchors, and timber. Spain claimed that it had never abandoned these wrecks and after a legal battle won ownership of both. The Spanish government later allowed the Park Service to exhibit some artifacts at Assateague Island NS (*Spain loans artifacts*, 2007).

Amrhein writes in *The Hidden Galleon* (2007) that he believes one of these wrecks was misidentified, and the true remains of *La Galga* rest in Chincoteague NWR. Amrhein asserts that *La Galga*, *en route* to Spain laden with horses and other valuables, ran aground in a small inlet, long since closed and forgotten. He believes that he has found its remains

The corrals bring mature stallions into close proximity, and dramatic battles often flare. Witch Doctor (left) and half-Arabian Copper Moose (right) were so intent on combat, the water troughs were no obstacle. Copper Moose was a 1996 Buyback, claiming the highest bid in the auction at $5,000.

in what is now an Assateague marsh, and he has appealed to the federal government to have the wreck recognized as a national historic site. The refuge now displays a scale model of *La Galga* in its visitor center. Was *La Galga* one of the wrecks found by the salvage company, or does it lie unexcavated in a marsh? And do the Assateague horses descend from its equine cargo?

We may never know for certain whether the first horses on Assateague arrived by way of shipwreck, but they almost certainly did not originate with the wreck of *La Galga*. First, by all indications, *La Galga* was not carrying horses (Shomette, 2008). Colonists used Assateague for grazing horses and other livestock from the late 1600s (*Assateague Island, Nature and Science*, 2010). If *La Galga* were really carrying undocumented horses back to Spain—which would have been highly unusual—and if these horses managed to swim to shore when *La Galga* wrecked, they would have found numerous other horses already on the island. Amrhein disagrees, and proposes that the earlier livestock was destroyed by a violent hurricane in 1749 that swept Assateague Island (Grey, 2014).

Escaping from such a ship would have been difficult. Spanish galleons were large and very sturdy. They were made for warfare and carried formidable armament. Horses in European ships were secured in solid compartments deep within the ship, typically 3 ft x 7 ft/0.9 m x 2.1 m, able to withstand the struggles of a panicked horse and the impact of its body in rough seas. Horses were tethered with ropes short enough to prevent entanglement in stalls too small to allow turning around or backing up. Often their front legs were tied together.

Slings were at times secured around the chest with a breastplate to prevent the animal from falling forward in the stall and breeching across the hindquarters to keep him from falling backwards. When the sea was fair, these hammocks allowed a horse to rest his feet, as the stall was too small to allow him to lie down (Hayes, 1902).

Baron Robert Baden-Powell, the founder of Scouting, was a young captain in the British cavalry when he advised (1885, p. 78),

> The hammock should be slung under the horse's belly loosely, not so tight as to raise him. The breast band and breeching should be securely fastened but not tightly. The object of the sling hammock is to enable the horse to rest himself without actually lying down; but this should only be permitted in fine, calm weather. When the ship is rolling, the hammock should hang quite loose below the horse, so that he will not throw his weight into it, because if he did so he would swing about with every motion of the ship, and so get bruised and chafed. The object of leaving the hammock under him at all in bad weather is to save him falling to the ground if his legs slip from under him; for this reason the fastenings of the suspending ropes of the hammock should be very secure.

> In bad weather ashes should be spread over the flooring of the stall to save the horse from slipping. If a horse falls he is very apt to trip up the horse in the next stall with his legs, and one may pass this on to another till a whole number are down.

> Should a horse fall, planks, which should always be placed in readiness when bad weather begins, should be run in along both sides of his stall to box him in, and prevent his legs from getting into the next stall.

> The men should stand by their horses' heads in rough weather, as the horses will not then be so frightened.

A 2005 buckskin mare, Kachina Grand Star, grazes in the marsh of the north pasture shortly after Pony Penning week. She is probably pregnant with next year's foal—on Chincoteague NWR, most mares foal annually.

Although some shipwrecks were disasters involving injuries, drownings, and loss of cargo, others were no more than minor strandings. The Lifesaving Service responded to 383 vessels in distress on the Eastern Shore between 1875 and 1915, of which only 174 became total losses. Horses have value, and it would be reasonable to think that owners made an effort to recapture any that escaped to the wild. When damage was minimal, horse-carrying ships were probably evacuated and unloaded with great difficulty, whereupon arrangements were made for another mode of transportation. Any equine castaways joining the wild herd would have most likely been strays that eluded recapture or those who had outlived their shipmates, their owners, or their usefulness.

Escaping from the carcass of a shattered ship and swimming to shore through monstrous storm-driven waves would have been an extraordinary feat for an unaided horse. It would be akin to a horse freeing himself from a narrow horse trailer overturned in a lake, except that the shipwrecked animal was below decks and restrained not only by a halter, head tie, and hobbles, but also perhaps by breastplate, breeching, and sling. The ship was typically afloat, aground, or sinking not in still water, but in violent waves.

Horses would be most likely to survive shipwrecks if someone freed them. On smaller ships, horses stayed on deck, legs hobbled together in fair weather. During storms, their bodies were tied down securely, but left uncovered and exposed no matter how harsh the weather (*History of Race Riding*, 1999). Horses on deck could be more easily released by passengers or crew, and perhaps were more likely to survive than their counterparts belowdecks.

Major shipwrecks typically occurred in storms. If horses got free of a stricken ship, they needed to reach shore to survive, and drowning was a strong possibility. Horses swim with their chins in or near the water, nostrils narrowed to slits just above the surface; high waves can easily overwhelm them. If they made it to shore, the violence of the surf could smash them on the beach with a force sufficient to break long, slender legs. Ponies standing on dry land sometimes drown when storm surge engulfs Assateague in a strong nor'easter or a hurricane.

In the hurricane of 1821 known as the "Great September Gust" the ocean pulled back, exposing bare sandbars, then leapt forward in a giant wave that engulfed both Assateague and Chincoteague "and in an unbroken mass swept across the low south marsh flats, carrying away men and ponies like insects; rushing up the island, tearing its way through the stricken pine woods" (Pyle, 1877, p. 743). A wire story about the Chesapeake-Potomac Hurricane of August 1933 reported that hundreds of ponies drowned and only three survived ("Chesapeake Storm Killed Hundreds of Wild Ponies," 1933, p. 6). Maryland Conservation Commissioner Swepson Earle predicted that Pony Penning might never recover. The 1933 reports were evidently exaggerated—Pony Penning seems to have occurred the next summer without incident (DeVincent-Hayes & Bennett, 2000, p. 104)—but all these accounts underscore the susceptibility of horses to drowning.

In the Ash Wednesday Storm of 1962, roughly half of the Assateague herd was swept into the sea and drowned—about 20 bodies later washed up (Ryden, 2005)—while almost 100 ponies drowned on Chincoteague (Mariner, 2003). A minor storm with high waves swallowed 12 horses on the Maryland end of Assateague in 1992. Sometimes horses disappear in storms and their bodies are never recovered; the assumption is that they were swept away and drowned.

To minimize the dangers of the Assateague coastline, the first lighthouse to stand guard over these waters became operational in 1833. By 1852, officials decided that this structure was neither tall enough nor bright enough. The present lighthouse, atop a 22-foot/6.7m hill on the Virginia end of Assateague, has been in service since 1867. The 142-foot/43.3-m structure was an unadorned brick column similar in appearance to the Currituck Beach Light in North Carolina until the 1960s, when it was painted with distinctive red and white stripes (U.S. Fish and Wildlife Service, 1999). In 2004, the U.S. Coast Guard transferred

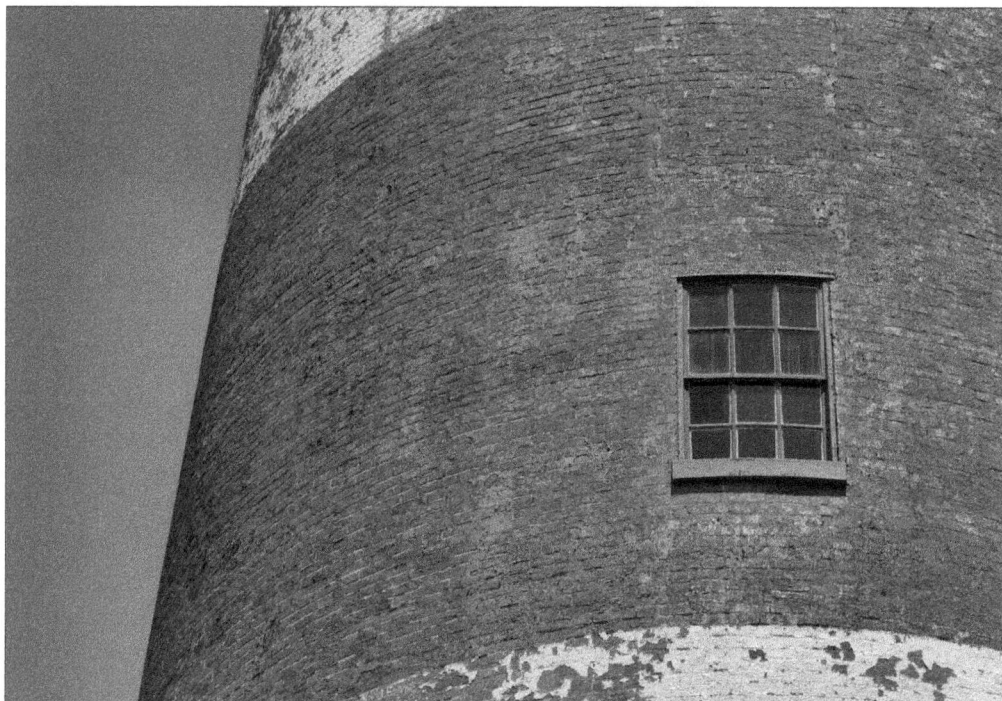

The Assateague Lighthouse still casts its light over an especially dangerous part of the Atlantic coast.

ownership of the Assateague Island Lighthouse to the Fish and Wildlife service, which opened it to the public for climbing.

Lanman (1856) wrote that there were only four families living on Assateague, those of the lightkeeper and three fishermen. Homes were constructed ruggedly to withstand the wrath of storms. Assateague islanders of the 1800s grew produce in gardens fenced to keep the ponies, sheep, and other animals out. They harvested the trees for firewood and building materials, collected wool from free-roaming sheep, and raised hogs in pens. White-tailed deer and huge flocks of migrating waterfowl provided game.

Eventually small, short-lived settlements appeared on Assateague. The villagers were mostly families of watermen, stockmen and, after 1875, employees of the U.S. Lifesaving Service. Assateague Village was the largest of these settlements, and at its peak it contained just over 200 people. In the 1920s factories were built in Assateague Village to process fish into fertilizer and oil. Evaporated seawater yielded salt.

Assateague Village was largely self-sustaining. Former resident Bill Jester (1977, p. 12) recalled in an interview,

We didn't get many supplies. . . . They could kill all the fowl they wanted to eat. . . .
They had their own hogs, all they had to do was turn 'em loose and get one. And if
they wanted beef they had cattle the same way that ran loose on the beaches.

Assateague Village petitioned the state to fund a school, but 40 students had to be enrolled. Assateague residents satisfied this requirement in 1890 by enrolling children as young as age 3. In 1919, a small church was added to the community—before this, religious services were held in the schoolhouse.

For centuries, free-roaming livestock of Chincoteague and Assateague were regularly corralled and branded to determine ownership, a practice that became known as Pony

Although it is possible for horses to swim to shore from a shipwreck in a stormy sea, the odds are not in their favor. Horses swim with their heads low in the water, and high waves are likely to drown them.

Penning (Wallace, 1897). In all probability, the practice began in the late 1600s or early 1700s. Pony and cattle penning has been practiced on Assateague since before records were kept, and sheep penning is said to have pre-dated them.

Said Refuge Inn founder Donald Leonard (2006, pp. 5–6) in an oral history interview,

Sheep was as much as a part of the livestock history as ponies or cattle was. . . .

It was a lot a cattle on Assateague before Fish and Wildlife bought it. In fact, there were far more cattle than there was ponies. . . .

And, of course, the beach goes from here to Ocean City. And it was a large area and supported a lot a cattle and horses. But when Fish and Wildlife bought it, of course, they forced the removal of all livestock except the ponies. And the fire company negotiated a grazing right from Fish and Wildlife, which stands today.

Assateague's sheep penning festivities took place in June, after the annual oyster harvest drew to a close. An 1891 article in a mainland newspaper reported, "The sheep penning at Assateague . . . was attended by some 500 or 600 people, the largest gathering ever known on a similar occasion" ("Chincoteague," 1891, p. 3). As the turn of the century approached, visitors could ride a steamship from the mainland to Chincoteague and back the following day for a dollar a head ("Local News," 1898). The final sheep penning was conducted in 1914. Thereafter, only ponies were gathered.

Up to 75 horses were owned by a single individual, and many people owned just one or two (Mariner, 2003). Pony Penning originated as a necessary chore, but over time it became formalized and ritualized, a cause for festivities (Mariner, 2003). There was one main annual summer penning, as there is today, and smaller pennings at other times of the year (Schoenherr, 2010).

An 1874 newspaper article reported,

At present the island is said to contain about five hundred of those diminutive horses, who travel in herds and bear the brands of various owners. Some thirty persons live on the island and claim to own all this wild stock in lots or herds of from ten to 100 head. ("Wild Horses in Maryland," 1874, p. 3)

Pony penning was apparently a long-standing tradition on Assateague when the practice was regularly adopted on Chincoteague (Mariner, 2003). Two Chincoteaguers kept large

On a gusty November day, the ocean smashes irritably into the Assateague beach. Waves this size could drown swimming ponies, yet they are minuscule compared to the surf of a violent storm.

herds of ponies on Wildcat and Piney Island marshes and in the mid-1800s began to pen them annually near the intersection of today's Beebe Road and Ridge Road. At one point, the stockmen attempted to capture the horses without the use of pens, allowing them to escape. When horses were killed in the process, the practice was abandoned in favor of penning. Holmes wrote, "The catastrophe ... did occur on Chincoteague island, of horses rushing into the sound, when indiscreetly attempted to be caught without pens, by driving detached portions of them upon narrow, projecting marshes; and some fine creatures were drowned" (1835, p. 419). (These drownings in shallow interior waters in controlled circumstances underscore how difficult it would be for horses to swim ashore from a wreck in a stormy sea.)

Holmes (1835, pp. 418–419) commented at length on the history and economics of pony penning:

> The horses of Assateague Island belonged principally to a company, most of whom resided upon the peninsula. No other care of them was required, than to brand and castrate the colts, and dispose of the marketable horses, all of which was effected at the period of their annual pennings (June), the whole, nearly, being joint stock. . . .
>
> The wild gang of Assateague horses were secured by driving them into pens, made for the purpose, of pine logs. The horses seized in the pens, (by islanders accustomed to such adventures, who pushed fearlessly into the midst of the crowded herd,) were brought to the main land in scows, and immediately backed, and broke to use; their wild, and apparently indomitable spirit deserting them after being haltered and once thrown, and subdued by man. More docile and tractable creatures could not be found.

Every summer, Chincoteaguers and men from all over would participate in the capture, branding, and sale of the Assateague ponies, and this occasion developed into a major event that drew crowds and spawned parties. Originally, the roundup involved the entire length of Assateague, then a peninsula, from Ocean City, Maryland, to Toms Cove, rather than gathering them from fenced sections of the Chincoteague refuge as is done today. Holmes (1835, pp. 417–418) described his experience with penning in the early 1800s:

The multitudes of both sexes that formerly attended these occasions of festal mirth, were astonishing. The adjoining islands were literally emptied of their simple and frolic-loving inhabitants, and the peninsula itself contributed to swell the crowd, for fifty miles above and below the point of meeting. All the beauty and fashion of a certain order of the female population, who had funds, or favorites to command a passage, were sure to be there. . . . It was a frantic carnival, without its debauchery. The young of both sexes, had their imaginations inflamed by the poetical narratives of their mothers and maiden aunts . . . the mad flight of wild horses careering away along a narrow, naked, level sand-beach at the top of their speed, with manes and tails waving in the wind before a company of mounted men, upon the fleetest steeds, shouting and hallowing in the wildest notes of triumph, and forcing the affrighted animals into the angular pen of pine logs, prepared to enclose them: and then the deafening peals of loud hurras from the thousand half-frenzied spectators, crowding into a solid mass around the enclosure, to behold the beautiful wild horse, in all his native vigor, subdued by man, panting in the toils, and furious with heat, rage and fright; or hear the clamorous triumphs of the adventurous riders, each of whom had performed more than one miracle of equestrian skill on that day of glorious daring—and the less discordant neighing of colts that had lost their mothers, and mothers that had lost their colts in the *melee* of the sweeping drive, with the maddened snorts and the whinnying of the whole gang—all, all together formed a scene of unrivaled noise, uproar and excitement, which few could imagine who had not witnessed it, and none can adequately describe.

But the play of spirits ended not here. The booths were soon filled, and loads of substantial provision were opened, and fish and water fowl, secured for the occasion, were fried and barbacued by hundreds, for appetites whetted to marvellous keenness by early rising, a scanty breakfast, exercise and sea air. The runlets of water and the jugs of more exhilerating liquor, were lightened of their burden. Then softer joys succeeded: and music and the dance, and love and courtship, held their undisputed empire until deep in the night.

It seems that little has changed in the past 200 years.

An 1874 newspaper article ("Wild Horses in Maryland," p. 3) offered a similar perspective:

On Chincoteague Island, the square in front of the Atlantic Hotel is used for a pen, but on Assateague a large pen has been built on the shore of stout pine logs. Men and boys mount tame ponies and start out to bring in the herds. They gallop to pasture grounds, and, after much yelling, fast riding, and some little swearing, they manage to drive one of the herds down to the shore. Nearly all the houses on Chincoteague are built along the sound, and the yards in front join each other, thus forming a continuous fence. When the herds get started down the shore the riders press close after, yelling and whooping, and there is a lively chase until the avenue that leads to the pen is reached. Here a crowd of men are standing, and they turn the head of the flying column into the square. Some of the ponies suspect treachery and run into the water, but the riders dash after them, and soon the whole herd is forced into the pen. The colts stick close to their dams and in all the rearing and plunging about through the pen they never become separated.

Descendants of the original settlers still gather descendants of the original ponies in the shadow of the Assateague Light. This tradition has persisted for hundreds of years with little change.

On the eve of the Civil War, Charles Lanman interviewed 82-year-old Rev. David Watts, then said to be the oldest resident of Horntown, Virginia, on the mainland near Chincoteague. Lanman wrote (1856, pp. 235–236),

> By far the most interesting circumstance connected with the wild horses of Assateague had reference to the annual festival of penning the animals for the purpose of, not only of bringing them under subjection, but of selling them to any who might desire to purchase. The day in question was the 10th of June, on which occasion there was always an immense concourse of people assembled on the island from all parts of the surrounding country; not only men, but women and children; planters who came to make money, strangers who wished to purchase a beautiful animal for a present, together with grooms or horse-tamers, who were noted at the time for their wonderful feats of horsemanship.
>
> But a large proportion of the multitude came together for the purpose of having a regular frolic, and feasting and dancing were carried on to a great extent, and that too upon the open sandy shore of the ocean, the people being exposed during the day to the scorching sunshine, and the scene being enlivened at night by immense bonfires, made of wrecked vessels or drift wood, and the light of the moon and the stars. The staple business of these anniversaries, however, was to tame and brand the horses, which were usually cornered in a pen, perhaps a hundred at a time, when, in the presence of the immense concourse of people, the tamers would rush into the midst of the herd, and not only noose and halter the wild and untamed creatures, but, mounting them, at times, even without a bridle, would rush from the pen and perform a thousand fantastic and daring feats upon the sand.

Few, if any, of these horsemen were ever killed or wounded while performing these exploits, though it is said that they frequently came in such close contact with the horses as to be compelled to wrestle with them, as man with man. But, what was still more remarkable, these men were never known to fail in completely subduing the horses they attempted to tame; and it was often the case that an animal which was wild as a hawk in the morning could be safely ridden by a child at the sunset hour.

Pyle (1877, p. 741) describes the process of capturing a single pony in the corral:

The momentous time arrives for casting the lasso; not as they do in the West, but by hanging it on the end of a long pole, and then dropping it skillfully over the pony's head. Uncle Ken takes the pole. Holding the noose well aloft on the top of it, so as not to frighten the intended prey upon which he has fixed his eye, he cautiously approaches the herd, around which the crowd has gathered. One of the ponies takes a sudden fright and a stampede follows, the spectators scattering right and left. For a moment the intended captive is wedged in the midst of the rest of the herd. Uncle Ken sees his advantage. He rushes forward, the noose is dropped and settles around the pony's neck. Immediately six lusty negroes, with glistening teeth, perspiring faces and glittering eyes, are at the other end of the rope. The animal makes a gallant fight. This way and that he hauls his assailants, rearing and squealing. Now he makes a sudden side dash and sends them rolling over and over, plowing their heads through the shifting sand till their wool is fairly powdered; still, however, "the boys" hold on to the rope. At length the choking halter commences to tell; the pony, with rolling eyes and quivering flanks, wheezes audibly. Now is the moment! In rush the negroes, clutching the animal by legs and tail. A wrestle and a heave, a struggle on the pony's part, a kick that sends Ned hopping with a barked shin like a crazy turkey, and Sambo plowing through the sand and stinkweed in among the spectators, and then over goes the pony with four or five lusty shouting negroes sprawling around him. The work is done: a running noose is slipped around the pony's nose, his forelock is tied to this by a bit of string, and soon his tantrums cease as he realizes that he is indeed a captive.

These cowboys were often black men, many of them freed slaves. By the 1880s, pennings were held on Chincoteague one day and Assateague the next in addition to the traditional Assateague sheep penning (Mariner, 2003).

Scott's Ocean House was a privately owned hotel and resort from about 1869 to 1894. It operated at Green Run Inlet on Assateague just north of the state line (Hall, Casey, & Wells, 2004). The inlet closed in 1880. The hotel was immensely popular, and it attracted an affluent clientele who feasted on local seafood, visited the beach, and enjoyed the ballroom and bowling alley. Nearby Ocean City attracted mostly local people while Scott's drew visitors from as far away as Pennsylvania and West Virginia. In time, Ocean City added homes, hotels, cottages, and boarding houses and blossomed into a fashionable resort area with a boardwalk that workers disassembled and stored during the off season.

Waterfowl are abundant on Assateague, but not nearly as numerous as they were before the arrival of Europeans. Through the 19th century, on Assateague and elsewhere, egrets, geese, and other birds were hunted relentlessly even during the breeding season and at night. The seemingly unreducible avian population dropped precipitously through the 1800s.

Pony pennings in the 19th century were action-packed events in which cowboys wrangled defiant ponies in a rousing East Coast rodeo. "Crossing to Assateague," by Howard Pyle (1877, p. 743), courtesy of Cornell University Library, Making of America Digital Collection.

One author observed of Chesapeake Bay in 1830,

> The quantity of fowl of late years, has been decidedly less than in times gone by; and the writer has met with persons who have assured him, the number has decreased one half in the last fifteen years. This change has arisen, most probably, from the vast increase in the destruction from the greater number of persons who now make a business or pleasure of this sport; as well as the constant disturbance they meet with on many of their feeding grounds, which induces them to distribute themselves more widely, and forsake their usual haunts. (Sharpless, 1830, p. 41)

Where once birds arose from the waters in amorphous clouds to be felled in great numbers by the most inept gunner, flocks shrank and individual species became endangered. Waterfowl and other birds were hunted commercially for meat and for feathers (which adorned women's hats) and for sport until the Migratory Bird Treaty Act of 1918 (16 U.S.C. §§ 703–712) limited the harvest to ducks, geese, and other game birds. It also gave the federal government the power to establish seasons for game birds and set other limits.

Hunters congregated in camps, cabins, and lodges and shot from blinds, watercraft, and the shore. Gunning shantyboats—shallow-draft, flat-bottom houseboats—were commonplace in Mid-Atlantic marshes from the 1880s to the early 1900s. Hunters could retreat to the marshes and shoot waterfowl for up to a week at a time.

In Virginia, where land on Assateague was unavailable to gun clubs, sportsmen found a loophole that allowed them access to the wild flocks. Clever hunters leased oyster beds, which granted them the right to build oyster watch houses, structures used by oystermen to monitor oyster grounds (Eshelman & Russell, 2004). When legislation prohibited the construction of new oyster houses, sportsmen placed trailer homes on oyster scows—floating watch houses that became readily available as the oyster industry declined. These scows were eventually prohibited because they lacked sewage holding tanks (Eshelman & Russell, 2004).

Another reason for the decline in waterfowl was the widespread practice of collecting wild bird eggs for food and as a social activity (Langley et al., 2009). "Egging" was popular on Assateague as early as the 18th century. Communities planned picnics for the purpose of harvesting delicacies from the nests of marsh and seabirds. Assateague "eggers" primarily foraged on two rookery islands in Sinepuxent Bay, Great Egging Beach, and Little Egging Beach, near the old ferry landing on the Maryland portion, but also visited Green Run Beach, just north of the state line, and North Beach, at the north end of the island (Eshelman & Russell, 2004).

Chincoteague was home to a number of year-'round residents. By 1835, the Island of Chincoteague supported in excess of 70 families, a number that more than doubled by 1860 (Mariner, 2003). In 1881 John Bunting built a fish factory on Chincoteague where Atlantic menhaden (*Brevoortia tyrannus*), small inedible fish, were converted to oil, which was used in cosmetics, paints, and lamps. The dried fish pulp was sold as fertilizer. Chincoteague residents grew potatoes, strawberries, and corn, but mostly made their living from the sea. The island was renowned for the distinctively flavored oysters that grew there in great quantities. Access to the mainland railroad increased local watermen's income by allowing them to supply New York and Philadelphia with oysters, trout, and channel bass.

Pyle wrote in 1877 (p. 738),

> There are two distinct classes of inhabitants upon Chincoteague: the pony-owners—lords of the land—and the fishermen. Your pony-owner is a tough, bulbous, rough fellow, with a sponge-like capacity for absorbing liquor; bad or good, whisky, gin, or brandy, so that it have the titillating alcoholic twang, it is much the same to him. Coarse, heavy army shoes, a tattered felt hat, or a broad-brimmed straw that looks as if it had never been new; rough homespun or linen trowsers, innocent of soap and water, and patched with as many colors as Joseph's coat; a blue or checked shirt, open at the throat, and disclosing a hairy chest,—these complete his costume. Your fisherman, now, though his costume is nearly similar, with the exception of shoes (which he does not wear), is in appearance quite different. A lank body, shoulders round as the bowl of a spoon, far up which clamber his tightly strapped trowsers; a thin crane-like neck, poking out at right angles from somewhere immediately between the shoulder-blades; and, finally, a leathery, expressionless, peaked face, and wiry hair and beard complete his presentment.

When Virginia communities took sides at the onset of the Civil War, Chincoteague voted almost unanimously to remain with the Union—mainly because islanders sold their seafood to northern markets (Langley et al., 2009). The Union, however, seldom acknowledged their fidelity, and it seized Chincoteague ships and cargo in northern ports.

As the natural bounty dwindled, watermen took to planting oyster beds as they would other crops. Starting around 1864, watermen selected "seed" oysters from the "rocks"—public beds—and distributed them in their private beds, growing them in shallow water for 12–18 months, then deeper water for the next year or so. The beds were then allowed to lie fallow for a year, and the process began again. Chincoteaguers leased the oyster beds from the state for 50 cents a year, and eventually much of the area surrounding Chincoteague Island was used for oyster farming. The flavor of the oysters was influenced by whether they

After tying her legs with rope and laying her flat, Chincoteague volunteer firemen trim the hooves of a feisty filly who is in no mood for a pedicure. Photograph circa 1940s, from the collection of Flickr user rich701.

were grown on mud, shell, or sand, and they were popularly deemed good for eating only during months with an *r* in their names, or September through April.

The industry boomed. During the 1879–1880 season, 318,113 bushels of oysters (roughly 11,200 m3 or 8,000 tons/7,300 metric tons) were harvested from the waters surrounding Chincoteague and exported by ship or rail (Mariner, 2003). In 1890, this number rose to 300–400 bushels a day. Chincoteague oystermen set the all-time record in 1889, sending out 1,600 bushels by rail in a single day. In 1913 alone, Chincoteague shipped 60,000 barrels/7,000 m^3 of in-shell oysters and 80,000 gallons/303,000 L of shucked oysters.

In the 1930s, the newly constructed Ocean City Inlet jetty system increased the salinity of the estuaries adjoining Assateague and increased the numbers of native oyster predators, such as starfish. Chincoteague oysters were overharvested, eelgrass (*Zostera marina*) was struck by a virus and all but disappeared, and in the 1950s two aggressive single-celled oyster parasites, MSX (*Haplosporidium nelsoni*) and dermo (*Perkinsus marinus*), infected oyster beds. These multiple insults sent the oyster population into sharp decline until the Chincoteague oyster was in danger of disappearing altogether (Hayward, 2007). Atlantic bay scallops (*Aequipecten irradians concentricus*) were also affected by these events, though with the resurgence of eelgrass, they have begun a modest comeback.

During the 1920s one man, Dr. Samuel Field of Baltimore, owned much of the south end of Assateague and denied others in the community access to the shellfish beds of Toms Cove. At that time about 25 families lived on Assateague, most in Assateague Village, facing Chincoteague near the lighthouse, where they maintained a school, a church, and two stores. Field employed a guard from Wyoming, Cooper H. "Cowboy" Oliphant, in 1921. Cherrix (2011, p. 74) says that Oliphant was hired "to be the manager of his stock farm and to collect rents from the people leasing the flats and oyster grounds. He was known to dress like a cowboy, ride a big horse, and carry a big gun." Finally, the frustrated residents floated their homes across the channel on barges and set up residence on Chincoteague.

Field owned the land, but not all the free-roaming horses on Assateague. Annual penning, sorting, and branding events had determined ownership every year, probably from the late 1600s. The centuries-old ritual came to an abrupt end in 1921 because Field would not let the stockmen cross his land to gather the horses. In 1922, the locals established a single penning on Chincoteague for both islands. Initially, ponies were ferried across from Assateague by boat, but soon stockmen began the tradition of swimming them across the channel, as is done today.

Around the turn of the 20th century, Chincoteague had become prosperous and had its own schools and hotels, a post office (established in 1854), and many homes. The island was home to two separate communities, "Up the Island" and "Down the Island," each with a church, a school, and general stores. Up the Island, or "Oysterville," even had a separate post office for a short time. These villages gave rise to several neighborhoods—Deep Hole, Downtown, Down the Marsh, Up the Neck, Snotty/Rattlesnake Ridge, and Chicken City (Hall, Reed, & Daisey, 2012; Waterhouse, 2003)—with subtle differences in speech and sometimes-testy relations: "They'd run you out ... if you tried to court a girl from one of these different neighborhoods. They'd chunk you with rotten eggs and brickbats and everything!" (Hall et al., 2012, p. 13).

Warren (1913, pp. 776–777) wrote of her experience with the Chincoteague people on a visit 15 years earlier:

> The people marry early, the girls sometimes at the age of fourteen, the men at eighteen, and they have large families. One woman is pointed to as the mother of eighteen children; another was a grandmother at thirty. . . . These people are encompassed by the poetry of life—by the three most ancient cries in the world: the cry of the sea-bird, the call of the wind, and the sighing of the sea. Yet they live according to a happy prose kept resolutely in their blood by the strong Anglo-Saxon strain in them, which has come down as unchanged perhaps as in any community in the world. And allowing for surface changes, they live much as their fathers did. . . . There was no mayor and no prison, and, after the first rage, people forgave easily whatever crime was committed.

Although the Fish and Wildlife Service worked to eradicate the ponies from the Chincoteague National Wildlife Refuge, Pony Penning continued on schedule, as shown in this photograph from the 1940s. From the collection of Flickr user rich701.

When Warren returned in 1913, the town was considerably more modern, with gas lamps, telephones, and two five-cent theaters. A chambermaid told Warren, "Things hain't like they were when you came before. We have a bathroom now; you can lie right down in the tub and let the water go all over you" (Warren, 1913, p. 777).

Livestock still ranged freely on the island, and herds of cattle and hogs regularly wandered into town. Mariner (2003) recounts that Clark Street was once called Madcalf Lane. It was so named after a boy walking his girlfriend home in the dark stumbled over what he thought was a log, which turned out to be a sleeping calf. The heifer leaped up and ran off with the boy on her back.

After reading about the island in national publications, visitors flocked to the town. In the late 1800s, a steamboat named *Chincoteague* carried passengers and freight between the mainland at Franklin City, Va., and the island. Chincoteague was incorporated as a town in 1908.

Free-roaming animals were outlawed on Chincoteague in the early 20th century, and any wandering stock was impounded—at the same address as the municipal jail! In the spring of 1920, workers began constructing a causeway linking Chincoteague to the mainland, a project completed in 1922. On the opening day, amid fanfare and parades, a rainstorm turned the causeway into a quagmire, stranding 96 cars in the dark.

The streets were narrow, and the wooden houses were close together. The people of Chincoteague feared fire because they knew that it could quickly wipe out much of their community. Their fears became reality in the early 1900s when a fire did considerable damage (Chincoteague Volunteer Fire Company, n.d.). Residents bought a hand-pump fire engine, then later a gasoline engine, and trained a team to use it. But when a serious fire struck on September 5, 1920, the equipment had fallen into disrepair and would not work properly. Twelve homes and businesses were lost, including the hotel, the post office, the shoe-repair shop, and the bank. As it turned out, the fire was set by 15-year old Etman Cherrix, who had been offered $10 to perform the deed by a resident trying to commit insurance fraud (Mariner, 2003).

The resilient town quickly recovered and rebuilt itself. Four years later, another fire took most of the buildings on the west side. Chincoteague residents vowed that this preventable tragedy would never recur. In 1924, residents formed the Chincoteague Volunteer Fire Company. To raise money for equipment, the fire company bought 80 of the ponies running free on Assateague.

Donald Leonard explained,

> I started [participating in] the roundups—I guess in the late 30's, penning not only on Assateague, but on Wallop's Beach.... (2006, p. 1)
>
> The ponies were owned by Mr. Joseph Pruitt [of Greenbackville, VA]. He was a very successful business man and he owned the livestock grazing rights on Assateague and at that time on Wallops Beach. At his death it became a problem for the fire company in that if the ponies went elsewhere were bought by someone else, the fire company went out of business....
>
> [T]hey were forced to buy as many of the ponies of Joseph Pruitt at the settling of his estate. And that's what put the fire company in the pony business. (2006, p. 5)

The fire company held its first Pony Penning in 1925. Residents organized the annual Firemen's Carnival, which included the roundup and auction of the ponies. Every year, most of the new foals went to the highest bidders, and the adult ponies returned to Assateague to live as wild. Thus the tradition of selling horses to raise money began.

The development of wetlands and the black market for waterfowl and their meat and feathers were putting pressure on many native species. Chincoteague NWR was established in 1943 as a breeding and wintering area for migratory and resident waterfowl. The refuge protected 9,000 acres/3,642 ha of coastal wetlands and wildlife. The land in question, however had been free-range grazing land for hundreds of years, and locals petitioned the U.S. Fish and Wildlife Service to continue this generations-old practice.

In 1943, with the formation of the refuge, the Service issued a permit to Wyle Maddox, allowing him to graze cattle and horses on part of the island. Three years later, the Service issued a special use permit, which allowed the fire company to graze up to 150 horses; since the early 1950s the fire company has owned all the horses on the refuge (Grey, 2014).

Rachel Carson, world-renowned marine biologist, environmentalist, and editor-in-chief for the Fish and Wildlife Service, wrote that when the refuge was created, the agency permitted residents of Chincoteague to graze 300 head of horses and cattle on the refuge, and noted no adverse effect on waterfowl. (Only 150 horses are permitted today.) "The presence of these grazing animals is not detrimental to the waterfowl for which the refuge was established," she said (1947, p. 17).

Later, the Fish and Wildlife Service removed the cattle and opposed the ponies as a nuisance that trampled vegetation and competed with the birds for forage. The refuge erected

Although many oyster watch houses were erected to monitor oyster beds or to allow hunters access to waterfowl, this house on Tom's Cove was built as a recreational property. In recent years it has grown unstable, and the primary residents are a nesting pair of ospreys that raised chicks in a nest built over the chimney.

fences to restrict their range to only 5% of the Virginia section of Assateague (Ryden, 2005). Almost all of this section was salt marsh, which provides plenty of food, but offers no way to escape the torment of insects and no high ground to climb in storms.

When the Ash Wednesday Storm of 1962 flooded the Assateague lowlands, about half the horses in the refuge drowned (as did nearly 100 horses on Chincoteague). In 1965, the fences were reconfigured to give the ponies access to high ground and to let them range more freely (Ryden, 2005).

On Chincoteague, the storm wrought great damage as well. In *Stormy, Misty's Foal* (1963), Marguerite Henry gives a fictional account that nonetheless accurately captures the horrors of the aftermath. Grandpa Beebe had died some years before the storm; but in Henry's story, he, Paul, and waterman Tom Reed combed the flooded pastures in the Deep Hole section of Chincoteague in a scow to flag the bodies of drowned ponies so that crews could remove them by helicopter. First the boat bumped into the body of Black Warrior, one of Grandpa's favorite stallions, then swirling crows led them to the bodies of Warrior's mares and foals. Grandpa surveyed the scene dejectedly. Henry wrote (1963/2007, 119–120),

> It was almost as if they were alive. Some were half-standing in the water, propped up by debris. . . .
>
> Then he took a good look, and he began to name them all, saying a little piece of praise over each. . . .
>
> "That Black Warrior was a good stallion. He died tryin' to move his family to safety, but . . . " his voice broke " . . . they just couldn't move."
>
> The heart-breaking work went on. . . . They found more stallions dead, with their mares and colts nearby. And they found lone stragglers caught and tethered fast by

twining vines. As the morning dragged into noon, and noon into cold afternoon, the pile of flags in the boat dwindled.

The storm surge swamped houses, ripped boats out of their moorings and hurled them through the stores on Main Street, and disinterred bodies. In one cataclysmic event, the once-thriving Chincoteague poultry industry was destroyed, never to recover. Water stood 6 ft/1.83 m deep in spots, and raging fires broke out. Dead fish, chickens, and livestock decomposed by the thousands in the soggy ruins, presenting a health hazard that necessitated the evacuation of the island.

Betts Devine, who grew up on Chincoteague, recalls,

> I spent my summers in my Grandfather's house on Peterson Street on Chincoteague, and I will always remember the five high tide marks on the wall paper in "the big kitchen." Not to mention the way the floor boards were warped in one corner, so the rocking chair was always in a reclined position.... The two highest marks were from 1938 hurricane and the Ash Wednesday Storm. (Personal communication, April 6, 2012)

Drowning still occasionally claims equine lives. A freak storm with high waves drowned 12 Assateague horses on the Maryland end in 1992. Wildlife biologist Jay Kirkpatrick wrote, "something that can only be described as a small tidal wave" swept across the island, engulfing the animals (1994, p. 141). The storm waters were powerful enough to wash them across the bay and deposit their bodies on the mainland. Some bodies were even found caught in trees.

Besides the danger of storms, wild island ponies are vulnerable to potentially fatal diseases spread by bloodthirsty insects. Eastern equine encephalitis, a neurological disease transmitted from birds by mosquitoes, is responsible for a significant number of pony deaths on Assateague. The virus can also be fatal to humans. In 1960, 30-odd Chincoteague Ponies succumbed to EEE in a span of 10 days. Most of them were foals that "had been held past the regular 'pony-penning' sale and were used in the filming, during July and August, of the motion picture, 'Misty'" (Byrne, 1968, pp. 357–358). The foal that portrayed young Misty herself was the first to contract the disease. Forty more horses in the Maryland herd died from EEE in 1989–1990 (Kirkpatrick, 1994). Another 40 died on Cumberland Island, Georgia, in 1990 (Goodloe, Warren, Osborne, & Hall, 2000). Those years saw an uncharacteristically large mosquito population. The fire company now vaccinates against EEE, but vaccinating the other free-roaming herds would involve stressful and expensive annual gathers—and on heavily forested Cumberland Island, it would be virtually impossible to gather every horse.

The salt marsh is an ideal breeding ground for several species of mosquito, but marshes are not necessary to ensure a large population of the insects. Mosquitoes can successfully breed larvae in water that collects in a broken bottle or a fallen leaf. The dried eggs can survive for up to five years, and a moistening rainfall is enough to trigger their growth. They can hibernate as eggs, larvae, or adults to survive any season, including severe winters. Drawn by the heat, moisture, carbon dioxide, and other chemicals released by warm-blooded species, mosquitoes are most active at dawn and dusk.

In addition to the usual behaviors such as tail-swishing, stomping, head-shaking, and mutual grooming, ponies often wade out into the bay or even the ocean to escape the onslaught of insects. Sometimes they wade out so far that only their heads can be seen. More typically, they move into water deep enough to discourage flies, but shallow enough to let foals nurse—sometimes as much as a half-mile (0.8 km) from shore.

Ponies are familiar with high spots where the wind blows more strongly and often congregate there when the insects are relentless. They plunge into deep brush and rub against trees to dislodge the pests. Horses also have a "fly-shaker" muscle over the shoulder area that can be twitched to remove insects.

In 1974, a colt sold to a family from New Jersey tested positive for equine infectious anemia (swamp fever), a fly-borne chronic equine disease. In 1975, almost half the Chincoteague NWR herd tested positive. The horses in question did not show signs of serious illness and were not likely to die from the disease, at least not in the short term, but state law mandated the quarantine or destruction of positive testers. For three years affected individuals were quarantined on the island, away from other horses, producing disease-free foals that were sold at auction. In 1978, the positive testers were euthanized to halt the spread of the disease.

Officials from the Maryland Department of Agriculture and the Park Service considered testing the Maryland herd for EIA and euthanizing positive reactors, but it would have been an enormous task to corral and test every individual—and very stressful to the horses. If the agencies missed just one horse that carried the disease, the whole herd could be reinfected. To implement such a plan, officials would have had to shoot any animals that evaded capture in order to ensure that the disease was not spread.

The Park Service finally concluded that the drawbacks outweighed the benefits and opted to let the herd remain untested. Because these animals are not sold to mainlanders like the Virginia ponies, the presence or absence of EIA poses no risk to other horse populations. Fences prevent the intermingling of herds. In theory, flies can carry the disease from Maryland ponies to Virginia ponies; but so far, there have been no problems, since horses must be in close proximity to transmit the disease.

In August 1978, a Maryland stallion with EIA crossed the fence at the state line and joined the Virginia herd (Mackintosh, 1982/2003). Over time, other potential carriers circumvented the fence, and, Park Service biologist John Karish laid plans to gather and test all Maryland ponies. The plans were never implemented, and there remains a small risk that Virginia ponies may someday contract EIA from a Maryland horse. The Park Service opens the beach to equestrians only after the fly season, so domestic horses would be unlikely to contract EIA from the wild herd.

Virginia, like most states, prevents the spread of EIA by requiring a Coggins test (antibody screen) of any horse attending an exhibition or sale or transported across state lines. Any horse that contracts EIA must be either quarantined for life or euthanized. In accordance with the law, Chincoteague Ponies must have a negative Coggins test before sale at the annual pony auction.

Epizootics are among the natural pressures that shape a free-roaming herd, and sometimes the medical and scientific communities are powerless to prevent them. Over the years, large numbers of ponies and horses of various breeds have been experimentally infected with equine infectious anemia in laboratories all over the world in hope of developing a vaccine against the dreaded lentivirus. This greatly anticipated vaccine, however, remains elusive.

For decades, professional and student zoologists, botanists, ecologists, oceanographers, geologists, and others have found Assateague ideal for original research (Frydenborg,

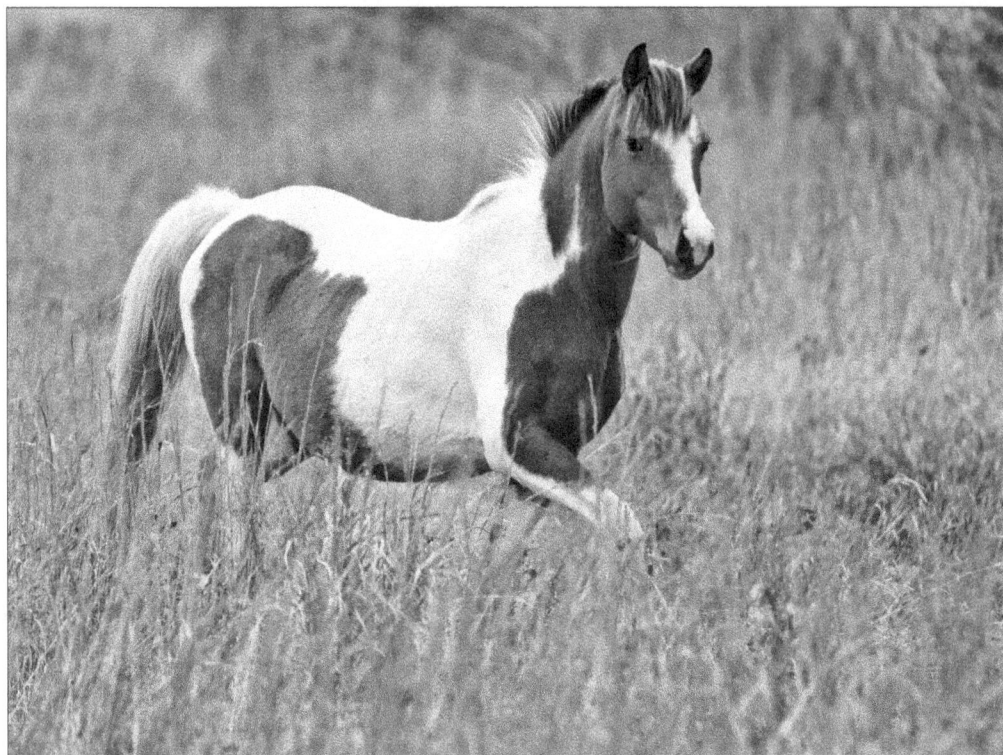

Even pregnant, Scotty ET, a 2002 pinto mare, shows beautiful Arabian features. She foaled a few days after the author took this photograph at the April 2011 roundup. Jean Bonde, who donated the mare to the fire company as breeding stock, settled on the name ET because, to her, the marking on her neck resembled the finger of the Hollywood extraterrestrial.

2012). Most of their work has increased understanding of the island or its wildlife without harming either. But in the 1950s and 1960s government-sponsored scientists experimented on Chincoteague Ponies from the Virginia herd with different agendas in mind.

For example, at least six teams of researchers studied equine and human influenza by infecting roughly 40–60 ponies and observing the results. The role of these experiments in evaluating influenza as a biological weapon (Wheelis, Rózsa, & Dando, 2006) is unclear, but they were underwritten by institutions that conducted bioweapons research, notably the U.S. Army. The following is a summary:

- Cameron et al. (1967) took 24 yearlings to the University of Maryland in College Park and infected them all with influenza H3N8, a virulent subtype that caused major outbreaks among horses in the United States in 1964–1965. The organism was also apparently responsible for the pandemics of 1889–1890 (Valleron et al., 2010) and 1898–1900 (Salmon, n.d.), and the economically devastating North American equine panzootic of 1872 (Morens & Taubenberger, 2010; Threlkeld, 2010). All infected ponies mounted an immune response. About half grew very ill, with fevers up to 105.2°F/40.7°C (normal is 99–101°F/37.2–38.3°C), respiratory rates as high as 80–90 per minute (normal is 12–24), and a dry cough. All survived, though three were treated for colic.
- Alford, Kasel, Lehrich, and Knight (1967) tested H3N8 on 33 prisoners and on four ponies used in Cameron et al. (1967).

- About a year later, Cameron, Kasel, and Couch (1974) gave eight ponies used in Cameron et al. (1967) an inactivated-virus influenza vaccine and checked their antibodies.
- Kasel, Byrne, Harvey, and Schillinger (1968) infected 12 Chincoteague Ponies of both sexes, all about 18–24 months old, with human influenza, apparently in College Park. It is not clear whether these ponies were veterans of earlier studies.
- Kasel, Fulk, Haase, and Huber (1968) infected an unknown number of Chincoteague Ponies with human influenza A and B in College Park.
- In late 1968, during a pandemic of Hong Kong influenza that killed about 34,000 people in this country and 700,000 worldwide (U.S. Dept. of Veterans Affairs, 2013), Kasel, Fulk, and Harvey (1969; see also Spaulding, 1968) infected 10 Chincoteague Ponies about 8 months old with the Hong Kong virus and five with A2/Rockville/65, then all 15 with equine influenza (H3N8). Around the same time, Couch, Douglas, Kasel, Riggs, and Knight (1969) infected 15 prisoners with equine influenza (H3N8). Then Couch, Fulk, Douglas, Kasel, and Knight (unpublished, cited in Kasel & Couch, 1969) gave those prisoners the Hong Kong virus. A Hong Kong influenza vaccine existed before this cluster of studies began (Health England, n.d.), so their purpose is unclear.

From 1957 or earlier till 1969 or later, these researchers and others in their network infected burros, domestic horses, and ponies of unknown origin with many other viruses, several of which had been or would be weaponized (Buescher et al., 1963; Byrne, 1963; Bailey et al., 1979; Cameron et al., 1967). Robert J. Byrne of the University of Maryland mentions "Stored, frozen brains of horses and ponies from which EEE virus has been isolated over the past 6 years" (Byrne, 1963, p. 6), but says nothing about the ponies' breed, manner of infection, or cause of death. The project received support from the Army Medical Research and Development Command at Fort Detrick, Md., which took interest in the bioweapon potential of these viruses.

Army-affiliated researchers also conducted research on or near Assateague in the 1960s. One team tested birds, mosquitoes, and 156 ponies from Chincoteague, Assateague, and the mainland for Eastern equine encephalitis (Byrne, 1963). There was even a small facility supported by several agencies on Assateague itself. What occurred at the Assateague Research Laboratory is unclear, though the Fish and Wildlife Service (USFWS, 1962, p. 22) described it as an "improvised temporary laboratory for study of encephalitis at Chincoteague Wildlife Refuge." It moved to Wallops Island by mid-1963.

The Army was well past its horse-drawn era when it became really interested in communicable diseases of horses and other livestock. When the United States entered World War II, Germany, Japan, and the Soviet Union already had biological warfare programs (Beck, 2003). In 1942, the civilian War Research Service hastily developed the first American bioweapons at Camp (later Fort) Detrick. Its successor, the Army Chemical Warfare Service, manufactured approximately 5,000 anthrax bombs, but did not deploy them during the war (Lindler, Lebeda, & Korch, 2005; Ryan, 2008). Other American factories geared up to produce vast quantities of anthrax, brucellosis, tularemia, Q fever, and VEE pathogens. A plant in Vigo, Indiana, could turn out *100 tons of anthrax spores a month* (Ryan, 2008).

After the Third Reich fell, Allied governments quietly recruited, rescued, or captured German scientists and engineers (Boyne, 2007; Hunt, 1991). As a result, American and

Soviet BW programs flourished during the Cold War. Researchers refined old bioweapons and developed new ones: bubonic plague, rabies, *Staphylococcus* (bacterial pneumonia, food poisoning, toxic shock syndrome, and flesh-eating infections), *Clostridium botulinum* (botulism), and several deadly hemorrhagic fever viruses (Ebola, Marburg, Lassa, and Rift Valley) (Beck, 2003; Greif & Merz, 2007). Some of this research helped develop vaccines, antisera, and therapeutic agents to protect troops and noncombatants from biological attack (Laird, 1970; Smith et al., 1997; Wheelis, Rózsa, & Dando, 2006), but its focus was offensive.

To investigate the real-world performance of bioweapons, the U.S. Army ran 239 or more clandestine tests of biological agents and proxies on American soil from the 1940s through the 1960s (Carlton, 2001). In one 1950 exercise, a Navy ship sprayed bacterial fog for 6 days over the San Francisco Bay area. The main agent, *Serratia marcescens*, was deemed harmless; but it evidently killed a 75-year-old civilian and sent others to the hospital, and it may have had delayed effects, such as a meningitis outbreak in 2001 (Carlton, 2001; Tansey, 2004). The Army tracked dispersal by adding the toxic, carcinogenic fluorescent compound zinc cadmium sulfide.

Army files released during 1977 U.S. Senate hearings document biological and chemical tests on unsuspecting subjects in most of the eastern and central United States and parts of Canada, in Manhattan subways, at Washington National Airport, even inside the Pentagon (Carlton, 2001; Cole, 1997; Mangold & Goldberg, 2000; Rubin, 2007; Subcommittee on Zinc Cadmium Sulfide, 1997; Wheelis, Rózsa, & Dando, 2006). Until information leaked out in the 1970s, the public knew nothing about the open-air tests or laboratory experiments on thousands more military and civilian subjects.

Safety concerns, political pressure, and advances in other weapons eventually made the American biological arsenal dispensable. President Nixon formally abolished this country's offensive BW capability in 1969 (Ryan, 2008; Tucker & Mahan, 2009). Coincidentally or not, experiments on Chincoteague Ponies apparently ended around the same time.

Other countries still possess or have acquired biological weapons, and the menace of bioterrorism is growing. Long after the Chincoteague-related research seemingly ended, several viruses studied remain high on the Army's threat list: "The viral encephalitides represent 15% (9 of 62) of the infectious diseases identified by the Armed Forces Medical Intelligence Center as being of U.S. military operational importance" (Hoke, 2005, p. 92).

Offensive and defensive BW programs seem equally porous. In 1977, a distinctive H1N1 influenza strain that had been extinct in humans more than two decades suddenly appeared in China, which officially labeled the event "mysterious." Western virologists maintained that the only rational explanation for the unchanged reemergence of a rapidly mutating virus is escape from a laboratory freezer. Anthrax caused 100 deaths near Sverdlovsk, USSR, in 1979 (Greif & Merz, 2007). Ft. Detrick was evidently the source of anthrax used in postal attacks that killed five and sickened 17 in 2001 (Meyer, 2008), and three vials of VEE virus that disappeared from there in 2009 are still missing (Shaughnessy, 2009).

In nature, viruses evolve to afflict specific hosts and typically cause benign infections in these reservoir species (Mandl et al., 2011; Seale, 1989). When a virus jumps to a novel host, it is likely to cause severe illness (Mandl et al., 2011). In humans, pathogens that originated in animals often cause pandemics, for example, "HIV (from chimpanzees), SARS coronavirus (from bats) and influenza A virus (from birds)" (Flanagan, Leighton, & Dudley,

A Chincoteague mare named Mystery wades through a shallow impoundment, keeping her body protectively between her newborn and the author. Nobody seems to know where she came from—she just appeared in the north refuge herd one day—and she did not come from the Maryland herd.

2011, p. 2; Sharp, Shaw, & Hahn, 2005). Smallpox and measles evolved from mutations of cowpox and rinderpest, pathogens that affect cattle (Diamond, 2009).

Influenza A continually reshuffles its genetic structure within its natural reservoir of wild birds (Taubenberger & Morens, 2013). Sometimes, in a process incompletely understood, it infects either domestic fowl or mammals, including humans, swine, horses, dogs, cats, and seals. Some virologists worry that introducing viruses into novel hosts in a laboratory encourages them to jump species and turn virulent (Taubenberger & Morens, 2013). Cameron et al. (1967) inoculated ponies with equine influenza virus grown in kidney tissue cultures from rhesus monkeys, green monkeys, and *human embryos*. Other researchers gave human subjects equine influenza. Many of the studies discussed here could have promoted cross-species transmission. One must wonder how they affected public health, how many other wild herds have been living laboratories, and whether any post-World War II disease outbreak among them was natural.

Researchers chose Chincoteague Ponies for influenza experiments because serum samples confirmed that this convenient herd was naive to influenza. Although Assateague is a tourist attraction, they deemed the herd "relatively isolated from humans" with "no known previous contact with other equines" (Cameron et al., 1967, p. 510). Most of the studies explicitly refer to Chincoteague Ponies from Assateague. The 28 horses living on the Maryland end when the Park Service took over in 1968 could not have supplied so many unique subjects. The Virginia herd is the only possible source, but there is no evidence that the refuge or the fire company knowingly provided ponies for experimentation. Nor is there evidence that researchers bid against tourists at the annual auction.

Still, the 15 weanlings studied by Kasel et al. (1969) must have come from the 1968 crop of foals, and the 24 ponies used by Cameron et al. (1967), with ages in a 6-month spread, probably belonged to an earlier cohort. Cameron et al. (1967) acknowledged Thomas J. Reed (1901–1993) of Chincoteague for "advice and assistance in securing the ponies for the experiments" (p. 515), and Kasel et al. (1968, p. 969) thanked him for unspecified "assistance." Reed was a prominent local figure—market gunner, waterfowl breeder, decoy carver, character in Marguerite Henry's *Stormy* (1963), and 24-year contract employee of "John Hopkins/Walter Reed" [sic] (Hall et al., 2012, p. 2)—yet his role in procurement is unknown.

Did research on Chincoteague Ponies cause the deadly outbreak of EEE on Chincoteague and Assateague in October 1960? More likely, it was a natural occurrence. EEE is endemic to East Coast barrier islands and follows natural cycles of resurgence and quiescence. The Fish and Wildlife Service (USFWS, 1962) says the Assateague Laboratory was created between July 1, 1960, and December 30, 1961, to study "whether birds are important in transmission of encephalitis to man and horses" (p. 21) and seems to suggest that the 1960 epizootic led to the laboratory, not vice versa: "The isolated refuge area was ideal for the virus transmission study because of the presence of wild ponies, a large bird population, and the known occurrence of the virus in the area" (p. 22). And though researchers in the Army network did infect ponies with EEE, there is no indication that these ponies were from or on Assateague. Many questions arising from this troubling episode remain unanswered or unasked.

Whenever an epizootic or natural disaster has devastated the Chincoteague refuge herd, the public has responded with donations and support as soon as the news hit the wire. In 1976, the equine infectious anemia crisis on Chincoteague caught the attention of Bob Evans, sausage maker and owner of a thriving restaurant in Ohio that would become a chain worth $1.7 billion. In 1972, Evans had acquired five mustangs with strong Spanish Barb characteristics captured in Utah and New Mexico. He freed them to roam his 1,000-acre/405-ha farm near Rio Grande, Ohio. Evans was distraught when he heard of the crisis on Assateague. "Here is an animal that was almost extinct," he said. "Something has to be done, and done quickly" ("Mustangs to cross breed with endangered pony herd," 1976, p. 14). He donated two young Spanish Barb stallions from his mustang herd to fortify the Chincoteague bloodlines.

Harry Bunting, a seafood dealer and chairman of the Chincoteague Volunteer Fire Company, applauded Evans's donation: "Centuries of inbreeding may have made the ponies more susceptible to disease and Spanish Barb blood should serve to strengthen the wild ponies of Assateague for future generations" ("Mustangs to cross breed with endangered pony herd," 1976, p. 14).

To revitalize the gene pool further and rebuild the population, the fire company imported 38 mostly solid-colored Western mustangs from a U.S. Bureau of Land Management

Surfer Dude, (left), born in 1992, is the direct descendant of Nevada mustangs introduced to the herd in 1978. He was born to a mare with a BLM freeze brand. His sire was reportedly a bay mustang named Pirate, who was later known as "Broken Jaw" after he fractured his mandible in a fight with another stallion (Szymanski, 2012).

adoption center near the California-Nevada state line in 1977. Bunting chose bays, blacks, and grays to bring the Chincoteague herd back to the solid colors he remembered from childhood. They "really look nice," he said. "I went out West to pick them out myself... they're having a problem getting rid of them out there and I thought here was 38 we could save from a can" ("Mustangs Not Accepted, Yet," 1977, p. 6).

According to Bunting, the fire company introduced new stock to the herd every 8 to 10 years, using mustangs, Arabians, and Morgan horses. "Otherwise you'd have a herd of idiots. They'd interbreed so," he said ("Breeding Horses Come East," 1977, p. 5).

At first the ponies shunned the mustangs, leaving them to form bands of their own. Challenged by the lower nutrient content of the forage and the relentless biting insects, most of these mustangs failed to adapt to the barrier island and died within a year (Keiper, 1985). Tim Ferry, a native of Alexandria, Va., recalls spotting mustangs during family vacations in the 1970s (personal communication, July 6, 2013).

> It was easy and fun to pick the remaining ones out because they were BIG horses, at least most of them. Some would dwarf even the largest Chincoteague pony standing at 14 hands plus. . . .
>
> [A] very special, beautiful wild mustang stallion [was] added to the Chincoteague herd in the 1970's and stayed around into the year 1982. . . . His foals for

A tall bay mustang towers over the native ponies as they rest in a pine forest on the south side of Beach Road. Many of the mustangs added to the herd were taller and heavier than the island ponies. Photograph by Tim Ferry, taken in the 1980s.

a few years went the highest at the auction. I believe the first $1,000 foal was his too. He had one of the largest herds on the island.

The fire company had also added mustangs in 1939, when it acquired 20 from Nevada for genetic diversity (Szymanski, 2007). It is unclear how many of these survived.

Over the years, the fire company has introduced other outside horses to the Chincoteague NWR herd to expand the gene pool and to improve quality, but it has not kept complete records of changes to herd composition. In the early 1900s, Shetland Ponies were added to promote pinto coloration, leaving this group with a greater proportion of pony genes that also decreased the average height (Keiper, 1985). A 1925 newspaper article notes, "The strain is not so pure now as it was in former years. The Shetland pony together with other breeds has been introduced, and the effect has wrought variety" ("Pony Round-Up at Chincoteague to Be the 'Wild West' Show of East," 1925, p. 3). Another article says that Samuel Field "improved" roaming horses "by the importation of valuable stallions" (Marinus, 1929). In 1945, when the National Advisory Committee for Aeronautics, the predecessor of NASA, purchased Wallops Island, the free-roaming ponies that inhabited it were gathered and moved to Assateague (DeVincent-Hayes, Bennett, & Hayes, 2001, p. 16).

Well-bred Arabian horses have also been added to the herd. Al-Marah Sunny Jim was a chestnut Arabian stallion donated to the fire company in 1965 ("Al Marah Sunny Jim," 2012). He was bred to Misty's daughter Stormy twice, producing Rainy and Misty II. He was also bred to Assateague mares and may have been released on the island.

Skowreym, a purebred 1953 gray Arabian stallion was leased to the fire company for two years. A grandson of the celebrated Raffles ("Skowreym," 2012), the 14.1-hand (57-in./1.45-m) gray was a seasoned endurance horse who often placed first or second in high-profile 50- and 100-mile (80- and 161-km) rides. On July 25, 1964. Linda

A graceful Arabian stallion crosses a marsh in the south pasture of the Chincoteague NWR in 1987. Arabians have been periodically outcrossed into the herd to infuse the resulting foals with exotic beauty, elegance, and stamina. Photograph by Tim Ferry.

Tellington-Jones, internationally acclaimed authority on animal behavior and originator of Tellington TTouch Training®, rode Skowreym (unsuccessfully) on the grueling Tevis Cup Ride— 100 rugged miles/161 km in the Sierra Nevadas. He began his sojourn on Chincoteague shortly thereafter.

Many of the outside horses introduced by the fire company were unable to adapt to the barrier island environment and died within a few years. In the mid-1990s, the fire company sent a group of Assateague mares to renowned trainer Stanley G. White at Grandeur Arabians in Florida to mate with purebred Arabian stallions. One filly and several colts (including the feisty buckskin Copper Moose) were kept as 1996 buy-backs and have lived long lives infusing the herd with the bloodlines of champions.

Premierre, a 1991 chestnut Arabian stallion with impeccable bloodlines, served mares on Chincoteague, then was released onto Assateague ("Premierre," 2012). He disappeared in 1999. Some believe he was stolen. Some think he died, though no body was recovered. He sired a number of half-Arabian foals, including island patriarch North Star. After Premierre's disappearance, the fire company kept a donated bay Arabian stallion at the carnival grounds with a group of mares. The registration papers of the resulting foals give his name as Striking Gold; but he was registered under a different name, and the Arabian Horse Association Registry and local sources are unsure of his true origins.

Old photographs can shed light on herd composition. An image taken at Pony Penning in 2000 and circulated on the Internet among Chincoteague Pony enthusiasts depicts an elderly bay mare in the Assateague corral with the south herd. She has a prominent BLM freeze brand on her neck indicating that she was born in 1973 in Nevada. Another image from the 1981 Pony Penning shows an Appaloosa mare and foal, probably mustangs. These

After the swim-back event, a flotilla of spectators in kayaks shadows the ponies back to their fenced enclosures on Assateague. Kayaking affords an optimal vantage point for the pony swim if the visitor is willing to paddle the distance to the crossing site.

two were the only known Appaloosas on Assateague, and it does not appear that this line contributed to the present-day herd.

The 2013 interim Chincoteague Pony management plan signed by Pony Committee Chairman Harry S. Thornton states "A wide variety of breeds such as Morgan, Welsh, Shetland, Arabian, and Mustangs were placed in the Chincoteague pony herd to increase genetic diversity and vigor among the present stock" (Grey, 2014, p. D-17). Other evidence for herd introductions is anecdotal. "Back then, record keeping was not high on their list of priorities, and some of the 'old heads' of the fire company have long since passed away," said Denise Bowden (personal communication, February 5, 2011). Szymanski (2007) writes that Quarter Horses were introduced in the 1960s, and at other times horses and ponies were turned out to run with the herd, details of their heritage lost to time. One former Eastern Shore resident remembers a 1960s news photograph purportedly portraying a Thoroughbred herd sire named "Red" leading a band of Chincoteague mares.

Betts Devine, who spent her childhood on Chincoteague and is a distant cousin of Maureen Beebe of *Misty* fame, recalled,

> My mother was born in 1921, and she remembers the ponies as being between 12 and 14 hands high (she was horse- and pony-crazy all of her life), and being mostly solid colors—bays, browns, chestnuts, and an occasional black. As she grew older, and western movies showed the "painted Indian ponies," the ponies on Assateague were bred with pinto stallions, to produce "splashy" paints. With the Chincoteague Volunteer Fire Company in charge of the breeding of the ponies, since the herds were decimated by the Ash Wednesday Storm, select mares (those with good conformation) have been bred to select stallions to improve the breed and keep from too much in-breeding. Mares have gone to Quarter horse, Morgan, Paint, and Arabian stallions. (Personal communication, April 6, 2012)

Pony Penning week provides children a unique opportunity to observe the behavior of wild horses and to join in the excitement of a roundup.

The fire company tried to maintain the vigor of the herd not only by importing Arabians and mustangs, but also by keeping natives. Sometimes the fire company agrees to accept a previously auctioned horse as a donation. Witch Doctor, a striking dark bay pinto, was sold as a foal to a farm in New York. After the young stallion escaped to cohabit with the wrong group of mares, the fire company agreed to take him back to his Assateague birthplace (Szymanski, 2012). Witch Doctor found his place among the island stallions and acquired a respectable harem.

In the past, ponies from the Maryland herd have been transferred to the Virginia part of the island. Over time, the Park Service relocated 44 Maryland horses to the Virginia herd before the start of immunocontraception in 1994 (Zimmerman, Sturm, Ballou, & Traylor-Holzer, 2006). Many of these horses were notorious for causing difficulties with campers by being too bold or by damaging property in their quest for human food. All horses tested negative for EIA before joining the Chincoteague herd.

Other returns are less compatible with island life. A young stallion was released on Assateague in 2008 and proceeded to engage in serious combat with the island stallions. He attacked with the fearlessness of a kamikaze, earning himself the name of Chaos. The Fire Company decided to sell him before the herd sires suffered serious injury, and he found a peaceful new home with a Maryland breeder (Szymanski, 2012).

The bloodlines of the introduced horses have blended with the native island stock to create a unique breed, the Chincoteague Pony. To keep lines pure, the fire company no longer introduces foreign stock into the refuge population. When outside genes would benefit the health of the herd, a genetically suitable foreign mare may be introduced to mate with a stallion and give birth. After weaning, the foal would remain on the refuge to continue the lineage, while the mare would return to the mainland (Grey, 2014).

Each of the ponies has a story worth telling, but some are legendary. The 1995 stallion Miracle Man started life as an orphaned foal found in Black Duck Marsh by carver, tour boat operator, and Chincoteague town councilman Arthur Leonard. Wearing dress clothes and shoes, Leonard waded into the marsh to rescue the newborn, but the foal was too quick. It took a team of volunteers hours to capture him. The foal had been motherless for several days and had abscesses on his eye and leg. The Leonard family coerced a lactating mare into nursing him and tended to his wounds. He wintered in Florida with a friend of the Leonards, who lavished attention on him and taught him how to bow and shake hands. The fire company released him onto Assateague as a yearling, and in time he became a successful and much-admired harem stallion (Szymanski, 2012).

Miracle Man was a smart stallion with an uncanny internal calendar. Just before the 2008 Pony Penning, the clever horse drove his mares into the channel and swam them to Memorial Park on Chincoteague four days ahead of schedule. (The palomino pinto stallion Prince did the same thing with his band in 2012.)

Another noteworthy Leonard performed a dramatic rescue of a black-and-white mare frozen in an Assateague pond, apparently dead. The late Donald Leonard was a lifetime member of the Chincoteague Volunteer Fire Company and former chairman of the Pony Committee. He and a friend extracted the unfortunate pony from the ice and found that she had a weak heartbeat. They rushed her to the fire station and rigged a sling to help her stand as she recovered. Icy, as she was named, survived to deliver a black colt in the firehouse, named "Little Icicle" (Szymanski, 2012).

The Chincoteague herd differs markedly in composition from other feral horse herds because of human interference. The majority of foals are sold to mainland homes a few months after birth. More colts are sold than fillies, and some of the fillies and an occasional colt return to the breeding herd every year. This results in a sex ratio of 4.6 mares for every stallion, nearly double the ratio seen in the herd on the north end of the island.

The mean age of each pony in the Chincoteague herd was greater, too. Sixty percent of the Virginia horses are adults; less than 40% were mature in the Maryland herd before birth control began. The foaling rate is also considerably higher in Virginia.

The health of the horses in the Virginia herd is enhanced by the management practices of the fire company, including treatment of injuries, annual deworming to improve use of the food they ingest, and vaccinations to prevent serious illnesses. High birth rates are advantageous in a population where there is no shortage of buyers for foals. With the pony swim and auction attracting new visitors every year and foal prices steadily climbing, the sales of each new colt or filly will boost the fire company's revenue.

On Chincoteague, every Pony Penning is at least as exciting as the one before as spectators pack the tiny island in hope of glimpsing the ponies swimming across the channel at slack tide. The details of the horses' origins have been lost to the passage of time, but one thing is clear—the free-roaming horses of Assateague have been in continual residence on the island for centuries, and with conscientious management they may remain in the centuries to come.

References

Al Marah Sunny Jim. (2012). *Pedigree Online All Breed Database*. Retrieved from http://www.allbreedpedigree.com/al+marah+sunny+jim

Alford, R.H., Kasel, J.A., Lehrich, J.R., & Knight, V. (1967). Human responses to experimental infection with influenza A/Equi-2 virus. *American Journal of Epidemiology, 86*(1), 185–192.

Amrhein, J., Jr. (2007). *The hidden galleon: The true story of a lost Spanish ship and the legendary wild horses of Assateague Island.* Kitty Hawk, NC: New Maritima Press.

Anderson, K. (1995). Feeding and care of orphaned foals (NebGuide G95-1237-A). Retrieved from http://digitalcommons.unl.edu/cgi/viewcontent.cgi?article=1233

Anderson, V.D. (2002). Animals into the wilderness: The development of livestock husbandry in the seventeenth-century Chesapeake. *William and Mary Quarterly, 3rd Series, 59*(2), 377–408. Retrieved from http://www.jstor.org/stable/3491742

Assateague Island, nature and science. (2010, November 21). Retrieved from http://www.nps.gov/asis/naturescience/index.htm

Baden-Powell, R.S.S. (1885). *Cavalry instruction. Course of lectures ordered by General Order 30, dated 1st of March, 1884, for instruction of cavalry, yeomanry, & c., & c.* London, United Kingdom: Harrison & Sons.

Beck, V. (2003). Advances in life sciences and bioterrorism. *EMBO Reports, 4* (Special Issue), S53–S56. doi: 10.1038/sj.embor.embor853

Beverley, R. (1705). *The history and present state of Virginia, in four parts.* London, United Kingdom: For R. Parker. Retrieved from http://docsouth.unc.edu/southlit/beverley/beverley.html

Bonetti, T. (Ed.), (2014) *Chincoteague and Wallops Island National Wildlife Refuges draft comprehensive conservation plan and draft environmental impact statement.* Chincoteague, VA: Chincoteague National Wildlife Refuge.

Boswell, R., & Mason, W. (2010, October 7). *Buyback Babes, island firemen have more to mend than fences.* Retrieved from http://wildponytales.info/archives/1337

Breeding horses come east. (1977, August 27). *Harrisonburg Daily News-Record* (Harrisonburg, VA), p. 5.

Bruce, P. A. (1907). *Economic history of Virginia in the seventeenth century: An inquiry into the material condition of the people, based upon original and contemporaneous records.* New York, NY: MacMillan.

Buescher, E.L., O'Dell, E.T., Scheider, F.G., Bourke, A.T.C., Eldridge, B.F., Thompson, E.G., . . . Suyamoto, W. (1963). Project 3A 0 12510 A 806: Military preventive medicine. (1963). In *Annual progress report, 1 July 1962–30 June 1963*, Vol. I, pp. 203–220. Washington, DC: Walter Reed Army Institute of Research.

Byrne, R.J. (1968). Sleeping sickness wakes again. In U.S. Department of Agriculture, *Science for better living: The yearbook of agriculture, 1968*, pp. 355–359. Washington, DC: Government Printing Office.

Cahill, A.E., Aiello-Lammens, M.E., Fisher-Reid, M.C., Xia, H., Karanewsky, C.J., Hae, Y.R., . . . Wiens, J.J. (2012, October 17). How does climate change cause extinction? *Proceedings of the Royal Society B: Biological Sciences.* doi: 10.1098/rspb.2012.1890 1471-2954

Cameron, T.P., Alford, R.H., Kasel, J.A., Harvey, E.W., Byrne, R.J., & Knight, V. (1967). Experimental equine influenza in Chincoteague Ponies. *Proceedings of the Society for Experimental Biology and Medicine, 124*(2), 510–515. doi: 10.3181/00379727-124-31777

Cameron, T.P., Kasel, J.A., & Couch, R.B. (1974). Persistence of antibody to envelope antigens of Heq2Neq2 virus in ponies after infection and vaccination. *Proceedings of the Society for Experimental Biology and Medicine, 146*(3), 658–660. doi: 10.3181/00379727-146-38166

Carlton, J. (2001, October 22). Of microbes and mock attacks—Years ago, the military sprayed germs on U.S. cities. *Wall Street Journal*. Retrieved from http://online.wsj.com/news/articles/SB1003703226697496080

Carson, R. (1947). *Chincoteague: A National Wildlife Refuge* (Conservation in Action 1). Washington, DC: U.S. Fish and Wildlife Service. Retrieved from http://digitalcommons.unl.edu/usfwspubs/1

Cherrix, M.J. (2011). *Assateague Island*. Charleston, SC: Arcadia Publishing.

Chesapeake storm killed hundreds of wild ponies. (1933, September 5). *New York Times* (New York, NY), p. 9.

Chincoteague. (1891, June 6). *Peninsula Enterprise* (Accomac, VA), p. 3. Retrieved from http://chroniclingamerica.loc.gov/lccn/sn94060041/1891-06-06/ed-1/seq-3/

Chincoteague Volunteer Fire Company. (n.d.). Chincoteague Volunteer Fire Company History, 1905–1950. Retrieved from http://cvfc3.com/about-us/history

Cohen, J. (2011, October 17). *Did Jamestown's settlers drink themselves to death?* Retrieved from http://www.history.com/news/did-jamestowns-settlers-drink-themselves-to-death

Cole, L.A. (1997). *The eleventh plague: The politics of biological and chemical warfare*. New York, NY: Henry Holt.

Couch, R.B., Douglas, Jr., R.G., Kasel, J.A., Riggs, S. & Knight, V. (1969). Letters to *Nature*. Production of the influenza syndrome in man with equine influenza virus. *Nature*, 224(5218), 512–514. doi: 10.1038/224512a0

Convention on the Prohibition, Development, Production, and Stockpiling of Bacteriological (Biological) and Toxin Weapons and on Their Destruction, April 10, 1972, 26 U.S.T. 583, 1015 U.N.T.S 163, retrieved from http://www.unog.ch/80256EDD006B8954/(httpAssets)/C4048678A93B6934C1257188004848D0/$file/BWC-text-English.pdf

Covington, H.F. (1915). The discovery of Maryland, or Verrazzano's visit to the Eastern Shore. *Maryland Historical Magazine 10*(3), 199–217.

DeVincent-Hayes, N., & Bennett, B. (2000). *Chincoteague and Assateague islands*. Charleston, SC: Arcadia Publishing.

DeVincent-Hayes, N., Bennett, B., & Hayes, J.R. (2001). *Wallops Island*. Charleston, SC: Arcadia Publishing.

Doney, S.C., Ruckelshaus, M., Duffy, J.E., Barry, J.P., Chan, F., English, C.A., . . . Talley, L.D. (2012). *Annual Review of Marine Science, 4*, 11–37. doi: 10.1146/annurev-marine-041911-111611

Dunbar, G.S. (1958). *Historical geography of the North Carolina Outer Banks*. Louisiana State University Studies, Coastal Studies Series 3. Baton Rouge: Louisiana State University Press.

Duncan, P., & D'Herbes, J.M. (1982). The use of domestic herbivores in the management of wetlands for waterbirds in the Camargue, France. In D.A. Scott (Ed.), *Managing wetlands and their birds: A manual of wetland and waterfowl management. Proceedings of the third Technical Meeting on Western Palearctic Migratory Bird Management* (pp. 51–56). Slimbridge, United Kingdom: International Waterfowl Research Bureau.

Eggleston, E. (1884, January). Husbandry in colony times. *Century Illustrated Monthly Magazine, 27*(3), 431–448.

Eshelman, R.E., & Russell, P.A. (2004, July 21). *Historic context study of waterfowl hunting* camps and related properties within Assateague Island National Seashore, Maryland and Virginia. N.P.: Eshelman and Associates. Retrieved from http://www.nps.gov/asis/parkmgmt/upload/AssateagueHuntingLodgesStudyFinalReport.pdf

Farrer, J. (1649). *A perfect description of Virginia*. London, United Kingdom: For Richard Wodenoth. Retrieved from http://etext.lib.virginia.edu/etcbin/jamestown-browse?id=J1080

Farrer, J., & Farrer, V. (Cartographers). (1667). *A mapp of Virginia discouered to ye Hills, and in it's Latt: From 35 dg: & 1/2 neer Florida, to 41 deg: bounds of new England....* Retrieved from http://memory.loc.gov/gmd/gmd388/g3880/g3880/ct000903.jp2

Flanagan, M., Leighton, T., & Dudley, J. (2011, June). *Anticipating viral species jumps: Bioinformatics and data needs* [Report Number OSRD 2011 020, Contract/MIPR Number 01-03-D-0017]. Ft. Belvoir, VA: Defense Threat Reduction Agency, Office of Strategic Research and Dialogues.

Force, P. (Ed.). (1846). *American archives: Fourth series. Containing a documentary history of the English colonies in North America ...* (Vol. 6). Washington, D.C.: M. St. Clair Clarke and Peter Force.

Fried, K. (2010, July 18). Where the wild ponies swim. *Parade*, 12.

Friedman, S.M. (n.d.). *The inflation calculator*. Retrieved from http://www.westegg.com/inflation/

Frydenborg, K. (2012). *The wild horse scientists*. Boston, MA: Houghton Mifflin Harcourt.

Giffin, J.M., & Darling, K. (2007). *Veterinary guide to horse breeding* (Kindle edition). Hoboken, NJ: Howell Book House. (Original work published 1999).

Goodloe, R.B., Warren, R.J., Cothran, E.G., Bratton, S.P., & Trembicki, K.A. (1991). Genetic variation and its management applications in eastern U.S. feral horses. *Journal of Wildlife Management, 55*(3), 412–421.

Goodloe, R.B., Warren, R.J., Osborn, D.A., & Hall, C. (2000). Population characteristics of feral horses on Cumberland Island, Georgia and their management implications. *Journal of Wildlife Management, 64*(1), 114–121. doi: 10.2307/3802980

Graham, J. (2007, August 7). *Geologic resources inventory scoping summary, Cumberland Island National Seashore, Georgia*. Washington, DC: U.S. National Park Service, Geologic Resources Division. Retrieved from http://www.nature.nps.gov/geology/inventory/publications/s_summaries/CUIS_gri_scoping_summary_2009-0807.pdf

Grey, E. (2014, May). 2013 interim Chincoteague Pony management plan. In T. Bonetti (Ed.), *Chincoteague and Wallops Island National Wildlife Refuges draft comprehensive conservation plan and draft environmental impact statement* (pp. D-1–D-54). Chincoteague, VA: Chincoteague National Wildlife Refuge.

Greif, K.F., & Merz, J.F. (2007). *Current controversies in the biological sciences: Case studies of policy challenges from new technologies*. Cambridge, MA: Massachusetts Institute of Technology.

Grimstad, P.R. (2001). Cache Valley virus. In M.W. Service (Ed.), *Encyclopedia of arthropod-transmitted infections of man and domesticated animals* (pp. 101–104). Wallingford, United Kingdom, and New York, NY: CABI Publishing.

Hall, D. (Interviewer), Reed, T., & Daisey, D. (Interviewees). (2012). *Tom Reed and Dave Hall, Chincoteague, VA March 24, 1989* [Interview transcript]. Retrieved from http://digitalmedia.fws.gov/cdm/singleitem/collection/document/id/911/rec/5

Hall, M., Casey, J., & Wells, D. (2004). A brief history of the Maryland coastal bays. In C.E. Wazniak & M.R. Hall (Eds.), *Maryland's coastal bays: Ecosystem health assessment* (DNR-12-1202-0009) (pp. 2-2-2-16). Annapolis, MD: Maryland Department of Natural Resources, Tidewater Ecosystem Assessment.

Harrison, F. (1927). The equine F F Vs: A study of the evidence for the English horses imported into Virginia before the Revolution. *Virginia Magazine of History and Biography, 35*(4), 329–370.

Hayter, E.W. (1963). Livestock-fencing conflicts in rural America. *Agricultural History, 37*(1) 10–20.

Hayward, L. (2007). *State of the parks: Assateague Island National Seashore, a resource assessment.* Washington, DC: National Parks Conservation Association.

Help preserve access to Assateague Island, VA. (n.d.) Retrieved from http://www.chincoteague.com/preserve-access/

Hening, W.W. (1823). *The statutes at large; Being a collection of all the laws of Virginia from the first session of the legislature in the year 1619* (Vols. 1–2). New York, NY: For the author.

Henry, M. (1947). *Misty of Chincoteague.* Chicago, IL: Rand, McNally.

Henry, M. (2007). *Stormy, Misty's foal.* New York, NY: Aladdin Paperbacks (Original work published 1963).

The history of race riding and the Jockeys Guild. (1999). Paducah, KY: Turner Publishing.

Hoke, Jr., C.H. (2005). History of U.S. military contributions to the study of viral encephalitis. *Military Medicine, 170*(4), 92–105.

Holmes, T. (1835, November). Some account of the wild horses of the sea islands of Virginia and Maryland. *Farmers' Register, 3*(7), 417–419.

Hooks, R.O. (2006). *Pine Ridge Horse Farm's illustrated guide to the wild pony auction at Chincoteague: A world famous attraction* (2nd ed.). Salisbury, MD: Pine Ridge Horse Farm.

Hunt, L. (1991). *Secret Agenda: The United States government, Nazi scientists, and Project Paperclip, 1945 to 1990.* New York, NY: St. Martin's Press.

Jester, W.F. (1977, December 5). Interview by Karen Croner. Retrieved from http://esplgenealogy.org/cohistory/transcripts/JESTER, BILL.pdf

Jones, H. (1865). *The present state of Virginia* (Original work published 1724). New York, NY: For Joseph Sabin.

Kasel, J.A., & Couch, R.B. (1969). Experimental infection in man and horses with influenza A viruses. *Bulletin of the World Health Organization, 41*(3), 447–452.

Kasel, J.A., Byrne, R.J., Harvey, E.W., & Schillinger, R. (1968). Experimental human B influenza virus infection in Chincoteague Ponies. *Nature, 219*(5157), 968–969. doi: 10.1038/219968b0

Kasel, J.A., Fulk, R.V. & Harvey, E.W. (1969). Susceptibility of Chincoteague ponies to antigenically dissimilar strains of human type A2 influenza virus. *Journal of Immunology, 103*(2), 369–371.

Kasel, J.A., Fulk, R.V., Haase, A.T., & Huber, M. (1968). Clinical investigations in viral infections and diseases. PHS-NIH individual project report, July 1, 1967 through June 30, 1968. Serial No. NIAID-14(c). In *Annual report of program activities, National Institutes of Health, 1967–1968: National Institute of Allergy and Infectious Diseases* (pp. 22–25). [Washington, DC: National Institutes of Health.]

Keiper, R. (1985). *The Assateague ponies.* Atglen, PA: Schiffer Publishing.

Keiper, R., & Houpt, K. (1984). Reproduction in feral horses: An eight-year study. *American Journal of Veterinary Research, 45*(5), 991–995.

Kirkpatrick, J. (1994). *Into the wind: Wild horses of North America.* Minocqua, WI: Northword Press.

Kirkpatrick, J.F., & Turner, A. (2003). Absence of effects from immunocontraception on seasonal birth patterns and foal survival among barrier island wild horses. *Journal of Applied Animal Welfare Science, 6*(4), 301–308.

Kirkpatrick, J.F., & Turner, A. (2007). Immunocontraception and increased longevity in equids. *Zoo Biology, 25,* 237–244. doi: 10.1002/zoo.20109

Kirkpatrick, J.F., & Turner, J.W. (1991). Compensatory reproduction in feral horses. *Journal of Wildlife Management, 5*(4), 649–652.

Kobell, R. (2006, July 28). Where men saddle up and ponies paddle over. *Los Angeles Times.* Retrieved from http://articles.latimes.com/2006/jul/28/nation/na-saltwater28

Laing, W.N. (1959). Cattle in seventeenth-century Virginia. *Virginia Magazine of History and Biography, 67*(2), 143–163.

Laird, M.R. (1970, July 6). *Memorandum for the president: National security decision memoranda 35 and 44.* Retrieved from http://www2.gwu.edu/~nsarchiv/NSAEBB/NSAEBB58/RNCBW22.pdf

Langley, S.B.M., & Jordan, B.A. (2007). *Archeological overview & remote sensing survey for maritime resources in Maryland state waters from the Ocean City Inlet to the Delaware Line, Worcester County, Maryland.* Crownsville, MD: Maryland State Historic Preservation Office. Retrieved from http://mht.maryland.gov/documents/pdf/ archeology_mmap_oceancity_survey_dnr.pdf

Langley, S.B.M., Van Driessche, P., & Charles, J. (2009). *Archeological overview and assessment of maritime resources in Assateague Island National Seashore, Worcester County, Maryland, & Accomack County, Virginia* (revised ed.). Crownsville, MD: Maryland State Historic Preservation Office. Retrieved from http://mht.maryland.gov/documents/PDF/Archeology_MMAP_AINS_Overview&Assess_optimized.pdf

Lanman, C. (1856). *Adventures in the wilds of the United States and British American provinces,* Vol. 2. Philadelphia: John W. Moore.

Leonard, D. (2006, April 25). Interview by Margo Hunt. Retrieved from http://espl- genealogy.org/cohistory/transcripts/LEONARD DONALD.pdf

Levin, P., Ellis, J., Petrik, R., & Hay, M. (2002). Indirect effects of feral horses on estuarine communities. *Conservation Biology, 16*(5), 1364–1371. doi: 10.1046/j.1523-1739. 2002.01167.x

Lindler, L.E., Lebeda, F.J., & Korch, G. (2005). *Biological weapons defense: Infectious disease and counterbioterrorism.* Totowa, NJ: Humana Press.

Local news. (1898, June 4). *Peninsula Enterprise* (Accomac, VA), p. 3. Retrieved from http://chroniclingamerica.loc.gov/lccn/sn94060041/1898-06-04/ed-1/seq-3/

Lucas, Z., Raeside, J.I., & Betteridge, K.J. (1991). Non-invasive assessment of the incidences of pregnancy and pregnancy loss in the feral horses of Sable Island. *Journal of Reproductive Fertility, 44*(Suppl.), 479–488.

Lynghaug, F. (2009). *The official horse breeds standards guide: The complete guide to the standards of all North American equine breed associations.* Minneapolis, MN: Voyageur Press.

Mackintosh, B. (2003, October 27). *Assateague Island National Seashore: An administrative history*. Washington, DC: National Park Service, History Division. (Original work published 1982.) Retrieved from http://www.nps.gov/asis/parkmgmt/upload/asisadminhistory.pdf

Mangold, T., & Goldberg, J. (2000). *Plague wars: A true story of biological warfare*. New York: Macmillan.

Mariner, K. (2003). *Once upon an island: The history of Chincoteague*. New Church, VA: Miona Publications.

Marinus, J. (1929, September 28). A little journey to Assateague. *Peninsula Enterprise* (Accomac, VA). Retrieved from http://eshore.vcdh.virginia.edu/node/1993

Maryland Department of Natural Resources, Resource Planning. (2005, October). *Assateague State Park land unit plan*. Retrieved from http://www.dnr.state.md.us/irc/docs/00011180.pdf

McCue, P.M. [2009]. *Foal heat breeding*. Retrieved from http://csu-cvmbs.colostate.edu/Documents/learnmares3-breed-foalheat-2009.pdf

Meyer, J., (2008, August 8). Inquiry sought into anthrax probe. *Los Angeles Times*. Retrieved from http://articles.latimes.com/2008/aug/08/nation/na-anthrax8

Migratory Bird Treaty Act of 1918, 16 U.S.C. §§ 703–712. (2012).

Morens, D.M., & Taubenberger, J.K. (2010). Historical thoughts on influenza viral ecosystems, or behold a pale horse, dead dogs, failing fowl, and sick swine. *Influenza and Other Respiratory Viruses, 4*(6). 327–337. doi: 10.1111/j.1750-2659.2010.00148.x

Move 'em out: Wild ponies herded for annual penning. (1981, July 30). *Annapolis Capital*, p. 4.

Mustangs not accepted, yet. (1977, October 5). *Radford News Journal* (Radford, VA), p. 6.

Mustangs to cross breed with endangered pony herd. (1976, July 22). *Bryan Times* (Bryan, TX), p. 14.

N.C. Office of Conservation, Planning, and Community Affairs. (2010, August 25). *North Carolina ecosystem response to climate change: DENR assessment of effects and adaptation measures* [Draft]. Retrieved from http://www.climatechange.nc.gov/pages/ Climate Change/Climate_Change_Ecosystem_Assessment_Summary.pdf

National Audubon Society. (2007). *Barrier island/lagoon system, Northampton and Accomack counties* (Audubon Important Bird Areas). Retrieved from http://web4.audubon.org/bird/iba/virginia/Documents/Barrier Island_Lagoon System.pdf

National Oceanic and Atmospheric Administration, National Climatic Data Center. (2012, August 21). *Global warming: Frequently asked questions*. Retrieved from http://www.ncdc.noaa.gov/cmb-faq/globalwarming.html

National Oceanic and Atmospheric Administration, National Climatic Data Center (2012, September). *State of the climate: Global analysis for August 2012*. Retrieved from http://www.ncdc.noaa.gov/sotc/global/2012/8

National Wildlife Refuge System Administration Act of 1966, 16 USC § 668dd (2000).

Nevada ponies imported. (1977, August 27). *Annapolis Capital*, p. 5.

Nieves, D.P. (2009, August 26). Application of the Sea Level Affecting Marshes Model (SLAMM 5.0.2) in the Lower Delmarva Peninsula (Northampton and Accomack counties, VA / Somerset and Worcester counties, MD). Arlington, VA: National Wildlife Refuge System Conservation Biology Program. Retrieved from http://www.slammview.

org/slammview2/reports/LDP_ChincoteagueFinal.pdf

Nova Scotia, House of Assembly. *Journal and Proceedings* (1892) at 119.

Peck, K.J. (2008). *Horse husbandry in colonial Virginia: An analysis of probate inventories in relation to environmental and social changes* (Unpublished honors thesis). College of William and Mary, Williamsburg, VA.

Pendleton, E.A., Williams, S.J., & Thieler, E.R. (2004). *Coastal vulnerability assessment of Assateague Island National Seashore (ASIS) to sea level rise* (U.S. Geological Survey Open-File Report 2004-1020, Electronic Book). Retrieved from http://pubs.usgs.gov/of/2004/1020/images/pdf/asis.pdf

Percy, G. (1625). *A trewe relacyon of the p[ro]cedeings and ocurrentes of momente w[hi]ch have hapned in Virginia....* Retrieved from http://www.history.org/foundation/journal/winter07/A Trewe Relation.pdf

Pilkey, O., Rice, T., & Neal, W. (2004). *How to read a North Carolina beach.* Chapel Hill: University of North Carolina Press.

Pippin, J. (2005, January 11). Head 'em up, move 'em out. *Jacksonville Daily News* (Jacksonville, NC). Retrieved from http://www.jdnews.com/news/horses-19054-herd-island.html

Pleasants, B. (1999). *Chincoteague pony tales.* Columbus, GA: Brentwood Christian Press.

Pony Round-Up at Chincoteague to Be the "Wild West" Show of East. (1925, August 3). *Lawrence Journal-World* (Lawrence, KS), p. 3. Retrieved from http://news.google.com/newspapers?nid=2199&dat=19250803&id=xNZkAAAAIBAJ&sjid=zHUNAAAAIBAJ&pg=2990,4305946

Premierre. (2012). *Pedigree Online All Breed Database.* Retrieved from http://www. all-breedpedigree.com/premierre

Proposed Comprehensive Conservation Plan (CCP) for the Chincoteague National Wildlife Refuge: Oversight Hearing before the Subcommittee on Fisheries, Wildlife, Oceans and Insular Affairs of the Committee on Natural Resources, U.S. House of Representatives, 112th Cong. 9 (2012a) (testimony of Wendi Weber).

Proposed Comprehensive Conservation Plan (CCP) for the Chincoteague National Wildlife Refuge: Oversight Hearing before the Subcommittee on Fisheries, Wildlife, Oceans and Insular Affairs of the Committee on Natural Resources, U.S. House of Representatives, 112th Cong. 14 (2012b) (testimony of Jack Tarr).

Proposed Comprehensive Conservation Plan (CCP) for the Chincoteague National Wildlife Refuge: Oversight Hearing before the Subcommittee on Fisheries, Wildlife, Oceans and Insular Affairs of the Committee on Natural Resources, U.S. House of Representatives, 112th Cong. 17 (2012c) (testimony of Wanda Thornton).

Protocol for the Prohibition of the Use of Asphyxiating, Poisonous or other Gases, and of Bacteriological Methods of Warfare, June 17, 1925, 26 U.S.T. 571, 94 L.N.T.S. 65, retrieved from http://disarmament.un.org/treaties/t/1925/text

Public Health England. (n.d.). *Influenza pandemics—history.* Retrieved from http://www.hpa.org.uk/Topics/InfectiousDiseases/InfectionsAZ/PandemicInfluenza/History/

Pyle, H. (1877, April). Chincoteague, the island of Ponies. *Scribner's Monthly, 13*(6), 737–746.

Rennicke, J. (2007, Fall). A climate of change. *National Parks, 81*(4), 26–31.

Rood, R.N. (1967). *Hundred acre welcome: The story of a Chincoteague pony.* Brattleboro, VT: Stephen Greene Press.

Rubin, J. (2007). *The living weapon: Program transcript*. Retrieved from http://www-tc.pbs. org/wgbh/americanexperience/media/uploads/special_features/download_files/ weapon_transcript.pdf

Rudman, R., & Keiper, R.R. (1991). The body condition of feral ponies on Assateague Island. *Equine Veterinary Journal, 23*(6), 453–456. doi: 10.1111/j.2042-3306.1991. tb03760.x

Ryan, C.P. (2008). Zoonoses likely to be used in bioterrorism. *Public Health Reports, 123*(3), 276–281.

Sallenger, A.H., Jr., Doran, K.S., & Howd, P.A. (2012). Hotspot of accelerated sea-level rise on the Atlantic coast of North America. *Nature Climate Change, 2*, 884–888. doi: 10.1038/nclimate1597

Salmon, R. (n.d.). *Swine flu: What next?* National Public Health Service for Wales, Communicable Disease Surveillance Centre. Retrieved from http://www.wales.nhs.uk/ sites3/Documents/882/R Salmon.pdf

Schoenherr, I. (2010, September 12). *"That miserable engagement"* [Web log post]. Retrieved from http://howardpyle.blogspot.com/2010/09/that-miserable-engagement.html

Seale, J. (1989). Crossing the species barrier—Viruses and the origins of AIDS in perspective. *Journal of the Royal Society of Medicine, 82*(9), 519–523.

Sharp, G.B., Kawaoka, Y., Jones, D.J., Bean, W.J., Pryor, S.P., Hinshaw, V., & Webster, R.G. (1997). Coinfection of wild ducks by influenza A viruses: Distribution patterns and biological significance. *Journal of Virology, 71*(8), 6128–6135.

Sharpless, J.T. (1830). Chesapeake duck shooting. *Cabinet of Natural History and American Sports, 1*, 41–46. Retrieved from http://www.scribd.com/doc/85925483/ The-Cabinet-of-Natural-History-and-American-Rural-Sports-Vol-1

Shaughnessy, L. (2009, April 22). *Army: 3 vials of virus samples missing from Maryland facility*. Retrieved from http://edition.cnn.com/2009/US/04/22/missing.virus. sample/ index.html

Shomette, D.G. (2008). The price of amity: Of wrecking, piracy, and the tragic loss of the 1750 Spanish treasure fleet. *The Northern Mariner/le marin du nord, 18*(3–4), 25–48.

Skinner, J.S. (1843). The horse, in England and America—As he has been, and as he is. In W. Youatt & J.S. Skinner, *The horse* (2nd ed.) (pp. 17–34). Philadelphia, PA: Lea and Blanchard.

Skowreym. (2012). *Pedigree Online All Breed Database*. Retrieved from http://www. all-breedpedigree.com/skowreym

Smith, J. (1624). *The generall historie of Virginia, New-England, and the Summer Isles*. London, United Kingdom: For Michael Sparkes.

Spain loans artifacts to Assateague Island National Seashore. (2007, September 12). Retrieved from http://www.nps.gov/history/archeology/sites/npsites/assateague.htm

Spaulding, J. (1968, December 21). Flu virus may first grow potent in animals. *Milwaukee Journal*, p. 19.

Spies, J.R. (1977). *The wild ponies of Chincoteague*. Cambridge, MD: Tidewater Publications.

Stewart, D.F. (July 24, 1977). Assateague ponies: A new look at their origin. *Baltimore Sun*, pp. I4–I7.

Subcommittee on Zinc Cadmium Sulfide, Committee on Toxicology, Board on Environmental Studies and Toxicology, Commission on Life Sciences, National Research

Council. (1997). *Toxicologic assessment of the army's zinc cadmium sulfide dispersion tests.* Washington, DC: National Academy Press.

Szymanski, L. (2007). *Out of the sea: Today's Chincoteague pony.* Centreville, MD: Tidewater Publishers.

Szymanski, L (with Emge, P.). (2012). *Chincoteague ponies: Untold tails.* Atglen, PA: Schiffer Books.

Taggart, J.B. (2008). Management of feral horses at the North Carolina Estuarine Research Reserve. *Natural Areas Journal, 28*(2), 187–195.

Tansey, B. (2004, October 31). Serratia has dark history in region/Army test in 1950 may have changed microbial ecology. *SFGate.* Retrieved from http://www.sfgate.com/health/article/Serratia-has-dark-history-in-region-Army-test-2677623.php

Tateo, A., Maggiolino, A., Padalino, B., & Centoducati, P. (2013). Behavior of artificially suckled foals. Journal of Veterinary Behavior, 8(3), 162–169.

Taubenberger, J.K., & Morens, D.M. (2013). Influenza viruses: Breaking all the rules. *mBio, 4*(4). doi: 10.1128/mBio.00365-13

Threlkeld, L. (2010, January 24). The Great Epizootic: Sick horses bring the economy to a square halt. *Eventing Nation.* Retrieved from http://eventingnation.com/home/the-great-epizootic-sick-horses-bring-the-economy-to-a-square-halt.html

Todd, J.D., Lief, F.S., & Cohen D. (1970). Experimental infection of ponies with the Hong Kong variant of human influenza virus. *American Journal of Epidemiology, 92*(5), 330–336.

Town of Chincoteague. (2010). *Beach access questionnaire.* Retrieved from http://www.chincoteague.com/preserve-access/Survey- Summary.pdf

Tucker, J.B., & Mahan, E.R. (2009, October). *President Nixon's decision to renounce the U.S. offensive biological weapons program* [Center for the Study of Weapons of Mass Destruction Case Study 1]. Washington, DC: National Defense University Press.

U.S. Department of Veterans Affairs. (2013, May 30). *Pandemic influenza (flu).* Retrieved from http://www.pandemicflu.va.gov/about/index.asp

U.S. Environmental Protection Agency. (2013, January 8). *Future climate change.* Retrieved from http://epa.gov/climatechange/science/future.html

U.S. Fish and Wildlife Service. (1999, December). *Assateague Island Lighthouse.* [Washington, DC]: Author.

U.S. Fish and Wildlife Service. (2012, August). *Chincoteague National Wildlife Refuge Comprehensive Conservation Planning Newsletter.* Retrieved from http://www.chincoteague.com/pdfs/usfws_newsletter_aug_2012.pdf

U.S. Fish and Wildlife Service, Bureau of Sport Fisheries and Wildlife. (1962, June). *Wildlife research progress* [Bureau of Sport Fisheries and Wildlife, Circular 146]. Washington, DC: U.S. Fish and Wildlife Service, Bureau of Sport Fisheries and Wildlife.

Upshur, T.T. (1901). Eastern-Shore history. *Virginia Magazine of History and Biography, 9*(1), 88–99.

Vallandigham, E.L. (1893, August 27). An island of ponies: American rivals of the little Shetlands. *Boston Sunday Globe,* p. 28.

Valleron, A.-J., Cori, A., Valtat, S., Meurisse, S., Carrat, F., & Boëlle, P.-Y. (2010). Transmissibility and geographic spread of the 1889 influenza pandemic. *Proceedings of the National Academy of Sciences of the United States of America, 107*(19), 8778–8781). doi: 10.1073/pnas.1000886107

Vavra, M. (2005). Livestock grazing and wildlife: Developing compatibilities. *Rangeland Ecology & Management, 58*(2), 128–134. doi: 10.2111/1551-5028(2005)58<128:LGA WDC>2.0.CO;2

Virginia Commission on Boundary Lines. (1873). *Report and accompanying documents of the Virginia commissioners appointed to ascertain the boundary line between Maryland and Virginia.* Richmond, VA: R.F. Walker, Superintendent of Public Printing.

Wallace, J.H. (1897). *The horse of America in his derivation, history, and development.* New York, NY: Author.

Warren, M.R. (1913, October). The island of Chincoteague. *Harper's Monthly Magazine, 127*(761), 775–785.

Washburn, R.M. (1877). *The peoples' condensed library: A compendium of universal knowledge.* Chicago, IL: Author.

Waterhouse, E. (2003). *Chincoteague summer of 1948: A waterman's childhood stories.* Lincoln, NE: iUniverse.

Wheelis, M., Rózsa, L., & Dando, M. (2006). *Deadly cultures: Biological weapons since 1945.* Cambridge, MA: Harvard University Press.

Wild horses in Maryland. (1874, February 19). *Indiana Progress* (Indiana, PA), p. 3.

The wild ponies of a Virginia island. (1910, July 21). *Arizona Republican* (Phoenix), p. 9.

Willis, D. (1995, July 25). Humane Society protests pony penning and sale. *Baltimore Sun* (Baltimore, MD). Retrieved from http://articles.baltimoresun.com/1995-07-25/news/ 1995206053_1_pony-humane-society-sale

Wise, J.C. (1911). *Ye Kingdome of Accawmacke or the Eastern Shore of Virginia in the seventeenth century.* Richmond, VA: Bell Book and Stationery.

Zimmerman, C., Sturm, M., Ballou, J., & Traylor-Holzer, K. (Eds.). (2006). *Horses of Assateague Island population and habitat viability assessment workshop: Final report.* Apple Valley, MN: IUCN/SSC Conservation Breeding Specialist Group.

Chapter 5

Currituck Banks, North Carolina

The End of the Road

Corolla
and Vicinity

NORTH CAROLINA–VIRGINIA

■ Highway	☐ Beach	
■ Road	☐ Marsh	
▮▮▮ ORV route	☐ Maritime forest	
▬ ▬ Trail	☐ Private	

0 1 2 3 4 5 6 km
0 1 2 3 4 mi

Virginia Beach

Sandbridge

BACK BAY
NATIONAL
WILDLIFE REFUGE

FALSE CAPE
STATE PARK
(No vehicular access)

BACK
BAY

Wash Woods

VIRGINIA
NORTH CAROLINA

615

Carova

MACKAY ISLAND
NATIONAL
WILDLIFE REFUGE

CURRITUCK
NATIONAL
WILDLIFE REFUGE

168

KNOTTS
ISLAND

Currituck

Ferry

N

MONKEY
ISLAND

NATURE
CONSERVANCY

CURRITUCK BANKS
NCNERR

158

Corolla

ATLANTIC OCEAN

Coinjock

CURRITUCK SOUND

12

Duck

DEWS
ISLAND

158

Southern
Shores

A round the time Paul Revere made his famous ride in Massachusetts on a Narragansett Pacer, a girl on a Banker Pony is said to have made a similar trek to become a heroine of the Revolutionary War. The legend, passed along through oral traditions, tells of a teenager named Betsy Dowdy who rode her mare from Currituck Banks to Hertford, North Carolina, on a cold winter night to inform the militia of the threat from British troops.

In the winter of 1775, John Murray, 4th Earl of Dunmore and royal governor of Virginia, had fled his palace in Williamsburg and begun making hostile raids on southeastern Virginia. He planned to best Colonel William Woodford of the 2nd Virginia Regiment at Great Bridge, then attack the Albemarle Sound region of North Carolina to obtain supplies and mounts for his soldiers. Locals were distraught, fearing that if Dunmore's campaign succeeded he would leave them destitute between harvests. Only the Perquimans Militia of General William Skinner had the power to help repel the attack, and it was too far away to respond in time.

The legend says that Betsy Dowdy, an Outer Banks teenager living opposite Knotts Island, heard of Dunmore's plans and resolved to gallop her rugged Banker mare, Black Bess, bareback through the night to alert the troops.

Creecy (1901, p. 6) tells her story:

> Down the beach she went, Black Bess doing her accustomed work. She reached the point opposite Church's Island, dashed into the shallow ford of Currituck Sound, and reached the shore of the Island . . . Through the divide, on through Camden, the twinkling stars her only light, over Lamb's old ferry, into Pasquotank, by the "narrows" (now Elizabeth City) to Hartsford's ford, up the Highlands of Perquimans, onto Yoepim creek, and General William Skinner's hospitable home was reached.

Betsy is said to have ridden 51 mi/82 km bareback through the dark winter night to Gen. Skinner's headquarters. Skinner dispatched his troops to meet Lord Dunmore's army, and they won the battle of Great Bridge. There is no documentation to verify the story, but the legend is a treasured part of Carolina coastal folklore.

In 2012, the author gained insight into what Betsy might have experienced. Riding with Steve Edwards of Mill Swamp Indian Horses in Smithfield, Va., who breeds domesticated Banker horses, she galloped a once-wild Corolla stallion named Manteo through a moonless Virginia wood in darkness so deep she literally could not see her hand in front of her face. He ran effortlessly along a pitch-dark forest trail over exposed roots and rocks, through deep mud and into pools of standing water up to his hocks. The little stallion, an athlete accustomed to 50-mi/80-km trail riding events, moved tirelessly at speed for the better part of an hour, never faltering, never distracted or disobedient. Effectively blinded by the night, she had no choice but to trust a wild-born Banker stallion that could see the way while she could not. It was a profound experience that she will remember always.

Free-roaming horses have inhabited the North Carolina Outer Banks for centuries. At one time, they numbered in the thousands, ranging over the 175-mile/282-km span of islands from Shackleford Banks north beyond False Cape in Virginia. As recently as 1926, a writer for *National Geographic* stated that 5,000–6,000 horses roamed the Outer Banks (Chater, 1926). They grazed primarily in the marshes, drifting out to the beach in the summer months to escape biting insects and catch the sea breeze. Like their human neighbors, these horses were rugged, tenacious, and independent. Until the 20th century, horses, cattle, hogs, sheep, and goats far outnumbered humans on this difficult-to-homestead barrier island chain. Once or twice a year, locals held roundups for branding to establish ownership of new calves and foals. The rest of the time, the animals were unattended.

On Currituck Banks, the northernmost reach of the Outer Banks, the surviving horses are strikingly Spanish in appearance—short-backed, deep-chested, and wide between the eyes—in shades of black, brown, bay, sorrel or chestnut. They are long-strided and very intelligent. There is no question that Banker Horses carry the blood of Spanish breeds brought to the New World in the 15th and 16th centuries, but how foundation stock reached the Banks is a topic of hot dispute. Some people believe that the first horses were left behind by the first explorers or swam ashore from early shipwrecks. Others assert that horses arrived with later colonists who placed them on the barrier islands to graze.

Throughout the 1500s, European explorers of several nationalities sailed the waters north of Cape Romain, South Carolina, including Chesapeake Bay, searching for places to establish a colony or for a passage linking the Atlantic with the Pacific. These scouting ships probably carried few, if any, horses. Although Spain mostly confined its colonizing efforts in the eastern United States to Florida, Georgia, and South Carolina, it did send expeditions farther north. The Eastern Seaboard, however, seemed relatively valueless to Spanish authorities because it offered little gold, no large cities to loot, and no concentrated populations to enslave.

Some speculate that Spaniards attempted to colonize the coast of North Carolina and left horses behind when the colonies failed. Most historians believe this is unlikely. In 1526 Lucas Vázquez de Ayllón established two short-lived settlements. The first may have been as far north as the mouth of the Cape Fear River, and the second was probably in South Carolina; but both are a long way from the Outer Banks. Spanish Jesuits from Florida established a mission on the lower Chesapeake Bay in 1570, but Indians killed all nine priests the next year. None of these brief settlements is a likely source for the Banker Horses.

Aside from Verrazzano's account of a short visit in 1524, the historical record offers few pre-settlement descriptions of the Outer Banks. On New Year's Day (March 25), 1584, Queen Elizabeth I granted Sir Walter Raleigh a huge tract of North America from South Carolina to Nova Scotia, territory claimed by Spain. Unwilling to undertake blindly the risks of colonization, which ranged from expensive failure to war with the most powerful country in Europe, Raleigh dispatched a reconnaissance mission. Two ships under Philip Amadas and Arthur Barlowe set sail on April 27 to explore the coast, evaluate potential settlement sites, and bring back news that might attract support. Barlowe's 1584 account is vague and contains obvious anachronisms, e.g., *Virginia*, a name not applied to the area till 1585, and it seems to include snippets of Verrazzano's 60-year-old narrative. These insertions suggest

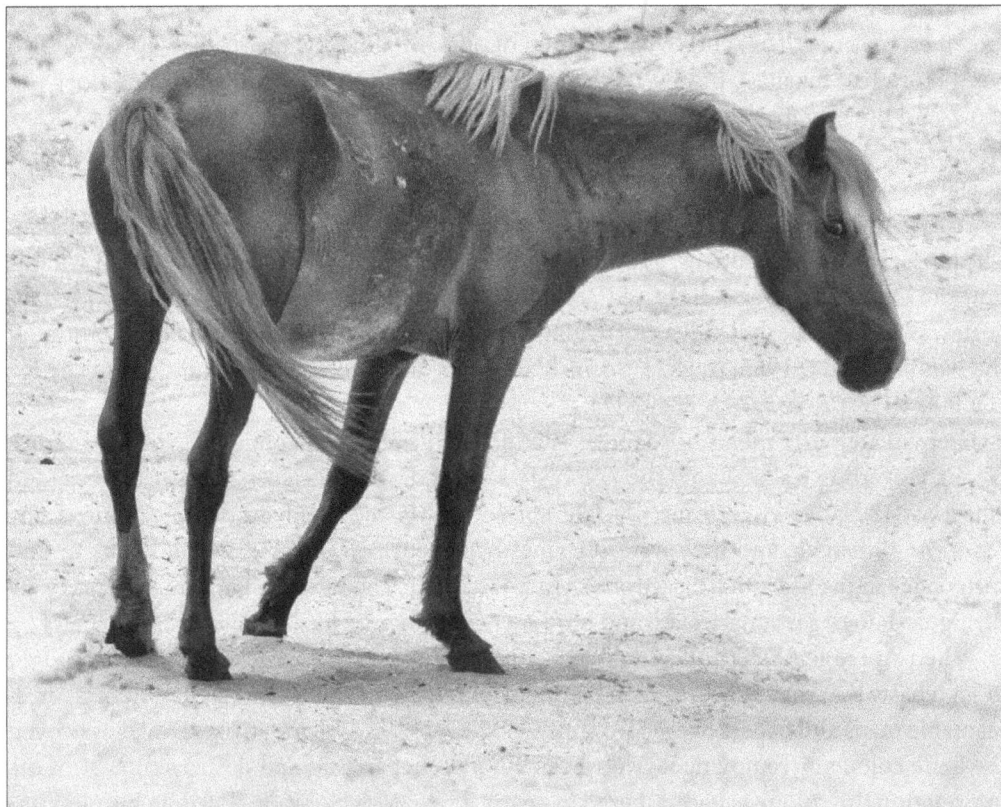

The wild Colonial Spanish Horses of Currituck Banks are a very old strain of a very rare American breed.

either that Barlowe wrote his account well after the mission had ended or that he, Raleigh, or someone else embellished it for publication.

There is no mention of livestock pre-dating colonization in any document from the period. Barlowe, in describing the bounty that he found on the Outer banks in 1584, wrote that it "had many goodly woods, full of Deere, Conies, Hares, and Fowle, euen in the middest of Summer, in incredible aboundance" (Quinn, 1955, p. 96). Francis L. Hawks misquoted this passage in his seminal *History of North Carolina* (1857), putting "horses" in one of two places (pp. 82, 88) where Barlowe had written "hares" (Prioli, 2007). This transcription error has been perpetuated by subsequent readers who cite it as evidence that horses were living on the Banks when English colonists arrived.

Barlowe was pressed to describe the New World in optimistic terms so that Raleigh might secure backing for colonization, and to justify the time the party had spent exploring a small fraction of Raleigh's grant (Quinn, 1955, p. 320). He extolled the abundant grapes, waterfowl, and timber and described the native people in great detail. One imagines that if horses had been present, Barlowe would have provided an enthusiastic description.

Records of early expeditions are seldom straightforward. The main source of information about the 1585 Grenville expedition is a spotty narrative of unknown authorship. The expedition left England on April 9. From May 12 to 23 or 24, the Englishmen laid over in temporary fortifications at Tallaboa Bay, Puerto Rico, which John White immortalized with a watercolor illustration depicting horses in a corral, among other things. The entry for May 22 describes a tense meeting with local Spaniards, during which Grenville "dispatched 20 footemen towards them, and two horsemen of ours, mounted vpon Spanish horses, which wee before had taken … on the Iland" (Quinn, 1955, p. 182). White's painting does not show, and the narrative does not mention, other livestock during the stay on Puerto Rico. Grenville tried to trade with the locals, perhaps for additional livestock; when they refused, he started a forest fire.

En route to Hispaniola to complete his provisioning, Grenville stopped to seize two Spanish ships, ransom his captives, and dig salt. In early June, he arrived at Isabela (in what is now the Dominican Republic) and traded for "horses [stallions], mares, kine [cows], buls, goates, swine, sheepe, bul hides, sugar, ginger, pearle, tabacco, and such like commodities of the Iland" (Quinn, 1955, p. 187). Spanish accounts add mules, calves, and dogs to the list (Quinn, 1955, pp. 735, 742, 786, 787).

Grenville left 107 men on Roanoke Island, the first English colony in North America. Under Governor Ralph Lane, the colonists explored, inventoried natural resources, and antagonized Native Americans. By late spring, 1586, they had probably devoured any remaining livestock, and they were starving. Lane sent half his men to the Banks to find sustenance in the sea. On a disastrous search for gold in the interior, Lane and his party ate their guard dogs, sassafras leaves, and ultimately nothing.

When Sir Francis Drake offered them passage to England a few weeks later, they accepted in such haste that they left three men behind, and impatient sailors threw the colonists' irreplaceable maps and papers overboard (Quinn, 1955, p. 293). In one stroke, Drake destroyed the Lane colony by removing it two weeks before relief arrived and destroyed most of the information the colonists had gathered in nearly a year of exploration. Further, his plundering tour of Spanish holdings to the south set into motion a series of events that ultimately prevented timely resupply of the 1587 colony. As in Ayllón's case, we can reasonably conclude that desperate settlers, natives, or the elements consumed any horses that had survived the trip before they could establish a self-sustaining herd.

In April 1587 John White sailed west again, this time as governor of a colony, including families and children, to be founded on Chesapeake Bay. He stopped at Roanoke Island in late July to confer with 15 men that Grenville had stationed to guard Lane's abandoned settlement. The men, however had apparently succumbed or fled; he found only bones and dilapidated buildings. Worse, his pilot unexpectedly refused to take the colonists farther, forcing White to establish his colony on Roanoke. Soon after the birth of his granddaughter Virginia Dare, the first English child native to the New World, White returned to England to obtain supplies and recruit new settlers.

Privateer William Irish led another expedition that apparently preceded White's. Alonso Ruiz, a seaman on a vessel that Irish had captured off Cuba in June 1587, later recounted a three-day stop in Virginia during which his captors "found traces of cattle and a stray dark-brown mule" (Quinn, 1955, p. 782). The location of this layover remains unclear. Unaware of White's diversion, Irish may have gone straight to Chesapeake Bay and discovered relics of the old Spanish mission, abandoned 15 years earlier. If Irish likewise took a detour to Roanoke Island and arrived before White, he would have found no settlers, but might have seen livestock. If he arrived after White, he would have seen White's ships, which were anchored offshore until August 28. A Spanish party from St. Augustine visited the Outer Banks in 1588 after searching Chesapeake Bay for the rumored English incursion, but it reported no settlers or livestock in either place (Quinn, 1955, p. 811).

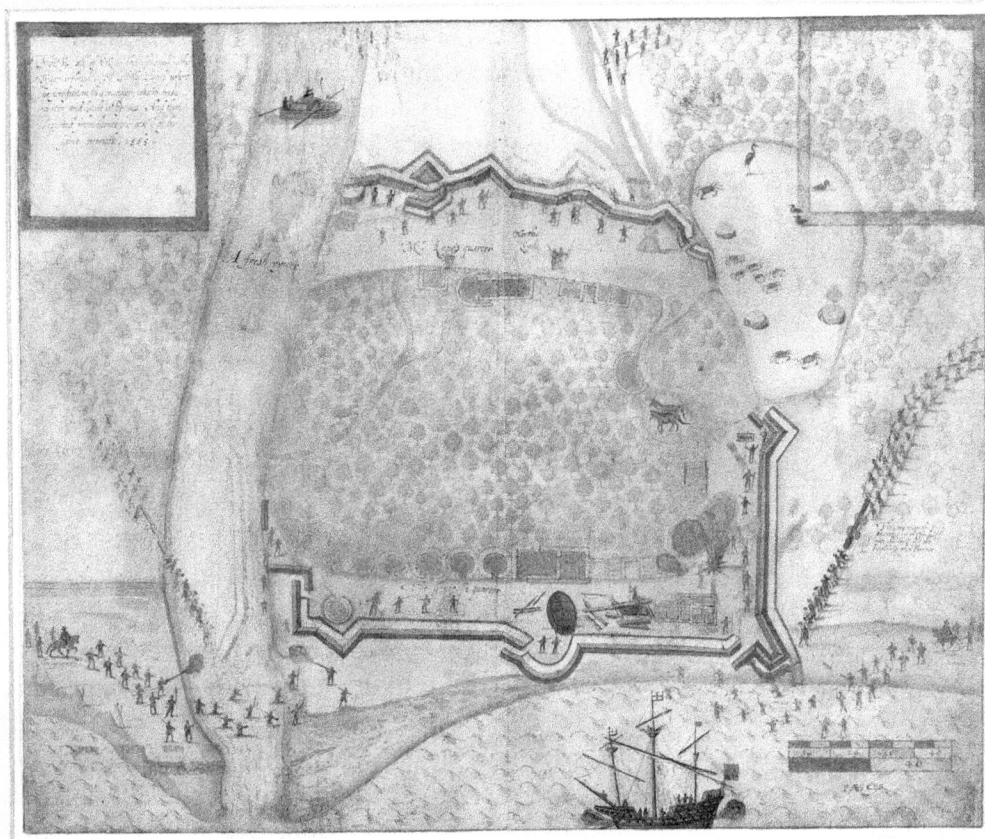

John White's graphite, ink, and watercolor depiction of the fortified camp at Tallaboa Bay, Puerto Rico (1585), shows horses that the Grenville expedition had taken from Spanish colonists. At bottom right, the flagship *Tiger* is anchored. In the bottom left corner, Grenville ("The General"), mounted on a stolen Spanish horse, returns with a party of foot soldiers. Courtesy of the Trustees of the British Museum.

Thomas Harriot, the Lane colony's chief scientist, seems to say in his *Briefe and True Report*, probably written in 1587, that one English colony had provisions for a year, and comparable rations should sustain successors for a similar interval:

> If that those which shall thit[h]er trauaile to inhabite and plant bee but reasonably prouided for the first yere as those that are which were transported the last, and beeing these doe vse but that diligence and care as is requisite, and as they may with eese: There is no doubt but for the time following they may haue victuals that is excellent good and plentie enough, some more Englishe sortes of cattaile also hereafter, as some have bene before, and are there yet remaining, may and shall bee God willing thit[h]er transported.... (1590, p. 32)

Because contemporary writers often referred to all livestock as cattle, it is unclear whether "Englishe sortes of cattaile" denoted bovids or other animals. It is also unclear which colonies Harriot meant, for no Roanoke colony seems to have been "reasonably prouided." The 1585 settlers fled after 11 months of hardship, and the 1587 colonists were so desperate for provisions that they sent Governor White himself back to England "for the better and sooner obtaining of supplies, and other necessaries" (Quinn, 1955, p. 533).

Does Harriot's report indicate that colonists in the 1580s brought animals with them from England? If so, it is unique. Did Ruiz's account indicate that livestock outlasted the 1585 colonists? Clearly, through the 1580s domestic animals were in short supply, and starvation threatened at every turn. Grenville incurred risk and expense acquiring livestock in a hostile region in 1585, and White was unsuccessful in his efforts to obtain more in 1587 (Quinn, 1955, p. 521). Their efforts would have been pointless if they already had English livestock on board. Further, the introduction of English livestock by Roanoke colonists would not explain the presence of predominantly Spanish horses on the Outer Banks in later centuries.

Handicapped by lack of funds and delayed by the attack of the Spanish Armada, White finally returned in 1590, without supplies or settlers, to find the site abandoned. He reported no signs of livestock. It appears unlikely that domestic animals remained on the Outer Banks after 1590. The fate of the Lost Colony is still a mystery.

Stick writes that when White sailed from the Banks, the English left the area "in the undisputed possession of the native Indians for another seventy-five years" (1958, p. 21). But by defeating the Spanish Armada in 1588, England cleared the way to colonize North America. Jamestown, founded in 1607, was the first of a new wave of English settlements along the coast. Some failed, but most succeeded.

The North Carolina coast has been inhabited for at least 10,000 years (Wiss Janney Elstner Associates & John Milner Associates, 2007). At the time of European contact, the East Coast was home to diverse tribes representing three different linguistic families. The native people encountered at Roanoke Island and later at Jamestown and Plymouth spoke Algonquian languages. Iroquoian-speaking tribes such as the Tuscarora lived to the west. Natives to the south spoke Siouan. Neighboring languages could be as different as English and Arabic, and linguistic barriers fostered hostility.

These native groups fought viciously against one another and among themselves. Barlowe, probably no stranger to combat, was surprised by the ferocity of their battles. Wars among the indigenous people, he noted, were "very cruell, and bloodie, by reason whereof, and of their ciuill dissentions, which haue happened of late yeeres amongst them, the people are maruelously wasted, and in some places, the Countrey left desolate" (Quinn, 1955, p. 113). The situation had not improved more than a century later, when John Lawson (1709, p. 225) linked the natives' continual hostility to communication problems:

> [T]he continual Wars these Savages maintain, one Nation against another, which sometimes hold for some Ages, killing and making Captives, till they become so weak thereby, that they are forced to make Peace for want of Recruits . . . and the Difference of Languages, that is found amongst these Heathens, seems altogether strange. For it often appears, that every dozen Miles, you meet with an Indian Town, that is quite different from the others you last parted withal. . . .

Fragmentation weakened the indigenous peoples and rendered them vulnerable to displacement and extermination by Old World invaders.

Upon encountering native cultures, Europeans, like any people, fit what they observed into their own intellectual and cultural framework. Consequently, much of what we know about these tribes is based on misconceptions and assumptions. Native languages often perplexed the European ear, and colonists often confused the names of people, places, and fragments of conversation. For example, the English believed that the name of the

Poney Penning on the Beach, near Oriental, N.C.

This postcard, produced around 1910, shows a roundup along the Neuse River near Oriental, N.C. Free-range horses and cattle still roamed much of Pamlico County and elsewhere on the mainland. These small horses appear to have Colonial Spanish characteristics and some coat colors typically found in the Banker herds. They are probably very similar to the horses originally released on the Outer Banks to graze. The ancestors of both groups probably included Chickasaw or Seminole Spanish horses. Note the high corral fence, suggesting that these small horses could jump a lesser barrier. Courtesy of University of North Carolina at Chapel Hill (*Poney* [sic] *Penning on the Beach*, ca. 1910).

territory they were exploring was *Wingandacon* until Harriot learned more Carolina Algonquian and concluded that the term meant "You wear fancy clothes." Later scholars suggested that it really related to evergreens. Likewise, John White's paintings offer honest portrayals of extinct cultures and vanished landscapes. But some of White's work survives only in engravings by Theodor De Bry, who lived in England only briefly and never crossed the Atlantic. What we see is filtered through White's and De Bry's perceptions and our own (Sloan, 2007).

Seventeenth-century settlers on the Eastern Seaboard came by long, difficult routes across the Atlantic, usually without livestock, and often purchased farm animals from Spanish ranches in the Caribbean. The cattle, hogs, sheep, goats, and horses raised there were of high quality, bred for hardiness, and much more likely to survive the shorter voyage to the colonies than animals shipped directly from Europe. As in Virginia, North Carolina colonists allowed their livestock to run free; and as in Virginia, roaming livestock soon became a nuisance, not only to other colonists, but also to the native people.

Lee wrote (2008, p. 48), "free ranging stock fostered local tensions. The lands into which colonists advanced were invariably occupied, and foraging cattle and hogs refused to abide by the spatial boundaries that humans created."

Virtually all authorities on Outer Banks history agree that early settlers, mostly small farmers in search of land, migrated down from the Chesapeake area in search of land on which to graze livestock (Dunbar, 1958). In July, 1653, Roger Green led a group from

Virginia to establish a settlement in what is now North Carolina (Wallace, 1897), alongside the Quakers and other refugees that had left Virginia to escape persecution of those who did not follow the Church of England. These refugees and colonists probably took horses with them. By the mid-1600s, most European settlement in the region that would eventually become North Carolina was in the lowlands between the Great Dismal Swamp and Albemarle Sound. By the time of the 1663 charter, which split Carolina from Virginia, about 500 whites, mostly subsistence farmers, "undesirables" from Virginia, runaway slaves, former indentured servants, religious dissenters, and debtors lived in the region with a small number of slaves (Barth, 2010). Whereas wealthy planters dominated government and society in Virginia and later in South Carolina, the people of the Albemarle were mostly small farmers of modest means who owned their own land.

Although the native tribes of northeastern North Carolina were in decline by the 1680s, before 1711 about 30,000 Tuscarora and Algonquian people remained in the region, and they often sold land directly to early white settlers (Barth, 2010). A 1672 pact with the Tuscarora stipulated that whites could only settle north of the Albemarle Sound—an agreement overturned through war and treachery. Where other white settlers resolutely subjugated blacks and Indians, Albemarle residents were more accepting of nonwhites, and free blacks often made their homes in this region. Indians and free blacks could vote until 1715 (Barth, 2010).

A French trader remarked in 1765 on the total absence of wealthy residents, and said that many homes in the Albemarle region were so utilitarian, a simple brick chimney was considered an extravagance (Barth, 2010). Geography impeded trade. While South Carolina and Virginia had easily accessible ports, ships arriving in North Carolina contended first with the shifting sand banks, then with the shallow sounds and widely scattered settlements behind them. At the treacherous inlets, goods were frequently transferred to and from smaller vessels. Smugglers sometimes evaded authorities in the challenging waters of North Carolina and found a lively market for ill-gotten goods among the lowland residents.

North Carolina stockmen followed the typical colonial practice of allowing their livestock free range, fencing them away from resources where they were not welcome, such as crops and gardens. Not until 1873 did the state General Assembly pass An Act Relating to Fences, which required confinement of livestock in five Piedmont counties. By 1880, all or parts of 30 or more counties were under the stock law (Johnson, 2006). Residents were obliged not only to fence in their own stock, but also to fence their crops to keep out roaming livestock from nearby areas not under the stock law. In 1901 several new laws allowed natural barriers, such as rivers and thick woods, to serve as fences.

When the British Parliament imposed a tax on fences in 1670, subsistence farmers in the colonies who had not already done so were motivated to move most of their livestock to islands and peninsulas, effectively penning them by water. A peninsula required at most only a fence at some narrow spot to contain the animals, and a barrier island required no fence at all. Islands had been used for grazing since the beginning of European settlement in the New World (Stewart, 1991). Eventually islands and necks all along the East Coast supported free-range livestock, including Boston Neck; Staten Island; Point Judith, Rhode Island; and many parts of Long Island. The first Georgia colonists likewise turned livestock loose in river swamps and "feeding marshes" as well as on barrier islands (Stewart, 1991).

Historically, Banker horses were generally short in stature and extremely rugged, like their Spanish ancestors. Drawing by the author.

Livestock typically grazed in marshy areas considered unsuitable for any other use. John Lawson, who traveled throughout the Carolina colony in the early 1700s, before its division into North and South, observed,

> The Country in general affords pleasant Seats, the Land (except in some few Places) being dry and high Banks, parcell'd out into most convenient Necks, (by the Creeks) easy to be fenced in for securing their Stocks to more strict Boundaries, whereby, with a small trouble of fencing, almost every Man may enjoy, to himself, an entire Plantation, or rather Park. (1709, pp. 79–80)

By the 1650s, farmers and stockmen had settled on the necks along the northern margin of Albemarle Sound, and some of them may have turned livestock loose on the Banks not long afterwards. Conant, Juras, and Cothran (2012, p. 53) wrote, "Feral horses are known to have existed on these islands once the mainland was settled, primarily by Englishmen, from 1650 until the present."

After a royal charter created the Carolina colony from Virginia in 1663, Europeans began to settle on the barrier chain. Lee (2008, p. 51) wrote that Sir John Colleton, one of eight Lords Proprietors, claimed Colington Island, on the west side of the Outer Banks. Colleton settled the island by proxy in 1664, when his agent Captain John Whittie "cleared a farm, planted corn, and turned cattle loose to graze" (2008, p. 51).

Not every settler was so well-connected, however:

> The land grants on the Banks during the colonial period were often very large, necessarily so for the purpose of stock raising . . . it should be remembered that the ownership represented by grants and deeds did not always mean control or occupation of the land. Presumably many of the early settlers acquired their property merely by settling on it. (Dunbar, 1958, p. 14)

These poor squatters fenced small homesteads and grew sand-tolerant vegetables such as sweet potatoes and collards on family farms. A small community of Native Americans, most likely of the same culture encountered by Verrazzano's expedition two centuries earlier, remained in the maritime forest near Cape Hatteras through the early 1700s (Impact Assessment, 2005, vol. 1, p. 10).

"By 1680 stock had been placed on the northernmost section of the Outer Banks" (DeBlieu, 1987/1998, p. 27). Stick explained,

> It was not long before the raising of cattle, horses, hogs and sheep was an important occupation on the Banks, though most of the stock seems to have been owned by the larger, non-resident property owners. When Sir William Berkeley sold a half interest in Roanoke Island to Joshua Lamb of Massachusetts in 1676, for example, it was specifically provided that Lamb should receive half of "all the Cattle, hogs, and other stock . . . thereon," and the first attempt to survey the boundary between Virginia and Carolina in 1692 was begun "at a place called Cowpenpoint on the north side of Corotuck River, or Inlet." (1958, p. 23)

Lee writes that early European arrivals lived in scattered settlements

> along the northern banks down to present-day Nags Head, below which Roanoke Inlet separated the islands. South of Roanoke Inlet a few landowners bought up large tracts, dividing nearly the entire outer banks [sic] between nine people by the early 1700s. (2008, p. 42)

Charles Felton Pidgin, a playwright and novelist from Massachusetts, described the Banker people as "distinctly strange, something of a cross between various nationalities; an unprincipled people, piratical, superstitious, uncleanly and ignorant; the substantials of life consisting of fish and wild hogs and cattle, with but scant provisions of bread and vegetables" (1907, p. 419).

The Outer Banks is one of several places known as the Graveyard of the Atlantic, having seen more than 1,000 shipwrecks since 1600. Unpredictable weather, dangerous shoals, strong currents, treacherous inlets, confusing topography, continuous change, and the late arrival of navigational aids made the place uncongenial to maritime traffic. Ships have come to grief not only on shoals, but also in deep water and on solid ground.

Legend has it that the Outer Bankers were opportunists who turned shipwrecks to their advantage. As the story goes, rather than waiting for ships to spill their cargo on the shore, the wily Bankers lured ships in to wreck on the shoals. It is said that they tied a lantern around the neck of a Banker pony and led him slowly up and down the high dunes now known as Jockeys Ridge so that the light was visible at sea. Any mariner to see the light would mistake it for a ship bobbling in a safe harbor and head for it, only to run aground. The land pirates would then loot and burn the stranded ship.

In the time of the Wright Brothers, Kitty Hawk appeared a veritable desert of barren, windblown dunes (Daniels, 1903). Courtesy of Library of Congress.

Pidgin (1907, p. 419) wrote,

> There are contradictory accounts relative to the name of this strip of sand. The early mariners say that the shore from Kitty Hawk, late the scene of the Wright brother's [*sic*] experiments with the flying machine, to the Oregon Inlet, presents the appearance of a nag's head, the ears made prominent by the high sand hills. Possibly this is true, but more probably may the name be accepted from the fact that the natives, a crude and lawless set of people, affixed torches on long poles, mounted their native banker ponies, and walked the beach stormy nights to allure the ships nearer the shore. This is the local acceptation; in those days there were no light-houses near, none save the stars of the universe, the light of the angels' eyes.

Dunbar (1958) disagreed with this stereotype. Like Stick (1958), he found only two instances of piracy, the looting of the grounded HMS *Hady* in 1696 and the exploitation, with outsiders, of three storm-damaged ships from the 1750 Spanish treasure fleet (Shomette, 2008). Dunbar concluded, "The nature of these cases and their rarity demonstrate that the early Bankers were undeserving of their reputation as wreckers or land pirates" (1958, p. 21). More typically, Banker people took shipwreck survivors into their homes and provided whatever assistance they could.

Even if they were not of nefarious bent, Banks dwellers showed a marked tendency to resist authority. In 1750, Colonial Governor Gabriel Johnston described them as "a set of People who live on certain sandy islands lying between the Sound and the Ocean, and who are very Wild and ungovernable, so that it is seldom possible to Execute any Civil or Criminal Writs among them" (Lee, 2008, p. 49).

Popular legends hold that the original Banker horses swam to the sandy islands from the wreck of a Spanish or English ship. While this scenario is possible, there is no proof. However, trade flourished from the earliest days of the North American colonies, and many of the ships sailing the Atlantic carried horses. Wrecks were commonplace and poorly documented, and storms often swept horses off the decks where they were carried even when the ships remained intact.

Horses were shipped to and from the British colonies from the early 1600s. The horses that rode the salty swells were drawn from a relatively small gene pool, and many outwardly diverse American breeds mostly derive from the horses cultivated in this colonial nursery.

These ancestral horses descended from European, Middle Eastern, and North African animals sent to Spanish, British, French, and Dutch colonies in the New World. Many colonists who settled in Virginia and Carolina initially obtained their horses in the West Indies. Most of these animals were Spanish Jennets bred on the islands to support Spanish exploration and exploitation of the New World. New England colonists also imported stock from England, and Dutch colonists in New York got some of theirs from the Netherlands. Horses turned out to forage for themselves multiplied prodigiously, especially in New England.

Settlers on the Outer Banks fenced their yards against the predations of the abundant free-roaming livestock that had the run of the island. Each village had its own identity, often characterized by the products of local industries (Lee, 2008, p. 107). Hatteras Village, "nearest to a deep and dependable inlet," was chiefly a fishing village and considered more sophisticated than other Hatteras Island communities. Kinnakeet (later called Avon), was known for boat building, the harvesting of eelgrass (*Zostera marina*) for mattress stuffing, and yaupon tea.

Dunbar (1958, p. 42) wrote that by 1776 the Banks were covered with cattle, sheep, and hogs, and "the few inhabitants living on the banks (were) chiefly persons whose estates consist in livestock." Roundups were held once or twice a year to divide and brand stock and remove certain animals to the mainland. Breeding stock remained to roam freely and multiply at will. These animals were often driven up the Banks to Virginia to be sold.

In 1900, *Chambers's Journal* published a piece that reads,

> In the state of North Carolina, along the shores of the Albemarle and Pamlico Sounds, lie miles of low sandy banks, the greater part covered with little vegetation but coarse grass, wild parsley, and other saltwater weeds. . . . On some of these banks are a breed of small wild horses, known in the neighbourhood as 'banker ponies.' They are quite untamed and uncared for, with rough shaggy coats, and are generally about twice the size of a Shetland Pony; now and again one even reaching the size of a small horse.
>
> Each year the herd-owners drive them into pens, where the foals are branded with the owner's mark, and those required are caught and sold to the dealers who attend the annual penning. The poor things are frightfully wild and—to apply the darky term for their state—'ignorant,' and have to be starved into eating grain and hay or grass. . . .
>
> In captivity they show equal intelligence, though seldom a reliable temper. They are tamed by darkness and semi-starvation, and make excellent draught animals, showing strength far beyond their size. They also eat voraciously, consuming as much as a full-sized horse.
>
> The foals bred from 'banker ponies' in captivity make valuable though small horses. They are strong, healthy, and intelligent, less vicious than their parents, and command good prices. One mare used for some years by the writer as a saddle-horse was sold for thirty pounds—a good price in those parts; her sire and dam had

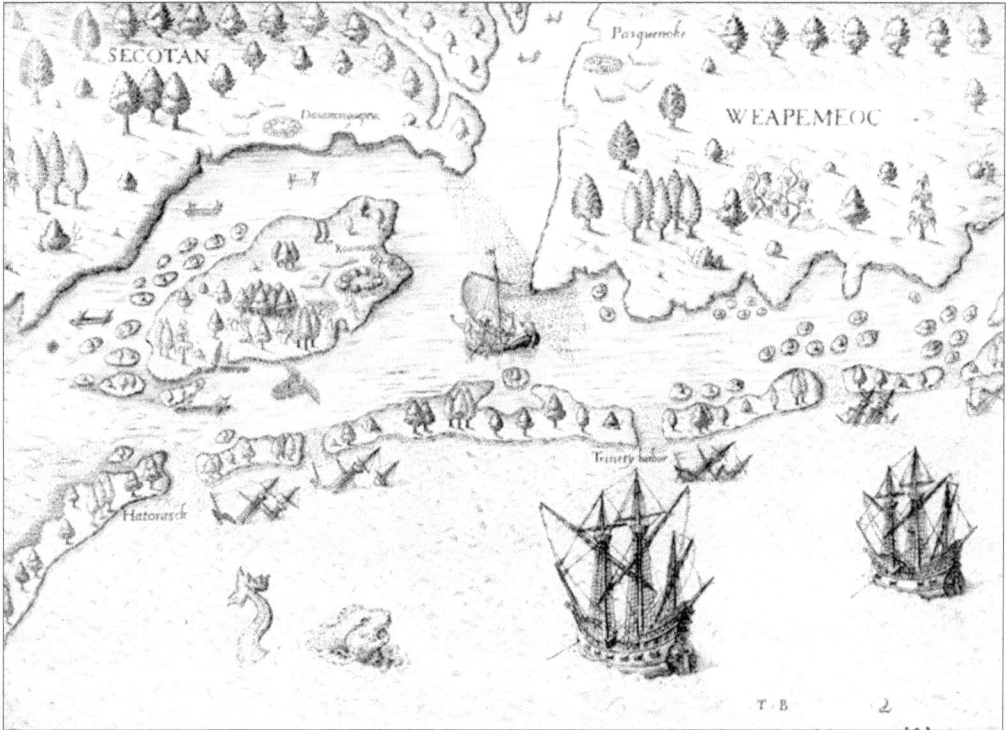

Theodor De Bry (1590) brought John White's paintings to the European public by converting them to reproducible engravings. De Bry never beheld the New World, but he apparently felt free to embellish White's faithful depictions. "The arriual of the Englishemen in Virginia," above, has no known surviving original and bears no clear connection to any of the Raleigh voyages. The tree symbols with which De Bry covered the Outer Banks do not appear on White's maps and may be as true to life as the sea monster at lower left.

cost respectively two and three pounds. She was a pretty little animal, could open any ordinary fastening with her teeth, and was frequently found with her head in the feed-bin. ("Fishing Horses," p. 493)

A *Washington Post* article after the turn of last century likened them to Chincoteague ponies, which at that time were similarly colored:

Currituck Sound has a race of ponies much like those of Chincoteague. They are not literally wild horses, since they have owners, and the colts are branded. The natives of the region use them for ordinary domestic purposes and the largest of the breed very well serve the needs of their owners.

Like the Chincoteague ponies, they are easily kept, and on the whole they furnish cheap horses for the poor farmers of the coast. The Chincoteague ponies can be bought at auction upon the island for $40 or $50 [roughly $1,000–$1,250 today (U.S. Bureau of Labor Statistics. (n.d.)], and the Curritucks are probably cheaper. ("Chincoteague Ponies," 1905, p. 6)

Marvin Howard of Ocracoke (1976, p. 26) gave this description:

The ponies of the Outer Banks did vary in weight from five hundred to eight hundred pounds. They lived on the range the year round as wild as deer or wild horses can ever be. For sustenance they had only the salt grass, the boughs of live oak and red cedar, and when the winters were severe, they dug in the sand hills with their hoofs to get the succulent roots of the sea oats. These ponies no doubt had strains of Arab steed for in numbers of them there was untold beauty in color and build. They were fleet of feet, hardy, well lined, and full of muscle. . . . None of these wild horses were ever large except the Pea Island pony, which came from the original quarter-bred horse.

A persistent legend holds that the horses of the Outer Banks were directly established by Spanish explorers through either shipwreck or colonization attempts. It is, however, unlikely that any Spanish horses brought to the East Coast of North America in the 1500s made it this far north or stayed long enough to leave progeny.

Farming was very small-scale—Bankers maintained small kitchen gardens beside their homes, fenced to keep the livestock out. Windmills ground corn that fishermen obtained in trade from mainland farmers (Senter, 2003).

In the early days, communities were usually established on the more protected mainland, leaving the Banks largely to livestock. The Banks were low and grassy, with wide beaches, scattered dunes to the west, and forested areas on the soundside. Over time, pilots and other watermen settled near the inlets, and villages were established in the parts of the islands stabilized by maritime forest. By the early 1800s, seasonal visitors began building cottages on the soundside. Hotels and other establishments followed. The first substantial oceanfront structure on the northern Banks was built in Nags Head just before the Civil War. After 1900, vacationers built cottages near the sea and moved them westward as the beach receded.

Oddly enough, commercial fishing was not a primary occupation or source of income on the Banks through the first two centuries of settlement. The 1850 census showed that fishing was the primary occupation on Hatteras Island, but commercial fishermen were in the minority on Portsmouth and Ocracoke. At that time, many mainland dwellers still lived close to waters that teemed with fish, so there was little incentive for them to import seafood from the Outer Banks.

This dynamic shifted abruptly between the Civil War and World War II, when Bankers turned to the sea and sound, not just to live, but also to make a living. They sold not only fish, but also porpoises, turtles (diamondback terrapins and loggerhead and green turtles), clams, scallops, oysters, crabs, waterfowl, and eelgrass.

Whales had been hunted off the coast of North Carolina since 1666 (Reeves & Mitchell, 1988). A porpoise fishery operated in Hatteras from the late 1700s until the Civil War (Senter, 2003). Watermen would trap and process entire pods of dolphins, including pregnant and lactating females and nursing calves, strip them of blubber, render the oil, and discard the flesh and skin. Up to 1,200 animals were taken in a season, and five or six porpoises could be processed into one barrel of oil (Mead, 1975). During the 1800s, free-roaming swine on Hatteras Island feasted on the scraps that remained after this butchery. Porpoise oil was burned in lamps and used in soapmaking, and from 1844 until around 1970 refined oil from the "melon" (an echolocation organ) was the favored lubricant for watches, clocks, and other delicate mechanisms. The Hatteras fishery harvested porpoises until about the early 1900s.

Loggerhead turtles were speared or tackled by fishermen who dived overboard onto the reptiles, turned their heads upward, and muscled them to shore. Eventually, the federal government stepped in to protect sea turtles from extinction.

"One of the great dunes at Kitty Hawk, anchored with brush fence and coarse grass" (Farrell, n.d.). The recent-looking plantings must have been made after the state proposed large-scale beach stabilization in 1933. Damage to the new highway in the foreground—note the water-filled gully and broken blacktop—may be from the hurricane of August 23 or September 15, 1933; the hurricane of September 8, 1934; or any of several nor'easters. Photograph courtesy of the State Archives of North Carolina.

Tourism and related development, now the economic underpinning of the Outer Banks, began on Ocracoke and around Nags Head before the Civil War (Mallinson et al., 2009; Stick, 1958). Because there were no roads, the mostly regional clientele arrived by steamer or private boat. The soundside at Nags Head soon had a hotel, a cluster of vacation houses, and docks up to a half mile (0.8 km) long, some supporting rails on which horses or oxen hauled wealthy visitors and their summer's baggage.

Battles and shipwrecks drew attention to the Banks, but the Wright brothers thrust the area permanently into the national limelight on December 17, 1903, when they made the first controlled flights in a self-propelled heavier-than-air machine near Kill Devil Hill. They had first visited three years earlier, after searching by mail for windy spots where they could experiment in privacy. Kitty Hawk postmaster William J. Tate (1900) wrote back, "our winds are always steady" and "you will find a hospitable people." His description of the terrain, flat and treeless in many places and free of rocks, may have sweetened the deal: "You could for instance get a stretch of sandy land 1 mile by five with a bare hill in center 80 feet high not a tree or bush anywhere to break the evenness of the wind."

At the time of the Wright Brothers' experiments, most of the Outer Banks were low, windswept dunes with wide beaches to the east and marsh and maritime forest to the west. For centuries, experts and laymen alike harbored the misconception that the banks were heavily forested before the arrival of Europeans. Engraver Theodor De Bry (1590) apparently added trees to John White's original maps, giving the false impression of forestation. Spears (1890) imagined the virginal Hatteras Island as "almost completely covered with a prodigious growth of trees" (p. 510). He blamed the change, and the sand wave burying the island, on overzealous logging.

The dense, heavy wood of live oaks (*Quercus* spp.) was favored in shipbuilding (Senter, 2003), but they were depleted before the Civil War (Bratton & Davison, 1987), Harvesting of loblolly pine (*Pinus taeda*) in Buxton Woods started in 1899 (Senter, 2003). From 1907 to 1911, the Foreman-Blades Lumber Company of Elizabeth City, N.C., aggressively logged parts of Hatteras Island, mostly cutting loblolly pine, oak, holly (*Ilex* spp.), and dogwood (*Cornus florida*) 8 in./20 cm or greater in diameter (Bratton & Davison, 1987). It spared the largest trees, many of which later succumbed to insect infestations. A sawmill on the sound-side in Nags Head processed trees from Nags Head Woods, and loggers cleared enough of what is now Southern Shores and Kitty Hawk to warrant a short narrow-gauge railroad ending at a sawmill on Kitty Hawk Bay.

Most of the islands' woodlands have recovered from the removal of old-growth trees in previous centuries and show few signs of exploitation. Logging, accidental and deliberately set fires, and land clearing, however, apparently have altered the composition of the maritime forest. Between roughly 900 and 170 years ago, Buxton Woods was 51% oak and 30% pine (Bratton & Davison, 1987). The balance began to shift around 1700. In 1987, Buxton Woods was 6% oak and 80% pine.

Around the beginning of the 20th century, persuasive commentators blamed free-roaming livestock, at least in part, for denuding the Banks. For example, Jay Bond of the U.S. Forest Service wrote (1908, p. 43) that hogs turned up the soil and exposed the roots of grasses, which cattle and sheep devoured, creating a "blowout" that could form a destructive sand wave. Bond recommended driving rows of planks into the ground to catch wind-borne sand; planting sea oats (*Uniola paniculata*) to hold the sand in place and catch more; and planting loblolly pines behind the artificial dune line to reestablish the forest. He also recommended forcing owners to fence in their livestock (Senter, 2003). His ideas took root a generation later.

Collier Cobb, a geology professor at the University of North Carolina, shared Bond's belief that the islands could be stabilized. He proposed "reforesting the sands" by systematic planting (1906, p. 317), which he said would turn Hatteras Island into "a subtropical garden." Cobb, however, did not see free-roaming horses as a problem, and he speculated that with the plantings, "the herds of wild ponies now dwindling away would again increase in numbers."

At the time, people did not understand barrier island dynamics. Now scientists realize that barrier islands are inherently unstable, with or without logging or livestock. Dunes appear to go through stages of migration that have little to do with human activity. They mobilize, engulfing and killing forests as overwash and the winds of nor'easters push them across the island. Then they restabilize when colonized by plants that form soil. Sometimes the dunes of the Outer Banks became mountains of sand much higher than the trees they smothered, reaching over 90 ft/27 m (Stick, 1958). The town of Seagull (later renamed Old Inlet) stood at the base of Penny Hill until the migrating dune completely buried the village in the 1950s.

By radiocarbon-dating the soil, geologists concluded that the large dunes on Currituck Banks have been live and migratory from 750 to 1000 CE, from 1260 to 1700, and from around 1830 to the present (Senter, 2003). There is also indirect evidence that sand waves on Hatteras Island pre-dated European contact. If Hatorask, on the 1590 White-De Bry map, comes from the Algonquian for "there is less vegetation" (Quinn, 1955, p. 864), it

On Currituck Banks, Penny (more correctly Lewark) Hill provides horses and other wildlife high ground during floods.

suggests sparse plant cover. Chicamacomico, an Algonquian-derived name for a cape at the north end of Hatteras Island, may mean "sinking-down sand" (Quinn, 1955, p. 864), which implies erosion or dune migration. In the late 16th century, this cape evidently had dunes and a maritime forest (Quinn, 1955). Less than a century later, the cape was gone, replaced by Wimble Shoals and a bend in the coastline (Comberford, 1657). Its rapid disappearance may have been part of changes that began centuries before the Raleigh colonists arrived.

Not all islands develop high dunes. On a natural barrier island, influences such as sand supply and transport and the direction of the prevailing winds give some areas a surplus of sand, resulting in higher dunes. Other islands, including most parts of the Outer Banks, naturally develop low dunes, wide beaches, and overwash flats. (Pilkey, Rice, & Neal, 2004). For example, sea-floor sediments 40 ft/12 m below the ocean surface determine the profile of Shackleford Banks. Where the nearby ocean floor is thick sand, Shackleford has high dunes. Where it is rocky with little sand, Shackleford has overwash flats (Pilkey et al.).

Under natural conditions, grasses can establish themselves on the dunes within a season. Sea oats grow in clusters, and between them, gaps form in the sandbanks that let fingers of sandy overwash probe the hollows (Pilkey et al., 2004). Once grasses stabilize a dune line, other plants can follow, though it may take a decade or two for shrubs to take hold and centuries to grow a maritime forest.

Dunes on Shackleford Banks stopped migrating sometime between 1917 and 1939, during an interval with fewer severe storms. From 1939 to 1971, sea-oat and scrub thicket cover expanded; the dune fields revegetated spontaneously despite the presence of large herds of free-roaming horses, cattle, sheep, goats, and hogs (Senter, 2003).

Photographs taken from the top of the Cape Hatteras Lighthouse in the 1920s show wide beaches and, behind them, a lightly vegetated overwash flat roughly 200 yards/183 m wide (Pilkey et al., 2004). These flats and beaches were important nesting habitat for native birds. Despite the presence of thousands of head of livestock, the island topography appeared to have changed very little over 300 years. Spared further human interference, it might have maintained this state indefinitely.

The Wright brothers last flew on the Banks in 1911, but their local legacy endured. As the 25th anniversary of aviation approached, W.O. Saunders, an Elizabeth City newspaper editor, and others prodded the federal government to erect a national monument atop Kill Devil Hill, near the site of the first successful flight. In 1928, Coast Guardsmen, residents, and paid workers planted vegetation and removed free-roaming livestock in hopes of anchoring the massive dune. Eventually the government mounted a 60-ft./18-m granite pylon at its apex (Chapman & Hanson, 1997). Completion of the monument in 1931 seemed to prove that impressive feats of stabilization were possible.

The Wright *Flyer* put the Outer Banks on the map, but it was the horseless carriage that made it a national resort. As automobiles became popular, adventurous visitors sometimes drove down the beach from Virginia or brought their cars to the Banks via private ferry. By 1925, Roanoke Island had a paved road, and 13 Hatteras Islanders owned automobiles (Chater, 1926). In 1927, after state engineers had declared that the Banks would need no road or bridge for "fifty years or more" (Stick, 1958, p. 244), Dare County Commissioner Washington Baum boldly started work on bridges and a causeway connecting Roanoke Island and Nags Head. Envisioning the automobile as a key to prosperity, he started this project without permission or funding at a time when few people owned automobiles and there were no paved roads on the Outer Banks.

The awkwardly-named Kill Devil Hill Monument National Memorial, near the site of the Wrights' first flight, drew visitors, and the expectation of more tourism justified expenditures for new roads and bridges. The privately operated 3-mi/4.8-km Wright Brothers Memorial Bridge, completed in 1930, granted vehicles access from the mainland. By 1931, it was possible to drive from the Currituck County mainland to Kitty Hawk, follow the new paved highway to Nags Head, and cross the bridge and causeway to Manteo. In 1932, the state of North Carolina assumed responsibility for maintaining all this new infrastructure. Investors from Elizabeth City and Norfolk eagerly built service stations, hotels, and restaurants on the newly accessible real estate.

These entrepreneurs based their dreams on an unstable barrier chain, and disaster was sure to follow. In 1933 alone, two hurricanes and a number of lesser storms flattened dunes and dwellings and breached or buried the new blacktop. Economic development, which had become imperative with the coming of the Great Depression, depended on preserving and extending the highway, which depended on stabilizing not just one hill, but the whole beach. The same year, the state Department of Conservation and Development proposed the North Carolina Coastal Development Project, which involved building dunes, restoring the banks to their "primal heavily forested condition," paving a Hatteras Island highway, and establishing a national coastal park (Senter, 2003).

In the early 1900s, while Bond and Cobb pondered stabilizing the Banks, H.C. Cowles of the University of Chicago introduced the concept of ecological succession, which became central to the emerging science of plant ecology (Senter, 2003). An ecosystem endures by maintaining a balance among interdependent species. These species make way for different species until a balanced ecosystem forms. For example, a meadow may emerge from a wetland, support the growth of shrubs and trees, and eventually turn into a forest. As succession progresses, the community becomes more diverse and the total mass of its organisms

The author took this photograph looking northeast from the top of the Currituck Beach Lighthouse in the late 1990s, well after the development boom had begun.

Same view, taken in 2012.

View to the southeast of the lighthouse, 2012.

increases. Ultimately, the ecosystem reaches a climax, a balanced state of maximal biodiversity and biomass. Cowles compared the climax community to the adult stage of an organism (Senter, 2003). It persists until, inevitably, some cataclysm such as a fire begins the process of secondary succession.

Stages of succession do not necessarily occur in a particular order, though, and sometimes species appear randomly and establish themselves in favorable conditions. Pioneer species change the environment, primarily by soil formation, and make it suitable for the plants that will replace them.

Sometimes succession occurs in response to climate change over millennia. Sometimes it occurs more quickly because the species within the ecosystem change the habitat with their presence. When grasses take root in open water, the ecosystem shifts to marsh. Some systems are beset so frequently by disturbances—fires, floods, overgrazing—that a climax community never appears.

These environmental dynamics were poorly understood in the early 20th century. Prevailing theories held that biological systems move inexorably "toward an unchanging plateau of stability where they will remain until their 'natural balance' is disturbed by, for example, human-induced deforestation" (Binkley, 2007, p. 201). Many feared that the deforestation of the Outer Banks caused shoreline erosion, an irrevocable loss of land that would end in the disappearance of the islands. In the 1930s, scientists, engineers, and politicians touted artificial dune construction as the way to save the Banks by artificially returning it to its "natural" state (Pilkey et al., 2004).

Between 1933 and 1940, the Civilian Conservation Corps, under direction of the Works Progress Administration and later the Park Service, installed more than 3 million ft (568 mi/914 km) of sand fencing in parallel rows to catch sand and build dunes as high as 25 ft/7.6 m, with bases nearly 300 ft/91 m across. This project planted 142 million ft² (3,260 acres/1,319 ha) of dune grasses and more than 2.5 million trees and shrubs along 115 mi/185 km of North Carolina coast. It laid the foundation for what was thought to be restoration of the prehistoric landscape (Senter, 2003).

Although the livestock industry was winding down, several thousand horses, cattle, sheep, goats, and swine still roamed the Banks. Some considered them a threat to beach stabilization and economic development. Advocates of wildlife refuges and a national seashore likewise considered them a threat to the natural balance of native species. MacNeill saw them as a health hazard. "Extinction of these wild cattle and ponies has been long advocated as a health measure. The cattle are virtually useless, and useful herds have died out of disease as soon as they were introduced" (MacNeill, 1938, p. 1).

In 1935, the North Carolina General Assembly passed legislation that required livestock owners to fence their animals (An Act To Place Certain Portions of Dare County under the State-Wide Stock Law, 1935). Two years later, lawmakers extended the stock law to Currituck Banks (An Act... 1937). Most affected stockmen did not own large tracts of land, and their animals traditionally grazed in marsh and woods owned by neighbors, hunt clubs, or absentees. They lacked the resources to maintain stock profitably on their own property. Suddenly families comprising generations of herders were forced to find another livelihood. Within a few years, the number of large livestock operations on the Outer Banks shrank from about 50 to seven or eight. The days of free-roaming herds north of Ocracoke had officially ended.

A live dune on Cumberland Island National Seashore engulfs and kills vegetation in its path. Sand dunes periodically mobilize and migrate, then stabilize and revegetate, a natural process that cannot be duplicated with earth-moving equipment and plantings.

In 1938, two local men equipped with "high-powered rifles and dum-dum bullets" methodically gunned down most of the remaining unfenced livestock in Dare County—"about 100 wild cows and upwards of 50 wild ponies" plus swine (MacNeill, 1938, p. 1). County officials paid the hunters a bounty of $5 a head (Federal Writers' Project, 1939) roughly $82 today (U.S. Bureau of Labor Statistics, n.d.). MacNeill (1958, p. 70) reported that First Lady Eleanor Roosevelt's younger brother Hall Roosevelt killed the last unfenced cow on Hatteras Island, apparently for sport, and gave the bounty away. Yet despite the sweeping eradication of domestic stock, small herds of horses remained on the islands below Hatteras Inlet and on the northern reaches of Currituck Banks.

By the late 1960s, ecologists began to realize that natural systems must constantly shift and adapt to changing environmental forces and can only remain "stable" under conditions that allow change (Binkley, 2007, p. 201). Natural dune systems maintain a state of dynamic stability, responding to changes in wave energy, water level, and sediment supply with continual changes in configuration (Gabriel & Kreutzwiser, 2000). Tsoar (2005) concluded that the main stressor limiting dune vegetation is erosion by wind.

Whereas the overwash passes and dunefields of a natural shoreface absorb the energy of storm-driven waves, the roiling ocean slams into continuous artificial dunes with full intensity. Waves erode the duneline and carry sand offshore, gradually narrowing the beach (Gabriel & Kreutzwiser 2000). Artificial dunes, jetties, groins, buildings, and other structures throw the system perpetually out of balance, so people must intervene to solve one problem after another. The beach erodes, so they add more sand. The dunes breach, so they rebuild them. A house, a business, a road washes away, so they rebuild to the west.

A pair of Banker Horses threads a maze of sand fences employed to slow the attrition of the primary dunes. In some places on Currituck Banks, natural dunes are in a "live" migratory phase, to the vexation of developers. Many blame free-roaming horses for destabilizing the dunes, but dune migration is an inevitable process.

Noting "how stabilization has changed the morphology and ecology of the beaches, dunes, and marshes," Dolan and Lins (1987, p. 76) wrote, "The paradox suggests that man-made structures do not merely fail to protect beaches but actually work to destroy them." Moreover, shoreline development, beach stabilization, and heavy use by off-road vehicles degrade habitats and have contributed to the decline of a long list of rare species (Weakley, Bucher, & Murdock, 1996, p. 11). The National Park Service gave up on dune stabilization in the early 1970s because of maintenance costs and negative geological and ecological effects (DeKimpe, Dolan & Hayden, 1991).

DeKimpe et al. (1991, p. 451) noted that beaches 328–410 ft/100–125 m wide before stabilization shrank to 230–328 ft/70–100 m by the mid-1940s and to an average of 39 m/128 ft by the early 1990s, when some were as narrow as 33 ft/10 m. The effects of stabilization, however, are not limited to the beach. Jockeys Ridge, the largest sand dune on the East Coast, has decreased in height from 138 ft/42 m above sea level in 1953 to 110 ft/33.5 m in 1974 and 87.5 ft/26.7 m in 1995 (Mallinson et al., 2009), probably because of stabilization of the surrounding area. The natural flow of sand is disrupted by artificial dunes, buildings, and opportunistic vegetation. Centuries of free-roaming livestock had little impact, but stabilization has caused rapid deterioration.

On Currituck Banks, largely ignored by officials and developers until the mid-1970s, horses survived. The small herd spent little time in the company of people, but when horses did wander into a village, they peacefully coexisted with the residents, who maintained deep emotional ties with them. A minimally manipulated line of natural dunes backed by maritime forest extends from Corolla north to the Virginia line. This area is essential to the survival of both free-roaming horses and the barrier chain. Spreading live oaks and other trees provide shelter during hurricanes, and their root systems hold soil in place when overwash sweeps the island. This and other maritime forests are the least stabilized, therefore most stable, parts of the barrier chain.

Corolla is the only survivor of several settlements in the area that developed around Lifesaving Service stations established in the 1870s. Three others also had post offices with unrelated names. Wash Woods lost its post office (Deals) in 1917, but had enough residents to remain a voting precinct in Virginia Beach until 1966.

Most of the other settlements on Currituck Banks were tiny and short-lived. The town of Seagull (later called Old Inlet) may have dated back to the days of commerce through New Currituck Inlet, which opened in the hurricane of 1713 and persisted after the 1730 closure of Old Currituck Inlet. Around the turn of the 19th century, the new inlet closed, along with Caffeys and Roanoke inlets to the south, causing a downturn in the shipping and fishing trades. Stock raising and subsistence farming continued, however, and the freshening of Currituck Sound caused a dramatic increase in populations of migratory waterfowl. After the Civil War, wealthy Northern sportsmen and local market gunners slaughtered waterfowl until populations collapsed and Congress outlawed commercial hunting in 1918. At its peak, this village had 35 houses, some of them occupied by families of the crew at Penny's Hill Lifesaving Station; two churches; a one-room school; and, from 1908 to 1924, a post office (Stick, 1958; Tennant, 2001).

When Penny Hill migrated in the 1950s, the village was smothered under the giant dune. The Ash Wednesday Storm, a devastating nor'easter, reopened New Currituck Inlet in 1962, and destroyed part of the dune, and federal and state agencies used some of the remaining sand hill to fill the breach.

The village of Corolla was originally called Whales Head, and then renamed Currituck Beach. When the post office opened in 1895, it was named Corolla. Eventually the name stuck on the little unincorporated town. Corolla was difficult to reach by land, but readily accessible by water.

In 1925, Edward Knight of Philadelphia, an industrialist who made his fortune primarily from the Pennsylvania Railroad and the American Sugar refinery, and his wife, Marie Louise Knight, completed the Whalehead Club, an opulent 21,000 ft^2/1,951 m^2 home in the Art Nouveau style. There they maintained their privileged lifestyle and hunted the abundance of fowl that wintered on Currituck Banks.

In 1939, shortly after it had absorbed the Lighthouse Service, the the U.S. Coast Guard dismissed the Currituck Beach Lighthouse keeper and automated the light. It ran on batteries charged by a generator till the Virginia Electric and Power Company ran lines to Corolla in 1955. Until 1973, the only land access from the south was by 15 mi/24 km of rutty, unpaved state road. Alternatively, one could use the beach, driving between the high and low tidemarks, but for the inexperienced this was often an unreliable and risky means of travel.

In 1973, developers paved a road from the Dare County line north to near the village, but restricted access until 1984. While access to Corolla from the south was limited, the Fish and Wildlife Service blocked access from the north through Back Bay National Wildlife

Refuge. Many Corolla residents and nonresident property owners found it difficult to leave or return legally. They railed against the restrictions with petitions and public hearings, and ultimately a child died when access to medical attention was unnecessarily delayed. In the 1980s, developers to the south of Corolla shifted their focus from maintaining exclusivity to accommodating tourists.

When the state took over responsibility for maintaining the road in 1984, the outside world descended on the little village. Beautiful Corolla was secluded no longer. In 1985, there were only 35 full-time residents. A decade later, that number had more than tripled, but the bulk of the population influx came from thousands of people eager to rent seasonal beach homes or just visit for the day.

Magazines and billboards touted Corolla as an "undiscovered paradise," and people came flocking to the "empty beaches" to soak up the solitude. Cars zoomed around the bends of Highway 12. Condominiums sprang up like the lesions of a fast-breeding virus. Price tags were high, but there was no shortage of wealthy vacationers eager to buy into this new retreat. The juxtaposition of sleek wild horses and expensive condominiums, nature and progress, inspired an emotionally charged war with powerful opinions on either side.

Free-roaming horses are strongly motivated to stay within their home ranges throughout their lives. Survival increases when they know which food source is available with the season, where to find water, and where to escape floods or biting insects. When developers placed a beach cottage or housing complex within this home range, the horses coped as best they could and foraged around the new obstacles. They had few options. The surrounding ranges were already claimed by other horse bands that would fight to repel intruders.

The dark horses that often crossed the road at night knew nothing about the impatient, lead-footed drivers that flew down the new roadways. By 1989, 17 horses had been killed in road accidents, six of them in a single incident. Residents and visitors united to create the Corolla Wild Horse Fund. Their mission was to guard the Corolla horses against the human invasion and attempt to preserve their wildness.

The horses needed government protection, but they were ineligible because government agencies deemed them not native or not wildlife. Hunt clubs complained that they ate the food planted for the waterfowl and asked the North Carolina Wildlife Resources Commission to remove the herd, but the horses were not considered native wildlife and were therefore not its concern. Currituck County looked to the state for help. The state bounced responsibility back to the county.

As Corolla residents wrestled with red tape, horses died on the highway. The Corolla Wild Horse Fund outfitted the horses with reflective collars to make them more visible to motorists, sprayed their bodies with glow-in-the-dark paint, and posted signs along Highway 12 warning motorists to slow down because of "horses on road at any time." These tactics helped to reduce, but not eliminate fatalities. Vacationers even stole some signs as souvenirs.

The fatalities were devastating. In 1990, a mare known as Bay Girl delivered her long-anticipated foal, who was to be named Freedom. The day after his birth, the newborn was found floating in a pond with the umbilical cord still attached. The wounds on his little body told the story—someone had hit him with a car and then thrown him into the water. Bay Girl's previous foal had also been killed by a vehicle.

Star, a black patriarch with a white star on his forehead, was a magnificent animal, the leader of a 23-member band. A photograph gracing the cover of *Outer Banks Magazine* showed him prancing down the beach, neck arched, a powerful animal in the prime of life. Star's son Midnight constantly fought his father for possession of the band, or at least a few of the mares. Their battles were frequent and intense. Midnight eventually acquired a mare from the north ranges, but returned to steal more from Star.

In May 1991, Star forced Midnight into a hasty retreat, which took the two battle-crazed stallions across a road at a gallop. Midnight made it across safely, but Star was hit by a fast-moving vehicle and killed. After witnessing the highway deaths of three of his mares and seven of his foals, Star joined the growing list of fatalities. Star was the 20th horse struck between 1985 and 1995. Midnight took over the band, but did not have the talent or experience to manage it. Other studs stole mares, and by 1993 the band was split into four.

In 1989, Currituck County adopted an ordinance that designated all of Currituck Banks a wild horse sanctuary. This made it unlawful for "any person to lure, attract, or entice a wild horse to come within 50 feet of any person" (*Currituck Outer Banks Wild Horse Management Plan*, n.d., p. 2) and protected the horses against trapping, taking, tormenting, injuring, or killing. In June 1995, Secretary Betty McCain of the North Carolina Department of Cultural Resources proclaimed, "The Corolla Wild Horses are one of North Carolina's most significant historic and cultural resources of the coastal area" (*Currituck Outer Banks Wild Horse Management Plan*, n.d., p. 3).

The county makes equine management decisions jointly with The Wild Horse Advisory Board, composed of representatives from the Corolla Wild Horse Fund, the Fish and Wildlife Service, the National Estuarine Research Reserve, along with two citizen representatives (USFWS, 2008). In 1994, Currituck County and the Corolla Wild Horse Fund assigned a management agreement that designated the Fund as the lead advisor in protecting the horses.

Currituck County officials assembled members of the Wild Horse Fund, the Currituck NWR, and the Currituck NCNERR to work out a 13-point management plan. As a result, federal and state agencies have incorporated the Currituck horses into their operations. Management practices now include maintaining a herd of fewer than 60 individuals; blocking wild horse access to the developed areas in North Carolina and Virginia and relocating horses that frequent populated areas; supervising their numbers and health status, maintaining enclosures for a few horses at the Whalehead Club or the Currituck Beach Lighthouse; and using private pasturage within the off-road area.

In a small herd like the one on Currituck Banks, it is important to maintain as much genetic variety as possible in order to avoid inbreeding. Sometimes this necessity is at odds with practicality. At one point, a large percentage of foals born happened to be male, and as they grew older, they presented a unique problem.

Usually colts are driven from the herd when they reach puberty. They form bachelor bands and keep one another company until about age 5 or 6, when they are mature enough to start their own bands. Lured by the lushness of lawns in Corolla, the Currituck colts had no incentive to head north to eat marsh grass. The youngsters stayed, chasing mares around in adolescent ardor and exhausting the mature stallions, which were compelled to continually interpose themselves between the mares and the competition.

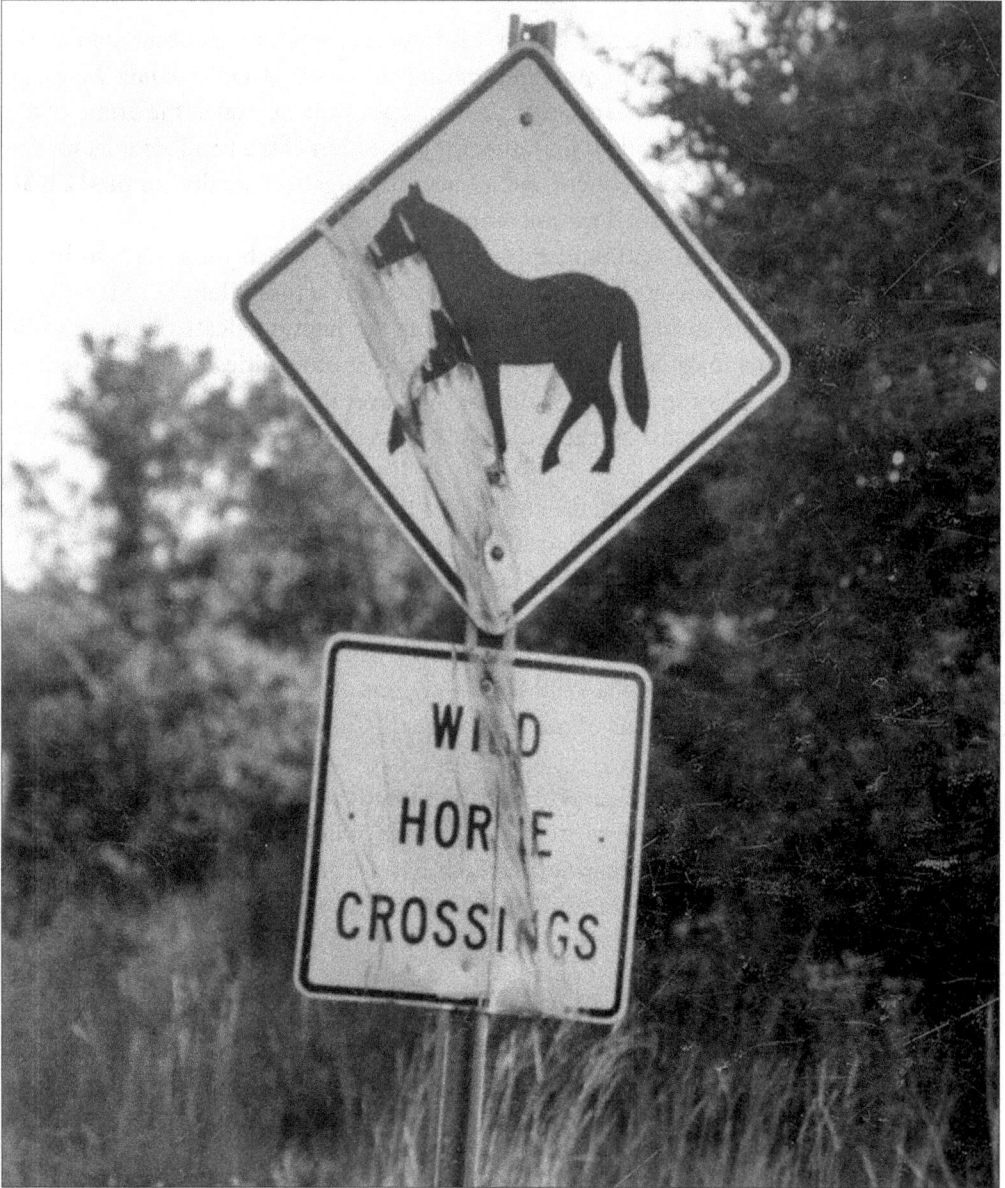

Through the 1990s, numerous signs along Highway 12 warned motorists of the wild horses in the road, but fatalities occurred regularly. In this case, it appears that a driver had run over the sign. Other visitors stole the signs as souvenirs, leaving the horses unprotected.

These colts ranged mostly in populated areas, and once they matured, there was sure to be intense fighting over the few available mares. Unaware tourists would surely get caught in the struggles. In 1990, the Corolla Wild Horse Fund proposed to geld bachelor colts, hoping that harem stallions would tolerate the young males if they no longer competed for reproductive rights.

This approach was controversial. Gelding would reduce the genes available for future generations. And what if something unfortunate happened to the herd stallions? The colts would then be the only hope for the continuity of bloodlines. With so many fatalities, this was certainly a concern. Opponents of the plan reasoned that wild horses should live without

A stallion inspects a newly constructed beach house in 1995, unfazed by the loud hammering of nearby carpenters working to complete the next house in the fast-growing complex.

human intervention, beneficial or detrimental, and a gelding is not a truly wild horse. The proposal to geld was defeated.

Aggressive attempts at public education did not adequately protect the horses. While most visitors respected the ordinances, some were incredibly reckless. The same vacationers that might be a little nervous mounting a well-trained rental horse for an amble down the beach actually felt safe in putting their children on the backs of these unbroken, unpredictable horses for photographs!

Unnecessary human contact eventually habituates much of the "wild" out of these horses by altering natural behavior. It can instigate aggression between horses as they squabble over proffered food. It can result in injury when the horse bites the hand that feeds it, or worse. Overly close contact can expose the visitor to vectors which carry diseases such as Lyme and encephalitis.

County ordinance required people to stay 50 ft/15 m from the horses—that is about the length of one and a half school buses. Too many tourists came in close to feed and pet them when they thought nobody was watching. Reports of kicks and bites were common.

People also did not stop to consider that a horse's digestive system is adapted to native grass. The animals may like the taste of pickles or peanut butter and jelly, but human food can make them very sick. Many well-meaning tourists fed the horses whatever was handy. Some even left plastic bags of apples and carrots out on their trashcans as treats for the horses. The animals did not understand that the bag was not intended to be part of the meal and consumed the entire offering, unaware that intestinal blockage was likely afterwards.

Domestic horses are prone to colic, but Banker Horses rarely develop digestive problems on their diet of grasses. When the horses were left to themselves, the rare instances of colic were usually due to sand ingestion. Once tourists appeared on the scene, colic became frequent and sometimes claimed lives.

One tourist actually lured a curious 2-year-old colt up onto the deck of a beach house with food. The animal fell and was almost killed. One woman trailered her mare to Corolla, hoping to breed her to a wild stallion, and exposed the herd to disease.

A normal horse, acting the way its instincts have commanded over millions of years of evolution, was justified in biting, kicking, or flattening the intruders. Horses are at their moodiest during the breeding season, which coincides with the tourist season.

Our society often has no sympathy for an animal being true to its nature. People want to interact with wildlife, and they are willing to cross boundaries and break rules to do it. They feed the Yellowstone bears. They pet the moose calf on the Vermont roadside. Wild animals accustomed to the presence of humans may appear docile, but remain unpredictable and can revert to dangerous instinctive behavior when stressed. Despite common sense and frequent press coverage, people are astonished and angry when the bear mauls, the moose charges, or the horse kicks. They demand retribution, and officials often typically inclined to hold the animal accountable. Animals that injure people are frequently destroyed, whether or not the action was appropriate from the animal's perspective.

Corolla Wild Horse Fund volunteers donated incalculable time to locating the herds and following nearby to keep tourists at a distance. But these volunteers had jobs and families vying for their attention, and there were too few of them to avert all potential tragedies.

In September 1994, the Wild Horse Fund attempted to block equine access to the heavily developed part of Corolla by building a fence from sea to sound at the North Beach Access ramp at the north end of N.C. 12. At first, the horses easily circumvented the fence and returned to their home ranges.

In 1995 the Fund completed an improved barrier reaching well out into the water to keep the horses north of town. It took two days to herd the horses beyond the fence, where they joined other bands that ranged as far north as Back Bay NWR in Virginia. Some observers scoffed at the fence, asserting that it would not deter horses known for their swimming ability. The average depth of Currituck Sound is only 5 ft/1.5 m, and few areas exceed 10 ft/3 m. Any enterprising horse, they argued, would easily find a way around.

Butterscotch, an ingenious lead mare, proved them right by persistently circumventing the sound side of the fence, leading her group back to the lush vegetation of the golf courses and green sod lawns. In one remarkable incident, she traveled north until she found a sandbar that extended out into the sound and reached 1,500 ft/450 m beyond the end of the fence. She and her friends sloshed through a length of foot-deep (30 cm) water and reentered Corolla well south of the barrier.

This adventure resulted in the death of Grecko, her black yearling son, in June 1995. Grecko wandered onto Highway 12 around 2 a.m. and was thrown 91 ft/28 m when struck by a vehicle. The 18-year-old driver was charged with possession of alcohol.

Sixteen horses in four separate bands regularly came around the fence to graze where the grass was greener. The public, having heard of the roundup to keep the horses north of the developed area, often did not expect to find them still in the middle of the highway.

The volunteers were shell-shocked from watching the horses die unnecessarily and horribly. They simply did not have the time, energy, or other resources to keep them safe on their home turf. In 1996, several horses that persistently returned to Corolla were removed from the Outer Banks and adopted. The rest of the horses seemed content to stay north of the barrier, living free on the mostly undeveloped land.

In the mid-1990s, a mare grazed in the driveway of a beach house off Highway 12 in Corolla. When developers built housing complexes within the home ranges of the horses, the animals adapted.

Then in 1999, reports of wild horses raiding yards, trash cans, and a produce stand made the news when a stallion named Little Red Man and his herd repeatedly came around the barrier to forage in Corolla. A black mare was killed on the road in 1999, orphaning her foal. Little Red Man was relocated with his mares to a 400-acre/162-ha hunt club on Dews Island in Currituck Sound (Williams & Page, 2002–2003).

The Currituck herd forages on about 7,544 acres/3,053 ha of the northern beach. About 70% of their range is privately owned, and the other 30% is public land specifically set aside for native wildlife preservation by entities that do not recognize horses as indigenous species. Individuals and corporations own 4,671.35 acres/1,890 ha, the Currituck National Wildlife Refuge administers 2,495.4 acres/1,010 ha, the North Carolina National Estuarine Research Reserve holds 326.5 acres/132 ha, and the nonprofit Nature Conservancy has 51 acres/21 ha (H.R. 306, Corolla Wild Horses Protection Act, 2011b).

Rheinhardt and Rheinhardt (2004) detailed seasonal habitat use by the Corolla horses. The researchers found that in late winter, horses spent much of their time in the maritime forest to escape from the wind and spent less time than expected in the cold, windy marshes. In spring, they used all available habitats. In summer, their preferred habitat was wet grassland—wet depressions dominated by plants such as cordgrass and rushes. They avoided the dry grassland of sand flats and dunes. Freshwater is always available to the horses in Currituck Sound and in ponds, puddles, and artificial canals.

In 2002, Currituck County and the Virginia Department of Conservation and Recreation built a second sea-to-sound fence 11 mi/18 km north of the Corolla barrier along the southern boundary of False Cape State Park at the Virginia state line to restrict horses from migrating to the developed areas of Virginia. Horses sometimes find their way through, around, or over this fence to range into the Back Bay NWR, where they graze on developing waterfowl food plants within impoundments (USFWS, 2010). The public

A hiker walks past driftwood sculptures crafted by some artistic visitor on the empty, wild beach at Back Bay NWR in Virginia. Wild horses sometimes range into the refuge through False Cape State Park despite the barrier erected to discourage them.

enjoys seeing wild horses on the refuge, and at present the equine incursions are few and cause minimal damage. But the refuge views them as a "potential nuisance animal problem" (U.S. Fish & Wildlife Service, 2010, pp. 2–3) that will become a concern if their population increases.

Like their brethren to the south, Currituck horses that expanded their range north through Back Bay NWR and False Cape SP were getting killed on the roads of Sandbridge. Donna Snow, former Virginia Beach animal control officer and Corolla Wild Horse Fund director, formed the Virginia Wild Horse Rescue, a nonprofit charity dedicated to protecting wild horses that come into Virginia from North Carolina. Working with the Sandbridge Civic League, she organized a response team to capture and return the wild horses to Currituck County. If horses migrate into the park, the refuge, or the neighborhood of Sandbridge, the Virginia Wild Horse Rescue is pressed into action to remove them (USFWS, 2010).

The development boom continues on Currituck Banks, and a proposed bridge that will connect Corolla directly to the North Carolina mainland will encourage more visitors. The northern reaches of Currituck Banks are becoming more appealing to developers. As of 2010, there were 3,090 platted lots (McCalpin, 2010, June 16) and more than 1,300 existing homes, including mansions with more than 20 bedrooms (McCalpin, 2010), in this area. The current herd management plan does not offer much hope for the future, stating that "as the northern Currituck Banks grow and develop ... confinement, relocation or other strategies may be necessary to maintain a viable herd" (County of Currituck, 2007). One very workable solution would inevitably recruit enthusiastic support from a

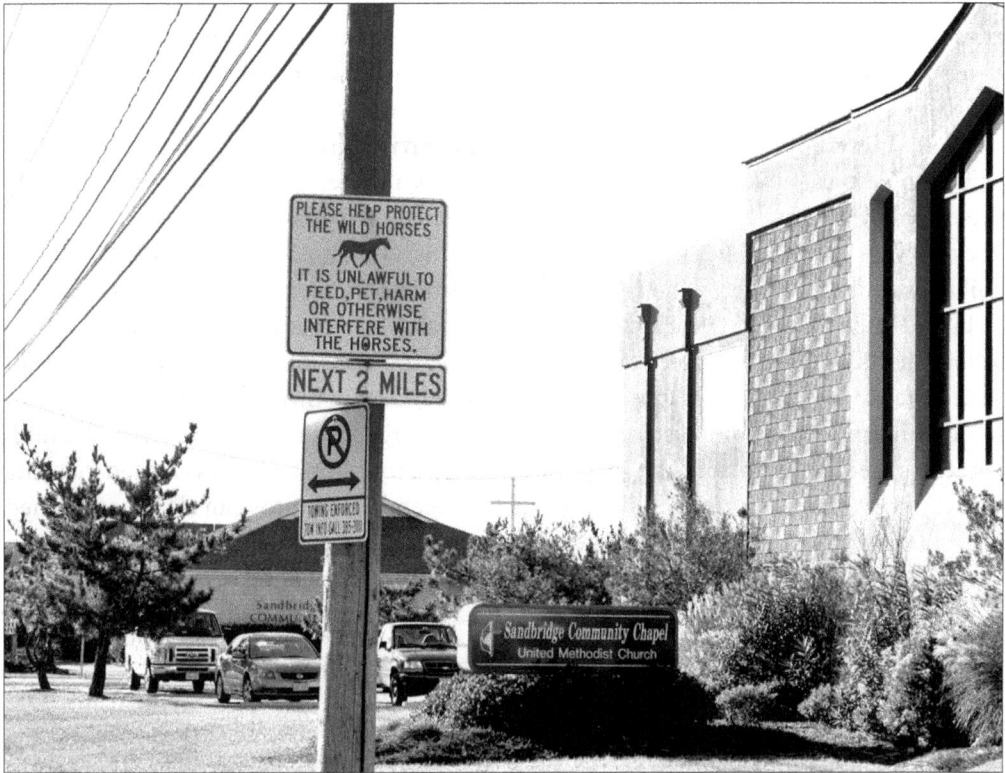

When horses wander through the refuge into the Sandbridge section of Virginia Beach, Virginia Horse Rescue mobilizes to capture them and return them to the safety of Currituck Banks.

large contingent of taxpayers and rabid resistance from state and federal officials and others: allow the Currituck horses to expand their range into the adjoining False Cape SP and Back Bay NWR and even into nearby Mackay Island NWR, which together cover about 22,000 acres/8,900 ha—an area almost half again as large as all of Assateague Island. In fact, federal and state governments could nearly treble the range available to the wild herds of North Carolina, Virginia, and Maryland by opening almost 73 mi^2/189 km^2 near and adjoining, but mostly *within*, preserves where the herds already live. This accommodation would not be quite as simple as the stroke of a pen or the turn of a key, and it would not be free of cost; but the greatest obstacles to its realization are bureaucratic, political, and attitudinal.

Commercial development is not permitted on the north beach, but property owners and developers have been trying to change its zoning. In 2010, Swan Beach Corolla, LLC (owned by developer Gerald Friedman), asked the Currituck County Planning Board to rezone 37.36 acres/15 ha in Swan Beach, part of the 4-wheel-drive area where the wild horses live, from residential to general business. Friedman proposed an inn with up to 302 units in Swan Beach. This complex would include a wellness center, indoor and outdoor pools, a fishing pier, stores, a helipad, a chapel, a fire and rescue station, and "a corral for the wild horses" (McCalpin, 2010, March 2). Ironically, the proposed name of this environmentally destructive hotel is Swan Beach Preserve. If these land tracts are developed, the horses will be restricted to even less land, and both horses and humans are more likely to suffer injury from the resulting overly close contact. Meanwhile, development arbitrarily ruled noncommercial continues. Although motels are banned, it is legal to cover a motel-size tract with large rental cottages that accommodate comparable numbers of people.

The species most damaging to the barrier island environment is not horses, but humankind. People tramp over, drive across, and slide down dunes that are easily degraded by these activities. They leave behind plastic bags, plastic and glass bottles, and other trash. They sully ground water with sewage, vehicular fluids, and other contaminants. To protect buildings and infrastructure, they disrupt natural processes and environmental balance.

In 2001, the Corolla Wild Horse Fund incorporated as a stand-alone nonprofit organization. In 2006, three full-time staff members took the reins of the organization. It converted the old Corolla schoolhouse to a Wild Horse Museum that offered activities such as wooden-horse painting and discussions. In 2012, the organization moved to a larger building across the street. Children can get personal with Corolla horses when a rescued mustang comes to visit regularly during the warm season. A membership program allows concerned citizens from all over the United States to become a part of the wild-horse preservation efforts. A Web site that receives more than 1 million hits annually serves as an information hub at www.corollawildhorses.com and includes a lively, poignant blog to keep enthusiasts up to date with the herd happenings.

Wesley Stallings was the manager of the Currituck herd for many years, and the author rode with him on several occasions. Almost every day he patrolled the area north of the fence, counting horses, looking for injuries and problems, monitoring trends, and evaluating habitat. Stallings tracked the bands of horses by stallion according to their range. He carried a notebook in which he recorded the characteristics and habits of each horse, but he did not name them, tag them, or brand them. He was well acquainted with every individual in the herd. "We don't know the specifics of family lines within the herd," said Stallings (personal communication, May 25, 2010). "The main thrust is land management."

Like the current manager Christina Reynolds, Stallings worked with the Fish and Wildlife Service and the Currituck NCNERR to protect the herd over the long term while maintaining ecological balance. "It is important that the work we do now is a permanent fix," explained Stallings. "Up until now, we had been putting Band Aids on our problems" (personal communication, May 25, 2010).

The herd managers work towards implementing a geographic information system to make his data accessible to lawmakers, scientists, and others. A GIS can help people collect, store, and analyze data associated with a particular location, merging cartography, statistical analysis, and database technologies. Stallings explained "GIS can create an evidence based model that people can relate to. When legislators want to implement some system with the horses, the GIS can illustrate where the horses range and how they utilize the environment" (personal communication, May 25, 2010).

One would think managing a wild horse herd would be an idyllic career. In reality, CWHF herd managers are on call around the clock and have as little downtime as a mother with a new baby. At home, the telephone is always ringing. A common scenario occurs when a horse lies down in someone's yard to take a nap. Unfamiliar with the habits of horses, the property owner calls 911 to report a dead horse. The emergency dispatcher calls the manager, and the manager talks to the person who made the report. Often by the time the manager has arrived, the horse has awakened and wandered off.

The beach north of Corolla is no longer an "undiscovered paradise." During the warm season, up to 3,000 vehicles per day cruise up and down the shorefront, and endless rows of trucks park with tailgates to the sea.

People call when they see someone feeding a wild horse. He gets frequent calls reporting a lame horse—the horse in question has a chronically locking patella and sometimes drags her toe. Tour guides show off for tourists by calling with insignificant questions, hoping to increase their tips.

A near-wilderness in the 1980s, the beaches north of Corolla are now subject to heavy vehicular traffic. Between 2006 and 2009, wild-horse tour companies proliferated, and the number of beach drivers mushroomed. The beach and unpaved roads became deeply rutted from vehicular overuse, the beach was like a parking lot, and inexperienced off-road drivers swerved unpredictably in a nerve-wracking semblance of urban rush hour. Karen McCalpin, director of the Corolla Wild Horse Fund, says that in July, up to 3,000 vehicles a day drive up and down the beach and behind the dunes (*H.R. 306, Corolla Wild Horses Protection Act*, 2011b).

According to the Wild Horse Fund, one company, operator of two "monster" buses and at least one large SUV, was planning to set up a commercial riding stable among the wild herds. In June 2009 a group of riders on three domestic horses led two other horses up the beach to the tour company's property to the north. A sanctuary patrol officer stopped them and explained that even vaccinated domestic horses can carry diseases to which the wild horses have no immunity and that the wild stallions are likely to charge and injure intruders.

The riders, at least one of whom had been drinking, were defiant and continued their ride, "allowing a bikini clad woman to step from the bed of her truck into the saddle and go for a pony ride of sorts" (McCalpin, 2009, June 29). A wild stallion crossed the dune line and charged the intruders to protect his harem. One of the horses reared and threw its rider. Sheriff's deputies repeatedly chased the stallion back. The rider remounted and the group

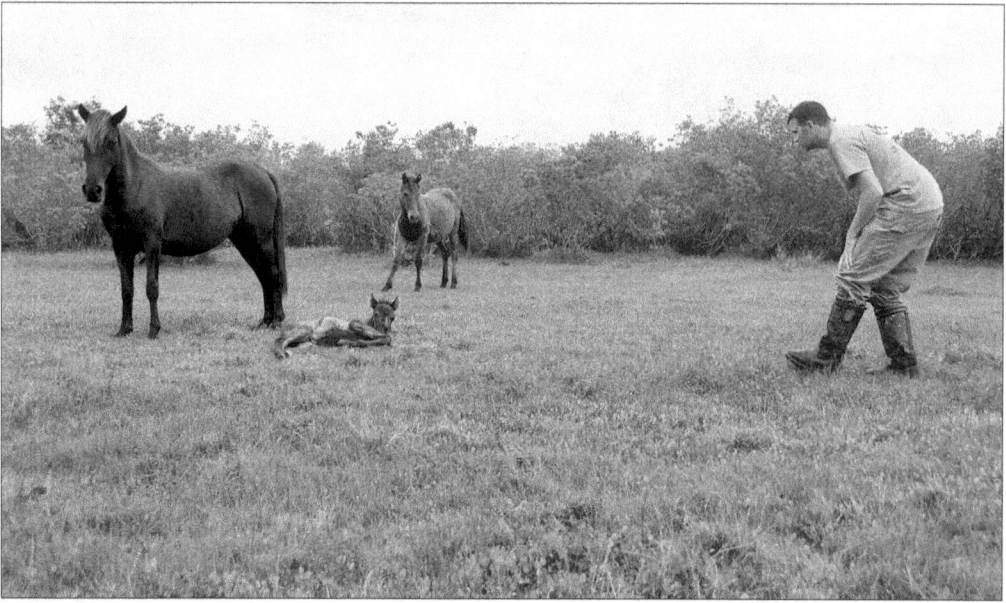

Herd manager Wesley Stallings assesses the health of a newborn colt.

continued up the beach, chatting with beachgoers along the way. Now a county ordinance prohibits riding on the beach or keeping domestic horses anywhere in the wild horse range.

As of 2010, any agency offering tours to see the wild horses must have permits, must include a guide, and must have signs on its vehicles. The county regulation will go a long way toward reducing traffic on the beach. In an attempt to make the tours more informative and less disruptive to the lives of the horses, the Corolla Wild Horse Fund educates guides in local history, equine behavior, and procedures for safe and legal horse-watching. The organization's Web site maintains a list of approved tour agencies that are compliant with regulations.

The best way to see the horses is to accompany the herd manager on his rounds. With the purchase of a Mustang Defender membership at the $250 level, two people can ride with the manager for about 4 hr, observing and even participating in census taking, record keeping, and other daily tasks. This trip is educational and individualized, it offers excellent opportunities for photographs, and every dollar goes to help the wild horses ("Take the trip of a lifetime!" 2009).

The Corolla herd represents one of the rarest strains of Colonial Spanish Horses. In the absence of proper management, they will become extinct. Horse breeds are always changing, and there is always a balance to be struck between keeping bloodlines pure and losing genetic diversity. Too much diversity, and the uniqueness of the population is lost; too little, and the population will collapse, and the uniqueness of the population will still be lost.

The horse has 64 chromosomes. To conceive a foal, the stallion and mare each contribute 32 chromosomes to the new individual, which come together at conception. Only half of each parent's genes are passed along to a given foal. These genes are passed randomly. A single foal can receive only half of each parent's genes; but if each horse has many foals, chances are that most of their genes will be perpetuated in the next generation. The larger the population, the better the odds. A coin flipped 10 times may yield eight heads; but if flipped

1,000 times the occurrence of heads will approach 50%. Horse groups become genetically uniform over time when certain genes are not passed on. In a small, closed population, only new mutations can increase genetic variability.

Large populations mating freely over generations tend to form homogeneous and stable herds with many genes circulating. Rare traits often linger silently within the herd's genome, manifesting episodically. Small populations, however, tend toward homogeneity and *instability*, either emphasizing or eliminating rare genes and new mutations. Slight differences in mortality or reproductive success can have dramatic effects on the population, causing certain traits to suddenly predominate or disappear.

The location of each gene is fixed on the chromosome. Each parent contributes one gene at each location, which at conception unites with its counterpart to form a pair of genes at each location. The two paired genes are called alleles. When the code for both alleles is identical, they are homozygous; if each allele is different, they are heterozygous. For example, a horse homozygous for the black gene will have two copies of it. If he is heterozygous, he will have one black gene and one gene for another color at that location.

A dominant gene expresses itself if the horse has either one or two genes for the trait. For example, black is dominant over chestnut; so if either gene is for a black coat, the horse's coat will be black. A recessive gene expresses itself only if the horse has two genes for the trait. A horse will have a chestnut coat only if he is homozygous—has two genes—for chestnut. If one of those two is for black, that gene will express itself as a black coat while the chestnut gene remains unexpressed. Many traits involve multiple alleles (such as the modifying genes that cause a bay horse to express black only on the mane, tail, and legs), and some traits are incompletely dominant.

When animals in a group are less closely related, healthier dominant genes usually balance rare, disease-producing recessive ones. Recessive genes are masked by dominant alleles and can skip many generations without making their presence known. When animals breed with close relatives, recessive genes are less likely to be balanced by dominant genes, and deformities and disease become likely. Many genetic diseases result from the coupling of hidden recessive genes, including cataract blindness, dwarfism, parrot-mouth, club-foot deformities, lethal white syndrome in pinto horses, and severe combined immunodeficiency in Arabians. A shallow gene pool can also decrease reproductive ability, size, and resistance to disease.

In wild herds with limited diversity, foals with genetic problems often simply die, usually before they can reproduce. Managers of wild herds can allow nature to take its course, or can intervene by increasing genetic diversity, removing diseased horses, or sterilizing horses with undesirable genes.

Isolated herds of fewer than 200 horses tend to lose dominant alleles through inbreeding (a loss of variability that occurs when close relatives frequently mate) and though genetic drift (horses in a group happen to mate and randomly pass along some genes but not others). For example, the once-present gene that causes a solid color to turn gray/white has evidently been lost from the genome of the Corolla herd. The graying gene is dominant; so if it remained, there would be some gray horses.

The original 1997 Currituck Outer Banks Wild Horse Management Plan, signed by representatives of the Corolla Wild Horse Fund, Currituck County, Currituck NCNERR, Currituck NWR, and the Fish and Wildlife Service, allowed 60 horses to remain free-roaming on the northern Outer Banks (*Currituck Outer Banks Wild Horse Management Plan, n.d.*). This number was not scientifically calculated (E.G. Cothran, personal communication, April 20, 2011), but was "merely a number upon which all parties were able to agree after prolonged and contentious debate" (Corolla Wild Horse Fund, 2008, p. 1). Cothran (personal communication, April 20, 2011) explained, "That is not a viable number, particularly for a population that has already had a great erosion of genetic variability. That population at one time was down below 20 individuals in the early 1990s."

Bhattacharyya, Slocombe, and Murphy write, "To broadly discuss all free-ranging horses in the region as though they were a homogenous group, and to generalize about their origins or management, demonstrates an oversimplified understanding of the heterogeneity of the horse populations and landscape" (2011, p. 620). Upon becoming the first executive director of the Corolla Wild Horse Fund in 2006, McCalpin immediately recognized that if the herd were managed at 60 animals, there would eventually be a complete genetic collapse similar to that experienced by the Ocracoke herd (Corolla Wild Horse Fund, 2008). The DNA testing performed in 1992 by Cothran (2008) collected DNA samples via dart gun to get current information about the 90-member herd's genetic health and showed that the Corolla horses have less genetic diversity than any other horse breed. He wrote that "much of the genetic diversity expected to be present in a horse population is gone, and this cannot be recovered" (2008, p. 2). Additionally, analysis of the mitochondrial DNA passed from mother to daughter showed that all the Corolla horses descend from a single maternal line (McCalpin, 2010, March 24).

McCalpin blamed inbreeding for a higher incidence of genetic disorders affecting the Corolla horses. A number of individuals are afflicted with locked stifles—an intermittent immobility where the hind leg joins the body. She said that if the number of horses is reduced to 60, Corolla would be seeing more horses with this problem, as well as overbites, crooked legs, and other conditions associated with inbreeding.

Conservation geneticists give the minimal sustainable effective population size as 50 individuals, and effective size should be ⅓ to ⅕ census size. (E.G. Cothran, personal communication, September 10, 2014). In other words, to be sustainable over the long term, a healthy herd should have 180–300 individuals. The number is higher for a genetically compromised group. Individually, none of the barrier islands have the forage to support herds of this size, but if the North Carolina herds are viewed as a single population, translocation of individuals between the groups can maintain genetic health.

Not all scientists agree. "The issue of what constitutes a genetically viable population has been turned into a bumper sticker issue," says reproductive physiologist Jay Kirkpatrick (personal communication, April 12, 2011). "There is no empirical evidence or data to support any of the numbers thrown around. They are theoretical and there are a lot of data to show that very low numbers are able to rebound with vigor." This seems to be the case with the sika elk of Assateague Island; the apparently robust population was founded by only seven individuals in the 1920's.

Vega-Pla, Calderón, Rodríguez-Gallardo, Martínez, and Rico (2006) studied the Retuertas horse or *Caballo de las retuertas*, a rare strain native to the Andalusia region of Spain. These horses closely resemble the horses that roamed that area in prehistoric times. Calderón noted that the present Retuertas horse population had passed though a genetic bottleneck of tens of horses and shows no sign of inbreeding depression. The team concluded that

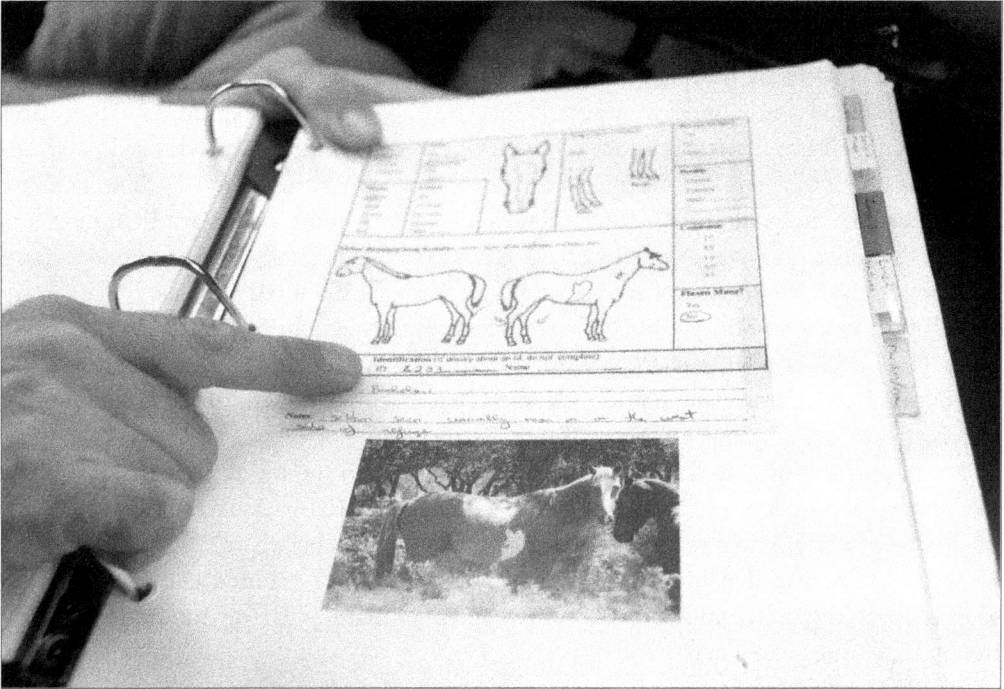

Wesley Stallings, former herd manager for the Corolla Wild Horse Fund, points out a data sheet describing the most colorful horse in the Currituck herd. This stallion has sabino coloration with a bold slash of white on his barrel. One of his sons carries a similar white marking. The rabicano gene, expressed by areas of roaning, is also present in the herd.

genetic variability and demographic and stochastic factors play a greater role in long-term survival than a specific numerical threshold. Likewise, the robust heard of 400-odd sika deer on Assateague Island arose from 7 animals released in 1923.

Kirkpatrick points out that the subject of minimum viable population, though important, has gotten entangled in nonscientific agendas.

> The MVP number for ASIS has been determined to be 48 and 52 for the Pryors, but these are only theoretical and have no empirical data to support them. Unless reproduction is failing, for physiological (genetic inbreeding depression) reasons, there is no genetic issue. (Personal communication, May 28, 2014)

Yet reproductive failure is a late sign of a shrinking gene pool. The herd north of Corolla, N.C., is still fecund, but genetic analysis shows genetic variability to be dangerously low. Its inbreeding expresses itself with an increase of recessive disorders, such as locking stifles that require surgical correction.

In her capacity as a certified nurse-midwife, the author serves a largely "plain" population in Lancaster County, Pennsylvania. These Amish and Mennonites descend from a small founder population, and tend to marry within their own groups. Over time they have experienced an erosion of genetic variability that reveals itself in a high rate of serious disorders, including severe combined immunodeficiency (SCID), microcephaly, many types of potentially lethal inborn errors of metabolism, and other conditions that are rare in the general population (Gura, 2012). For example, Ellis-van Creveld Syndrome, an often-lethal form of dwarfism, occurs once in about 60,000 live births in the general population. Among Old Order Amish, the rate is greater than 1 in 200 live births. Yet contraction of the gene pool has not diminished fertility. It is common for families to have 8, 12, or even more children. So far, most of them are born healthy.

When isolated herds undergo a population crash, they tend to reproduce rapidly and soon regain their former size. But each crash inevitably creates a genetic bottleneck, weakening the health of the herd and making all members more uniform. In a small, closely related population like the horses of Currituck Banks, diversity is more important than numbers alone if they are to survive for the long haul. A herd of 60 genetically diverse horses is more likely to to thwart the specter of extinction than a herd of 150 inbred animals.

In 2009, there were only about 100 breeding mares between the Shackleford and Corolla herds. This decline prompted the Equus Survival Trust (2011), a nonprofit organization committed to protecting heirloom equine breeds, to change the status of the Banker Colonial Spanish Mustang to Critical/Nearly Extinct.

The population of Corolla horses bloomed well beyond the 60-horse legal maximum. The census-takers counted 119 horses in 2006, 94 in 2007, 101 in 2008, 88 in 2009, 115 in 2010. Forty-two horses were removed from Currituck Banks between 2006 and 2008, and 33 of these were placed in adoptive homes. The census of 2011 counted 121 horses. In that year, 16 foals were born, and Stallings contracepted 53 mares with the immunocontraceptive porcine zona pellucida (PZP).

In 2014, at about the time this book went to press, the population was 100 horses, and McCalpin had administered contraceptive vaccines to all the mares with the help of retired police officer E.T. Smith. Only 2 foals were born in 2014. One, a spunky colt named Vivo, was removed from the wild with his mother for the correction of congenitally contracted tendons which forced him to walk on the tips of his hooves. Only one 2014 foal remains in the wild, a leggy filly named Primavera.

The National Academy of Science, in its 2013 report on the genetic health of Western herds, recommended that the smaller populations should be regarded as a single herd, and managers should exchange horses between groups to maintain genetic diversity:

> The committee recommends routine [genetic] monitoring at all gathers and the collection and analysis of a sufficient number of samples to detect losses of diversity. The committee also recommends that BLM consider at least some animals on different HMAs as a single population and use the principles of metapopulation theory to direct management activities that attain and maintain the level of genetic diversity needed for continued survival, reproduction, and adaptation to changing environmental conditions. . . . Although there is no minimum viable population size above which a population can be considered forever viable, studies suggest that thousands of animals will be needed for long-term viability and maintenance of genetic diversity. Few HMAs are large enough to buffer the effects of genetic drift and herd sizes must be maintained at prescribed AMLs, so managing HMAs as a metapopulation will reduce the rate of reduction of genetic diversity over the long term. (NRC, 2013, p. 9)

In July 2014, McCalpin and Cothran collected DNA samples from horses in the Cedar Island, N.C., herd via dart gun to evaluate possible candidates for genetic exchange. Although there have been horses on Cedar Island for more than 100 years, many in the current population were translocated from Shackleford Banks after most of the original herd was euthanized for carrying equine infectious anemia. Both the Corolla and Shackleford/Cedar herds

Banker horses transition gratifyingly from beautiful wild creatures to loyal domestic companions. At Mill Swamp Indian Horses in Smithfield, Va., Lydia Barr and Steve Edwards train Edward Teach, a wild-born stallion removed from Currituck Banks after receiving a life-threatening neck wound that has completely healed.

appear to have descended from the same original population, and mating between the two groups would not be considered crossbreeding. Shackleford Banks horses, however, are genetically healthy, with three maternal lines where Corolla has only one.

A sound stallion, given the opportunity, can impregnate many mares simultaneously and shift the herd genome toward better health more rapidly than a mare, which almost never carries more than one foal at a time. But for stallions, reproductive success is far from certain. Before he gets a chance to breed, a mature male must fight seasoned stallions to win mares and retain them in his harem. Many young stallions do not have the skills to entice mares and fight or bluff to keep them. If the Corolla Wild Horse Fund introduces an outside stallion who is unable to assemble a harem, he will have no genetic impact. And artificial insemination is not an option, as on a panicky wild mare, the procedure would require capture and anesthesia.

Cothran analyzed the DNA of the two most promising prospects, then rejected a gorgeous silver dapple colt because he carried the gene for multiple congenital ocular anomalies. Ultimately, on November 20, 2014, the Corolla Wild Horse Fund introduced a stallion from the Cedar Island herd. The veterinarian found him healthy, and analysis of his DNA proved that he is a Colonial Spanish horse without identified genetic disease. Named for Gus Cothran, who was instrumental in this revitalization project, the stallion will breathe new life into a dying gene pool if he successfully mates with Corolla mares. The Corolla Wild Horse Fund hopes to introduce Cedar Island mares in the future. Cedar Island horses, most of which carry Shackleford Banks bloodlines, are genetically similar to the Corollas and will keep the breed pure, but are different enough to add new alleles.

Low genetic diversity and low numbers are the two greatest hreats to the Corolla herd. When the census is very low, the herd is vulnerable to destruction by disease, drought, fire, flood, or hurricane. This risk is very real. On Cumberland Island NS, a 1990 outbreak of mosquito-borne Eastern equine encephalitis killed about 40 horses (Goodloe et al., 2000). On Assateague Island, at least 15 horses died of EEE in a 5-year span, and 12 drowned in a single storm.

The critically endangered Colonial Spanish Horses on Great Abaco Island in the Bahamas numbered an estimated 200 in the mid-20th century, but, excluded from habitat, gunned down for sport and from antipathy, and attacked by dogs, dwindled to a low of three individuals in the 1970s (Kohn, 2004). Wild horse advocates intervened to save the remaining animals, and by 1992, the herd had increased to 35. But when Hurricane Floyd struck in 1999, storm damage and resulting fires pushed the horses onto a citrus plantation. The rich forage on the plantation caused obesity, decreased fertility, and crippling laminitis, and herbicides, pesticides, and poisonous plants caused extensive liver damage. As of this writing, the once-thriving herd has been reduced to a single aging mare named Nunki. According to Milanne Rehor of the Wild Horses of Abaco Preservation Society, reproductive specialists are attempting to harvest ova from Nunki, in hopes of fertilizing them with sperm from genetically appropriate stallions and gestating them in surrogate mares (personal communication, February 17, 2014).

In the Dzungarian Gobi of southwest Mongolia, a critically endangered population of Przewalski horses was thriving and rapidly increasing its numbers until a harsh, stormy winter devastated it. Frequent blizzards and subzero cold reduced the Przewalski population from 137 horses in December 2009 to 48 in April 2010. Whereas the herd typically increased by 28–36 foals annually, a solitary foal was born in 2010. It will take many years for the herd to reach its projected minimum viable population of 200–300, and another disaster could easily wipe it out entirely (Kaczensky, Ganbaatar, Altansukh, & Enkhsaikhan, 2010). Even at the minimum size for genetic sustainability, the herd will remain vulnerable to natural disaster. The world population of the species, including zoo stock, is only about 1,400 animals.

Over the long term, the Corolla herd is likely to experience genetic collapse and extinction within a few generations if managed at 60 animals. As Sponenberg wrote to McCalpin in 2008,

> In this, as in other cases, the competing interests need to somehow come to an effective compromise. I don't know what that will look like, but I do know that if a genetically isolated horse population is to be genetically secure for the future, then the total population must be much closer to 100 than 60. (H.R. 306, Corolla Wild Horses Protection Act, 2011b)

The Corolla population's best chance for survival lies in increasing its numbers and introducing outside horses occasionally to maintain genetic diversity over the long term. To that end, U.S. Representative Walter B. Jones (NC 3) introduced the Corolla Wild Horses Protection Act (H.R. 5482) to the 111th Congress on June 8, 2010. This legislation was to

Newborn foals are becoming an uncommon sight on Currituck Banks with the implementation of the immunocontraceptive program. Colonial Spanish Horses are endangered as a breed, but the Bankers are especially rare, with only about 100 breeding mares between the Corolla and Shackleford herds.

establish a partnership between the secretary of the Interior and the Corolla Wild Horse Fund similar to that between the Park Service and the Foundation for Shackleford Horses, which allows for a target population of 120–130 horses and occasional introductions of horses from Shackleford Banks. After it died in committee, Jones introduced a new bill (H.R. 306) with the same title to the 112th Congress on January 18, 2011. It was approved by the House Natural Resources Committee on October 5, 2011, despite a written dissent by six members from northern and western states (H.R. Rep. No. 112-210, 2011), and was unanimously passed by the House on February 6, 2012. It died there when the 112th Congress went home. On January 3, 2013, the first day of the 113th Congress, Jones reintroduced the bill once more as H.R. 126. It once again passed the House unanimously on June 3, 2013, and is now waiting to pass the Senate, with chance of being enacted that one source estimates at 20% (Corolla Wild Horses Protection Act, H.R. 126, 2013).

Passage of this bill would allow horses to roam anywhere on the refuge unless credible scientific evidence shows that they threaten survival of an endangered species for which the refuge is critical habitat. The bill is supported by the Wild Horse Fund, former North Carolina Governor Beverly Perdue, Marc Basnight (former president *pro tempore* of the North Carolina State Senate), the Humane Society of the United States, the County of Currituck, the Animal Welfare Institute, the Foundation for Shackleford Horses, Saving America's Mustangs, American Wild Horse Preservation Campaign, and Equus Survival Trust.

But a healthy herd of sustainable size would raise anew the old question, "Where will these horses live?" Detractors maintain that the purpose of the refuge is to support what they consider wildlife, including waterfowl, migratory birds, and endangered species. McCalpin pointed out that at Shackleford Banks on Cape Lookout NS, the National Park Service maintains 120–130 horses on a 3,000-acre/1,200-ha range. Over the past 12 years, there has been no documented negative impact to the seashore. Yet the comparable number of Corolla horses stand accused of damaging the 7,544 acres/3,053 ha where they roam, most of it private land (*Re: Denial of request*, 2008).

The Fish and Wildlife Service opposes this bill, as do The Wildlife Society, the National Wildlife Refuge Association, Defenders of Wildlife, Ducks Unlimited, the Public Lands Foundation, and the North Carolina Wildlife Federation (Hutchins et al., 2012). To the National Wildlife Refuge Association, the leading nonprofit support group for the refuge system, wild horses are as unwelcome as feral cats or hogs. Its Web site opines,

> Like the kudzu vine that ravages southeastern states, or the Burmese python in Florida, feral horses are an introduced species that can wreak havoc on native plants and wildlife. . . .
>
> The so-called "Corolla Wild Horses" that roam the Outer Banks venture onto Currituck National Wildlife Refuge and trample a fragile ecosystem that many native species depend on. (Woolford, 2012)

The declaration is at odds with the fact that the horses were resident for hundreds of years before the refuge was established, including during the 19th century, when waterfowl populations were at their zenith. Although the Association does not necessarily speak for the Fish and Wildlife Service, the agency seems to have made no effort to distance itself from this inflammatory statement.

The Congressional Budget Office (2011) calculated that managing the horses on the refuge would cost $200,000 a year. The nonprofit Corolla Wild Horse Fund would absorb all costs of horse management, including census-taking, record keeping, health inspection, removal of animals unfit for the wild, and population control. The Fish and Wildlife Service (2008) has expressed doubt that the refuge can support additional horses without compromising the environment, and it asserts that the bill is at odds with the refuge's mission. The Corolla Wild Horse Fund says, however, that if the bill were signed into law, the Fish and Wildlife Service's role in horse management would not change. Because the Currituck Refuge is a satellite of Mackay Island NWR with no permanently assigned staff, it appears the only active role the agency has taken in horse management has been pressing to have the horses thinned or removed.

While today the Fish and Wildlife Service (USFWS, 2008, p. 194) considers horses "exotic and potentially damaging to vegetation under active management," in the 1940s the agency thought otherwise. Rachel Carson, internationally celebrated marine biologist, environmentalist, and editor-in-chief for the Fish and Wildlife Service, observed that the presence of 300 head of cattle and horses grazing the Chincoteague NWR was "not detrimental to the waterfowl for which the refuge was established"(1947, p. 17).

Michael Hutchins, former Executive Director and CEO of The Wildlife Society, maintains that feral horses trample soils and vegetation, change the distribution and abundance of native plant species and nutrients, selectively consume palatable plants, and disturb avian nesting sites (*H.R. 306, Corolla Wild Horses Protection Act, 2011a*). He testified at a hearing in April 2011 the legislation would make it more difficult for the Fish and Wildlife Service to manage the feral horses on Currituck NWR and hamper the refuge system's mission. It would put the Fish and Wildlife Service in the difficult position of being legislatively required to manage for the conservation of native wildlife and habitat on the one hand and to support what it deems a non-native invasive species on the other (*H.R. 306, Corolla Wild Horses Protection Act,* 2011a).

Do wild horses cause permanent environmental damage to their home ranges? Research has shown that they do—and that they do not. Promoters and detractors alike can draw on a large body of evidence to support their positions. Taken as a whole, these studies appear to demonstrate that in the absence of overpopulation, horses and other grazers harmonize with and even benefit the rest of the natural world, and species disrupted by their activities typically rebound quickly from damage.

Numerous studies have documented or suggested detrimental or potentially harmful effects of free-roaming herbivores (Barber & Pilkey, 2001; De Stoppelaire et al., 2004; Oduor, Gómez, & Strauss 2010; Seliskar, 2003; Turner, 1987). Nimmo and Miller (2007, p. 409) compiled a list of adverse environmental effects of wild horses, which the author has adapted:

- Soil loss, compaction, and erosion (Andreoni, 1998; Beever & Herrick, 2006; Dyring, 1990)
- Trampling of vegetation (Beever & Brussard, 2000b; Dyring, 1990; Rogers, 1991; Turner, 1987)
- Reduced plant diversity (Beever & Brussard, 2000)
- Killing native trees by chewing bark (Ashton, 2005; Schott, 2002)
- Damage to bog habitat (Dyring, 1990; Rogers, 1991)
- Damage to water bodies (Dyring, 1990; Rogers, 1991; Rogers, 1994)
- Facilitated weed invasion (Rogers, 1991; Taylor, 1995; Weaver & Adams, 1996)
- Altered community composition and reproductive habitat of birds (Levin, Ellis, Petrik, & Hay, 2002; Zalba & Cozzani, 2004), fish, crabs (Levin et al., 2002), small mammals, reptiles (Beever & Brussard, 2004), and ants (Beever & Herrick, 2006)
- Fewer birds (Levin *et al.*, 2002)
- Reduced plant biomass and salt marsh vegetation (Turner, 1987)
- Lower diversity and density of fishes (Levin *et al.*, 2002)

Conversely, many researchers have demonstrated the beneficial or potentially beneficial effects of the grazing of large herbivores on ecosystems, especially wetlands and grasslands (Anderson, 1993; Benot, Bonis, Rossignol, & Mony, 2011; Duncan & D'Herbes, 1982; Keiper, 1981; Levin et al., 2002; Menard, Duncan, Fleurance, Georges, & Lila, 2002; Plassmann, Laurence, Jones, & Gareth Edwards, 2010; Stroh, Mountford, & Owen, 2012; Vavra, 2005; Wylie, 2012). Documented beneficial environmental effects of feral horses include

- Dispersing seeds of desirable native plants (Hobbs, 1996; Severson & Urness, 1994; Stroh, Mountford, & Owen, 2012)
- Benefiting endangered reptiles (Tesauro & Ehrenfeld, 2007)
- Promoting biological diversity, accelerating succession, and encouraging a diverse mosaic of desirable plants (Bakker, 1985; Bazely & Jefferies, 1986; Hobbs, 1996; Jensen,1985; Menard et al., 2002; Severson & Urness, 1994; Vavra, 2005)
- Increasing nutritional value of forage (Hobbs, 1996; Severson & Urness, 1994)

A black stallion saunters through a community on the "isolated" reaches of the north beach that consists of a development of opulent beach houses and a small park, fenced to keep horses out.

- Increasing browse vital to winter survival of deer and elk (Reiner & Urness, 1982)
- Providing feeding, nesting, dusting, and display sites for upland and passerine birds (Vavra, 2005)
- Breaking up homogeneous grass stands, producing a patchy, open cover with a diversity of forbs (USFWS, 1999)
- Enhancing regrowth of forbs beneficial to wildfowl (Evans, 1986)
- Greatly increasing diversity of bird species (Levin *et al.*, 2002)
- Altering community composition of birds, increasing foraging habitat for willets, least sandpipers and other birds that prey on small invertebrates (Levin et al., 2002)
- Greatly reducing height and density of invasive *Phragmites* (Duncan & D'Herbes, 1982)

Horses beneficially remove excess vegetation from the most productive plant communities and break up dense, tall stands of *Phragmites australis*, the common reed, giving desirable plants a competitive advantage (Duncan & D'Herbes, 1982; Menard et al., 2002). *Phragmites* has been present in North America for tens of thousands of years and is considered a native species in both inland and coastal areas (Bertness, Silliman, & Jefferies, 2004). Pieces of native *Phragmites* were found in the fossilized dung of a Shasta ground sloth dating to 40,000 years ago (Lellis, 2008).

Over the past two centuries, an introduced strain, likely from Europe or Asia (Lellis, 2008), has overtaken wetlands throughout the United States. When developers removed woody vegetation bordering the marshes from Maine southward and ditched tidal wetlands for mosquito control, nitrogen-rich freshwater runoff inundated coastal marshes, lowering salinity and giving *Phragmites* a competitive edge. About 90% of the *Phragmites*

Coastal States Herds and Ranges (2014 statistics)

Herd	Population	Range (acres/ha)	Acres/ha per horse
Assateague Island			
Maryland	~100	9,761/3,950	~97.6/39.5
Virginia			
North herd	~100	2,695–3,399/ 1,091–1,376	~27–34/ ~10.9–13.8
South herd	~50	547/221	~10.9/4.4
Mt. Rogers National Recreation Area/ Grayson Highlands State Park (Va.)			
Mt. Rogers NRA	~100	140,000–200,000/ 56,656–80,937	~1,400–2,000/ ~567–809
Grayson Highlands SP	~45	4,822/1,951	~107.2/42.4
Currituck Banks (N.C.)	100	7,544/3,053	63.9/25.9
Ocracoke (N.C.)	18	180/73	6.7/2.7
Shackleford Banks (N.C.)	~107	2,990/1,210	~24.9/10.1
Rachel Carson NCNERR (N.C.)	~30	1,073/434	~35.8/14.5
Cedar Island (N.C.)	~39		
Cumberland Island (Ga.)	~143	25,734–25,802/ 10,414–10,442	~238.3–238.9/ 96.4–96.7
Western HAs & HMAs	~15,000–31,500	~28,896,000/ 11,694,000	~917.3–1926.4/ ~371.2–779.6

Herds shrink and grow continually with or without artificial population control, and census-taking methods are diverse. Ranges also change because of management practices and natural processes. In the West, for example, federal agencies have reduced the area designated for mustangs by one third since 1971. In the East, shoreline erosion has taken valuable range from the barrier island herds. On the Virginia end of Assateague, the Chincoteague Fire Company seasonally adjusts the north herd's range. Published range figures disagree because of differences in calculation, definition, and record keeping and because of various methods of accounting for surface water, easements, and inholdings. Further, fragmented, partly developed ranges such as Currituck Banks are not exactly equivalent to contiguous, undeveloped ones such as Shackleford Banks. Despite the fog, it is clear that all the wild herds of the East Coast except the ponies of Grayson Highlands are much more crowded than Western mustangs. Some of the Eastern herds inhabit lush, well-watered environments that potentially support more animal units per acre than arid Western lands. Like mustangs, most of the East Coast herds have been arbitrarily banned from government land near, sometimes within, the preserves they inhabit. The most notable example in the East is Ocracoke Island, where the Park Service owns 4,578 acres/1,853 ha, but restricts the herd to 180 acres/73 ha of fenced pasture, about 3.9% of the space available, and most of the horses spend most of their time in smaller paddocks near the barn.

dominance is attributable to shoreline development (Bertness et al., 2004). When nitrogen is abundant, tall plants such as *Phragmites* capture much of the available sunlight and squeeze out their shorter rivals, sharply decreasing the diversity of marsh vegetation and decreasing productivity of fish and shellfish. Once *Phragmites* has taken hold, it is difficult and expensive to subdue.

The invasive reed is nutritious and palatable to horses, and grazing pressures weaken its unwanted foothold on estuarine margins. When horses bite off *Phragmites* shoots below water level, the plant often starts to decompose and may die. Under heavy grazing, *Phragmites* declines quickly in height and density, allowing sunlight to reach competing native plants (Duncan & D'Herbes, 1982). *Phragmites* grows tall and dense in ungrazed exclosures, but it is low and sparse in heavily grazed areas (Bos, Bakker, de Vries, & van Lieshout, 2002). In France, Duncan and D'Herbes (1982) found that by breaking up *Phragmites* stands, horses increase the area of open water and the abundance of submerged plants.

Refuge officials maintain that horses spread the seeds of non-native plants through their manure; this accusation is true. Free-roaming horses live most of their lives in well-defined home ranges and eat problem vegetation such as *Phragmites* with desirable species such as *Spartina*. In one study, non-native species did sprout in dung piles, but then so did native species, which were often at a competitive advantage. The study concluded that since horses avoid grazing where they have defecated, these dung piles might represent "refuges for palatable species," including natives, to take hold (Loydi & Zalba, 2009, p. 107).

Of course, horses can only spread the seeds of plants that exist within their range. In 1985, Davison found exotic plant species had become established in the Ocracoke pony pen, such as Bermuda grass (*Cynodon dactylon*), annual buttercup (*Ranunculus sardous*), and some pasture grasses, introduced when the Park Service fed hay in winter (Bellis, 1995). The author concluded that these exotic plant species did not endanger the native plants beyond the enclosure. Seeds are spread by other herbivores as well and by wind and water. Wild birds are more numerous than horses, travel much farther, and distribute many more seeds in their droppings.

Horses do alter the environmental balance of the salt marsh, but not necessarily to its detriment. On Shackleford Banks, horse grazing changes the marshes from nesting habitat for gulls and terns to a bountiful feeding ground used by a diverse community of foraging shorebirds (Levin et al., 2002). Willets and least sandpipers *prefer* grazed marshes and are nearly absent from ungrazed marshes. Marshes used by ungulates support large populations of crabs, which in turn attract birds that feed on them. Although the number of individual birds is greater in marshes that were not grazed by horses, grazed marshes show much greater diversity. One study showed that twice as many avian species foraged in the marshlands used by horses—20 species versus 10 (Levin et al., 2002). Likewise, Zalba and Cozzani (2004) examined the impact of wild horses on bird communities in the Pampas tall grasslands in Argentina. They found that diversity and total abundance of birds were greatest with moderate grazing, decreased in ungrazed areas, and greatly decreased under high grazing pressure although some species preferred intensely grazed areas.

Disturbance by large herbivores *increases* the diversity and quality of wildlife habitat, creating a patchwork of food resources (Lamoot, 2004; Vavra, 2005). Research has demonstrated that light to moderate grazing does damage plants and disrupt other wildlife, but this damage is temporary. Wood, Mengak, and Murphy (1987) studied the effects of horse grazing alongside goats, cattle, and sheep on Shackleford Banks and observed evidence of damage by these species, but expected the resilient island to recover and reestablish *Spartina* in denuded areas.

Rheinhardt and Rheinhardt found that horses on the Currituck Banks "consume few forb [herbaceous plant] species and graminoid [grass] species seem to recover from grazing by early summer when primary production is highest. . . . Because rooting impacts of feral hogs may be more severe than horse grazing impacts on Currituck Banks, exclosure experiments would have to be designed to separate horse grazing from hog rooting" (2004, p. 258).

The Currituck NWR is concerned that grazing may threaten certain endangered plants: "The effect horses have on sea beach amaranth (if any) needs to be determined, as does interdune grasslands and marshes [sic]" (p. 194, USFWS, 2008).

Seabeach amaranth (*Amaranthus pumilus*) is a rare plant that grows on Atlantic coastal overwash flats and on beaches above mean high tide (Marion, 2010). To thrive, this species appears to need large areas of naturally functioning barrier island beaches and inlets, and when people develop shorelines to prevent overwash, its habitat shrinks. Seabeach amaranth virtually disappeared from Cape Hatteras north after construction of the artificial dune barrier in the 1930s (Marion, 2010), and at Cape Lookout it is not found in areas used by off-road vehicles (Altman, 2009). Its numbers fluctuate dramatically from year to year in response to environmental conditions, thriving when overwash and severe storms bring seeds to the surface to germinate.

Plant counts vary widely from year to year and bear no relationship to the presence or absence of horses. On ungrazed Hatteras Point, Hatteras Inlet, and North Ocracoke, researchers counted 15,828 seabeach amaranth plants in 1988, only one in 1995, and 34 in 2003 (Marion, 2010). On Cape Lookout National Seashore in 2003, the Park Service counted 206 plants on ungrazed Core Banks, and 1,354 plants on the grazed Shackleford Banks (Altman, 2009)—but the following year the census was 137 plants total through all of Cape Lookout NS. Interestingly, in 1995, when the herd was at its peak size of 184 horses (the Park Service had overestimated by 30% and put the population at 240), the seabeach amaranth count on Shackleford Banks was a bountiful 1,155 plants.

The Fish and Wildlife Service blames grazing animals for "reduction of vegetation, encouraging the formation of sand sheets and sand hills, destabilizing much of Currituck Banks" (U.S. Department of Interior, 2010, p. 18) and points to H.F. Hennigar, Jr.'s, master's thesis (1979) for corroboration. More recently, Havholm et al. (2004) used ground-penetrating radar, modern dating techniques, and soil analysis to study dune activity and stabilization along the North Carolina-Virginia coast. The team noted that "overgrazing has not had significant impact" on dune dynamics (2004, p. 993). As in other locations, dunes form where there is an adequate supply of sand on the shoreface, winds are strong enough to mobilize the sand, and there is enough moisture to support vegetation that can stabilize it (Havholm et al., 2004). The presence of grazers does not alter this ancient process.

Plassmann et al. (2010) found that dune vegetation is influenced primarily by rainfall, and large herbivores had minimal effect on compaction and moisture retention. They also noted that vegetation appeared to benefit more from year-'round grazing than from seasonal grazing (Plassmann et al., 2010). Additionally, physical disturbance of large sand flats by horses helps to provide nesting habitat for least terns, piping plovers, and other shorebirds by delaying their colonization by perennial grasses and shrubs.

Barrier islands maintain dynamic stability by rolling over themselves toward the mainland. They can adapt to violent storm surges and sea level rise, but are weakened by cumulative stresses introduced by people, such as shore protection (Gabriel & Kreutzwiser, 2000).

Clearly, the problem species is humans, not horses. Residential development not only changes the appearance of the landscape, it also interferes with the natural function of the barrier island system. Developers bulldoze roads, flatten home sites, and replace natural dune-building vegetation with lawns. Buildings block the flow of sand-laden wind and cause it to eddy, altering the growth and migration of dunes (Gabriel & Kreutzwiser, 2000). Anthropogenic modifications—roads, buildings, dune ridges, planted vegetation, artificial inlet closures, and inlet stabilization disrupt natural processes that maintain the islands (Smith et al., 2008). Even the marsh impoundments constructed on federal refuges to provide habitat for waterfowl and other wildlife disrupt the natural marsh-building process and may hinder the normal migration of salt marshes in the face of rising sea levels (Gabriel & Kreutzwiser, 2000).

The undeveloped northern part of Pea Island grew wider between 1852 and 1998 at a rate that did not change when livestock were removed in 1935 (Smith et al., 2008). The southern part, along with the entire Avon-Buxton area, eroded steadily over the same period. Smith et al. (2008) concluded, "Attempts to protect the barrier islands through construction and maintenance of artificial barrier dune ridges and through rapid closure of inlets (e.g., Buxton Inlet in 1963 and Isabel Inlet in 2003) promote the opposite result" (p. 80).

The dunes north of Corolla were built by nature, up to 35 ft/11 m in height, and they are riddled with overwash passes. Horses lived in and around these dunes for centuries at a much greater population density than they have now. Despite the grazing pressures from a multitude of horses, sheep, goats, cattle, and hogs, the dunes grew high and maintained their natural ever-changing equilibrium with waves and wind as they migrated westward.

In the absence of large-scale human manipulation of the environment, horses become part of this balance. Sometimes the balance results in overwash flats and island migration, opening up habitat for birds such as the piping plover. Sometimes the balance results in high dunes as in Corolla and at the west end of Shackleford Banks. Both are natural processes. If horses were invariably detrimental to dunes, mature dune systems would not have developed in Corolla and Shackleford Banks, where horses have roamed for centuries.

Detractors argue that long-domesticated horses introduced from Europe did not co-evolve with the estuarine ecosystem of the New World, and as exotics they do not belong on federal lands at all. Many wildlife scientists, however, consider the horse a reintroduced native. Although most domestic animals are subspecies of ancestral animals, *Equus caballus* remains genetically identical to its Pleistocene progenitors. Thousands of years of selective breeding have wrought changes in horses that are no more radical than those caused naturally by genetic drift (Kirkpatrick & Fazio, 2010).

Recently, DNA molecules were recovered from core samples of Alaskan permafrost, showing that woolly mammoths and horses persisted in interior Alaska until 7,600 years ago—or less (Haile et al., 2009). They may have lasted longer in locations distant from the Bering land bridge, where no permafrost, and hence no buried DNA record, exists. The horse coevolved with North American ecosystems for more than 50 million years, lived here as *E. caballus* for 3 million years, and then was absent for fewer than 8,000 years.

Pleistocene horses grazed the North American marshes into an ecological balance very similar to that produced by modern free-roaming horses (Russell, Rich, Schneider, & Lynch-Stieglitz, 2009). *Spartina alterniflora* evolved under the teeth of grazers, not only horses, but also proboscidians (relatives of modern elephants) and camelids (relatives of camels and llamas). So did native *Phragmites,* which evolved as a harmonious component

Lush grass and healthy lactating mares indicate that the Currituck horses remain in balance with their habitat.

of the primordial world. It would appear that grazing herbivores are necessary to re-create a truly natural state.

The continent was teeming with horses when the non-native Paleo-Indians expanded into the Americas. There was an abundance of other horses besides *E. caballus*, so similar in form that paleontologists distinguish the various species mainly by pattern variations of the grinding surfaces of the cheek teeth (Sanders, 2002).

At the peak of the North American glaciation about 18,000 years ago, nearly 5 million mi²/12.9 million km² of the earth was covered by an ice sheet 2 mi/3.2 km thick. Ocean levels were about 400 ft/122 m lower than they are at present, and what is now submerged continental shelf was exposed as much as 60–90 mi/97–145 km east of the present shoreline (Carroll, Kapeluck, Harper, & Van Lear, 2002). The climate of the area that would someday become North Carolina was arid-cool during the height of the Ice Age, arid-hot 7,500–5,000 years ago, and thereafter warm-humid (Carroll et al., 2002). This flat coastal plain contained bogs and swamps interspersed with scrublands of pine and deciduous species (Carroll et al., 2002). Megafauna grazed plentifully in an open, park-like landscape of arid scrub, prairie, and savanna that probably looked much like modern-day Wyoming, rather than the dense, humid closed-canopy woodlands encountered by Europeans in the 1500s (Carroll et al., 2002).

Human cultures made extensive changes to the environment, ecosystem functions, and evolution (Donlan et al., 2006). Paleo-Indians intentionally burned native vegetation to drive prey, to create a favorable habitat for game species, and later to clear land for agriculture. The modern dominance of oaks, hickories, and pines in the Southeast reflects more than 12,000 years of regular burning by human inhabitants (Carroll et al., 2002). Many scientists believe that early North American tribes contributed to the extinction of Pleistocene megafauna through overhunting.

Not long after the arrival of people, horses disappeared, and it is quite possible that Paleo-Indians helped to eradicate them. After 15,000–20,000 years of continuous residency,

American Indians were naturalized as "native people." Non-native Europeans then arrived, repatriating the native wild horse and partially restoring the balance that was lost when most large ungulates died out. "Horses, like cattle, can serve as convenient proxies for the set of large mammals that inhabited the Americas during and before the Pleistocene. Horses, in fact, are more than proxies: they *are* Pleistocene megafauna" (Barlow, 2000, p. 32). A few scholars have gone so far as to suggest "Pleistocene rewilding," or attempting to re-create the ecological balance that existed before people arrived, using extant species, including wild elephants for mammoths (Donlan et al., 2006). Others doubt that we can restore the continent's virginity. Pleistocene rewilding might jeopardize existing native species and long-standing ecosystems without reestablishing the ancient balance, and it would inevitably clash with roads, fences, buildings, and people.

At a hearing for H.R. 306, Hutchins testified, "It's all about values. Do we want to protect our native wildlife, or turn our national refuges into theme parks for exotic animals?" (*H.R. 306, Corolla Wild Horses Protection Act*, 2011a). McCalpin (2011, April 13) replied, "We asked for wild horses—not elephants!.... These horses were on this land long before we were and certainly long before USFWS purchased it."

Public land managers are reluctant to encourage horse grazing because of biases held by many professional wildlife biologists and wildlife interest groups (Vavra, 2005). Clearly, overgrazing is detrimental to the environment, but the absence of large herbivores may be harmful in ways that we cannot yet comprehend. When the herd is maintained at a healthy population level, environmental harm is usually temporary, and compromised species recover quickly. As elsewhere, these free-roaming horses have acquired a bad reputation that they do not deserve.

Free-roaming horses, like any other species, change the environment with their presence. Innumerable species write in the diary of the wetlands. On any given day, broken stems mark where a band of horses dined on *Spartina* or *Phragmites*. Impressions in the mud and a scattering of feathers announce that a hawk put an end to a songbird on a meadow's rim. Vegetation is compressed in a thicket where a deer bedded down for the day. Irregularly mounded litterfall marks where a sounder of feral hogs rooted in the forest floor. Every goose, raccoon, muskrat, duck, plover, bat, beetle, and so on down to the tiniest single microbe affects the environment by living its life as part of the web.

The magnitude of horse impact depends on many factors—population density, growth and composition, type of habitat, climatic variation, animal behavior, anthropogenic influences, and other disturbances such as fire and flood. Mitchell, Gabrey, Marra, and Erwin (2006) found that light to moderate grazing probably has little effect, but with more intense grazing, impacts accumulate.

No definition of *overgrazing* is universally accepted. Range managers, forest managers, ecologists, wildlife biologists, conservationists, politicians, journalists, and private citizens have different definitions based on their attitudes, habits, and agendas, so they cannot always agree on whether or not an area is overgrazed. Wilson and Macleod (1991, p. 481) suggest overgrazing is relatively uncommon. "Although there are some notable exceptions, there are reasonable grounds to suspect that many rangelands are not overgrazed to the extent that is frequently claimed."

Mysterud (2006) argues that the term *overgrazing* is often invoked to support management agendas:

The U.S. Fish and Wildlife Service fenced a 135-acre/55-ha exclosure in Currituck NWR to ascertain whether horses were causing environmental damage. From this photograph, taken in November 2012, it is impossible to determine which side of the fence permits equine access, indicating that horses cause minimal impact. In the September 2012 census, only eight horses were counted on the 3,000 acres/1,200 ha of refuge property accessible to them. The remainder of the 121-member herd were counted on private property. Photograph courtesy of the Corolla Wild Horse Fund.

> The term 'overgrazing' (including related processes such as . . . browsing and trampling) is much used and abused in scientific literature . . . and it is usually value-laden as it implies grazing at a higher level than wanted relative to a specific management objective. . . . it has been used to describe almost any kind of . . . negative impact of grazing. Few use an explicit definition of overgrazing related to a specific ecological pattern or process. (p. 130)

Indications of overgrazing include loss of the original vegetation coverage to where recovery is progressively slower or irreversible, increasing areas of bare ground, local extinction of seed sources, and fundamental changes to successional pathways (Mysterud, 2006). But it is possible to make too much of such signs:

> In virtually any grazing system, spots of bare ground can be found around water holes, salt licks, and along fences . . . that may be attributed to trampling or might be considered evidence of overgrazing, even though they have no serious effect on the system. Overgrazing is an almost meaningless concept unless the spatial scale is considered. (Mysterud, p. 135)

Using Eberhardt's model (2002), as the herd reaches the carrying capacity of its habitat, sequential changes will occur in the population. First, foal mortality increases. Next, young mares produce their first foals at increasingly older ages. Third, reproductive rates decline. Finally, in extreme cases, adult mortality increases. As food resources become scarce, equine bodily condition and fat reserves decline, particularly for adult mares (Scorolli, 2012), which require more calories than stallions to gestate and nurse foals (which they often do simultaneously). Scorolli noted that when food supplies are limited, a higher percentage of mares will show poor body condition. After a two-year study of an Argentine herd, Scorolli (2012, p. 92) concluded, "Adult males, stallions, were in good condition . . . and had higher values than adult females in all months of both years." It should follow that if the majority of lactating mares in a herd are well-fleshed, the rest of the herd is at least equally healthy, forage is abundant, and adverse environmental effects are negligible.

The body condition of lactating mares is a sensitive, easily observed, and easily quantified indicator of the health of a herd and its habitat. When food is abundant, lactating mares have more body fat, which can be assessed from a distance using the Henneke scale. As the herd approaches carrying capacity, competition for food increases, and horses lose body fat and become more vulnerable to diseases and parasites. Lactating mares require more calories than the other horses in the herd and are the first to lose condition when food becomes scarce. If the majority of lactating mares in the herd are significantly underweight, it is likely that the food supply is insufficient to support a population of that size. This would be true whether the decline in plant production is the result of overgrazing or some other process such as development, storm damage, or drought. The author proposes using a lactational body condition index to estimate a herd's proximity to carrying capacity, discussed comprehensively in Chapter 7.

Disease can cause decline in condition, but disease generally involves individuals, while insufficient food production affects all lactating mares. Likewise, intestinal parasites can cause an individual's condition to decline markedly, but widespread heavy parasitic infestation tends to accompany overpopulation.

While equine overpopulation is undeniably damaging to any ecosystem, small bands that graze lightly and move on can be surprisingly beneficial. Elsewhere, land managers recognize the potential benefits of grazing. Horse grazing in the Camargue of France has proved a useful instrument for management of marshes for waterfowl by opening up the emergent vegetation, especially where the water level is controlled by the manager (Duncan & D'Herbes, 1982). Conservation groups in northern Europe have encouraged grazing in wetlands to improve habitat for ducks, coots, wigeon, and other waterfowl (Gordon & Duncan, 1988). Wylie (2012) found that non-native herbaceous cover increased at the expense of native herb cover on San Clemente Island grassland sites after feral herbivores were removed.

In the 1960s, the scenic balds of the Mount Rogers National Recreation Area and the adjacent Grayson Highlands State Park were becoming choked with brushy overgrowth, and fir and hardwood seedlings threatened to someday obstruct the scenic vistas. The U.S. Forest Service introduced first sheep and then cattle to feed on the brush, but they died after eating toxic plants. In 1974, the Forest Service invited Bill Pugh of Sugar Grove, Va., to graze his herd of 50+ Shetland Ponies on the mountain peaks. These ponies have successfully performed this function for nearly 40 years and are now managed by the Wilburn Ridge Pony Association. On the other hand, it can be argued that the scenic balds are not natural features, and the Mount Rogers ponies are helping to conserve an engineered environment.

Steve Edwards communes with Tradewind, a once-wild Corolla stallion who enjoys his new career as an long distance trail horse. Edwards, who during the work week is an attorney for Isle of Wight County, Va., uses natural horsemanship methods to train Corolla and Shackleford horses, readying them for adoption or placement in the off-site breeding program. Mill Swamp Indian Horses stands Corolla stallions at stud— service is free to any Corolla mare.

The herd currently numbers roughly 150 animals. The Association gathers ponies each September and sells foals at auction (Tennis, 2009). Proceeds fund veterinary care for the herd and support the Rugby Volunteer Rescue Squad and Fire Department.

The Fish and Wildlife Service uses carefully managed grazing rotations as a tool to maintain healthy habitat for Attwater's prairie chickens. Roughly 1 million of these birds populated the Texas and Louisiana Gulf coastal prairie at the time of first European contact, their odd booming mating calls resonating over the endless grasslands each spring (USFWS, 2012). The prairie chicken evolved alongside the vast herds of bison, horses, and other large herbivores that ruled the primordial tall-grass prairie. When the large grazers vanished, so did the birds. Overgrazing by cattle compounded the problem. By the 1930s, the species had dwindled to about 1% of its former population. Today, rotational grazing has helped to reestablish the bird in its native ranges (USFWS, 1999).

Historically, land managers have tried to preserve natural communities by protecting them from physical disturbance (Hobbs & Huenneke, 1992). They have come to understand that many forms of disturbance benefit the environment and maintain the natural balance.

The Fish and Wildlife Service may grant ponies a role in the management of impoundments on the Chincoteague NWR. Currently the staff uses disturbances such as burning, discing with a tractor, and mowing to enhance the habitat for migrating birds. Lou Hinds, former refuge manager, said,

We are considering . . . moving some portion of the herd into one of the impound-ments to allow the ponies to graze on the plants for a while, to clear away under-growth and use their hooves to punch in seed. The waste will be broken down by other invertebrates and microorganisms, and that becomes nutrients for worms and other things shorebirds feed on. . . . It is something that needs to be looked at and studied, but it might be possible to use the ponies in my wildlife management practices. (Personal communication, May 21, 2010)

People like easy explanations and cause-effect relationships. Ecology, however, involves a complex, ever-changing web of relationships, many of which are poorly understood. A change in one part of the web can cause unexpected problems in another part.

The marsh snail (or marsh periwinkle, *Littoraria irrorata*) is found in salt marshes all along the Atlantic coast and is most abundant—as many as 1,500 snails per square meter—in "die-back" zones, where large areas of *Spartina* are dead or dying (Bertness et al., 2004). This predilection led ecologists to conclude that *Littoraria* is a detritivore that feeds on microorganisms that live on decomposing plants. As it turns out, the snails were actually causing the die-offs to supply themselves with food. Grazing *Littoraria* damage the stalks of live cordgrass, creating lesions that become infected by the fungus that is their primary food. In this way the snails cultivate crops of nutritious fungi, fertilizing it with their dung. Snail grazing and fungus growth kill the cordgrass and causes marsh die-offs. Crabs preying on the snails keep the population in check, minimizing die-offs. Overharvesting blue crabs may be an important contributor to die-offs. Horses also eat cordgrass, but without destroying it, and their grazing thwarts die-offs by increasing the population of crabs (Levin et al., 2002) and limiting the population of Littoraria snails (Turner, 1987).

When one component of an ecosystem veers off-balance, other components are affected. Until recently, the grazing of lesser snow geese (*Chen caerulescens caerulescens*) strongly ben-efitted south-central salt marsh communities. These geese spend their summers grazing the Hudson Bay lowlands, then each fall migrate along the Mississippi and Central flyways to winter in Texas and Louisiana. When the geese were surveyed 30 years ago, 600,000 of them (Bertness et al., 2004) grazed these southern marshes and fertilized them with their nitro-gen-rich excrement, causing plants to quickly regenerate. Goose population size was limited by the availability of food in the Gulf salt marshes. When farmers along the migration route began using nitrogen fertilizer and high-yield crops in the 1960s, lesser snow geese stopped to feed on corn, soybeans, and wheat, and their reproductive rate sharply increased. Many of these birds began to winter in Arkansas and Missouri, foraging on crops, and many more survived the winter. Today, the census counts 3 million birds, which is probably half the actual total. After gorging on grains all winter, the birds return to the Hudson Bay area to breed alongside Canada geese and destroy the vegetation in the marsh above and below ground. A single goose can strip a square meter (10.8 ft^2) of marsh in an hour, creating a barren mudflat that is unlikely to recover while occupied by geese (Bertness et al., 2004).

Historically, the Atlantic flowed directly into Currituck Sound via inlets, creating a high-salinity estuarine habitat favorable to oysters and eelgrass and a nursery for saltwater fish (Lloyd, 2006). When the three closest inlets—New Currituck, Caffeys, and Roanoke—closed between 1795 and 1828, Currituck Sound became an estuary of increasingly fresh water fed by the tannin-rich rivers draining the Great Dismal Swamp. Shellfish beds disap-peared, and sunlight filtered down through the shallow waters and encouraged the growth

At Mount Rogers National Recreation Area and the adjacent Grayson Highlands State Park, the U.S. Forest Service and the state of Virginia use Shetland Ponies to maintain scenic grassy vistas by limiting the growth of brushy plants.

of aquatic vegetation such as wild celery and wigeon grass. Migrating waterfowl, always plentiful, flocked to this banquet until they literally blackened the sky in a phenomenon known by the locals as "smoke" (Lloyd, 2006). Recreational hunters were secretive about the whereabouts of their hunting grounds. An exclusive invitation-only organization known as the Currituck Hunting Club purchased more than 3,000 acres/1,214 ha of waterfowl habitat by 1857 (Lloyd, 2006). Following the Civil War, word of the bountiful waterfowl had spread, and affluent Northerners established more than 100 hunt clubs within a 50-mi/80-km radius of Currituck Sound. Wealthy and powerful men such as J.P. Morgan and Andrew Carnegie made annual pilgrimages to Currituck Banks to hunt waterfowl, as did hunters and fishermen of lesser means. The economy centered on the waterfowl hunting business, with locals working as guides, offering lodging to hunters, and guarding hunting grounds from poachers.

Currituck's waterfowl population, once seemingly infinite, has taken many devastating hits over the past few centuries. Market hunters shot great numbers of wild ducks for commercial sale, using enormous guns up to 10 ft/3 m in length mounted on flat-bottomed "punt" boats propelled by poles (Lloyd, 2006). One shot could slay as many as 100 ducks. Working in a team of about 10, each gunner could kill up to 700 ducks in a single day. Around the turn of 20th century, the seemingly inexhaustible flocks grew sparse, not only around Currituck Sound, but all along the Atlantic Flyway.

Thousands of adult snowy egrets were killed in the breeding season, when their plumes were most beautiful—leaving their chicks to starve in the nest. After Congress tightened restrictions, birds were still shot for the black market and shipped in trunks, suitcases, butter firkins, egg crates, horse trailers, and the carcasses of other animals. In 1902, a raid on one cold-storage house in New York yielded 8,058 snow buntings, 7,607 sandpipers, 5,218 plover, 7,003 snipe, 788 yellow legs, 288 bobolinks, 96 woodcock, 7,560 grouse, 4,385 quail, and 1,756 ducks. Most of these birds were illegally hunted, and fines would have totaled $1,168,315 had they been imposed (Hornaday, 1913).

Fashion-conscious women and their hatmakers created a seemingly insatiable demand for feathers, and the rarest and most exotic species fetched the highest prices. Species facing extinction brought the highest prices of all. Even hummingbirds were hunted relentlessly to be sold to European hat manufacturers. In 1911, three London millinery-trade auctions included the plumage from 129,168 egrets, 13,598 herons, 20, 698 birds of paradise, 41,090 hummingbirds, and 9,464 eagles and condors put on the block (H.R. Doc. No. 1447, 1913b). Before April 1911, when Governor Dix signed the Bayne law and halted the sale of wild native game in the state of New York, Currituck County saw the slaughter of about 200,000 wild fowl annually (H.R. Doc. No. 1447, 1913a). Members of sportsmen's clubs could shoot without limit, often bagging 150 or more birds on a two-day hunting trip. Coveting the money that wealthy sportsmen brought in and fearing the consequences of antagonizing them, local and state governments made no move to limit their excesses.

By 1913, Currituck County had earned a reputation as "the bloodiest slaughter-pen for waterfowl that exists anywhere on the Atlantic Coast" (Hornaday, 1913, p. 292). Hunters plied their trade without bag limits, even during the nesting season, and shipped vast quantities of birds to Northern restaurants and dealers.

Habitat loss and wanton hunting caused the American passenger pigeon (*Ectopistes migratorius*) population to collapse from 5–6 billion birds to ignominious extinction in less than 100 years. It appeared that the waterfowl of the Atlantic Flyway would soon follow suit. When President William McKinley signed the Lacey Act (16 U.S.C. §§ 3371–3378) into law in 1900, it became illegal to trade or sell wildlife, and many commercial killing machines were silenced, though illicit market gunning continued for decades. Hunt clubs encouraged the avian population to recover by limiting hunting to strictly defined seasons, creating refuges on key portions of their property, prohibiting spring hunting, and imposing bag limits on game. The National Audubon Society incorporated in 1905 as a grassroots conservation agency. In 1918, the Migratory Bird Treaty Act (16 U.S.C. §§ 703–712) protected over 800 species by making it illegal to possess these birds, alive or dead, intact or in part, including feathers, eggs, and nests. From his 7,000-acre/2,800-ha estate on Mackay Island in Currituck Sound, Joseph Palmer Knapp organized international conservation efforts for migratory waterfowl that became Ducks Unlimited in 1937. Knapp's estate later became Mackay Island NWR.

Besides the challenge of heavy hunting pressures, waterfowl populations have suffered large, mysterious die-offs. By the middle of the 20th century 3,000 tons/2.7 million kg of expended lead shot was piling up in the marshes every year, and it was ingested by waterfowl in great quantities. Roughly 2 million ducks succumbed to lead poisoning every year, and many more became chronically ill and slowly wasted away (Bolen, 2000). The birds of Currituck Sound and Back Bay were afflicted in great numbers—yet hunters opposed legislation that would require nontoxic shot, fearing that it would be more expensive or damage their weapons. Ammunition manufacturers, reluctant to retool to accommodate new materials, also resisted a shift away from lead.

After initial problems, the sea-to-sound barrier has been generally effective in keeping wild horses north of Corolla. Yet careless motorists continue to maim and kill horses in vehicular accidents on the beach.

The connection between lead ingestion and waterfowl die-offs was recognized in 1874. Lead shot remained entirely unrestricted until 1975 when hunters were required to use steel shot at locations where more than 5% of the ducks harvested were positive for lead ingestion. In 1986, the secretary of the Interior finally imposed a nationwide ban on lead shot for waterfowl hunting in the United States, which took effect in 1991. This ban reduced mortality from lead poisoning in mallards by 64% in the Mississippi Flyway alone and probably saved about 1.4 million ducks annually in the United States in the first years after its enactment (Bolen, 2000).

In the late 19th century, epizootics also devastated the duck population in California, Utah, and other western states, sometimes leaving more than 1,000 dead birds to the acre (2,500/ha) (Bolen, 2000). Outbreaks also occurred in North Carolina and the Chesapeake Bay region. Some killed more than a million birds, and mortality of half that number was not uncommon (Locke & Friend, 1989). The best minds of science could not find a cause, though it appeared that toxins such as the chlorides of calcium and magnesium were responsible. Decades later, in 1930, researchers determined that botulism was the true culprit, and it was somehow related to the multitude of aquatic invertebrates dying on mud flats (Bolen, 2000). As it turns out, *Clostridium botulinum* lives in the wetlands as spores, ready to proliferate under the right conditions. When something shifts the balance and causes a large die-off of invertebrates and other small fauna, *Clostridium* spores germinate and produce toxins. When waterfowl feed on the toxin-ridden remains, they die. The maggots that feed on the carcasses also contain high levels of *Clostridium* toxins. Sudden fluctuations in water level, pollution, and agricultural runoff can precipitate an outbreak of avian botulism

Today, waterfowl are challenged by environmental contaminants, invasive plants, and the insidious creep of development (Pease, Rose, & Butler, 2005). Despite improvements in several water quality elements, submerged vegetation has apparently declined in Currituck Sound and Back Bay since tide locks were removed from the Albemarle and Chesapeake Canal, allowing brackish water from Chesapeake Bay to enter through the Elizabeth River.

Bhattacharyya et al. write,

> Interviews with some field managers—resident and nonresident in the area— indicate a tendency to rationalize land use and range management decisions as though the manipulation of a single species (in this case horses) could rectify a century of environmental degradation that has occurred as a result of multiple disturbance factors. In reality, horses are one of many elements impacting ecosystems, any or all of which may require active management in areas where the land becomes degraded. (2011, p. 624)

Some managers of federal lands reject free-roaming horses out of hand as an invasive exotic species that can only cause damage. Many public land managers are pressured by professional wildlife biologists and wildlife interest groups who themselves are heavily biased against wild horses (Vavra, 2005). Others are more open to solutions that accommodate horses as a *resident* species, native or not.

Although horses and hogs are vastly different species that occupy different ecological niches, the Currituck NWR conservation plan makes no distinction between the two, stating unequivocally that "feral hogs (*Sus scrofa*) and horses (*E. caballus*) have overgrazed areas near Carova Beach to the elimination of habitat for native mammal species" (USFWS, 2008, p. 36). Later in the document, under the heading of "Pest Animals," the agency appears less certain: "Animals such as feral horses, feral hogs, and nutria *may* [emphasis added] have an impact on habitat and other species, but the Service does not currently staff or fund the refuge to investigate that impact" (p. 88).

In fact, ongoing research has demonstrated that the horses cause no lasting damage to the refuge. Rheinhardt and Rheinhardt (2004) studied the Currituck Banks herd as it foraged on the refuge and found that the herd (including two donkeys) was well below the carrying capacity of the range. They concluded that horses do affect native vegetation "primarily via cropping and trampling" (2004, p. 258), but any damage is temporary. Horses consumed few forb (herbaceous) species, preferring instead graminoid (grass) species, which "seem to recover from grazing by early summer when primary production is highest" (2004, p. 258).

Photographs presented by McCalpin at the July 27, 2010, hearing on H.R. 5482 (the Corolla Wild Horses Protection Act) showed no evidence of overgrazing in the vicinity of six 16 ft x16 ft/4.9 m x 4.9 m exclosures on refuge property (McCalpin, 2010, July 27). At the 2009 census, only eight members of the 121-horse herd were counted foraging on federal land—a population clearly too small and diffuse to impact the environment to any measurable degree. The Wild Horse Fund argued that over a 3-year period, there were never more than 26 horses counted on Fish and Wildlife Service property (CWHF, 2008). Her testimony in 2011 (*H.R. 306, Corolla Wild Horses Protection Act*, 2011b), included the following figures:

- Before 2006, no official census records were found in CWHF archives. Beginning in 2006, aerial counts were conducted by the CWHF Herd Manager and the CNWR Manager . . .
- 2006—119 horses . . .

A mare crosses a tangle of sand fencing in the village of Carova. The sure-footedness of Banker horses is legendary, largely because they learn to negotiate endlessly varied obstacles in their life in the wild.

- 2007—94 . . . 26 horses on CNWR property; 68 on private property; 0 on NCNERR
- 2008—101 . . . 23 horses on CNWR property; 74 on private property; 4 on NCNERR
- 2009—88 . . . 0 horses on CNWR property; 84 on private property; 4 on NCNERR.
- 2010—115 . . . 35 horses on CNWR property; 71 on private property; 9 on NCNERR

In March 2010, the Currituck NWR fenced two large areas with electrified high-tensile wire to determine the impact horses have on the barrier island environment—143 acres/58 ha in Swan Beach, and 135 acres/55 ha in North Swan Beach. Thirteen horses grazed the North Swan Beach tract at the time the fence was erected. Refuge staff literally built the fence around these horses, trapping them, and then asked the Wild Horse Fund to extricate the animals. Stallings complied, but the evicted horses were displaced into the home ranges of other existing harems (McCalpin, 2010, July 27).

For days, the expatriate horses battled with the resident band and resisted its efforts to drive them away. Harem stallions clashed violently over mares, sustaining cuts and bruises, and one pregnant mare from the displaced group miscarried a foal that was very close to full term (McCalpin, 2010, July 27). Another mare, apparently healthy when removed from the exclosure, dramatically lost condition over the next month and was euthanized after an heroic attempt to save her life.

Feral horses occupy a home range and will attempt to return to it if moved (Goodwin, 2002). Horses have been known to travel more than 9 mi/15 km to return to a home range.

If enough people work together to protect them, wild horses will continue to roam free on the Outer Banks.

The displaced horses wandered up and down the fence, eating grass near the border of their former home ground, trying to get back in, and competing with other horses for resources in the area to which they were banished.

The Currituck NWR built the exclosures to evaluate how the ecosystem functions in the absence of horses. In actuality, the area surrounding the exclosure was used much more heavily and irregularly by horses than it would have been without the fence. Stallings expressed doubt that the results of this experiment are meaningful: "A 12x12 or 16x16 foot enclosure would provide more accurate results, because it does not displace horses from their home ranges" (personal communication, May 25, 2010). "If a stallion has raised a family in a home range, when removed from the 100 acres he will stay as close to the excluded land as possible and possibly overgraze the borders, whereas if the exclusion site were open to horses, the grazing would be more evenly distributed."

Meanwhile, development continues to boom on the northern Banks. Horses graze amid real estate signs that mark where the next homes will be built. Stallings maintains that development poses a greater threat to sensitive species than do horses. "Each house has a septic field and a well," he explained. "The water table is high here. Eventually water quality will become affected, and plants adjacent to septic systems will hold toxins detrimental to the environment" (personal communication, May 25, 2010).

"Horses don't destroy grass," Stallings says. "They mow it to a height of 4–6 inches, which is no different than trimming a shrub at your house to a proper level" (personal communication, May 25, 2010). Moreover, light grazing stimulates plants to grow lower and denser (Taggart, 2008).

The fence keeps the horses out of the paved and thickly settled village of Corolla, but the problems it was meant to solve remain. Horses are still being hit by vehicles and left to die slowly, only now the vehicles are 4x4s and ATVs on the beach. Three horses were struck by vehicles in 2009. A young foal was hit on the beach by a reckless driver and was euthanized while his mother stood nearby, unwilling to leave him. "Most of the incidents are centered around Carova," says Stallings (personal communication, May 25, 2010).

T-Rex, a Corolla stallion in the prime of his life, was struck by a hit-and-run driver in March 2009. The driver knew that she had hit him, but did not report the accident, leaving him to suffer for many hours, in agonized terror, shaking from the effort to remain upright, until the vet finally came to euthanize him. The driver—a recent college graduate whose parents own a beach house in Carova—owned up to the crime more than *two months* after the incident.

Two months later, Spec, another proud stallion featured in a Wild Horse Fund brochure, was struck and left to suffer. It appears that ATVs were being used to chase the horse around the beach at 1 or 2 a.m. before one hit him from the side with force enough to snap his leg in two. With his leg dangling by a piece of skin, in unimaginable pain, Spec dragged himself almost a mile (1.6 km) across sand and dunes to return to his band. Spec suffered for hours before he could be euthanized. McCalpin wrote, "Spec did not want to die and he fought and struggled long and hard" (2009, May 24). "It was gut wrenching. It was a waste. It was sickening. He was terrified."

In her first four years as director of the Corolla Wild Horse Fund, McCalpin experienced the deaths of 19 horses. In 2009, a lactating mare almost died of agonizingly painful colic after being fed by a resident. This mare was lucky—she lived. Horses are exceptionally vulnerable to digestive disturbances and cannot vomit. A horse's intestines contain great lengths of loopy, mobile, narrow, winding passages that can easily twist, shift, or become obstructed. Consuming unusual foods, even those enjoyed by domestic horses, can kill a wild horse. A lactating mare died of colic in 2006. Attempts to bottle-feed her newborn colt were unsuccessful, and he died shortly after she did. Then in 2008 another horse died after eating moldy alfalfa hay put out by some well-meaning person. Every death subtracts rare and possibly irreplaceable genes from the herd.

Unintentional killings are heartbreaking enough, but, incomprehensibly, there have also been premeditated shootings. From 2001 to 2008, seven horses were gunned down in cold blood and left to decompose (Hampton, 2001; Owens, 2008). The shooters are still at large despite a $12,550 reward for information leading to their arrest.

Although it is illegal to approach closer than 50 ft/15 m to a Corolla horse, and a $500 fine awaits anyone who violates the statute, people still ignore the boundaries. In 2009, a mother sat by while her two young boys walked up to a stallion and patted him on his hindquarters, oblivious to the danger of even a casual kick. A mother and father attempted to place their child on the back of an unbroken wild horse for a photograph (Hampton, 2009). A woman walked down the beach alongside a group of mares. A family renting a cottage was charged by a stallion when they approached for pictures. A tour group operator prevented a group of children from shooting at a harem of horses with paintball guns (McCalpin, 2011, March 21).

Although the Corolla horses are maintained as wildlife, and herd managers do not typically interfere with natural sickness and death, the Wild Horse Fund rescues and rehabilitates wild horses injured by encounters with people. Horses are treated in the wild whenever possible. Some problems are fairly minor, such as the colt with a can stuck on his hoof or the filly with a fishhook caught in her leg. One mare was trapped in a deep rut where she would have drowned if not extricated. Another mare managed to ensnare her head in a tomato cage.

As lethal as a gun, but much slower acting, concentrated grain pellets can kill a wild horse geared to digesting high-fiber, low-energy grasses. Here somebody dumped a bag of grain on the ground in Swan Beach, presumably to attract horses and bring them within easy viewing range. The herd manager, with the help of the author, scooped the hazard into plastic bags before any horses were poisoned.

Another horse was trapped in a deep, flooded canal till she was saved by Kimberlee Hoey, president of the board of directors for the Wild Horse Fund. Hoey ran from dock to dock, climbing fences and clawing through heavy brush, pushing the mare with a lunge whip. Determined that the horse would not drown, she forced her to swim north toward the nearest break in the bulkhead and guided her to safety.

As if to intensify pressure on the horses from development and traffic, in December 2013, the Fish and Wildlife Service erected 2.9 mi/4.7 km of 15-gauge, 4-strand, 4-point barbed-wire fence, which connects with an existing 142-acre parcel surrounded by high tensile wire electric fence. The combination effectively transformed the entire Swan Beach Unit of Currituck NWR, 1,390 acres/563 ha, ocean to sound, into an exclosure that restricts access by horses and deer, but freely admits destructive feral hogs (Corolla Wild Horse Fund, 2013; Hampton, 2013; U.S. Dept. of Interior, 2010). The agency's reasoning is elusive. Barbed wire is difficult to see and frequently maims or kills not only horses, but also other wildlife, notably deer and large birds such as owls. Kline (2005) says authoritatively that it "should never be used for horse fencing." The Service strung the fence on uncapped metal t-posts, which pose a risk of impalement. Because the refuge has no regularly assigned staff, this fence cannot be monitored. It is easy to foresee that entangled horses, deer, or off-roaders might suffer for hours or days awaiting help. The Corolla Wild Horse Fund posted a photograph of this fence on Facebook, and within days hundreds of outraged people barraged the agency and local politicians with demands to remove it. Yet the fence still stands.

Volunteers must permanently remove some injured horses from the wild. Once a horse relies on humans for its care or is exposed to the diseases carried by domestic horses, it

Horses feed and nap on the lawn surrounding a community of beach houses near Carova. When development transformed their home ranges into suburban outposts, the animals adapted by using the manicured lawns as a food source. Some of the visitors taking horse tours to view the animals are disappointed to find them lounging in back yards and carports rather than galloping wildly down an empty beach.

can never rejoin its wild brethren. Veterinary care is expensive, as are boarding fees. Horses unable to return to the beach must be gentled and trained for months before they are suitable for adoption.

Rescues are resource-intensive. When capturing a Corolla horse, volunteers use steel panels to set up a "chute" to funnel the horse into a trailer. The group surrounds the horse and closes the circle toward the trailer. The horse then takes a 2-hr ride to the Dominion Equine Clinic in Suffolk, Va., a hospital with the staff and experience to manage wild horses (CWHF, 2008). In the past few years, many horses have been saved from certain death, including Uno, Tresie, Sunny, Hope, Croatoan, Manteo, Pomiac, Suerte, Tradewind, Valor, Barb, and Rainbow.

A young black stallion, Edward Teach (a.k.a. "Blackbeard"), was rescued after a bite wound laid open his neck during a fight with a rival stallion. *En route* to Dominion Equine Clinic he kicked 15 dents in the trailer. After more than $2,500 worth of veterinary care, the stallion was healed. Unable to return to the herd, he was trained to saddle by Steve Edwards of Mill Swamp Indian Horses in Smithfield, Va., and was eventually adopted.

Rainbow, a black yearling filly, sustained a puncture wound above her chest. Infection traveled up her neck to within 0.25 in./6 mm of her jugular vein and blew out a hole near her jowl almost the size of a tennis ball. Stallings and CWHF volunteers captured her. Dominion Equine Clinic brought her back to health, improved her nutrition, and used physical therapy to return flexibility to her neck. Visitors can see her at the Island Farm, a living-history site near Manteo.

In 2012, the Corolla Wild Horse Fund rescued a curious young horse that ascended a flight of stairs and was unable to get down. It also rescued one horse who got a tomato cage stuck on its head and neck, potentially life threatening—or blinding—for the horse. Photographs courtesy of the Corolla Wild Horse Fund.

Suerte, from the Spanish for "lucky," was rescued near death after he had ingested something toxic, perhaps antifreeze. Sunny, another young foal, was orphaned when her mother was stolen by another stallion. Young foals must be bottle-fed every 2 hr, so Stallings took the night shift, sleeping in his truck between feedings. McCalpin says, "These little horses have the strongest will to live that I have ever seen" (2009, August 24).

Sometimes horses are removed for their own good. Croatoan, Red Feather, and Swimmer were all escape artists who would not stay between the fences of the north beach and were likely to be struck by vehicles in town. Croatoan was so thin and weak when he was captured in March 2007, a well-meaning bystander pleaded for his euthanization because he was "old and hungry" (Edwards, 2008). As it turned out, he was only 11 years old, and with medical attention and good nutrition he achieved robust health. Edwards found him surprisingly easy to train. A few months after his capture, young children were using him in riding lessons, and a 10-year-old student rode him in the 2007 Smithfield Christmas Parade. In 2009, the Horse of the Americas Awards Committee gave 8-year-old Sarah Kerr-Applewhite its Buckaroo Award for her long-distance riding accomplishments on Croatoan.

In 2006 a young stallion was removed from the herd and gelded after he begged a resident for food, then knocked her down and injured her when no food was forthcoming. People had been feeding him, so he grew bold and aggressive. A mare named Baton Rouge was removed because she would bite people who came too close, and she gained a new career as a lesson horse.

An adoption system places Corolla horses in loving homes from Texas to Maine. Some are maintained in off-site breeding programs so that the Banker Horse might not become extinct if a disease or disaster devastates the island herds. Some are halter-trained before adoption, and some are broken to ride. Martin Community College in Williamston, N.C., trains adoptable Corolla horses as part of its degree program in equine technology. The goal is to place horses in good "forever" homes.

Adopters must demonstrate that they have the knowledge and skills to care for a horse. They are required to provide a shelter or box stall and an outside corral that is at least 20 ft x 20 ft / 6 m x 6 m and at least 5 ft / 1.5 m high, made of approved, sturdy materials such as pipes or planks. Barbed wire is forbidden because of the horrific injuries a horse may sustain if he becomes entangled. It is easy to adopt a horse in an impulsive emotional rush, but much harder to commit to the daily reality of feeding, watering, cleanup, expense, and meaningful interaction over the long term. A horse can live 20, 30, or even 40-plus years—as much as half a human lifetime.

The legend of Betsy Dowdy's midnight marathon underscores the stamina and heart of the North Carolina Banker Pony. Feats of endurance are well within the repertoire of a marsh pony. Edwards (2011) wrote,

> Exactly what can one expect from a Corolla Colonial Spanish mustang? I can only speak from experience. They are the easiest horses to train with which I have ever worked. They are strong, easy-keeping horses with incredible endurance. Many of our horses have completed rides of 50 miles in a day.
>
> Tradewind, the 2011 Horse of the Americas Registry's National Pleasure Trail Horse of the Year, is a 12.2 hand stallion, weighing 626 pounds in peak condition. In 2011 he carried me 206 hours in the woods, the vast majority of those hours either trotting or cantering. This does not include the many hours that others rode

him on the trails. At the time my weight was from 212 to 222 pounds. He did so even though he was captured because he was utterly crippled with founder. He is now wonderfully recovered and has produced two beautiful colts.

The Corolla horse is one of the oldest and rarest strains of Colonial Spanish Horse in the United States. They can trace their ancestry through centuries of free-roaming horses that probably descend from the initial Spanish horses to set hoof in the New World. Other breeds have contributed their genes over the years, but their conformation gives little hint of these introductions. Mill Swamp is one of the off-site breeding farms that raise domestic Banker Horses in an attempt to continue this rare, ancient lineage. Edwards (2011) writes,

> We rehabilitate and train these horses and breed them domestically, not as a replacement for the wild herd, but as a safety net in the event that the wild herd is destroyed by bureaucrats, developers, or a natural catastrophe. The off-site breeding program is designed to insure that these horses, which are the state horse of North Carolina and are among the rarest and oldest distinct genetic grouping of American horses, will always be with us.

It is equally important, however, to preserve this breed as free-roaming. Wesley Stallings, herd manager, believes that a designated wild horse sanctuary would be an optimal solution. He envisions a fenced area that limits the wild horse range to the marsh, maritime forest, and high ground such as Penny Hill. The marsh would meet the majority of the horses' nutritional needs. The high dunes would provide the same benefits as the beach—strong breezes and insect relief. The horses would remain safely out of the way of traffic, and the prime beachfront property valued by developers would be available for vacation homes. A greenway or boardwalk could allow people to observe the horses in their natural environment from a safe distance. Admission fees would enable the Corolla Wild Horse Fund to better preserve and maintain the herd.

The Fund works tirelessly to raise public awareness and support of the Corolla wild horse. More than 1,000 schoolchildren petitioned to have the Colonial Spanish Mustang designated the North Carolina state horse, a dream that was realized in 2010.

These horses and their ancestors have ranged freely over the dunes of Currituck County for hundreds of years, and cling to existence in the windswept wilds north of Corolla. "It is their last stand—all that remains of their habitat," says McCalpin (2011, February 28).

These animals owe their liberty to the advocates who have battled so relentlessly on their behalf. With continued providence and careful protection, these beautiful horses can remain free on the shifting sands that cover the bones of their ancestors; but maintaining a viable herd will require deliberate conservation efforts. Says Edwards (2011), "Extinction lasts forever and the clock is ticking."

References

An Act Relating to Fences, and for the Protection of Crops. (1873). NC Sess L 1873 ch 103.
An Act To Place Certain Portions of Currituck County under the State-Wide Stock Law. (1937). NC Sess L 1937 ch 389.
An Act To Place Certain Portions of Dare County under the State-Wide Stock Law. (1935). NC Sess L 1935 ch 263.

Altman, J. (2009). *Cape Lookout National Seashore seabeach amaranth* (Amaranthus pumilus*): 2009 report.* Retrieved from http://www.nps.gov/calo/parkmgmt/upload/Cape Lookout National Seashore Sea Beach Amaranth-2009.pdf

Anderson, E.W. (1993). Prescription grazing to enhance rangeland watersheds. *Rangelands, 15*(1), 31–35.

Anderson, V.D. (2002). Animals into the wilderness: The development of livestock husbandry in the seventeenth-century Chesapeake. *William and Mary Quarterly, 3rd Series, 59*(2), 377–408. Retrieved from http://www.jstor.org/stable/3491742

Andreoni, F. (1998). *Evaluating environmental consequences of feral horses in Guy Fawkes River National Park: A report to National Parks and Wildlife Service.* NR 490 Project. Armidale, New South Wales, Australia: University of New England.

Ashton, A. (2005). *Bark chewing by the wild horses of Guy Fawkes River National Park, NSW: Impacts and causes* (Unpublished B.Sc. honors thesis). University of New England, Armidale, New South Wales, Australia.

Bakker, J.P. (1985). The impact of grazing on plant communities, plant populations and soil conditions on salt marshes. *Vegetatio, 62*(1–3), 391–398.

Barber, D.C., & Pilkey, O.H. (2001). *Influence of grazing on barrier island vegetation and geomorphology, coastal North Carolina.* Paper No. 68-0 given at the Geological Society of America Annual Meeting, November 6, 2001. Retrieved from https://gsa.confex.com/gsa/2001AM/finalprogram/abstract_28327.htm

Barlow, C. (2000). *The ghosts of evolution: Nonsensical fruit, missing partners, and other ecological anachronisms.* New York, NY: Basic Books.

Barth, J.E. (2010). "The sinke of America": Society in the Albemarle borderlands of North Carolina, 1663–1729. *North Carolina Historical Review, 87*(1), 1–27.

Bazely, D.R., & Jefferies, R.L. (1986). Changes in the composition and standing crop of salt-marsh communities in response to the removal of a grazer. *Journal of Ecology, 74*(3), 693–706.

Beever, E.A., & Herrick, J.E. (2006). Effects of feral horses in Great Basin landscapes on soils and ants: Direct and indirect mechanisms. *Journal of Arid Environments, 66*(1), 96–112. doi: 10.1016/j.jaridenv.2005.11.006

Beever, E.A., and Brussard, P. F. (2000). Examining ecological consequences of feral horse grazing using exclosures. *Western North American Naturalist, 60*(3), 236–254.

Bellis, V.J. (1995, May). *Ecology of maritime forests of the southern Atlantic coast: A community profile* (Biological Report 30). Washington, DC: U.S. National Biological Service.

Benot, M.L., Bonis, A., Rossignol, N., & Mony, C. (2011). Spatial patterns in defoliation and the expression of clonal traits in grazed meadows. *Botany, 89*(1), 43–54. doi: 10.1139/B10-082

Bertness, M., Silliman, B.R., & Jefferies, R. (2004). Salt marshes under siege. *American Scientist, 92*(1), 54–61.

Bhattacharyya, J., Slocombe, D.S., & Murphy, S.D. (2011). The "wild" or "feral" distraction: Effects of cultural understandings on management controversy over free-ranging horses (*Equus ferus caballus*). *Human Ecology, 39*(5), 613–625. doi: 10.1007/s10745-011-9416-9

Bill Summary & Status, 112th Congress (2011–2012), H.R.306: All Congressional Actions. (2012). Retrieved from http://thomas.loc.gov/cgi-bin/bdquery/z?d112:HR00306:@@@X

Bill Summary & Status, 113th Congress (2013–2014), H.R.126: All Congressional Actions. (2013). Retrieved from http://thomas.loc.gov/cgi-bin/bdquery/D?d113:13:./temp/~bdWbBw:@@@X|/home/LegislativeData.php?n=BSS;c=113|

Binkley, C. (2007, August). The creation and establishment of Cape Hatteras National Seashore: The Great Depression through Mission 66. Atlanta, GA: U.S. National Park Service, Southeast Regional Office, Cultural Resource Division. Retrieved from http://archive.org/stream/creationestablis00bink#page/n1/mode/2up

Bolen, E.G. Waterfowl management: Yesterday and tomorrow. Journal of Wildlife Management, 64(2), 323–335.

Bond, J.F. (1908). Report on an examination of the sand banks along the North Carolina coast. In Biennial Report of the State Geologist, 1907-1908 (pp. 42–48). Raleigh, NC: E.M. Uzzell.

Bos, D., Bakker, J.P., de Vries, Y., & van Lieshout, S. (2002). Long-term vegetation changes in experimentally grazed and ungrazed back-barrier marshes in the Wadden Sea. Applied Vegetation Science, 5(1), 4–54.

Bratton, S.P., & Davison, K. (1987). Disturbance and succession in Buxton Woods, Cape Hatteras, North Carolina. Castanea, 52(3), 166–179.

Burney, D.A., & Burney, L.P. (1987). Recent paleoecology of Nags Head Woods on the North Carolina Outer Banks. Bulletin of the Torrey Botanical Club, 114(2), 156–168.

Carroll, W.D., Kapeluck, P.R., Harper, R.A., & Van Lear, D.H. (2002, September). Background paper: Historical overview of the southern forest landscape and associated resources. In D.N. Wear & J.G. Greis (Eds.), Southern forest resource assessment (General Technical Report SRS-53) (pp. 583–605). Asheville, NC: U.S. Department of Agriculture, Forest Service, Southern Research Station.

Carson, R. (1947). Chincoteague: A National Wildlife Refuge (Conservation in Action 1). Washington, DC: U.S. Fish and Wildlife Service. Retrieved from http://digitalcommons.unl.edu/usfwspubs/1

Chapman, W.R., & Hanson, J.K. (1997, January). Wright Brothers National Memorial historic resource study. Atlanta, GA: U.S. National Park Service, Southeast Field Area.

Chater, M. (1926). Motor-coaching through North Carolina. National Geographic, 49 (5), 475–523.

Chen, H. (2013, June 26). Ellis-van Creveld Syndrome. Medscape. Retrieved from http://emedicine.medscape.com/article/943684-overview#a0199

Chincoteague ponies: Good points of diminutive draft horses of the coast islands. (1905, September 17). Washington Post, p. E2.

Cobb, C. (1906). Where the wind does the work. National Geographic, 17(6), 310–317.

Comberford, N. (1657). The south part of Virginia, now the north part of Carolina [Map]. Retrieved from http://digitalgallery.nypl.org/nypldigital/dgkeysearchdetail.cfm?trg=1&strucID=744285&imageid=ps_mss_cd18_271&total=1&e=w

Conant, E.K., Juras, R., & Cothran, E.G. (2012). A microsatellite analysis of five colonial Spanish horse populations of the southeastern United States. Animal Genetics, 43(1), 53–62. doi: 10.1111/j.1365-2052.2011.02210.x

Congressional Budget Office. (2011, November 9). Cost estimate: H.R. 306, Corolla Wild Horses Protection Act. Retrieved from http://www.cbo.gov/sites/default/files/cbofiles/attachments/hr306.pdf

Corolla Wild Horse Fund. (2008, November). Shackleford 127—Corolla 60. *Wild and Free*, *1*(7), 1. Retrieved from http://www.corollawildhorses.com/Images/Newsletter/November%202008%20Newsletter.pdf

Corolla Wild Horse Fund. (2013, September 23). *YOUR tax dollars at work on the Outer Banks*. Retrieved from http://www.corollawildhorses.com/tax-dollars-work-outer-banks/

Corolla Wild Horses Protection Act. H.R. 5482, 111th Cong. (2010).

Corolla Wild Horses Protection Act, H.R. 306, 112th Cong. (2011).

Corolla Wild Horses Protection Act, H.R. 126, 113th Cong. (2013).

Cothran, E.G. (2008). *Analysis of genetic diversity in the Corolla feral horse herd of North Carolina*. Retrieved from http://www.corollawildhorses.com/Images/News/genetic-diversity-analysis.pdf

County of Currituck. (2007). Currituck Banks wild horse management plan: *Final cooperative plan among Currituck County, Corolla Wild Horse Fund, North Carolina Department of Environment and Natural Resources and U.S. Fish and Wildlife Service*. Currituck, NC: Author.

Creecy, R.B. (1901). The legend of Betsy Dowdy: An historical tradition of the Battle of Great Bridge. *North Carolina Booklet, 1*(5).

Currituck County Wild Horse Advisory Board. (2008, October 16). *Currituck County Wild Horse Advisory Board Meeting, October 16, 2008* [Minutes]. Retrieved from http://co.currituck.nc.us/pdf/wild-horse-advisory-board-2008-2011/wh-minutes-08oct16.pdf

Currituck Outer Banks wild horse management plan. (n.d.). Retrieved from http://www.corollawildhorses.com/wp-content/uploads/2012/08/wild-horse-management-plan.pdf

Daniels, J.T. (Photographer). (1903, December 17). *First flight, 120 feet in 12 seconds, 10:35 a.m.; Kitty Hawk, North Carolina*. Library of Congress, Prints & Photographs Division, LC-DIG-ppprs-00626.

De Bry, T. (Engraver). (1590). The arriual of the Englishemen in Virginia. In T. Harriot, *A briefe and true report of the new found land of Virginia* (plate 2). Frankfurt-am-Main, Germany: Johann Wechel.

DeKimpe, N.M., Dolan, R., & Hayden, B.P. (1991). Predicted dune recession on the Outer Banks of North Carolina, USA. *Journal of Coastal Research, 7*(2), 451–463. http://www.jstor.org/stable/4297850

De Stoppelaire, G.H., Gillespie, T.W., Brock, J.C., & Tobin, G.A. (2004). Use of remote sensing techniques to determine the effects of grazing on vegetation cover and dune elevation at Assateague Island National Seashore: Impact of horses. *Environmental Management, 34*(5), 642–649. doi: 10.1007/s00267-004-0009-x

DeBlieu, J. (1998). *Hatteras journal*. Winston-Salem, NC: John F. Blair (Original work published 1987).

Dolan, R., & H. Lins. (1987). Beaches and barrier islands. *Scientific American, 257*(1), 68–77. doi: 10.1038/scientificamerican0787-68

Donlan, C.J., Berger, J., Bock, C.E., Bock, J.H., Burney, D.A., Estes, J.A., . . . Greene, H.W. (2006). Pleistocene rewilding: An optimistic agenda for twenty-first century conservation. *American Naturalist, 168*(5), 660–681. doi: 10.1086/508027

Dunbar, G.S. (1958). *Historical geography of the North Carolina Outer Banks*. Louisiana State University Studies, Coastal Studies Series 3. Baton Rouge: Louisiana State University Press.

Duncan, P., & D'Herbes, J.M. (1982). The use of domestic herbivores in the management of wetlands for waterbirds in the Camargue, France. In D.A. Scott (Ed.), *Managing wetlands and their birds: A manual of wetland and waterfowl management. Proceedings of the third Technical Meeting on Western Palearctic Migratory Bird Management* (pp. 51–56). Slimbridge, United Kingdom: International Waterfowl Research Bureau.

Dyring, J. (1990). *The impact of feral horses* (Equus caballus) *on sub-alpine and montane environments* (Unpublished master's thesis). University of Canberra, Australia.

Eberhardt, L.L. (2002). A paradigm for population analysis of long-lived vertebrates. *Ecology, 83*(10), 2841–2854.

Edwards, S. (2008, November 24). Croatoan. *Mill Swamp Indian Horse Views.* Retrieved from http://msindianhorses.blogspot.com/search?q=croatoan

Edwards, S. (2011, December 3). Wild Spanish mustangs in Corolla. *Mill Swamp Indian Horse Views.* Retrieved from http://msindianhorses.blogspot.com/2011/12/wild-spanish-mustangs-in-corolla.html

Equus Survival Trust. (2011). *Equus Survival Trust 2011 equine conservation list.* Retrieved from http://www.equus-survival-trust.org/documents/equineconservationlist.pdf

Evans, C. (1986). *The relationship of cattle grazing to sage-grouse use of meadow habitat on the Sheldon National Wildlife Refuge* (Unpublished master's thesis). University of Nevada, Reno.

Farrell, C.A. (n.d.) (Photographer). *One of the great dunes at Kitty Hawk, anchored with brush fence and coarse grass.* PhC9_2_72_21, Charles A. Farrell Photograph Collection, State Archives of North Carolina, Raleigh, NC.

Federal Writers' Project. (1939). *North Carolina: A guide to the Old North State.* Chapel Hill: University of North Carolina Press.

Fishing horses. (1900). *Chambers's Journal, 6th Series* (3), 493.

Friedman, S.M. (n.d.). *The inflation calculator.* Retrieved from http://www.westegg.com/inflation/

Gabriel, A.O., & Kreutzwiser, R.D. (2000). Conceptualizing environmental stress: A stress-response model of coastal sandy barriers. *Environmental Management, 25*(1), 53–69.

Goodloe, R.B., Warren, R.J., Osborn, D.A., & Hall, C. (2000). Population characteristics of feral horses on Cumberland Island, Georgia and their management implications. *Journal of Wildlife Management, 64*(1), 114–121. doi: 10.2307/3802980

Goodwin, D. (2002). Horse behaviour: Evolution, domestication and feralisation. In N. Waran (Ed.), *The Welfare of Horses* (pp. 1–18). Dordrecht, Netherlands: Kluwer Academic Publishers.

Gordon, I., & Duncan, P. (1988). Pastures new for conservation. *New Scientist, 117*(1604), 54–59.

Green, P. (1937). *The Lost Colony: A symphonic drama in two acts (with music, pantomime, and dance).* Chapel Hill: University of North Carolina Press.

Gura, T. (2012). Genomics, plain and simple. *Nature, 483*(7387), 20–22. doi: 10.1038/483020a

H.R. 306, Corolla Wild Horses Protection Act . . . *Legislative hearing before the Subcommittee on Fisheries, Wildlife, Oceans and Insular Affairs of the Committee on Natural Resources, U.S. House of Representatives,* 112th Cong. 19 (2011a) (testimony of Michael Hutchins).

H.R. 306, Corolla Wild Horses Protection Act . . . Legislative hearing before the Subcommittee on Fisheries, Wildlife, Oceans and Insular Affairs of the Committee on Natural Resources, U.S. House of Representatives, 112th Cong. 23 (2011b) (testimony of Karen H. McCalpin).

H.R. Doc. No. 1447 at 5307 (1913a).

H.R. Doc. No. 1447 at 5317 (1913b).

H.R. Rep. No. 112-210 (2011).

Haile, J., Froese, D.G., MacPhee, R.D.E., Roberts, R.G., Arnold, L.J., Reyes, A.V., . . . Willerslev, E. (2009). Ancient DNA reveals late survival of mammoth and horse in interior Alaska. *Proceedings of the National Academy of Sciences of the United States of America, 106*(52), 22352–22357. doi: 10.1073/pnas.0912510106

Hampton, J. (2001, November 27). Gunshots kill 3 wild horses. *Virginian-Pilot* (Norfolk, VA), p. B1.

Hampton, J. (2009, July 12). New N.C. law designed to keep visitors, wild horses safe. *Virginian-Pilot* (Norfolk, VA). Retrieved from http://hamptonroads.com/2009/07/new-nc-law-designed-keep-visitors-wild-horses-safe

Hampton, J. (2013, September 11). Currituck refuge wants more fencing to block wild horses. *Virgininan-Pilot* (Norfolk, VA), p. B5.

Harriot, T. (1590). *A briefe and true report of the new found land of Virginia*. Frankfurt-am-Main, Germany: Johann Wechel.

Harrison, F. (1927). The equine F F Vs: A study of the evidence for the English horses imported into Virginia before the Revolution. *Virginia Magazine of History and Biography, 35*(4), 329–370.

Havholm, K.G., Ames, D.V., Whittecar, G.R., Wenell, B.A., Riggs, S.R., Jol, H.M., . . . Holmes, M.A. (2004). Stratigraphy of back-barrier coastal dunes, northern North Carolina and southern Virginia. *Journal of Coastal Research, 20*(4), 980–999.

Hawks, F.L. (1857). *History of North Carolina: With maps and illustrations* (2nd ed., vol. 1). Fayetteville, NC: E.J. Hale & Son.

Hennigar, H.F. (1979). *Historical evolution of coastal sand dunes on Currituck Spit, Virginia/North Carolina* (Unpublished master's thesis). College of William and Mary, School of Marine Science, Williamsburg, VA.

Henning, J. (1985). *Conquistadores' legacy: The horses of Ocracoke*. Ocracoke, NC: Author.

Hobbs, N. (1996). Modification of ecosystems by ungulates. *Journal of Wildlife Management, 60*(4), 695–713.

Hobbs, R.J., & Huennecke, L.F. (1992). Disturbance, diversity, and invasion: Implications for conservation. *Conservation Biology, 6*(3), 324–337. doi: 10.1046/j.1523-1739.1992.06030324.x

Hornaday, W.T. (1913). *Our vanishing wild life: Its extermination and preservation*. New York, NY: Charles Scribner's Sons.

Howard, M. (1976). Ocracoke horsemen. In C. O'Neal, A. Rondthaler, & A. Fletcher (Eds.), *The story of Ocracoke Island: A Hyde County bicentennial project* (pp. 25–27). Charlotte, NC: Herb Eaton.

Hutchins, M., et al. (2012, May 1). [Letter to Sen. Barbara Boxer and Sen. James Inhofe.] Retrieved from http://www.publicland.org/08_current_past_news/2012/120501_wild_horse_protection_act.pdf

Impact Assessment, Inc. (2005). *Ethnohistorical description of the eight villages adjoining Cape Hatteras National Seashore and interpretive themes of history and heritage: Final technical report*. Manteo, NC: Cape Hatteras National Seashore.

Jensen, A. (1985). The effect of cattle and sheep grazing on salt-marsh vegetation at Skallingen, Denmark. *Vegetatio, 60*(1), 37-48.

Johnson, K.T. (2006). Fences. *NCPedia*. Retrieved from http://ncpedia.org/fences

Kaczensky, P., Ganbaatar, O., Altansukh, N., & Enkhsaikhan, N. (2010, August). *Winter disaster in the Dzungarian Gobi—Crash of the Przewalski's horse population in Takhin Tal 2009/2010*. Retrieved from http://www.takhi.org/media/forschung/2010_Winter-disaster-in-Dzungarian-Gobi-2009_10.pdf

Keiper, R.R. (1981). *Ecological impact and carrying capacity of ponies*. Contract number 51570–0055. Chincoteague, VA: Chincoteague National Wildlife Refuge.

Kennish, M.J. (2001). Coastal salt marsh systems in the U.S.: A review of anthropogenic impacts. *Journal of Coastal Research, 17*(3), 731–748.

Kirkpatrick, J., & Fazio, P. (2010, January). *Wild horses as native North American wildlife*. Retrieved from http://awionline.org/content/wild-horses-native-north-american-wildlife

Kline, K. (2005, April 18). *Safe fencing for horses*. Retrieved from http://www.livestocktrail.illinois.edu/horsenet/paperDisplay.cfm?ContentID=6727

Kohn, E. (2004). Can we save the historic Abaco wild horses? *Natural Horse Magazine, 6*(3). Retrieved from http://www.naturalhorse.com/archive/volume6/Issue3/article_8.php

Lacey Act of 1900, 16 U.S.C. §§ 3371–3378 (2006 & Supp. II 2008).

Lamoot, I. (2004). *Foraging behaviour and habitat use of large herbivores in a coastal dune landscape*. Brussels, Belgium: Research Institute for Nature and Forest. Retrieved from http://www.inbo.be/files/bibliotheek/14/167914.pdf

Lawson, J. (1709). *A new voyage to Carolina; Containing the exact description and natural history of that country....* London, United Kingdom. Retrieved from http://docsouth.unc.edu/nc/lawson/lawson.html

Lee, F.G. (2008). *Constructing the Outer Banks: Land use, management, and meaning in the creation of an American place* (Unpublished master's thesis). North Carolina State University, Raleigh.

Lellis, K.A. (2008). *Native Phragmites australis in national parks: A three-step approach to locate, identify, and verify native stands of the common reed in the northeast* (CIIP White Paper). Retrieved from http://www.ci.uri.edu/ciip/WhitePaper/2008/Lellis.pdf

Levin, P., Ellis, J., Petrik, R., & Hay, M. (2002). Indirect effects of feral horses on estuarine communities. *Conservation Biology, 16*(5), 1364–1371. doi: 10.1046/j.1523-1739.2002.01167.x

Lloyd, J. (2013, February 13). *The story of the rise and fall of the Currituck hunt clubs*. http://www.ncbeaches.com/Features/Wildlife/HistoryofCurrituckHuntClubs

Locke, L.N., & Friend, M. (1989). *Avian botulism: Geographic expansion of a historic disease* (Fish and Wildlife Leaflet 13.2.4). Washington, DC: U.S. Fish and Wildlife Service.

Loydi, A., & Zalba, S.M. (2009). Feral horses dung piles as potential invasion windows for alien plant species in natural grasslands. In A.G. van der Valk (Ed.), *Herbaceous Plant Ecology* (pp. 107–116). doi: 10.1007/978-90-481-2798-6

Lynghaug, F. (2009). *The official horse breeds standards guide: The complete guide to the standards of all North American equine breed associations*. Minneapolis, MN: Voyageur Press.

MacNeil[1], B.D. (1938, June 14). Dare sharpshooters begin Banker Ponies' extinction. *News and Observer* (Raleigh, NC), p. 1.

MacNeill, B.D. (1958). *The Hatterasman.* Winston-Salem, NC: John F. Blair.

Mallinson, D.J., Culver, S.J., Riggs, S.R., Walsh, J.P., Ames, D., & Smith, C.W. (2008). *Past, present and future inlets of the Outer Banks barrier islands, North Carolina.* Greenville, NC: East Carolina University.

Mallinson, D.J., Riggs, S.R., Culver, S.J., Ames, D., Horton, B.P., & Kemp, A.C. (2009). *The North Carolina Outer Banks barrier islands: A field trip guide to the geology, geomorphology, and processes.* Retrieved from http://core.ecu.edu/geology/mallinsond/IGCP_NC_Field_Trip_Guide_rev1.pdf

Marion, J.L. (2010). Management, monitoring, and protection protocols for seabeach amaranth at Cape Hatteras National Seashore, North Carolina. In *A review and synthesis of the scientific information related to the biology and management of species of special concern at Cape Hatteras National Seashore, North Carolina* (Open-File Report 2009–1262) (pp. 89–100). Reston, VA: U.S. Geological Survey.

McCalpin, K. (2009, August 24). Sunny and Suerte. *Wild and Free Weekly.* Retrieved from http://corollawildhorses.blogspot.com/2009/08/sunny-and-suerte.html

McCalpin, K. (2009, June 29). Killing them softly. *Wild and Free Weekly.* Retrieved from http://corollawildhorses.blogspot.com/2009_06_01_archive.html

McCalpin, K. (2009, May 24). Another senseless death. *Wild and Free Weekly.* Retrieved from http://corollawildhorses.blogspot.com/2009/05/another-senseless-death.html

McCalpin, K. (2010, March 2). Horses or hotel? *Wild and Free Weekly.* Retrieved from http://corollawildhorses.blogspot.com/2010/03/horses-or-hotel.html

McCalpin, K. (2010, March 24). HB 4867—The Corolla Wild Horse Protection Act. *Wild and Free Weekly.* Retrieved from http://corollawildhorses.blogspot.com/2010/03/hb-4867-corolla-wild-horse-protection.html

McCalpin, K. (2010, June 16). Our fragile home. *Wild and Free Weekly.* Retrieved from http://corollawildhorses.blogspot.com/2010_06_01_archive.html

McCalpin, K. (2011, February 28). They paved paradise and put up a parking lot . . . *Wild and Free Weekly.* Retrieved from http://corollawildhorses.blogspot.com/2011/02/they-paved-paradise-and-put-up-parking.html

McCalpin, K. (2011, March 21). The height of stupidity. *Wild and Free Weekly.* Retrieved from http://corollawildhorses.blogspot.com/2011/03/height-of-stupidity.html

McCalpin, K. (2011, April 13). We asked for wild horses—not elephants! *Wild and Free Weekly.* Retrieved from http://corollawildhorses.blogspot.com/2011/04/we-asked-for-wild-horses-not-elephants.html

McCalpin, K.H. (2010). *Saving the horses of kings: The wild horses of the Currituck Outer Banks.* Kitty Hawk, NC: Outer Banks Press.

McCalpin, K.H. (2010, July 27). *Hearing on H.R. 5482 Corolla Wild Horses Protection Act, U.S House of Representatives, Subcommittee on Insular Affairs, Oceans, and Wildlife: Testimony of Karen H. McCalpin, Executive Director, Corolla Wild Horse Fund, Inc. Corolla, North Carolina.* Retrieved from http://naturalresources.house.gov/UploadedFiles/McCalpinTestimony07.27.10.pdf

McGowan, C.P., & Simons, T.R. (2006). Effects of human recreation on the incubation behavior of American oystercatchers. *Wilson Journal of Ornithology, 118*(4), 485–493.

Mead, J.G. (1975). Preliminary report on the former net fisheries for *Tursiops truncatus* in the western North Atlantic. *Journal of the Fisheries Research Board of Canada, 32*(7): 1155–1162. doi: 10.1139/f75-136

Menard, C., Duncan, P., Fleurance, G., Georges, J.Y., & Lila, M. (2002). Comparative foraging and nutrition of horses and cattle in European wetlands. *Journal of Applied Ecology, 39*(1), 120–133. doi: 10.1046/j.1365-2664.2002.00693.x

Migratory Bird Treaty Act of 1918, 16 U.S.C. §§ 703–712 (2006).

Mills, D.S., & McDonnell, S.M. (Eds.). (2005). *The domestic horse: The evolution, development and management of its behaviour.* Cambridge University Press.

Mitchell, L.R. Gabrey, S., Marra, P.O., & Erwin, M.R. (2006). Impacts of marsh management on coastal-marsh bird habitats. *Studies in Avian Biology, 32,* 155–175.

Mysterud, A. (2006). The concept of overgrazing and its role in management of large herbivores. *Wildlife Biology, 12*(2), 129–141.

Nimmo, D.G., & Miller, K.K. (2007). Ecological and human dimensions of management of feral horses in Australia: A review. *Wildlife Research, 34*(5), 408–417. doi: 10.1071/WR06102

Oduor, A.M.O., Gómez, J.M., & Strauss, S.Y. (2010). Exotic vertebrate and invertebrate herbivores differ in their impacts on native and exotic plants: A meta-analysis. *Biological Invasions, 12*(2), 407–419. doi: 10.1007/s10530-009-9622-1

Outer Banks Conservationists. [2008]. *Currituck Beach Lighthouse: Chronological list of keepers, 1875–2006.* Retrieved from http://www.currituckbeachlight.com/ ChronologicalListofCBLHkeepers.pdf

Owens, G. (2008, January 21). *Reward offered in wild horse shootings.* Retrieved from http://www.wral.com/news/local/story/2334908/

Pease, M.L., Rose, R.K., & Butler, M.J. (2005). Effects of human disturbances on the behavior of wintering ducks. *Wildlife Society Bulletin, 33*(1), 103–112.

Pendleton, E.A., Thieler, E.R., & Williams, S.J. (2004). *Coastal vulnerability assessment of Cape Hatteras National Seashore (CAHA) to sea-level rise* (U.S. Geological Survey Open-File Report 2004-1064). Retrieved from http://pubs.usgs.gov/of/2004/1064/images/pdf/caha.pdf

Pidgin, C.F. (1907). *Theodosia: The first gentlewoman of her time.* Boston, MA: C.M. Clark.

Pilkey, O.H., Neal, W.J., Riggs, S.R., Webb, C.A., Bush, D.M., Pilkey, D.F, . . . Cowan, B.A. (1998). *The North Carolina shore and its barrier islands: Restless ribbons of sand.* Durham, NC: Duke University Press.

Pilkey, O., Rice, T., & Neal, W. (2004). *How to read a North Carolina beach.* Chapel Hill: University of North Carolina Press.

Plassmann, K., Jones, M.L.M., & Edwards-Jones, G. (2010). Effects of long-term grazing management on sand dune vegetation of high conservation interest. *Applied Vegetation Science, 13*(1), 100–112. doi: 10.1111/j.1654-109X.2009.01052.x

Poney Penning on the Beach, Near Oriental, N.C. (ca. 1910). In Durwood Barbour Collection of North Carolina Postcards (P077), North Carolina Collection Photographic Archives, Wilson Library, University of North Carolina-Chapel Hill.

Prioli, C. (2007). *The wild horses of Shackleford Banks.* Winston-Salem, NC: John F. Blair.

Quinn, D.B. (Ed.) (1955). *The Roanoke voyages, 1584–1590: Documents to illustrate the English voyages to North America under the patent granted to Walter Raleigh in 1584.*

London, United Kingdom: For the Hakluyt Society.

Re: Denial of request to allow Corolla wild Colonial Spanish Mustangs to remain at a genetically healthy level. (2008). Retrieved from http://www.corollawildhorses.com/Images/News/genetic-crisis-rev.pdf

Reeves, R.R., & Mitchell, E. (1988, March). *History of whaling in and near North Carolina.* NOAA Technical Report NMFS 65. Washington, DC: National Oceanic and Atmospheric Administration, National Marine Fisheries Service.

Reiner, R.J., & Urness, P.J. (1982). Effect of grazing horses managed as manipulators of big game winter range. *Journal of Range Management, 35*(5), 567–571.

Rheinhardt, R.D., & Rheinhardt, M.C. (2004). Feral horse seasonal habitat use on a coastal barrier spit. *Journal of Range Management, 57*(3), 253–258.

Rogers, G. (1994, August). *Kaimanawa feral horses: Recent environmental impacts in their northern range.* Landcare Research Contract Report: LC 9495121. Hamilton, New Zealand: Manaaki Whenua—Landcare Research.

Rogers, G.M. (1991). Kaimanawa feral horses and their environmental impacts. *New Zealand Journal of Ecology, 15*(1), 49–64.

Rubenstein, D.R., Rubenstein, D.I., Sherman, P.W., & Gavin, T.A. (2006). Pleistocene Park: Does re-wilding North America represent sound conservation for the 21st century? *Biological Conservation, 132*(2), 232–238. doi: 10.1016/j.biocon.2006.04.003

Russell, D.A., Rich, F.J., Schneider, V., & Lynch-Steiglitz, J. (2009). A warm thermal enclave in the Late Pleistocene of the South-eastern United States. *Biological Reviews, 84*(2), 173–202. doi: 10.1111/j.1469-185X.2008.00069.x

Sanders, A.E. (2002). Additions to the Pleistocene mammal faunas of South Carolina, North Carolina, and Georgia. *Transactions of the American Philosophical Society, 92,* Part 5.

Schott, C. (2002). Ecology of free-ranging horse in northern Guy Fawkes River National Park—Research progress report. Armidale, New South Wales, Australia: University of New England.

Scorolli, A.L. (2012). Feral horse body condition: A useful tool for population management? In *International Wild Equid Conference, Vienna 2012: Book of abstracts* (p. 92). Vienna, Austria: Research Institute of Wildlife Ecology, University of Veterinary Medicine. Retrieved from http://www.vetmeduni.ac.at/fileadmin/v/fiwi/Konferenzen/Wild_Equid_Conference/IWEC_book_of_abstracts_final.pdf

Seliskar, D.M. (2003). The response of *Ammophila breviligulata* and *Spartina patens* (Poaceae) to grazing by feral horses on a dynamic Mid-Atlantic barrier island. *American Journal of Botany, 90*(7), 1038–1044

Senter, J. (2003). Live dunes and ghost forests: Stability and change in the history of North Carolina's maritime forests. *North Carolina Historical Review, 80*(3), 334–361.

Severson, K.E., & Urness, P.J. (1994). Livestock grazing: A tool to improve wildlife habitat. In M. Vavra, W.A. Laycock, & R.D. Pieper (Eds.), *Ecological implications of livestock herbivory in the West* (pp. 232–249). Denver: Society for Range Management.

Shomette, D.G. (2008). The price of amity: Of wrecking, piracy, and the tragic loss of the 1750 Spanish treasure fleet. *The Northern Mariner/le marin du nord, 18*(3–4), 25–48.

Sloan, K. (2007). *A new world: England's first view of America.* Chapel Hill: University of North Carolina Press.

Smith, C.G., Culver, S.J., Riggs, S.R., Ames, D., Corbett, D.R., & Mallinson, D. (2008). Geospatial analysis of barrier island width of two segments of the Outer Banks, North Carolina, USA: Anthropogenic curtailment of natural self-sustaining processes. *Journal of Coastal Research, 24*(1), 70–83. doi: 10.2112/05-0595.1

Spears, J.R. (1890, October). Sand-waves at Henlopen and Hatteras. *Scribner's Magazine, 8*(4), 507–512.

Stewart, M.A. (1991). "Whether wast, deodand, or stray": Cattle, culture, and the environment in early Georgia. *Agricultural History, 65*(3), 1–28.

Stick, D. (1958). *The Outer Banks of North Carolina, 1584–1958*. Chapel Hill: University of North Carolina Press.

Stroh, P.A., Mountford, J.O., & Hughes, F.M.R. (2012). The potential for endozoochorous dispersal of temperate fen plant species by free-roaming horses. *Applied Vegetation Science, 15*(3), 359–368. doi: 10.1111/j.1654-109X.2011.01172.x

Taggart, J.B. (2008). Management of feral horses at the North Carolina Estuarine Research Reserve. *Natural Areas Journal, 28*(2), 187–195.

Take the trip of a lifetime! (2009). Retrieved from http://www.corollawildhorses.com/trip_of_a_lifetime.html

Tate, W. (1900, August 18). Letter to Wilbur Wright. Retrieved from http://www.bsu.edu/eft/wright/p/library/letter6.html

Taylor, U. (1995). *Seed dispersal from feral horse manure at Guy Fawkes River National Park*. NR 490 Project. Armidale, New South Wales, Australia: University of New England.

Tennant, D. (2001, December 2). Swept away: Shifting sands cover what once was Seagull, N.C. *Virginian-Pilot* (Norfolk, VA), p. E1.

Tennis, J. (2009, September 13). *Home on the range*. Retrieved from http://www.tricities.com/news/article_09b7b5d3-bd47-5d18-80e4-5b20bc9cbf52.html

Tesauro, J., & Ehrenfeld, D. (2007). The effects of livestock grazing on the bog turtle [*Glyptemys* (=*Clemmys*) *muhlenbergii*]. *Herpetologica, 63*(3):293-300. doi: 10.1655/0018-0831(2007)63[293:TEOLGO]2.0.CO;2

Tsoar, H. 2005). Sand dunes mobility and stability in relation to climate. *Physica A: Statistical Mechanics and its Applications, 357*(1), 50–56.

Turner, M.G. (1987). Effects of grazing by feral horses, clipping, trampling, and burning on a Georgia salt marsh. *Estuaries, 10*(1), 54–60.

U.S. Bureau of Labor Statistics. (n.d.). *CPI inflation calculator*. Retrieved from http://www.bls.gov/data/inflation_calculator.htm

U.S. Department of Interior. (2010, March). *Draft environmental assessment, 2010 sport hunting plan amendment for Currituck National Wildlife Refuge, Currituck County, North Carolina*. Retrieved from http://www.fws.gov/currituck/images/D4_Envir_Assess.pdf

U.S. Fish and Wildlife Service. (1999, May). *Attwater Prairie Chicken National Wildlife Refuge*. Retrieved from http://library.fws.gov/Refuges/attwater.pdf

U.S. Fish and Wildlife Service. (2008, November). *Currituck National Wildlife Refuge comprehensive conservation plan*. Atlanta, GA: USFWS Southeast Region.Retrieved from http://digitalmedia.fws.gov/cdm/ref/collection/document/id/397

U.S. Fish and Wildlife Service. (2010, September). *Back Bay National Wildlife Refuge comprehensive conservation plan*. Retrieved from http://www.fws.gov/northeast/planning/Back%20Bay/pdf/FinalCCP/BACKBAYNWRFinalCCP9_2010.pdf

U.S. Fish and Wildlife Service. (2012, November 23). *Attwaters prairie-chicken*. Retrieved from http://www.fws.gov/refuge/Attwater_Prairie_Chicken/wildlife/APC.html

Vavra, M. (2005). Livestock grazing and wildlife: Developing compatibilities. *Rangeland Ecology & Management, 58*(2), 128–134. doi: 10.2111/1551-5028(2005)58<128:LGA WDC>2.0.CO;2

Vega-Pla, J.L., Calderón, J., Rodríguez-Gallardo, P.P., Martínez, A.M., & Rico, C. (2006). Saving feral horse populations: Does it really matter? A case study of wild horses from Doñana National Park in southern Spain. Animal Genetics, 37, 571–578. doi: 10.1111/j.1365-2052.2006.01533.x

Wallace, J.H. (1897). *The horse of America in his derivation, history, and development.* New York, NY: Author.

Weakley, A., Bucher, M., & Murdock, N. (1996). *Recovery plan for seabeach amaranth* (Amaranthus pumilus) *Rafinesque.* Atlanta, GA: U.S. Fish and Wildlife Service. Retrieved from http://pbadupws.nrc.gov/docs/ML0719/ML071970324.pdf

Weaver, V., & Adams, R. (1996). Horses as vectors in dispersal of weeds into bushland. In R.G. Richardson & F.J. Richardson (Eds.), *Eleventh Australian Weeds Conference Proceedings* (pp. 383–387). Frankston, Victoria, Australia: Weed Science Society of Victoria.

White, J. (Artist). (ca. 1585). [Plan of the Grenville expedition's camp at Tallaboa Bay, Puerto Rico] [Watercolor]. London, United Kingdom: British Museum.

Williams, B., & Page, M.P. (2002–2003). Where the feral horses roam. *Roanoke Colonies Research Newsletter, 8*(1 & 2). Retrieved from http://www.ecu.edu/rcro/RCRONLvol-8WhereFeralHorsesRoam.htm

Wilson, A.D., & MacLeod, N.D. (1991). Overgrazing: Present or absent? *Journal of Range Management, 44*(5), 475–482.

Wiss, Janney, Elstner Associates & John Milner Associates. (2007). *Portsmouth Village cultural landscape report.* Atlanta, GA: U.S. National Park Service, Southeast Regional Office.

Wolfe, M. (2003, September 20). (Photographer). [Inlet cut through Hatteras Island by Hurricane Isabel]. Retrieved from http://commons.wikimedia.org/wiki/File:FEMA_-_8414_-_Photograph_by_Mark_Wolfe_taken_on_09-20-2003_in_North_ Carolina. jpg

Wolfe, M. (2003, September 23). [(Photographer). Damage from Hurricane Isabel in Kitty Hawk, NC]. Retrieved from http://commons.wikimedia.org/wiki/File:FEMA_-_8411_-_Photograph_by_Mark_Wolfe_taken_on_09-23-2003_in_North_ Carolina.jpg

Wood, G.W., Mengak, M.T., & Murphy, M. (1987). Ecological importance of feral ungulates at Shackleford Banks, North Carolina. *American Midland Naturalist, 118*(2), 236–244.

Woolford, P. (2012, May 9). *At Currituck NWR: No horsing around when it comes to protecting native wildlife.* Retrieved from http://refugeassociation.org/2012/05/at-currituck-nwr-no-horsing-around/

Wylie, D.D.J. (2012). *Vegetation change on San Clemente Island following the removal of feral herbivores* (Unpublished master's thesis). San Diego State University, CA.

Zalba, S.M., & Cozzani, N.C. (2004). The impact of feral horses on grassland bird communities in Argentina. *Animal Conservation, 7*(1), 35–44. doi: 10.1017/S13679430030010

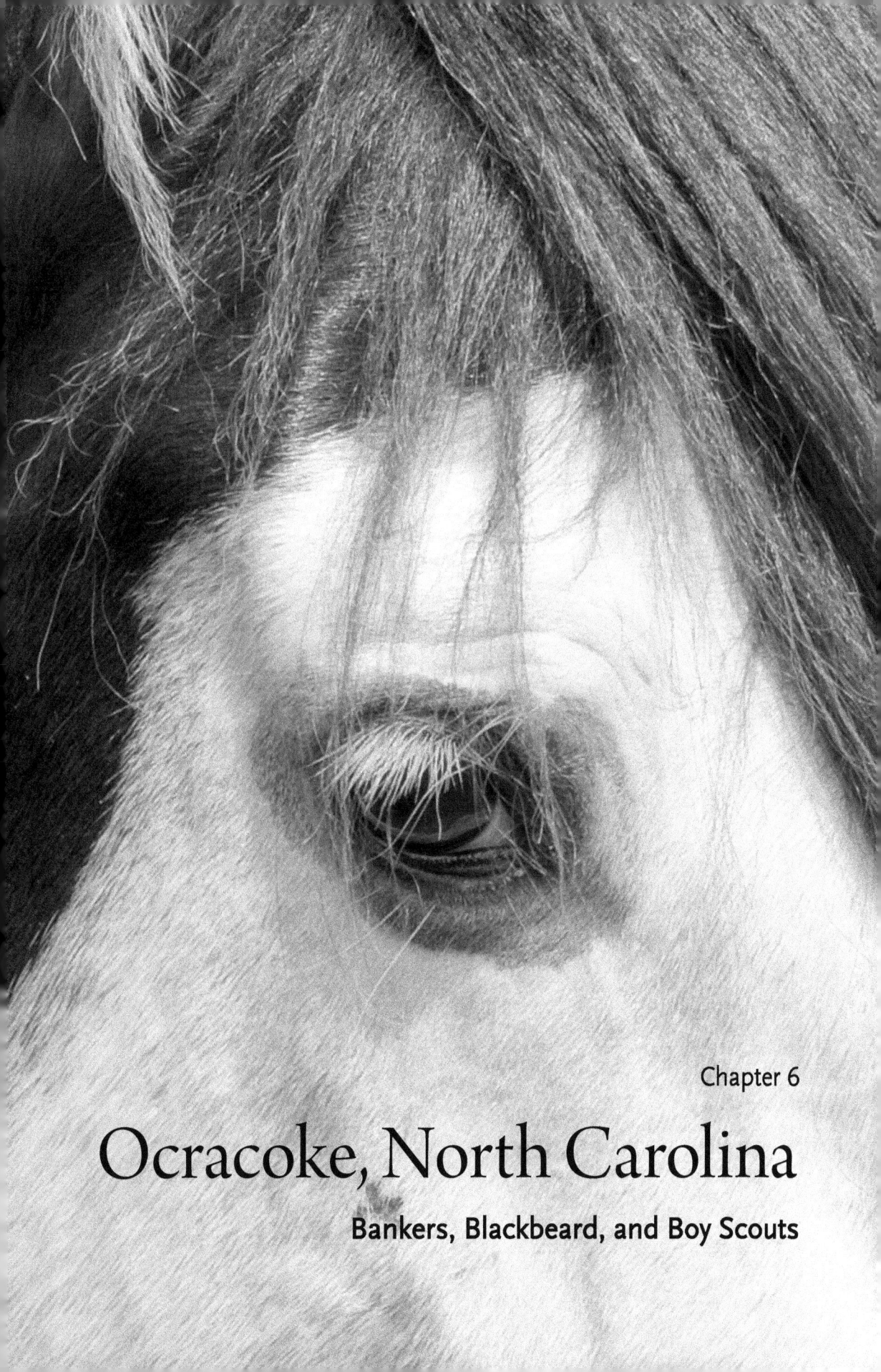

Chapter 6

Ocracoke, North Carolina

Bankers, Blackbeard, and Boy Scouts

Ocracoke Island

NORTH CAROLINA

Legend:
- Highway
- Road
- ORV route
- Trail
- Beach
- Marsh
- Maritime forest
- Private

0 1 2 3 km
0 1 2 mi

PAMLICO SOUND

N

Hatteras Island

Horse Pen

Nature Trail

Cape Hatteras NS Visitor Center

Ferry to Swan Quarter (Toll)

Ocracoke

12

Campground

Ferry to Cedar Island (Toll)

Ramp 70

CAPE HATTERAS NATIONAL SEASHORE

Passenger Ferry (Toll)

Airstrip

Ramp 72 (ORV access only)

PORTSMOUTH ISLAND

ATLANTIC OCEAN

OCRACOKE INLET

CAPE LOOKOUT NATIONAL SEASHORE

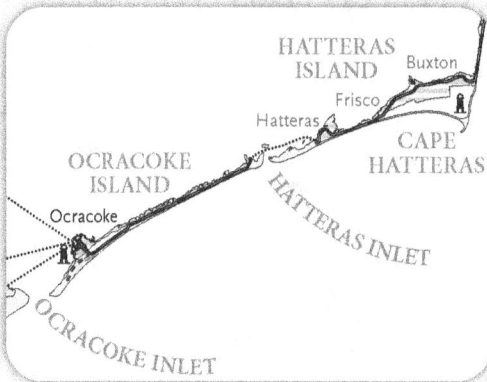

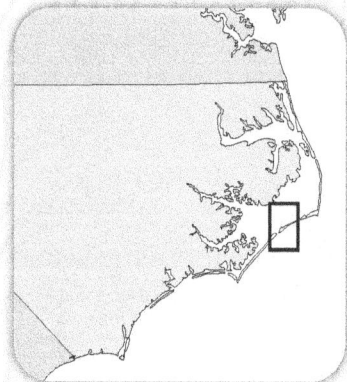

HATTERAS ISLAND

Buxton

Frisco

Hatteras

OCRACOKE ISLAND

CAPE HATTERAS

Ocracoke

HATTERAS INLET

OCRACOKE INLET

O n March 22, 2010, a celebrity came to Ocracoke Island. She arrived secretly in the dark of night, and on discovering her, the paparazzi plastered her photographs all over the Internet. The starlet was oblivious to the fanfare, being a newborn filly with nothing on her mind but frolic and sunshine.

Paloma, as she was later named, captivated the audience by capering around her patient dam, Spirit. Her base coat was bay, accented with four high, white stockings and a large diamond on her forehead. Like her dam, she displayed a white pinto marking, looking for all the world as if some troublemaker had upended a paint bucket over her back.

Paloma's instant celebrity went well beyond the public's fascination with whiskery new foals. Before 2009, the Ocracoke herd mostly comprised elderly horses and those unsuitable for breeding. It appeared that this historic herd was spiraling toward inevitable extinction. Under the advice of Dr. Sue Stuska of Cape Lookout National Seashore, two stallions and two mares from Shackleford Banks had arrived to invigorate the genome with new bloodlines from a very old breed. When Spirit became romantically involved with Wenzel, a diminutive bay Shackleford stud, the result was Paloma.

Ocracoke Island lies east-southeast of the mouth of the Pamlico River, about 30 mi/38 km from the nearest point on the mainland. A pearl in the barrier island necklace, Ocracoke is separated from Hatteras Island by Hatteras Inlet to the northeast, and from Portsmouth Island by Ocracoke Inlet to the southwest. An earlier Hatteras Inlet, which closed around 1755, once crossed the barrier chain about 7.5 mi/12.2 km farther east, just beyond Quork Hammock, across the lower end of the present Tar Hole Beach (Engels, 1942). When Old Hatteras Inlet was open, what is now the northeastern part of Ocracoke was part of Hatteras Island. When the inlet closed, *all* of Ocracoke became part of Hatteras Island. Ocracoke did not regain its independence until September 7, 1846, when New Hatteras Inlet opened in the same storm that created Oregon Inlet.

Ocracoke remains an appealing Outer Banks outpost, all the more endearing for its inaccessibility. Because it is still a village of working watermen, it has escaped some of the artificiality of other resort communities. Most visitors arrive by one of three ferry routes—the one from Hatteras is free. The Park Service manages most of this island, and as a result it remains devoid of homes and businesses along the entire 12-mi/19-km drive from the ferry dock to the village.

As elsewhere on the Banks, Highway 12 has been repeatedly relocated in response to beach recession, and it remains in jeopardy. The artificial dune ridges are unstable and are affected by overwash in every severe storm, in spite of the fact that free-roaming livestock were removed more than 50 years ago.

Barrier islands are always new, constantly shifted and rearranged by the elements, yet a feeling of timelessness awaits the visitor who hikes beyond the developed areas of Ocracoke.

Parts of the island probably look about the same as they did centuries ago, when there were hundreds of free-roaming horses and cattle over the next dune. This timelessness is one of the reasons people keep coming back to Ocracoke. Schedules and deadlines are less important here, and many who spend time on the island gradually subside into the ageless rhythm of tides, moons, and seasons.

Horses once ran free on Ocracoke, but their liberty came to an abrupt halt soon after the island was designated a national seashore. Paloma and her family reside in the Ocracoke Pony Pen, a handicapped-accessible facility that allows visitors a chance to get a close look at the horses. In all seasons, the Pony Pen is a popular attraction for thousands of visitors. A brief synopsis of their supposed origins stands mounted on a plaque in front of the paddock.

As with all Banker Horses, legends, theories, and a few definite facts vie to explain their history. Misinformation appears in print so often that it passes for fact. Much is open to speculation. To understand how Spanish-blooded herds came to run wild on such unlikely outposts as North Carolina barrier islands, one must consider how and when Europeans brought horses to the North American continent, how these animals were used and maintained, and the trends and events that gave rise to unique American breeds.

The first horses reintroduced to the New World came with Columbus's second expedition in December 1493, when 1–3 dozen were unloaded in Hispaniola. For the following 30 years, almost every fleet sailing from Spain to New Spain carried horses, and Hispaniola became a major equine breeding hot spot. Only about half the horses survived the 2–3-month trans-Atlantic voyages. Once the advantages of following the Atlantic currents were better understood, later colonists would typically sail from England to the West Indies before heading north to the colonies, giving them the opportunity to trade for livestock that was much more likely to survive the shorter voyage.

Thornton Chard (1940) describes many instances of early Spanish explorers bringing horses from Hispaniola to Florida, but makes a good case that none of these animals remained to reproduce; some died, some were eaten, and, some returned to the starting point. Numerous early Spanish explorers brought horses from the West Indies to what is now the United States, it does not appear that any horses started herds..

The first Spanish horses in what is now the United States were landed in Florida by Juan Ponce de Léon, who arrived in 1521 with 2 ships, 200 men, and 50 horses from Puerto Rico. They were beaten back by Calusa warriors, and de Léon was struck by a poisoned arrow. Horses were valuable at the time—Chard (1940) wrote that a single horse typically cost an astonishing 3,000–4,000 pesos (worth $18,680 –24,684 in his era). His estimate is probably very low. Taking inflation into account, each horse might be worth roughly $500,000–$600,000 today, a major investment comparable with a good racehorse today (Bank of England, n.d.; *Purchasing Power of British Pounds*, 2013; Turner, n.d.; U.S. Bureau of Labor Statistics, n.d.). Survivors would have reclaimed any serviceable horses when they withdrew to Cuba, where de Léon died of his wound.

One popular hypothesis holds that the horses on Ocracoke and elsewhere descend from stock imported for the Spanish settlement established by Lucas Vázquez de Ayllón. The evidence is sketchy at best, and virtually all credible sources disagree with one another on prominent details.

In 1521, Ayllón, a lawyer and senator in Hispaniola, aspired to discover a new territory and establish a prosperous colony there. He acquired the necessary license and dispatched

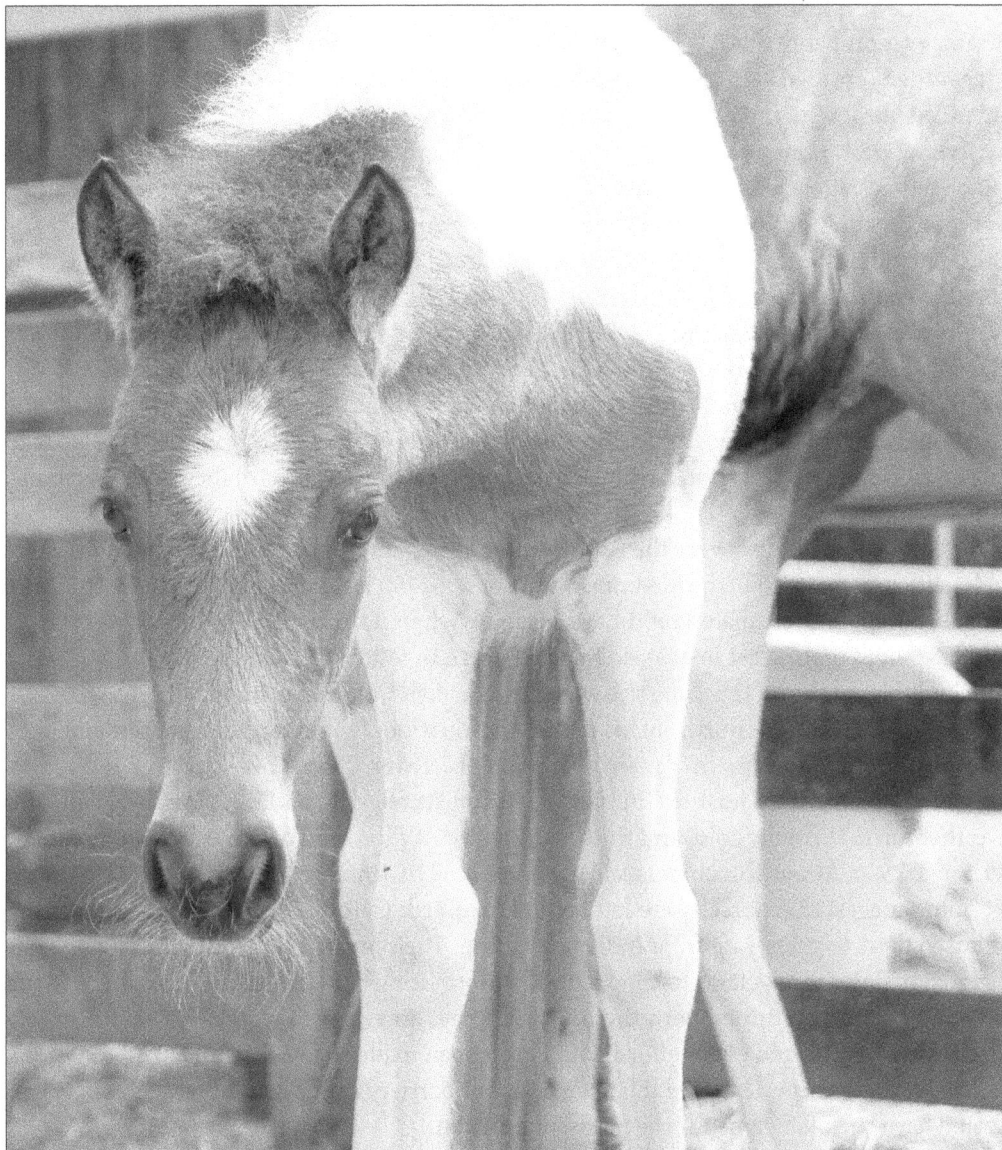

Paloma, born in 2010, holds the promise of the future in her genes. Her pinto coloration, however, as well as her dam's graying coat result from introductions of outside horses within the past five or six decades.

Francisco Gordillo to sail north through the Bahamas to reach continental North America, instructing him to cultivate good will with the natives. (Other accounts say that Ayllón sent the ships to search for slaves.) He expected to find a legendary Indian land called Chicora, inhabited by gigantic fair-skinned natives with long, blonde hair that reached their heels.

On his journey Gordillo encountered a kinsman, Pedro de Quejo, who was on a mission to capture Indians to sell into slavery. (Or Ayllón hired Quejo to capture slaves.) Gordillo and Quejo traveled north for 8 or 9 days until they reached the mouth of a very large river, which they named for John the Baptist, *San Juan Bautista*. They believed that they found this river in latitude 33° 30' (Winsor, 1886). Many modern scholars suppose that it was Winyah Bay, in the vicinity of present-day Myrtle Beach, South Carolina, if the reported latitude is accurate (it probably is not). They landed 20 men on shore and presented the Indians with gifts—then captured 70 natives and sailed away without further exploring the coast. (Or the intended purpose of their venture was to secure slaves for Ayllón.)

Ayllón was granted license to explore and occupy this territory. Legal disputes and other setbacks delayed his expedition, but in June 1526 he set sail with three large ships (or four or six), carrying "six hundred persons of both sexes" (Winsor, 1886, p. 240) (or 300), including clergymen, physicians, black slaves, goats, hogs, chickens, and 100 (or 90 or 80) Spanish horses.

They reached the mainland not at the mouth of San Juan Bautista, but at another river flowing directly into the ocean, which they believed to be at 33° 40' N (Winsor, 1886). Both Winyah Bay and the Cape Fear River in North Carolina fit this description; but the actual location is widely disputed, and various sources place it anywhere from the Chesapeake region to Georgia. Soon after, Ayllón's flagship ran aground in the entrance to the river and was destroyed.

Ayllón named his discovery the River Jordan. He explored upriver but did not find a suitable settlement site, so he headed down the coast 40–50 leagues. (Because the length of a league varied, that estimate could have been anywhere from 40 to 150 mi/64–241 km.) Other sources report that he instead traveled north to follow Chesapeake Bay into Virginia and built his colony on the same site where the English would later build Jamestown.

Someplace south (or north) of his original destination, Ayllón founded the settlement of San Miguel de Gualdape in a marshy area on the River Canaan. The natives refused to feed them, then turned hostile. The starving colonists fell sick, and the slaves rebelled. The weather turned brutally cold, and the survivors fled. Ayllón died of exposure and disease in October 1526. About 150 colonists made it back to Hispaniola.

Some suggest that horses were left behind on North Carolina barrier islands after being released there to graze, safe from Indian predation. Their presence would be like living graffiti advertising Spain's claim to the land. It is more likely that any horses not eaten by the starving colonists returned with them. At the time, horses were scarce and valued highly in the New World, and every effort would have been made to keep them. Further, Ayllón's colonization attempts involved the mainland, not barrier islands, and horses left anywhere would have been fair game for the native population.

But even if this expedition did, improbably, leave horses on islands, the mouth of the Cape Fear River is about 150 mi/241 km from Ocracoke. If his colony was established at the mouth of the Pamlico or Neuse River, Ocracoke would have been closer, but records indicate that Ayllón's colony was located by a river or estuary that emptied directly into the ocean or a major arm of the ocean. The Pamlico and Neuse do not fit this description.

Historically, free-roaming horses have grazed on many of the islands of the east coast. While opening and closing of inlets might eventually allow horses placed anywhere on a barrier chain to migrate elsewhere on that chain, horses could not have traveled from Cape Fear to Ocracoke without the assistance of people.

Rather than leaving stock on barrier islands, Ayllón's sick, starving colonists probably would have slaughtered any remaining horses for sustenance. Other settlers were certainly not above eating their horses. In 1527, Narvaez landed 600 colonists at Tampa Bay, Florida; every last horse was eventually slaughtered to sustain the starving people (Chard, 1940). In Jamestown, Virginia, after all the cattle, hogs, and horses had been consumed, a few planters ate an Indian (or several) during the Starving Time. One man even ate his own wife, according to John Smith (1624) and others.

Other early explorers imported horses to the North American continent. In 1538, Hernando De Soto sailed from Spain to Santiago, Cuba, with nine vessels and 600 men, then brought a contingent of horses to his encampment on Tampa Bay. On August 1, 1539, he set out to explore what is now Florida, Alabama, and Mississippi with 500 men and 200 horses, most of them probably purchased in Cuba (Chard, 1940). Most were killed in battles with indigenous people. Chard writes, "As all seem to be accounted for it is safe to say that they left no progeny east of the Mississippi" (1940, p. 93). In 1558 a Spanish expedition from Mexico attempted to colonize present-day Parris Island, South Carolina. Of the 240 horses loaded onto the ships at the outset of the voyage, only 130 survived. Ultimately, there was insufficient food to sustain the colony, and both horses and humans became weak and emaciated. Here again, the colonists survived by eating their cattle, hogs, and horses.

Similarly, Pedro Menéndez de Aviles established permanent settlements in St. Augustine and San Mateo in Florida, which included many horses; but by 1572, colonists were starving and had eaten them (Chard, 1940). St. Augustine was resupplied with horses as the colony rallied. The Timucua and other tribes probably acquired their horses from colonists of St. Augustine, and these horses apparently gave rise to the Seminole breed, named for an ethnic group that did not exist till the 18th century. Seminole and Chickasaw Horses remained for centuries pure lines derived from the original Jennet.

None of the English settlers who filled the vacuum that Spain left in North America mentioned finding horses or any other familiar livestock grazing in the marshes when they alighted. In the early 1950s David Quinn compiled and annotated nearly all the known English and Spanish records of the Raleigh colonies in *The Roanoke Voyages* (1955), which remains the standard work on the subject. Even in this definitive form, the 16th-century sources are incomplete and often puzzling. The surviving documents say little about livestock anywhere and almost nothing about horses in and around the Roanoke Island colony. One can only speculate about the numbers and kinds of domestic livestock introduced by the colonists and what happened to those animals during and after occupation.

Thomas Harriot was the scientific advisor for the Lane colony of 1585. His *Briefe and True Report of the New Found Land of Virginia* (1590) was the first remotely scientific treatise on North America, and it influenced European thought and exploration for two centuries. He describes native animals and plants in detail, from cacti to the now-extinct Carolina parakeet, but never mentions horses (Berenger, 1771; Harriot, 1590).

In 1585 Sir Richard Grenville set sail for Roanoke Island in command of a flotilla laden with colonists and some of the supplies that they needed to survive in the New World. Along the way, they acquired many tropical plants, such as pineapples, which they expected to cultivate in the warm climate of Roanoke Island, but were thwarted by the unexpectedly cold winters that the region endured during the Little Ice Age. Two decades later, John Smith mused, "The sommer is hot as in *Spaine*; the winter colde as in *Fraunce* or *England*.... The colde is extreme sharpe.... In the yeare 1607 was an extraordinary frost in most of *Europe*, and this frost was founde as extreme in *Virginia*" (1612/1910, pp. 47–48).

Stopping at Puerto Rico, Grenville built a replacement for the pinnace he had lost, reunited with one of his missing ships, and obtained the horses shown in John White's watercolor of the fortified camp. Continuing to Hispaniola for more livestock, including additional horses, Grenville captured two Spanish ships.

According to the log of Grenville's flagship, the *Tiger*, while in Hispaniola Grenville traded for "horses [stallions], mares, kyne [cattle], buls, goates, swine, sheepe . . . and such like commodities of the island" (Quinn, 1955, p. 187). Evidently any livestock that survived the voyage were not enough to adequately supply the colony. On September 3, 1585, the governor of the new colony, Ralph Lane, wrote, "if Virginia had but Horses and Kine in some reasonable proportion . . . no realme in Christendome were comparable to it" (Quinn, 1955, p. 208).

Grenville himself sailed on the *Tiger*, so he may have assigned the chore of transporting livestock to other vessels in the group. Unfortunately, his ship ran aground repeatedly at Wococon (probably Ocracoke Inlet or one of its predecessors), necessitating repairs. There is some indication that plants and seeds that the colonists intended to grow in Virginia were damaged or destroyed aboard the *Tiger* during its eventful passage through the inlet. Some writers suggest that livestock were thrown overboard to lighten the foundered ship; the animals then swam to the island and escaped to run wild on Ocracoke. Documents do not mention livestock on the *Tiger* or any other vessel, however, and there is no evidence to support (or undermine) this supposition.

Like the rest of the Outer Banks, Ocracoke was long used for grazing livestock. Because horses and cattle roamed rest of the Outer Banks from 1660s to the 1930s, it is likely that livestock was also established on Ocracoke long before records reflect their introduction. At various times, Cape Hatteras and Ocracoke were connected by land, and herds on the two islands probably mingled.

Local herdsmen grazed sheep, cattle, pigs, goats, and horses free-range in the marshes, fencing their homesites against the depredations of hungry livestock. They bought, sold, and bred horses, transported them by land and water, and let them roam and mix at will. Sometimes they added outside horses to the islands, largely by introducing animals from other islands or the local mainland. Historically, the Ocracoke horses were genetically most similar to those at Currituck Banks, Shackleford Banks, Carrot Island, and Cedar Island. A Currituck Banker born in 1913 described cattle drives from Oregon Inlet to Wash Woods, almost 60 miles, as late as the interwar years of the 20th century (Eaton, 1989). Horse penning was probably of a similar scope, and the gathering likely mixed bands, disrupted harem leadership, and redefined home ranges. The dynamic nature of barrier islands, always separating and reuniting, also shuffled the genetic deck by allowing horses to migrate and find new mates.

The Banks from Chicamacomico (now the villages of Waves, Salvo, and Rodanthe) to Ocracoke were united from 1755 to 1846, save one short-lived inlet. It was possible to walk from Hatteras to Ocracoke; in fact, one of the early Cape Hatteras light keepers commuted over land from Ocracoke to tend the light. Often, there was only one shallow inlet between Portsmouth and Shackleford. Currituck Banks had as many as three open inlets at a time before 1828. Since the early 19th century, shoreline has extended in an unbroken sandy corridor from Cape Henry at the mouth of Chesapeake Bay to Oregon or New Inlet (Mallinson et al., 2008).

Barrier islands are strong, flexible barricades that mitigate ocean waves and lessen their effect on the mainland. The islands change shape to absorb the large waves of a storm and

A Currituck Banks mare navigates around tree stumps at the water's edge, remnants of a forest that once grew on the sound side.

respond to sea-level rise by migrating westward, remaining more or less intact and maintaining about the same distance from the mainland (Pilkey, Rice, & Neal, 2004).

The beaches maintain a dynamic equilibrium with sea-level change, wave energy, the availability of sand, and the shape of the beach—when one element shifts, the others adapt to keep the balance. A powerful storm can literally cut an island in half, creating an inlet where there was solid ground the day before. Conversely, sediment can fill in old inlets to create solid ground. These banks continuously merge and separate, connect and divide.

Barrier islands, about 2,500 in all, are found on all continents except Antarctica. They are adaptive, sometimes exhibiting characteristics more befitting a life form than a land form. Dunes suddenly wake and migrate across an island, engulfing everything in their path, then return to dormancy. Inlets open and close regularly; some migrate, others remain stable.

Evidence of this constant movement is observable by any alert visitor. For example, a walk along the beach of Currituck Banks, especially after a storm, will reveal hundreds of tree stumps. These are the remains of forests, 400–1,000 years old, that once grew on the west side of the island. Overwash moved the island westward and buried the forest, as someday the forest growing to the west today will be buried if nature has its way.

Storm surge can raise the water level on the North Carolina coast by as much as 20 ft/6 m (Pilkey, Rice, & Neal, 2004). During severe weather, dramatic changes in features can take place over days or even hours. Within the Pea Island National Wildlife Refuge, New Inlet is now an area of flat ground where a series of inlets first appeared on maps in 1738 (Mallinson et al., 2009). These inlets were small and shallow and filled in quickly because the larger Roanoke Inlet provided a more convenient outlet for the northern sounds until the early 19th century. By the time Oregon Inlet opened in 1846, New Inlet was choked

Hurricane Isabel slammed into the Outer Banks of North Carolina, washing out Highway 12 and breaching Hatteras Island. Photograph by Mark Wolfe (2003) courtesy of the Federal Emergency Management Agency.

with sediment. It closed in 1922, but reopened around 1932 and remained until around 1945 (Mallinson et al., 2008).

Inlet/outlet systems function as self-adjusting safety valves (Mallinson et al., 2009). They are created during storms or floods when storm surge washes a corridor through the sand. Hurricane storm surge is created by the forces of low atmospheric pressure, winds and surface currents circling around the eye, and the piling up of water pushed from the deep ocean to the shallows (Pilkey, Rice, & Neal, 2004). The mounded water carries sand and whatever else it can pick up across the island toward the mainland.

As the storm moves away, the wind reverses direction, and the water rushes back to the sea. The sounds have a total surface area of roughly 3,000 mi^2/7,770 km^2; and can also produce a powerful storm surge. This swollen, unruly tide escapes by smashing through the barrier island into the ocean, producing an inlet by finding an outlet. If the water had no outlet, the barrier islands would function as dams, holding the excess water in the sound and increasing mainland storm damage due to flooding (Mallinson et al., 2009). Between storms, sand tends to fill the inlets.

The barrier island environment was challenging for livestock and humans alike. Herds and homesteads were equally vulnerable to the wrath of storms. Violent storms drowned countless Banker Horses, but others survived and contributed their hardy genes to their progeny. To cope with the changes in geography, island livestock migrated to whatever solid ground existed, becoming separated from their brethren whenever an inlet blocked an old route. After horse penning, animals were often sold from island to island, remixing the herds.

Since colonial times, between Cape Lookout and the Virginia line more than two dozen inlets have existed long enough to appear on printed maps, and dozens more have appeared

Ocracoke village *circa* 1950s. Courtesy of the Ocracoke Preservation Society.

suddenly and closed rapidly (Stick, 1958). Roanoke Inlet, which Sir Walter Raleigh's colonists may have used, closed before 1819. It was probably located in the vicinity of Whalebone Junction, the intersection of U.S. Route 158, U.S. 64, and N.C. 12 in Nags Head. Once there were numerous inlets north of Roanoke Island, but now the entire drainage of Albemarle and Currituck sounds and their tributaries must flow around the island to exit through Oregon Inlet (Mallinson et al., 2009). Only Ocracoke inlet may have remained open since European contact. As of this writing, six inlets exist north of Cape Lookout.

Oregon Inlet, the northern boundary of Hatteras Island, formed during an 1846 hurricane about 2.5 mi/4 km north of its present location. The inlet has been moving south at the rate of about 100 ft/30 m a year for more than 160 years. It destroyed the first Bodie (pronounced "body") Island lighthouse, built to the south on Pea Island (now the northern part of Hatteras Island) in 1848. Confederates destroyed its 1859 successor, also on Pea Island. The present-day Bodie Island Lighthouse was built on the north shore of the inlet, therefore *on* Bodie Island, in 1872. Bodie Island, however, had been an extension of Currituck Banks, a peninsula, since Roanoke Inlet closed. Oregon inlet continued its journey southward and left the lighthouse standing watch farther and farther away. The Inlet migrated at rates as great as 300 ft (91 m) annually, demolishing campgrounds, parking lots, and roads and finally threatening the Coast Guard station at the south end of the Herbert Bonner Bridge, completed in 1963 (Mallinson et al., 2009). Under natural conditions, Oregon Inlet would continue to move south, but at the expense of infrastructure. In 1989 the North Carolina Department of Transportation built a rock jetty to prevent the inlet from migrating further. This has temporarily arrested the south bank but not the north, causing Oregon Inlet to narrow and clog. Dredging is necessary to keep a channel open for boats under the highest part of the bridge.

Another good illustration of island movement is the Cape Hatteras lighthouse. When the present structure was commissioned in December 1870 (replacing one built in 1803), it stood 1,500–1,600 ft/457–488 m inland. Through the last few decades, the beach moved west to meet it. Finally the lighthouse stood right on the beach. From the 1960s, preservation attempts vied with the forces of nature: workers built three groins, rebuilt them when they failed, added reinforcements of large rocks and sandbags, repeatedly nourished the beach, and planted beds of artificial seaweed. Asphalt from the adjoining parking lot crumbled and piled onto the beach. One plan involved reinforcing the lighthouse to let it stand as an island in the advancing sea. In October 1998, Congress appropriated $9.8 million to save this national landmark by moving it to a safer location about a half-mile (0.8 km) inland. The move was accomplished in the summer of 1999 amid much fanfare, and the famous lighthouse stands well protected in its new location—until nature catches up with it again.

Most of South Nags Head, in an inlet-formation zone, was developed since the 1980s and 90s, when all the houses were required to be built a designated distance from the beach. Since then, some oceanfront cottages have been washed away, and the second-tier cottages behind them have become oceanfront, awaiting their turn to topple in. The beach at South Nags Head currently recedes at a rate of 10–16 ft/3–5 m a year. Sandbagging and bulldozing may buy the houses a little more time, but the trade-off is loss of the public beach, debris from ruined homes littering the beach, and septic tanks exposed and broken during each major storm (Mallinson et al., 2009). After septic tanks are repaired and reburied, the houses are sold or rented again.

Farther south, the beach at Rodanthe, the northernmost community on Hatteras Island, erodes at a rate of about 16 ft/4.9 m a year. The house filmed for the 2008 movie *Nights in Rodanthe* was nearly swept into the ocean not long after the movie premiered. The structure was moved to a new lot, then was walloped by Hurricane Bill in August 2009.

Nearly every hurricane that comes through inflicts severe property damage. Extratropical storms, or nor'easters, are more common and last longer, however, and they cause most of the damage along the Outer Banks. Each year, 30 to 40 nor'easters churn up high waves and storm surges in this area.

There are many instances of wild island horses drowning in storms. On September 18, 2003, Hurricane Isabel, a Category 2 storm when it came ashore, descended with record waves nearly 40 ft/12 m high and sustained winds of approximately 85 knots/157 kph. Isabel swept five horses from Carrot Island, off Beaufort, N.C., and carried their bodies 3 mi/5 km east to Harkers Island (Taggart, 2008). Three additional horses survived after they were carried 1 mi/1.6 km south to Shackleford Banks. Between 1989 and 1993, 12 horses died in storms on Assateague Island. The nor'easter of November 2009 that threw tires on the beach at Assateague did not kill any horses, but ripped across the Outer Banks with three days of rain and 40–55 mph/64–89 kph winds. In some places, half to three quarters of the dunes were flattened, and houses were damaged so badly that officials condemned them.

State Highway 12 is a boon to the area, but it has been a costly, and some would say quixotic, venture to maintain a fixed roadway on an unstable island. When sand and water destroy the road, workers rebuild it to the west and stabilize the artificial dunes with bulldozers, sand fencing, sandbags, and grasses. But when a significant storm blows in, the dunes are violated yet again. There are six locations in and near Cape Hatteras NS where the dunes are frequently destroyed by storms. When overwash breaches the dunes, they are usually

The Outer Banks has seen thousands of wrecks. This line engraving was published in *Harper's Weekly* in 1863, depicting the USS *Monitor* sinking in a storm off Cape Hatteras on the night of December 30–31, 1862. Courtesy of the United States Naval History and Heritage Command, #NH 58758 via Wikimedia Commons.

repaired right away, a practice that interferes with natural barrier overwash processes, potentially increasing vulnerability to storm damage (Pendleton, Thieler, & Williams, 2004). Government agencies collaborate to fix the dunes to limit the loss of property and maintain access, but holding dunes in place is like holding back the tide—which is in fact another project elsewhere on the island. It may be folly to continue this expensive battle against the elements with no prospect of long-term success; but coastal tourism is economically vital to North Carolina, and the highway is the only vehicular hurricane evacuation route.

Despite more than 75 years of concerted effort to maintain a high, stable artificial dune line on Pea, Hatteras, and Ocracoke islands, the great dunes persistently encroach on the highway with every significant storm. The roads are impassible until the government clears or replaces them. Hurricane Isabel created a new inlet—Isabel Inlet—just north of Hatteras Village, leaving Highway 12 in crumbles on either side. The U.S. Army Corps of Engineers hastily filled the breach before a major flood-tide delta could form. It appears that Hatteras Island would become an archipelago again if left entirely to natural processes.

If Highway 12 is to remain, it must be relocated to the west ahead of the migrating dunes. Building to the west would disrupt fragile wetlands essential to the health of the environment and the missions of two federal agencies. Because erosion has taken its toll, in places the islands are already too narrow to support a new highway. Soon the state will need to build a causeway to carry the road west of the overwash flats or implement a system of water taxis or ferries to move people and vehicles between communities.

It will be hard for state and federal governments to commit funds to a soundside cause-way system, but such a project may become politically necessary. Meanwhile, funds for temporary fixes are hemorrhaging from the treasury. Another option is to harmonize with nature and maintain a temporary gravel road across the most vulnerable portions of the island (Mallinson et al., 2009). An unpaved roadway would allow vehicles to pass but would be much easier to repair and relocate when overwashed. And islands with ferry-only access such as Ocracoke, Nantucket, Block Island, and Martha's Vineyard prosper despite lack of a blacktop connection to the mainland.

Ocracoke appears on various 1500s maps as *Wococon, Woccocock,* and other similar-sounding variations. This is probably because the name was an Indian word, and it was spelled phonetically by mapmakers who may never have heard the original pronunciation. These gave way to *Occocock,* also spelled various ways) and later to *Ocracoke.*

In 1715, the North Carolina Assembly passed "An Act for Settling and Maintaining Pilots at Roanoke and Ocacock Inlett," but because the land was privately owned at the time, these early settlers were squatters (Stick, 1958). A community originally called Pilot Town flourished along Cockle Creek (the harbor known today as Silver Lake). In 1760, they were allotted 50 acres/20 ha for the settlement, which became Ocracoke Village.

By the end of the colonial period, the only two sizable settlements on the Outer Banks were Ocracoke and Portsmouth. Most ships traveling to and from the major seaports of New Bern, Washington, and Edenton were obliged to pass through Ocracoke inlet, and these two towns stood on either side of it (Olszewski, 1970).

Ocracoke Inlet was a major trade route through the Outer Banks, but it was fairly shallow and treacherous, and most ships could not pass through without a pilot. Some large or heavy-laden ships could not pass through it at all.

The first mention of settlement at present-day Portsmouth "on the south side of Ocracoke Inlet" dates to 1685 (Wiss Janney Elstner Associates & John Milner Associates, 2007, p. 15). Portsmouth was founded in 1753 for the purpose of providing a site at Ocracoke Inlet for warehouses and wharves. A gristmill, windmill, and several residences soon followed. By 1770, Portsmouth was the largest settlement on the Outer Banks. Livestock, including horses, roamed this island, too.

Portsmouth became a major port for lightering. Merchants would hire shallow-draft boats, lighters, or barges to carry their goods over the bar or move cargo between ships on opposite sides of the inlet. A town grew up around the inlet, but few wanted to settle there. In 1755, Governor Arthur Dobbs found Portsmouth Harbor "so exposed that every privateer sailing along the coast could from their mast head see every vessel in the harbor, and go in and cut them out, or destroy them" (Olszewski, 1970). Dobbs recommended the construction of a fort on Beacon Island, a 20-acre/8 ha islet west of the inlet in Pamlico Sound. The resulting stronghold, Fort Granville, was garrisoned by 53 officers and men in 1757, but only 5 in 1763. When the Treaty of Paris ended the French and Indian War in 1764, the fort was closed.

In 1789 merchants John Gray Blount and John Wallace established another lightering port on a half-mile-long (0.8 km) aggregate of oyster beds—"Old Rock"—in Ocracoke Inlet (Norris, 2006). They named the settlement Shell Castle, and in time the surprisingly sturdy base supported wharves, warehouses a windmill, a gristmill, a store, a lumber yard, a lighthouse, and a tavern. Optimally situated to transport goods to and from the mainland, Shell

Castle prospered for decades until storm damage and shoaling put the little port out of business around 1812.

The census of 1800 showed that Core Banks—mainly Portsmouth Island and Shell Castle Island—supported a population of 165 whites and 98 blacks, as well as a lumber yard, a tavern, a ship chandlery, a porpoise fishery, and warehouses (Olszewski, 1970). In 1841, a marine hospital was built to tend to sick and injured seamen. By 1830, the white population increased to 342, and the slave population had reached 120.

Stock raising was well established on the Outer Banks by the late1600s. Stick wrote that when Richard Sanderson died in 1733, he owned all of Ocracoke, as well as a large "Stock of Horses, Cattle, Sheep and Hoggs" (1958, p. 33).

In an 1810 letter to the editor of the *Raleigh Star*, an unknown writer describing Portsmouth said,

> Seven years ago an inhabitant of the Island of his own mark, Sheared 700 head of sheep—had between two hundred & fifty, & three hundred head of cattle & near as many horses. . . . It is believed the Island at present is overstocked & much benefit would result from diminution of one third the present number. (Newsome, 1929, p. 401)

A half-century later, Edmund Ruffin pointed out,

> The rearing of horses is a very profitable investment for the small amount of capital required for the business. There are some hundreds of horses, of the dwarfish native breed, on this part of the reef between Portsmouth and Beaufort harbor—ranging at large, and wild, (or untamed,) and continuing the race without any care of their numerous proprietors. (1861, p. 130)

Portsmouth stockmen would attend roundups on Ocracoke, and vice versa, to acquire new horses for their herds. On occasion a Banker Horse sold from Ocracoke to Portsmouth was so eager to return to its herd that it would swim successfully across the inlet. Ocracoke Inlet is a wide, fast-moving stretch of water. Even on a huge ferry burdened with numerous automobiles and passengers, one can feel the powerful grab of current while crossing. Many small boats are incapable of navigating it.

In one well-documented instance, an Ocracoke horse named Old Jerry, who had a taste for straw hats and achieved notoriety by consuming tourists' headgear left within his reach, was sold to someone on Portsmouth Island. Shortly thereafter, he was found grazing contentedly, back on Ocracoke. The horse had swum across 1.5 mi/2.4 km of surging tidal current (O'Neal, Rondthaler, & Fletcher, 1976). A syndicated story about the 1946 Ocracoke pony penning published around the country, in some instances with embellishment, describes a feat even more remarkable, though likely fictional. This version is representative:

> It is said that a few years ago a Carteret county man who made a hobby of training the ponies bought a young one at the Ocracoke island penning and bought him to the mainland. . . .
>
> But—it did not like the mainland[.] One night it escaped and was not seen again until his owner revisited the island. There he saw his duly branded pony happily back on his windswept sand bar. To get there the pony had performed an almost incredible feat—he had waded and swum six miles to Shackleford Banks, and after a long walk had crossed Barden Inlet, then after rounding Cape Lookout, had travelled up lonely Core Banks and plunged into Drum inlet. From there he followed

the beach to Portsmouth and finally breasted the dangerous currents of Ocracoke inlet, a journey which might have fazed even a well-equipped seafarer. (Munsell, 1946, p. 11)

In 1839, a powerful hurricane inundated Portsmouth and Ocracoke and swept away most of the livestock (Olszewski, 1970). In 1846, the same storm that carved both Hatteras and Oregon inlets demolished much of the town. Quoting native pilot Reuben Quidley, Welch wrote in 1885, "there were several families living where the inlet [Hatteras] is now . . . but to their great surprise, in the morning they saw the sea and sound connected together, and the live oaks washing up by the roots and tumbling into the ocean" (pp. 6–7). The new inlet at Hatteras was deeper for a while, so the shipping business that had kept Portsmouth prosperous shifted north. The population of Portsmouth Island declined from 685 in 1860 to 17 in 1956. In 1971 the last two residents moved to the mainland. Today Portsmouth stands as a ghost town, complete with houses, church, U.S. Lifesaving Service station, school, and cemeteries carefully restored by the Park Service. It is listed on the National Register of Historic Places and maintained by the Cape Lookout NS.

The waters surrounding Ocracoke have long been a challenge to sailors. Stick (1958) wrote that during the 16th century, Spanish fleets and Spanish armies swarmed Mexico, Central America, and the Caribbean and transported commodities from the New World along a route that passed near Cape Hatteras on the way to Europe. Ships traveling from Europe to the East Coast colonies shortened the voyage by sailing south to the Canary Islands, riding the Equatorial Current and the Trade Winds to the West Indies, and sailing up the North American coast on the Gulf Stream. Spanish mariners discovered that they could save substantial time in their travels from the Caribbean to Spain if they rode the Gulf Stream current northward along the coast until they were in the vicinity of Cape Hatteras, then headed northeast to cross the Atlantic, aided by the prevailing Westerlies.

Aptly referred to as the Graveyard of the Atlantic, the treacherous Diamond Shoals of Cape Hatteras extend up to 8 mi/13 km into the ocean, far enough that land is often invisible, even in ideal conditions. Many a sailor who believed he was miles out to sea has run on the outer shoals without warning. Strong currents and extreme weather add to the difficulties. Southbound sailing vessels were obliged to hug the coast in order to avoid fighting the Gulf Stream, and sometimes the passage between current and the shoals was only a mile or two (1.6–3.2 km) wide, leaving little margin for error (Chard, 1940).

In the past few hundred years, there have been more than 1,000 recorded shipwrecks around Cape Hatteras. Even the famous Union ironclad *Monitor* wrecked off Cape Hatteras in a storm. The actual number of losses is probably much higher. Before the 19th century, records were inconsistent and fragmentary. Wrecks became commonplace, and people who could write were not always moved to write about them.

As with many of the Atlantic coastal wild herds, many people believe that the horses that gave rise to Ocracoke's wild herd were shipwreck survivors who swam to the island though stormy seas. Some or all of the foundation animals could have arrived by shipwreck, but as with the other herds, there is no proof. It is clear, however, that ships carrying horses often passed close to the Outer Banks and its dangerous shoals.

Beginning in the mid-1600s, New England was so well supplied with horses, merchants turned a tidy profit exporting them in large numbers to power cane-crushing mills on sugar plantations in the Caribbean and as far away as the Dutch colony of Surinam, in South

America. Horses could be shipped to Barbados in about 37 days, roughly half the time required for a trip from England (Avitable, 2009). From 1683 to 1794, colonists, mostly in New England, sold over 33,000 horses to Surinam. In fact, Surinam would trade only with vessels that brought horses as part of their cargo.

Merchants realized a 50% profit margin on the sale of horses, or exchanged them for slaves, which they sold to other colonies (Du Bois, 1896). Trade was brisk and highly profitable. Logs account for 1,583 horses shipped from Connecticut to Jamaica between 1762 and 1768, but there were apparently many more. Another record counts 27,809 horses and cattle shipped from New London, Connecticut, from 1785 to 1788 (Phillips, 1922, p. 927). In the years 1771 and 1774, 7,130 horses were imported into the British Isles from "North America" (Phillips, 1922, p. 922). Horses were carried on the decks of ships bound for the West Indies. Even Jamaica and Santo Domingo imported horses from New England despite their abundance of livestock. As the planters became wealthy, they began to import elite saddle horses for comfort and prestige, especially Narragansett Pacers from Rhode Island. New England also traded horses to other colonies on the Eastern Seaboard. In 1642 Massachusetts Bay shipped horses to Lord Baltimore's colony in Maryland and in 1665 to Virginia. New England merchants even sold horses to the Dutch living on the Hudson River (Phillips, 1922, p. 902).

In 1648, Governor John Winthrop of the Massachusetts Bay colony wrote in his journal of a ship "lying before Charlestown [now part of Boston] with eighty horses on board bound for Barbados"—probably an eyewitness account of the first recorded exportation of horses from New England to the West Indies (Phillips, 1922, p. 902). It is also evidenced that New England ships sometimes hugged the coast *en route* to the West Indies. Ships carrying horses could have wrecked anywhere along the way.

Despite patchy records, accounts of equine losses along the Atlantic coast are numerous. During the colonial period, livestock were typically transported on deck, and they often fell victim to storms and rogue waves (Avitable, 2009). On Christmas Day in 1770, a storm struck a small fleet from Connecticut bound for the West Indies and drowned more than 100 horses and a number of sailors. In February 1725, another storm drowned 16 of 18 horses on a sloop from New London and a total of 21 horses on three different Rhode Island ships (Avitable, 2009). In September 1766, a Connecticut vessel bound for Grenada encountered a series of hurricanes that drowned its entire shipment of horses and oxen. The captain sought shelter in Bermuda and saw the bodies of innumerable dead horses floating in the ocean around the island. Another gale on January 7, 1758, struck a vessel in New Haven Harbor awaiting departure to the West Indies. Ten of 34 horses drowned or were battered to death against the rocks (Avitable, 2009). The list of mishaps is long.

Numerous ships carrying horses skirted the Carolina coast, and many horses were lost in maritime disasters. Although it is more likely that the horses of the Outer Banks originated from stockmen who used the islands as a grazing commons, horses fleeing a maritime disaster might have swum to shore and started or contributed to some of the wild herds. Some residents of the Outer Banks may be descendants of shipwreck survivors, and it is possible their horses are, too (Stick, 1952).

Horses are good swimmers and possess great endurance. One news item ("Battling Stallions," 1951, p. 3B) describes an altercation involving "two or more banker ponies," in which the contenders chased each other "off Ocracoke Island into the waters of Hatteras

Inlet, finally swimming through the dangerous current and reaching the safety of Hatteras." In 2012, an Arabian named Air of Temptation demonstrated incredible stamina when he bolted into the ocean at Summerland Beach, California. He swam for *three hours* before he was rescued 1 mi/1.6 km from shore (Almendrala, 2012).

Horses do, however, frequently drown when overwhelmed by storm-driven waves, swimming as they do with their heads in or near the water. Near shore, swimming becomes more hazardous as breakers slam them onto sandbars or the beach. There are many accounts of barrier island horses drowning in storm surge, usually in hurricanes but sometimes during minor nor'easters. Most of them were standing on high ground when the surge occurred.

Even if horses managed to make their way to shore from wrecked ships, horses not acclimated to the Banks have often perished, as hardy mustangs did on Assateague. Edmund Ruffin (1861) commented that horses introduced to Banker herds often succumbed to the harsh conditions and the poor-quality forage.

Any genes contributed to the herds by shipwreck survivors would have been very similar to those of the horses that safely arrived by other means. The horses traveling along Atlantic shipping routes were the same horses found in the colonies: Spanish Jennets; English, Dutch, Irish, and French horses; and the horses developed in the colonies through the interbreeding of these types.

Stud books, newspaper articles, and stallion advertisements from the 16th, 17th, and 18th centuries give insight into the breeds that were popular in the colonies. Wallace found that the horses of colonial Virginia, Pennsylvania, and Maryland were similar in size and type because of "constant intercourse and trade" among those colonies (1897, p. 141).

The earliest farm animals in the English colonies of North America were probably Spanish livestock bought (or, by some accounts, stolen) from the Spanish ranches in the Caribbean. Records are vague, incomplete, and confusing—at the time, writers often used the term "cattle" to mean any livestock. In 1609, six or seven ships from England brought many domestic animals to Jamestown, including a stallion and six mares, presumably Spanish horses from the Caribbean. But during the "starving time" the following winter, the colonists butchered them all for food. Thomas Dale brought 17 horses, probably Spanish, in 1611 (Anderson, 2002); Samuel Argall obtained French horses in a raid on Nova Scotia in 1613 (Lynghaug, 2009); and the Virginia Company supplied English horses in the 1620s (Harrison, 1927). Some of these horses, their descendants, and later imports seem to have avoided the cookpot.

The next wave of livestock appears to have originated in Europe. In 1625, the Dutch West India Company shipped horses from the Netherlands to New York, and in 1635 Dutch draft horses arrived in the form of 27 Flanders mares and three stallions. Francis Higginson shipped about 50 mares and stallions to Massachusetts Bay, perhaps from Leicestershire, in 1629; but most died *en route* or shortly after, and only one stallion and seven mares survived (Browne, 1854).

Swedes settled southeastern Pennsylvania with their Old World horses, which were "undoubtedly pacers" (Phillips, 1922, p. 904). New Jersey acquired horses not from Europe, but from New York, Pennsylvania, and Virginia, and, like other colonies, allowed them to forage free range. By the mid-1600s, Connecticut and Massachusetts supported enormous herds of free-roaming horses, which increased faster than traders could sell them to the West Indies. By the 1700s the other colonies were in similar straits. Phillips (1922, p. 903) quotes

Along the Atlantic Seaboard, purely Spanish horses occurred in two distinct types—the Seminole or Creek horse originating in Florida, "small in size and capricious in nature," and the Chickasaw or Choctaw horse, "larger and more docile," originating in the plains west of the Mississippi River (Harrison, 1931, p. 170). The Corolla Banker herd shows evidence of both lineages.

William Harris, who had been sent out by the Board of Trade in 1675, and observed that the country had so many horses "that men know not what to do with them."

Phillips (1922, pp. 894–895) wrote,

> In view of the lack of any direct evidence to the contrary, it is fair to assume that the first shipments were mainly from England and of the small nondescript type which at that time made up the bulk of the English horses. There was, however, some admixture of other blood. In the primary importation into the Massachusetts Bay colony in 1629, three at least are mentioned specifically as "having come out of Leicestershire," which at that time was the source of a more or less distinct type of horse of a sort better than the average. The importation of Flemish mares also has been noted. Wallace contends that these latter were not Flemish but were rather of a Dutch type, but his conclusion is based merely on the fact that the vessel cleared from a Dutch port—which does not seem a very valid reason for controverting Winthrop's specific statement as to their Flemish origin, especially since Flemish horses were well known at that period as a distinct type.
>
> There is one other possible source of some of the New England horses which deserves consideration, especially because it may tend to explain in some measure the persistently small size of these horses, even when carefully bred—as later they were in Rhode Island and Connecticut—and, further, the constant occurrence among them of individuals possessed of a natural pacing gait. This possible progenitor is to be found in the Irish hobbies, a race of small, hardy, wild ponies existing in Ireland during the first part of the 17th century. These horses were in great demand in England for saddle purposes, and were exported thence in such quantities that they are said to have become practically extinct in Ireland before the year 1634. They were well known in England, and their natural pacing gait made them

especially desirable in any place where travel was of necessity on horseback; it is not at all improbable, therefore, that some of them found their way to New England, where they would have been especially serviceable. There seems to be no direct evidence to this effect, but any comparison of such fragmentary descriptions of the two as are available discloses a rather striking similarity between these Irish hobbies and the famous Narragansett pacers which were later developed in Rhode Island.

In colonial times, horses were not categorized as breeds, but rather types or families recognized by one progenitor (such as Snip) or an owner's name (for example, The Godolphin Arabian, Byerley Turk, Justin Morgan). These horses were grouped primarily by ability, usefulness, and individual accomplishment (Westvang, 2008), and from these sires and mares arose may diverse types that evolved into what we now know as breeds.

The Narragansett Pacer of Rhode Island was the first unique American breed. It was probably developed from English, Irish, Spanish, and Dutch horses imported in the early 1600s. The Galloway, the rugged little pacing horse of Scotland, probably figured prominently in its heritage. William Robinson, the Lieutenant Governor of Rhode Island, improved the lines by importing a stallion named Snip, who was variously described in writings of the time as an Andalusian, a Thoroughbred, a Barb, and an Arabian. The Hazard family of Rhode Island (Phillips, 1922), who inherited Robinson's estate, maintains that Snip was a fine Spanish stallion that Robinson imported directly from Andalusia.

The Narragansett Pacer quickly became popular throughout colonial America. Also known as "Rhode Islands" or "New England Pacing Horses," this breed was present in South Carolina as early as 1682 and probably in North Carolina, Virginia, and Maryland around the same time. Narragansetts probably reached North Carolina not only by trade, but also with immigrants from New England, for example, "Quakers from Newport, Rhode Island, who had settled in the Albemarle region of North Carolina by 1700, then gradually spread into the counties south of Albemarle Sound" (Little, 2012, p. 18). However Narragansetts arrived, they were the dominant saddle horse until the early 1740s, when the Colonial Spanish-blooded Chickasaw (or Choctaw) surpassed it in popularity (Dunbar, 1961).

Early mainland roads were poor, little more than paths through the forest. The Narragansett Pacer was a fast, sure-footed horse with an uncommonly smooth gait and great endurance, making it possible for the rider to comfortably travel the many miles between communities. He could pace 12–14 mph and could sustain that speed without undue fatigue to either himself or his rider (Harrison, 1931). A Narragansett Pacer could carry a man on a journey of 800 miles at 100 miles per day. During the 1700s, colonists prized these horses and bred large numbers of them. Newspaper advertisements for stud service from around the turn of the 19th century used descriptions such as "stout and handsome"; "high-spirited ... and perfectly sure-footed"; "a chestnut colour, inclined to a sorrel, 14 hands 3 inches high, round made, and very strong"; and "tractable, and free from vicious habits" (Harrison, 1931, pp. 163–164).

James Fenimore Cooper described the Narragansett Pacer in a footnote to the second edition of *The Last of the Mohicans* that reads, in part,

> They were small, commonly of the colour called sorrel in America, and distinguished by their habit of pacing. Horses of this race were, and are still, in much request as saddle horses, on account of their hardiness and the ease of their movements. As they were also sure of foot, the Narragansets were greatly sought for by

females who were obliged to travel over the roots and holes in the 'new countries.'" (1831, p. 14)

Phillips (1922, p. 926) quotes an article by Robert Livingston of New York from the 1832 edition of the *Edinburgh Encyclopedia*:

They have handsome foreheads, the head clean, the neck long, the arms and legs thin and taper; the hindquarters are narrow and the hocks a little crooked, which is here called sickle hocked, which turns the hind feet out a little: their color is generally, though not always, bright sorrel; they are very spirited and carry both head and tail high. But what is most remarkable is that they amble with more speed than most horses trot, so that it is difficult to put some of them upon a gallop. Notwithstanding this facility of ambling, where the ground, requires it, as when the roads are rough and stony, they have a fine easy single footed trot. These circumstances, together with their being very sure footed, render them the finest saddle horses in the world; they neither fatigue themselves nor their rider. It is generally to be lamented that this invaluable breed of horses is now almost lost by being mixed with those imported from England and from other parts of the United States.

In Virginia, breeders crossed the Narragansett, English Thoroughbreds, and Chickasaw horses. Horsemen also imported Scottish Galloways and Irish Hobbies to Virginia from England or from the New England states (Wallace, 1897). Many of the horses of Colonial Virginia could pace, but they were smaller and usually capable of trotting and cantering. Wallace wrote:

When the Rev. Dr. McSparran, of Rhode Island, made a trip in Virginia and rode the Virginia pacers some hundreds of miles, early in the last [18th] century, he seems to have observed them closely and spoke very highly of them, but he said they were not so large and strong as the Narragansett's, nor so easy and gliding in their action. It might be suggested that this opinion was the natural result of esteeming one's own as better than those of a neighbor, but he was certainly right in the matter of size. In 1768 the Rhode Island horses averaged fourteen hands one inch, while the Virginia horses averaged (1750–52) thirteen hands one and three-quarter inches, making a difference of three and one-quarter inches in height. In the matter of gait they were not all natural pacers. . . . From this it may be inferred that breeders, in order to increase the size, had incorporated more or less of the blood of the early Dutch importations (1897, p. 134).

Through much of the 18th century, "nothing was accepted as a 'saddle horse' that could not take the pacing gait and its various modifications" (Wallace, 1897, p. 115). But as roads improved, Americans came to prefer horse-drawn conveyances to riding astride and favored trotters as carriage animals. Breeders lost interest in the Narragansett Pacer; Herbert described the breed as "extinct" (1857, p. 112) before the Civil War. Wallace wrote:

After more than a hundred years of faithful service, of great popularity, and of profitable returns to their breeders, the little Narragansetts began to disappear, just as their ancestors had disappeared a century earlier. Rhode Island was no longer a frontier settlement, but had grown into a rich and prosperous State. Mere bridle paths through the woods had developed into broad, smooth highways, and wheeled vehicles had taken the place of the saddle. Under these changed conditions, the little pacer was no longer desirable or even tolerable as a harness horse, and he was supplanted by a larger and more stylish type of horse, better suited to the particular kind of work required of him. This was simply the "survival of the fittest," considering the nature of the services required of the animal (1897, p. 182).

The pacers of the American colonies were popular for long-distance travel, sport, and general use. Spanish planters in the West Indies, notably Cuba (Wallace, 1897), paid premium prices for the Narragansett Pacers, stocking their stables with the most valuable and elite examples of the breed. In so doing, they depleted the nursery until they grew scarce, and suppliers in Rhode Island were breeding fewer every year. Eventually the breed collapsed.

As the Narragansett Pacer faded into obscurity, the Spanish horse moved into the limelight. D. Phillip Sponenberg, DVM, PhD, a professor at the Virginia-Maryland Regional College of Veterinary Medicine, noted that the earliest Spanish horses imported to North America, including the populations in the southeastern United States, were typically brought from Mexico rather than directly from the Caribbean (Sponenberg, 1992). After these original introductions, it was uncommon for planters to import horses directly from Spain to cross into these local populations (Sponenberg, 1992, p. 337). In South America, however, the earliest horses were introduced from the Antilles, followed by repeated importations of horses from the Caribbean and directly from Spain (Sponenberg, 1992). This divergence of bloodlines gave rise to the numerous distinct types of pure Colonial Spanish Horse that persist today.

Around 1700, the purely Spanish horse was found in an arch that stretched northward from Florida; through the Carolinas, Tennessee, and the Great Plains; and into the Western mountains (Sponenberg, 1992). These horses averaged just under 13.2 hands, and many were gaited—able to amble, pace, trot, and canter.

By the mid-1700s, great numbers of untended Spanish horses roamed the Southeast, the descendants of Jennets introduced by Spanish explorers, probably after 1580 (Conant, Juras, & Cothran, 2012). The first Europeans to settle the Carolina mountains found large numbers of wild horses roaming the open forests (Gregg, 1867). Native American tribes, both Western and Eastern, acquired Spanish Jennets by trading with or stealing from European settlers or by capturing feral animals.

In the 18th century, two strains of Spanish horse were common along the Atlantic Seaboard—the Seminole or Creek horse originating in Florida, "small in size and capricious in nature," and the Choctaw or Chickasaw horse, "larger and more docile," originating in the plains west of the Mississippi River (Harrison, 1931, p. 170). These types persist in Colonial Spanish horses today. Steve Edwards (2010), a breeder of Colonial Spanish horses, observed that some Banker stallions, like the legendary Red Feather, are smaller and more aggressive, often amassing large herds of mares by virtue of their superior fighting ability. The larger stallions, like his adopted Croatoan, tend to be calm and very easy to train.

Smyth (1784, p. 139) described the Chickasaw horses as "named from a nation of Indians who are very careful in preserving a fine breed of Spanish horses they have long possessed, unmixed with any other." This stock was widely distributed and bred along the Atlantic Seaboard throughout the 18th century.

In time, a wide variety of Indian horses were termed "Chickasaws." Harrison (pp. 167–168) commented on the blurring of distinctions:

At the end of the eighteenth century the name "Chickasaw" had spread north from Carolina to connote, if not a breed, a *type* of desirable saddle horse derived from

A band of Bankers climbs Penny (Lewark) Hill on Currituck Banks. Over the centuries Banker horses have developed unique hooves adapted for running in sand. This band shows the two most typical hoof shapes: the sorrels have wide hooves that spread the horse's weight over a wider surface area and prevent sinking. The black has hooves with a strong, steep wall and a deeply cupped sole that let them sink in and provide traction. These unique hoof shapes are present at birth.

Spanish America. John Davis and other travelers heard and recorded the name as then current in Virginia and Maryland in the same sense that the name 'mustang' was later used.

As evidence he quotes an interesting 1801 stud advertisement from Norfolk, Va., that refers to Spot, a "pure Chickasaw horse. . . . brought from South America" (p. 168).

In the early 19th century David Ramsay, physician, public official, and prominent historian of the American Revolution, apparently referred to all Spanish-blooded horses as Chickasaws when he wrote,

Before the year 1754 the best horses for the draught or saddle in Carolina were called the Chickesaw breed. These were originally introduced by the Spaniards into Florida, and in the course of time had astonishingly increased. Great numbers ranged wild. . . . Many of them were caught and tamed by the Indians, and sold to the traders. They made use of them for pack horses to bring their peltry to market, and afterwards sold them in the low country. These horses in general were handsome, active, and hardy, but small; seldom exceeding thirteen hands and a half in height. The mares in particular, when crossed with English blooded horses, produced colts of great beauty, strength and swiftness. Before the year 1754 these Chickesaw horses were the favorite breed. (1809, p. 403)

Until the mid-18th century, wealthy planters typically invested in expensive pacers, while poor subsistence farmers acquired Spanish ponies that Indians conveyed from the south and west to traders in Virginia and Carolina.

Harrison (1931, p. 168) wrote,

Unmistakable Spanish conformation is evident in the horses of the Outer Banks, such as this Shackleford Banks stallion . . .

[T]he contemporary evidence is that early in the eighteenth century the Carolina planters, wedded to the comforts of the Narragansett pacer, despised the pack horses brought in by the Indian traders as 'tackies'. There is no evidence at all that they esteemed any Indian horse as fit for the saddle and turf until after 1740.

Before 1740, traders would frequently sell the little Indian horses to poor farmers. Lowland stockmen used the local islands and necks as pastures, and freed their horses to graze the marshes and prodigiously increase their numbers. The free-roaming horses on barrier islands in North and South Carolina and Georgia were termed "Marsh Tackies," though it is unclear whether this term was used dismissively by the social elite or to denote Spanish Chickasaw heritage.

In 1922, Hervey Allen immortalized the lineage of these hardy little horses in his poem "Marsh Tackies" (Heyward & Allen, 1922, pp. 112–113).

Browsing on the salty marsh grass,
Barrel-ribbed and blowsy-bellied,
With a neigh as shrill as whistles
And their mouths red-raw from thistles,
I have seen the brown marsh tackies,
Hiding in the swamps at Kiawah,
With the gray mosquito patches
Gory on their shaggy thatches.

Balky, vicious, and degenerates,
They are small as Spanish jennets,

But their sires were with El Tarab,
When he conquered Andalusia
For the Prophet and the Arab;
And they came with Ponce de Leon,
When the Spaniard made a peon
And a Christian of the Carib.
Peering from palmetto thickets
At some fort's coquina wickets,
Startled Indians saw them grazing,
Thunder-stamping and amazing
As the beasts from other stars,

. . . as well as the Pryor Mountain horses. The herds live live on opposite sides of the continent, but closely resemble each other. Both herds trace back to Spanish horses—probably those bred on the islands of the Caribbean to supply Spanish explorers and colonists. Additionally, certain individuals in both herds carry the rare Q-ac genetic marker denoting old Spanish bloodlines.

When they galloped down savannas,
And their masters seemed centaurs
With the new white metal blazing.

Thus they came, these little beasts,
With the men-at-arms and priests,
In the west with Coronado
When he reached the Colorado,
In the east with bold De Soto
In the search for El Dorado,

And they packed the bells and toys
That the chieftains loved like boys;
Struggling through the swamps and briars
After dons and tonsured friars;
Dying in the forests dismal,

Till the shrill of silver clarion
Brought the buzzards to the carrion
Round the smoke of lonely fires
In a continent abysmal.

So De Soto left them dying,
Heedless of their human crying;
Here he turned them loose to die
Underneath a foreign sky;
But they lived on thicket dross,
On the leaves and Spanish moss—

And I wonder, and I wonder,
When I hear the startled thunder
Of their hoofs die down the reaches
Of these Carolina beaches.

Despite the widespread prejudice against small horses and those ridden by Indians, Spanish horses quickly garnered respect among horsemen. Men would show up at match races with nondescript Spanish horses and best the competition again and again, stealing purses from under the noses of elite Thoroughbreds. Southern breeders soon began crossing readily available native Spanish horses with expensive imports. Harrison observed (1931, p. 55) that "the longest Carolina racehorse pedigrees all extend to a taproot in the Chickasaw stock." Even in New York and New England, which esteemed English, French, and Dutch horses, breeders used North American Spanish horses to improve bloodlines.

Wallace (1897, p.114) lamented the lack of detailed information about native Spanish horses, but offered his own impressions:

> We here have a stock of horses that the people of Virginia have bred and ridden and raced for a hundred years, and we know comparatively nothing about them. They seem to have been specially adapted to the saddle, but they could run four miles, or they could run a quarter of a mile, like an arrow from a bow. They were not a breed, although selecting and crossing and interbreeding for a hundred years would make them quite homogeneous. There is a romantic interest attaching to these little horses... this old stock furnished half the foundation, in a vast majority of cases, for the triumphs of future generations of the Virginia race horse, and the same may be said of the old English stock upon which the eastern blood was engrafted.

Culver (1922, p. 39) also made note of the homogeneity of "Chickashaw" [sic] horses in the Carolinas prior to 1754 and the scarcity of imported stock during that period. He described the horse as "of small size but well formed and active, and when covered with imported thoroughbreds produced animals of great beauty, strength and speed."

The Spanish horse persisted in the Seminole and Chickasaw strains well into the 18th century. In the new British colony of East Florida, novelist and playwright Oliver Goldsmith (or his brother John) described the local horses as "of the Spanish breed, of great spirit, but little strength; they are seldom above fourteen hands high: the Indians here, by mixing the Spanish breed with the Carolina, have excellent horses, both for service and beauty" (1768, p. 361).

In most of colonial America, horse racing had always been popular for sport and status. By 1748, races at the pace, trot, and gallop had become so commonplace in New Jersey colony, racing was outlawed as a public nuisance (Wallace 1897). Then, as now, fine horses were viewed as status symbols. Men of the upper class talked about their horses and admired them in the same way that they would themselves or one another (Peck, 2008). As colonists became more prosperous, they imported English Thoroughbreds and Arabians to increase the size, speed, and quality of Colonial horses and matched them over 6-mi/9.7-km courses or in three to five 4-mi/6.4-km heats.

The American Quarter Horse originated in the Carolinas and Virginia. Early settlers found Virginia and the Carolinas covered with old-growth forest. Cleared land represented an enormous investment in time and sweat and was necessary for agriculture. Unable to spare the valuable cleared land for racetracks, people who lived in heavily forested regions raced horses in pairs along parallel quarter-mile (0.4-km) tracks hewn through the woods, often separated by trees or a fence. Until around the time of the Civil War, horse racing was the primary American sport, and outcomes often involved civic pride in addition to personal wealth.

Smyth (1784, pp. 22–23) described the early quarter racing horse as running

> with astonishing velocity, beating every other, for that distance, with great ease; but they have no bottom. However, I am confident that there is not a horse in England,

Ocracoke's patriarch stallion, Santiago, is a brilliantly marked black and white tobiano. Some members of the Ocracoke herd carry two dominant genes not found in any of the other Banker herds, for tobiano pinto and gray coloration. Because these colors appeared suddenly in the 20th century, their presence can be attributed to outcrossing.

nor perhaps the whole world, that can excel them in rapid speed: and these likewise make excellent saddle horses for the road.

The Virginians, of all ranks and denominations, are excessively fond of horses, and especially those of the race breed. The gentlemen of fortune expend great sums on their studs, generally keeping handsome carriages, and several elegant sets of horses, as well as others for the race and road: even the most indigent person has his saddle-horse, which he rides to every place, and on every occasion; for in this country nobody walks on foot the smallest distance. . . . In short, their horses are their pleasure, and their pride.

From 1682 to 1740, the Narragansett Pacer was the dominant breed in the Carolinas; from 1740 to 1786, the Chickasaw; and from 1755 onward, horses descending from English stock (Chard, 1940). Colonists had an ardent enthusiasm for fast horses, and as their fortunes grew, they began to import English Thoroughbred and Arabian horses from Europe. Samuel Patton and Samuel Gist imported Bulle Rock, an esteemed son of the Darley Arabian and grandson of the Byerley Turk (two of the three primary foundation sires of the American Thoroughbred) to Virginia in 1730 at the age of 21, and offered him at stud to local mares. His English owner noted that the American colonies were passionate about quarter-racing on Chickasaw horses and correctly speculated that they would enthusiastically bring their mares to an English Thoroughbred. Bulle Rock was the first of many European and Middle Eastern race horses imported to "improve" the existing gene pool in hopes of producing faster, larger horses. Despite arriving in America late in life, Bulle Rock was probably the first stallion who could have been described as a leading sire in the Colonies.

Wallace (1897, pp. 116–117) explained,

> From about 1750 to 1770 seems to have been a period of great prosperity in Virginia and, notwithstanding the general improvidence of the times, many of the large landholders and planters were getting rich from their fine crops of tobacco and their Negroes. This prosperity manifested itself strongly in the direction of the popular sport of horse racing and improving the size, quality, and fleetness of the running horse. England had then been selecting, importing Eastern blood, and "breeding to the winner" for a hundred years, with more or less intelligence and success, while the colonists had rested content with the descendants of the first importations from the mother country.

The fervor for imported racing stock took hold first in Virginia and Maryland and spread to the Carolinas by around 1754. Many of these English Thoroughbreds were stallions who stood at stud in Virginia and the Carolinas. The historical record is shaped by the snobbery of the upper class—the blooded stallions were advertised and celebrated, while the local mares were scarcely acknowledged.

Mares are vital to any horse breeding operation. Clearly, the dam provides 50% of the foal's genes and 100% of its environment during gestation, as well as strong influence in shaping the mind and body of a young foal. Themselves a product of a sexist culture that favored males, however, horsemen of the time concluded that stallions made a far greater genetic contribution and often regarded mares as little more than carriers for the get of a sire. Moreover, a mare crossed with an inferior horse was somehow tainted, and thereafter incapable of producing a "pure" foal:

Colonial Spanish Horse Type Matrix

	Most typical—score 1	Least typical—score 5
Head Profile	Concave/flat on forehead and then convex from top of nasal area to top of upper lip (subconvex) OR Uniformly slightly convex from poll to muzzle OR Straight	Dished as in Arabian OR Markedly convex
Head Front View	Wide between eyes (cranial portion), but tapering and "chiseled" in nasal/facial portion (A very important indicator, and width between eyes with sculpted taper to fine muzzle is very typical.)	Wide and fleshy throughout head from cranial portion to muzzle
Nostrils	Small, thin, crescent-shaped; flare larger when excited or exerting	Large, round, and open at rest
Ears	Small to medium length with distinctive notch or inward point at tips	Long, straight, with no inward point at tip OR thick, wide, or boxy
Eyes	Vary from large to small (pig eyes); usually fairly high on head	Large and bold, low on head
Muzzle Profile	Refined, usually with the top lip longer than the bottom lip	Coarse and thick with lower lip loose, large, and projecting beyond upper lip
Muzzle Front View	Fine taper down face to nostrils, slight outward flare, and then inward delicate curve to small, fine muzzle that is narrower than region between nostrils	Coarse and rounded OR heavy and somewhat square as the Quarter Horses, rather than having the tapering curves of the typical muzzle
Neck	Wide from side, sometimes ewe-necked, attached low on chest	Thin, long, and set high on chest
Height	Usually 13.2–14.2 hands (54–58 in./1.37–1.47 m) high; horses over 15 hands (60 in./1.53 m) are not typical	Under 13 hands (52 in./1.32 m) or over 15 hands (60 in./1.53 m)
Withers	Pronounced and obvious—"sharp"	Low, thick, and meaty
Back	Short, strong	Long, weak, and plain
Croup	Angled from top to tail. Usually a 30° slope, some are steeper.	Flat or high
Tail Set	Low; tail follows the croup angle so that tail "falls off" the croup	High; tail up above the angle of the croup
Shoulder	Should be long and 45–55°	Short and steeper than 55°
Chest Profile	Deep, usually accounting for half of height	Shallow, less than half of height
Chest Front View	Narrow and "pointed" in an "A" shape	Broad with chest flat across
Chestnuts	Small, frequently absent on rear, and flat rather than thick	Large and thick
Rear Limbs Rear View	Straight along whole length OR inward to have close hocks and then straight to ground ("close hocks") OR slightly turned out from hocks to ground ("cow hocks"), but not extreme. Legs very flexible. At trot the hind track often lands past the front track.	Excessive "cow hocks." Heavy, bunchy gaskin muscle, tight tendons.
Feathering on Legs	Absent to light fetlock feathering, though some have long silky hair above ergot and a "comb" of curled hair up back of cannon. Some horses from mountain areas have more feathering than others and lose this after moving to other environments.	Coarse, abundant feathering as is seen in some draft horse breeds
Rear	Contour from top of croup to gaskin has a "break" in line at the point of the butt	Contour from top of croup to gaskin is full and round ("apple butt") with no break at the point of the butt
Hip Rear View	Spine higher than hip, resulting in "rafter" hip Usually no crease from heavy muscling	Thickly muscled with a distinct crease down the rear
Hip Profile	Long and sloping, well angled, and not heavy	Short, poorly angled
Muscling	Long and tapered	Short and thick ("bunchy")
Front Cannon Bones	Cross-section is round (Best to palpate this below the splint bones.)	Cross-section is flat across the rear of the bone

Adapted from Sponenberg and Reed (2009, pp. 140–142).

Pony Penning, July 4, 1956, at the Sam Jones Corral. Although Ocracoke supported hundreds of wild horses though the 18th and 19th centuries, the census fell sharply after World War II and dwindled further with the arrival of the National Park Service. In 1956 there were only about 70 free-roaming horses on the island. The following year, the Park Service had eliminated all but the 35 owned by the Boy Scouts and helped them build a pen to confine them. This photograph may show the last Pony Penning on Ocracoke. Photograph by Aycock Brown, courtesy of Outer Banks History Center.

> A mare once crossed with a sire of different blood, not only produces, but *becomes* herself, a cross; and is incapable of ever again producing her own strain. Thus a thorough mare, once stinted to a coldblooded horse, could never again bear the pure colt, even to a pure sire; while a cold-blooded mare, having once foaled to a thorough horse, would always be improved as a breeder by the change produced in her own constitution. (Herbert, 1857, p. 113)

While much was written about Thoroughbred sires, the mares they served often faded into obscurity. The majority were probably Spanish or Spanish crosses. Some of them bore piebald or skewbald coats, which Harrison says were "a characteristic of the plains horses, as noted in 1804 by Lewis & Clark" (1934, p. 39). Apparently some had Appaloosa markings as well. Harrison wrote (1934, p. 39), that in the late 18th century "the quarter racing 'Chickasaw' or 'Opelousas' stock still abounded in the Roanoke valley." Elsewhere he notes, "there were a number of mares 'from Old Spain' in Virginia and Maryland before 1750" and "for some generations prior to 1750 Virginia and Maryland had absorbed a steady inflow of mares 'from New Spain', brought back by [I]ndian traders, from their contacts with the horse owning [I]ndians of the Southwest" (1929, p. 23).

Before 1730 the horses of the English colonies were either Spanish-blooded Chickasaw stock or descendants of horses imported from Spain and Spanish colonies, England, France, or the Netherlands (Culver, 1922). The Indian ponies of the West descended directly from the earliest Spanish horses. But through the 1700s the East Coast Spanish horses were

Edward Teach, a.k.a. Blackbeard, often intimidated victims into submission with his wild appearance and reputation for ruthlessness. He frequently wore his long, black beard in beribboned braids and stuck burning cannon fuses under his hat brim. In his two years of piracy, he and his men took more than 50 ships.

crossed with Narragansett Pacers, English Thoroughbreds, heavy French and Dutch draft horses, and Andalusian horses of a more modern derivation.

In 1747, Samuel Ogle, the resident lieutenant governor of Maryland, imported Spanish, Barb, and Arabian horses to his Belair Stud, the first prominent stud in America. Another renowned breeding facility, John's Island Stud, was established in South Carolina in 1750 by Edward Fenwick. The founder of the South Carolina Jockey Club at Charleston in 1758, Fenwick imported many descendants of the Godolphin Arabian and often crossed them with Chickasaw Spanish ponies.

Wild ponies follow the Ocracoke shoreline in the 1950s. Note the low dunes and the grasses cropped short by grazing. Photograph courtesy of Ocracoke Preservation Society.

The Thoroughbred gained popularity as a racehorse, especially among the social elite, and upper-class horsemen came to regard this breed as the *only* horse breed of exceptional quality. Wealthy breeders considered quarter racing a primitive sport fit only for frontiersmen and incorrectly concluded that the small Spanish horses raced at that distance because they had no "bottom" or endurance. Thoroughbreds were raced on 3-or 4-mile tracks, requiring both speed and stamina to win (Harrison, 1934).

As with most breeds, the exact origins of the Quarter Horse are open to speculation. One of the foundation sires was Janus, an English Thoroughbred imported to Virginia as a 10-year-old in 1746. All modern Thoroughbreds can trace their lineage to at least one of three foundation stallions of Arabian, Barb, and Turkoman heritage imported into England from the Middle East: the Byerley Turk (1689), the Darley Arabian (1703), and the Godolphin Arabian (1729). Janus, a grandson of the Godolphin Arabian, stood at stud for 24 years in North Carolina and southern Virginia. Janus was not built like today's Thoroughbred or even a modern Quarter Horse, but was a compact, muscular 14.1 hands (57 in. / 1.45 m) tall, "a small but beautiful horse. . .. chestnut; speckled on the rump as he grew old; a small blaze in the face, and hind foot white" ("D.," 1832, p. 272). Wallace (1897, p. 95) observed, "Janus became the progenitor of a tribe of very fast quarter horses, and although he did not found that tribe, which had been in existence for a hundred years on the border line between Virginia and North Carolina, he doubtless improved it."

Theodore Dodge (1892, p. 671) observed that horse breeding in the American colonies was often a haphazard affair:

> A farmer had a stanch mare. The only available stallion was in the neighboring village—perhaps on circuit. All he could see was that there were good qualities present in both, and he believed that these would be transmitted. . . . Often the mare was not bred from until she was unfitted for work by something which equally unfitted her for breeding. No doubt the average produce of this lack of method may have been of excellent service in its way, but it was none the less "nondescript."

Colonial Spanish horses were used for riding, driving, and draft. When taller, heavier horses gained popularity, the cobby Spanish horse fell out of fashion and its numbers rapidly decreased. Racism also played a role in its loss of popularity. The small, stocky

Barrier island horses can swim from their first day of life and frequently use water to escape from biting insects or the oppressive heat of summer. These Ocracoke horses were apparently undisturbed by the activities of working boats in Silver Lake around 1957. Photograph by Hugh Morton, from the Hugh Morton Collection of Photographs and Films #P0081, © North Carolina Collection, University of North Carolina at Chapel Hill Library.

Spanish horse was associated with Indians, Mexicans, and the poor, and because of this association was perceived as lower in quality than the larger breeds of the dominant English culture (Sponenberg, 1992). In the late 1800s, Spanish Indian ponies were systematically exterminated by order of the U.S. government in an effort to demoralize and subjugate the native people. The once-ubiquitous Spanish horse was suddenly teetering on the brink of extinction.

Numerous lines of Colonial Spanish horses descend from old Spanish bloodlines, and, according to Sponenberg, "are a direct remnant of the Iberian horses of the 1500s" (1992, p. 335). The Colonial Spanish horse strains of today include the Banker, the Belsky, the Cerbat, the Choctaw, the Florida Cracker, the Marsh Tacky, the New Mexico, the Santa Cruz, the Sulphur, the Wilbur-Cruce, and some strains of the Pryor Mountain. All these breeds are classed as critically endangered, both collectively and individually (American Livestock Breeds Conservancy, 2009). There are numerous breed registries for horses of the Colonial Spanish type: Spanish Mustang Registry, Spanish Barb Breeders Association, Southwest; Spanish Mustang Association, Florida Cracker Horse Association, and American Indian Horse Registry. Each of these strains is similar to the others, but each registry emphasizes different characteristics. The Horse of the Americas registry accepts all Colonial

In the 1950s, Ocracoke was home to the only mounted Boy Scout troop. Marvin Howard is mounted at the far right. Photograph courtesy of the Ocracoke Preservation Society.

Spanish strains. Sponenberg says that it is important to conserve these North American horses of old Spanish bloodlines "since they probably most closely represent the original, less selected type of Iberian horse brought to the New World" (1992, p. 337).

Sponenberg writes (1992) that through centuries of divergent selection, the modern Iberian Spanish horse is quite different from the ancient Jennet, and certain New World Spanish breeds are closer in type to the historic horse of the Golden Age of Spain. He points out (2011) that the old Spanish horse type was variable. Some individuals were compact and heavy, and others were lightly built. Some horses had higher- or lower-set tails or broader or narrower chests. Ryden (2005) writes that Roman-nosed Barb-type and dished-faced Arabian-type individuals often can be found within in the same herd of Spanish-blooded mustangs. These differences are observable in Banker horses. When overfed in domesticity, Bankers typically develop a cobby, pony-like appearance with fat over the neck, back, and hindquarters that obscures the lines of Spanish conformation (Ives, 2007).

There is great disagreement about which breed is the most direct descendant of the original Spanish Jennet. Andalusians, Lusitanos, Lipizzaners, Cerbat and Pryor Mountain Mustangs, Paso Finos, Peruvian Pasos, and the free-roaming horses of the Banks all descend from Spanish stock; but that blood has mixed, lines have diverged, and all have developed into unique breeds.

In 2007, inspectors from the Horse of the Americas breed registry evaluated the conformation of Corolla and Shackleford Banks horses. They found that all the horses of both herds had the conformation, gaits, and skeletal structure of Colonial Spanish Horses (Ives,

The Ocracoke Scouts would select a pony, often a stallion, from the wild herds and pay its nominal owner the price of purchase. Working together, they trained the horses to accept saddles and bridles. Photograph courtesy of the Ocracoke Preservation Society.

2007). Banker Horses are accepted by the Horse of the Americas registry as Colonial Spanish horses based on their physical characteristics, even though their origins are obscure.

In the 1970s, certain members of the Ocracoke herd were admitted to the Spanish Mustang Registry as purebreds, and in 1982, two Shackleford horses were admitted. There have been no further inclusions from East Coast feral populations. Some present-day Ocracoke horses, however, descend from an Andalusian stallion outcrossed into the herd in the 1970s. Andalusians are of Spanish heritage, but are a different breed entirely. The Ocracoke horses resulting from these matings are regarded as crossbreeds; as such, the Spanish Mustang Registry considered them ineligible for registration.

Breeds persist for as long as people esteem their unique qualities, or, in the case of some wild herds, when people stop interfering with them. When a breed is repeatedly crossbred with outside horses, it loses its distinctiveness. Breeds become extinct when they are no longer valued, when their performance niche becomes obsolete, or when another breed becomes better at the job for which they were bred. At least 107 equine breeds are known to have become extinct (Hall, 2005, p. 26).

E. Gus Cothran of Texas A&M University has studied the DNA in blood samples from horses all over the globe and has offered insights on the relationship of Banker Horses to other breeds. His work provides a scientific basis for discussing the origins of these horses. He says that Corolla horses are a unique population with low genetic variability—only 29 alleles—and show no close resemblance to any known horse breed. This lack of diversity is probably the result of genetic loss caused by a shrinking breeding population (Cothran, 2008). This analysis indicates that they have bred only to each other for quite some time, to the point of inbreeding. He wrote. "Rather than being feral horses with a mixture of domestic breeds, they are in effect a breed unto themselves" (Ives, 2007, p. 13).

The horses of the North Carolina coast—Corolla, Shackleford Banks, Carrot Island, and Ocracoke—have more genetically in common with one another than with other breeds. Shared ancestry could mean that they descended from a single population that was placed on the island or arrived by shipwreck. Conversely, it could indicate that each herd originated from horses on the nearby mainland that were then of similar type and breeding. Their genes and conformation show Spanish ancestry, but do not clarify just what that ancestry was or when and how ancestors arrived on the islands.

Cothran (personal communication, April 20, 2011) commented on the presence of a Spanish genetic marker in the Shackleford herd:

> It can be difficult to determine whether there is direct Spanish ancestry in a lot of the wild populations, because breeds like the Narragansett Pacer were important in the development of some of the North American breeds. . . . in a feral herd, you don't know whether the marker is from old Spanish bloodlines, or is from these North American breeds which have some Spanish heritage.

There is variation within any breed, and even the original Jennets showed differences in conformation and attributes. Sponenberg and Reed (2009) developed a 5-point matrix that scores traits related to Colonial Spanish type. Horses are graded from 1 to 5 on specific physical attributes. The examiner scores each feature of the horse, adds the results, and then divides the total by the number of items. A score of 1 indicates strongly Iberian conformation, 2 is acceptable, and 3 is marginal. Ratings of 4 and 5 indicate significantly non-Spanish features. In a purely Spanish herd, scores should cluster around 1 and 2, with very few in 4 and none in 5.

An individual horse of a non-Spanish breed may have a low score, but when more than 80% of horses in a given population have low scores on this scale, the herd is likely to be Iberian in origin (Sponenberg & Reed, 2009). Herds with 50% or fewer Iberian types either do not have Spanish heritage, or interbreeding with unrelated breeds has diluted the gene pool until Iberian genes are no longer widely expressed.

As a species, all horses have five or six lumbar vertebrae or show a partial fusion of the fifth and sixth. Many or most Colonial Spanish horses (but not all) have five lumbar vertebrae, but so do many other breeds. This trait is very common in Arabians and regularly occurs in other breeds as well, such as Thoroughbreds. Indeed, the Colonial Spanish Horse possibly owes his vertebrae count to the ancient proto-Arabians in his lineage.

Just because a horse has Colonial Spanish characteristics does not mean it can be considered part of the breed. Sponenberg (2011) writes that the frame overo pattern (a base of a solid color like black, bay, or chestnut splashed with irregular white patches which rarely extend to the back, lower legs, and tail) "is almost entirely limited to North American Colonial Spanish horses or their descendants" (Sponenberg, 2011). The author's registered Paint Horse, Chics War Eagle (Leonardo), demonstrates the frame overo pattern. His forebears include paints, Thoroughbreds, Quarter Horses, and, apparently, Spanish ancestors that bestowed his flashy coloration. Leonardo's pedigree traces to many of the horses mentioned in this book—Snip, Janus, Bulle Rock, the Godolphin Arabian, the Darley Arabian, the Byerley Turk. These great foundation horses are also prominent in the pedigrees of Morgans, American Saddlebreds, Tennessee Walking Horses, Standardbreds,

The rebellious element of Ocracoke Island: teenagers riding bareback on fast horses, cigarettes in hand. Even though the Scouts weren't immune to the usual vices, Howard often mentioned the lack of juvenile delinquency on Ocracoke, which he saw as a direct result of involvement with the ponies. Photograph courtesy of the Ocracoke Preservation Society.

Quarter Horses, and Thoroughbreds. The gene pool was smaller in colonial days, and wealthy breeders drew heavily on the services of a handful of elite stallions, resulting in kinship among outwardly diverse American breeds.

Virtually all the Corolla and Shackleford horses are sorrel, chestnut, bay, brown, or black. The graying gene was once present in the Corolla herd, but it is unknowable whether this color was native or introduced. The Shackleford herd contained duns, buckskins, and palominos before 1996, when North Carolina health officials euthanized 74 horses that tested positive for equine infectious anemia. A few Corollas exhibit the white ticking of the rabicano trait, and at least two have spashy white sabino spots on their bellies. Many of the Carrot Island horses are line back red duns, and the Cedar Island herd has two buckskin mares, mother and daughter. All are small horses, mostly 12–14 hands (48–56 in./1.22–1.42 m), with Colonial Spanish conformation.

Historical photographs and written accounts of Banker Horses describe them much as they are today in color, size, and conformation. The modern herd at the Ocracoke Pony Pen, however, shows evidence of crossbreeding. The Ocracoke herd has two dominant colors not found in any other Banker herds, tobiano pinto and gray. As a dominant trait, tobiano is expressed in nearly every generation, sometimes so minimally that the white markings may go unnoticed (D.P. Sponenberg, personal communication, December 4, 2012). Gray is also a dominant trait—every gray horse has at least one gray parent. When grays or tobianos suddenly appear in a herd of horses where they have been historically absent, the trait has been brought in by an outside horse (D.P. Sponenberg, personal communication, December 4, 2012). It appears that gray and tobiano both appeared in the herd in the 20th century. Dunbar, (1958) gives evidence of this crossbreeding when he wrote that the Ocracoke ponies had been "recently improved and should not be considered Banks ponies or Tackies."

Two gray stallions are known to have been introduced to Ocracoke. Phillip Howard, a native Ocracoker, relates how his great-grandfather, James Howard, brought a two-year old gray Arabian stallion to the island "from somewhere on the mainland" in the 1880s (Howard, 2002). Jim Howard kept White Dandy in Ocracoke Village and may have bred him to Banker mares. Because White Dandy was an Arabian, any foals sired by him would be half-Arabian crossbreds. In the 1980s, Cape Hatteras National Seashore used the services of a purebred gray Andalusian stallion named Cubanito "to breed with Banker mares to insure an adequate gene pool" (Henning, 1985, p. 3). Unfortunately, the Andalusian breed is only distantly related to the Colonial Spanish horse, and Cubanito's introduction was essentially a crossbreeding that undermined the genetic integrity of the herd. Before horses were brought in from Shackleford Banks and Corolla to increase genetic variability, the average horse in the Ocracoke herd was about 15 hands (60 in./1.5 m), considerably larger than a Colonial Spanish Horse.

So what exactly are Banker Horses? We know that they descended from Spanish lineage, probably through the Chickasaw Horses brought from the West by Indian traders, Seminole Horses brought north from Florida, and Spanish horses from the West Indies introduced by British colonists. It is possible that the Narragansett Pacer, the Galloway, the Irish Hobby, and other light European saddle horses contributed their genes to the Banks herds centuries ago, and there is some evidence that Quarter Horses were added to the Pea Island herd more recently.

If outside genes were added episodically to "improve" the bloodlines, a common practice among stockmen, these contributions would likely have come from breeds or types already found in, or frequently imported to the Carolinas. Outcrosses appear to have been infrequent, however. Although there is no way to know for sure, it appears that these horses have bred primarily with one another, or with horses of similar breeding introduced from the mainland since the late 1600s or even earlier, and have developed into a unique and rare strain of Colonial Spanish Horse. The Carolina Banker Horse represents a critically endangered strain of perhaps the oldest surviving American horse breed. Conant et al., writing about the Colonial Spanish horses of the Southeast, commented that "these relic populations are worth preserving, both for their genetic as well as their historical heritage as descendants of the first modern horses in the Americas" (2012, p. 62). In 2010, North Carolina designated the Colonial Spanish Mustang as the official state horse.

Shipwrecks were commonplace on the hazardous shoals of the Carolina coast, many of them were undocumented. It is possible that horses swimming from wrecked ships might have originated or augmented the North Carolina barrier island herds. Any horses reaching shore from a shipwreck would most likely have been bred in the New World. Before the mid-1700s, they would have been Spanish horses from the West Indies (or, less often, European horses) bound for the Eastern Seaboard. Thereafter, any shipwrecked horses

Hitched to the porch rail, horses wait patiently for their riders to return. Photograph courtesy of the Ocracoke Preservation Society.

would have most likely originated in New England or another British colony on the North American mainland.

To warn ships of dangerous shoals, the Ocracoke Lighthouse was constructed in 1823. The white brick structure remains one of the oldest functioning navigational aids in the United States. Confederates extinguished the light in 1861, and Union forces resurrected it in 1864. After 1874, the federal government sought to improve marine safety by extending coverage by the U.S. Lifesaving Service to the Outer Banks. Based at stations positioned various distances apart along the coast, its men monitored the beaches and struggled to rescue sailors from stormy seas. Where he could, a surfman walked half the distance to the next station, scanning as he went; turned a key in a clock or exchanged a token with another surfman; then walked back, no matter whether nature offered raw, biting winds, blazing heat, or hurricane. The more horrendous the storm, the more likely their services would be needed.

The Ocracoke Pony Pen confines the ponies to a tiny fraction of their historic habitat.

Unflappable, loyal, sensible Banker Horses worked in partnership with the men of the Lifesaving Service. In the very early days, the Lifesaving Service expected its men to patrol the beach on foot and to push or pull the surfboats and beach apparatus carts unaided. Each of those burdens weighed half a ton/450 kg or more, and the stations were miles apart, so the crews had a strong incentive to provide their own mounts and draft animals at no cost to the government. When the service officially authorized horses and their accommodations, keepers or district inspectors may have bought horses from local stockmen. The tireless horses hauled equipment over the sand to shipwreck sites and often entered surging surf in the most violent tempests as their riders rescued people from the waves. In one famous act of heroism, Rasmus Midgett, a 48-year-old surfman from the Gull Shoal Station patrolling the beach around 3 a.m. at the height of the Category 3 San Ciriaco Hurricane of 1899 (Barnes, 2007), pulled 10 men from the wreck of the barkentine *Priscilla* aided only by his Banker horse Tom Creef (Henning, 1985).

One reason Portsmouth was established and a fort was built nearby was to protect the area against attacks by Spanish pirates and privateers in the mid-18th century (Wiss Janney Elstner Associates & John Milner Associates, 2007). Throughout the 1740s, Spanish privateers attacked ships off the Outer Banks. (In the spring of 1741, two Spanish ships captured six vessels in 10 days.) They established camp in Ocracoke, burned houses, and killed cattle. In 1747, they briefly seized Beaufort. The next year, they sacked the town of Brunswick, on the lower Cape Fear River, which never recovered.

The short, mostly fictional golden age of piracy on the Outer Banks occurred a few decades earlier, between 1713 and 1718, when brigands such as Christopher Moody, John Cole, Robert Deal, Charles Vane, Richard Worley, "Calico Jack" Rackham, Stede Bonnet, and Edward Teach, the infamous Blackbeard, terrorized—and occasionally visited—the

Visitors can usually watch horses grazing and interacting in the front paddocks of the Ocracoke Pony Pen. Here, Little Doc, the 1986 son of Cubanito and Old Paint, crops the spring grass within his enclosure. He is a full brother to Lindeza.

area (Stick, 1958). They ranged all over the western Atlantic, but sometimes used the barrier islands as hideouts and preyed on cargo-laden ships around Ocracoke Inlet and the Cape Hatteras bottleneck.

Blackbeard, probably an Englishman who served as a privateer during Queen Anne's War, turned to piracy in 1716, and by 1718 he employed about 400 men on four ships. He commandeered a 14-gun French slave ship, the *Concorde*, and made her his flagship, increasing her armament to 40 guns and renaming her *Queen Anne's Revenge*.

Blackbeard arrived in North Carolina in June 1718, sought and received a royal pardon from Governor Charles Eden (who apparently benefitted from the buccaneer's spoils), and settled in the colony's capital, Bath, where Eden lived. His presence there, as well as his alliances with other lawbreakers, created a state of alarm among regional maritime interests. When Eden ignored their concerns, frustrated Ocracokers and other coastal residents sought help from Governor Alexander Spotswood of Virginia. Spotswood seized the opportunity to eliminate a criminal while embarrassing a corrupt colleague.

Governor Spotswood hired two light, fast sloops that arrived off North Carolina's Outer Banks on November 21. It was Lieutenant Robert Maynard, aboard the *Ranger*, that killed Blackbeard on November 22, 1718, in a furious battle at Ocracoke Inlet. The pirate allegedly received five musket-ball wounds and more than 20 sword lacerations before he succumbed. In only two years of pirating, Blackbeard had taken more than 50 ships. Maynard sailed up the Pamlico River to Bath with Blackbeard's severed head swinging from his bowsprit. On November 21, 1996, a wreck that may be *Queen Anne's Revenge* was discovered by the private firm Intersal near Beaufort Inlet under 20 ft/6 m of water.

Some believe that the wild horses originated with pirates who brought them to Ocracoke as a source of mounts or even food if necessary, but scholars discredit this idea. Though pirates probably did kill and eat free-roaming livestock, the herds were evidently well established before pirates arrived. Contrary to popular lore, most pirates were not native to the Banks, and the local population had little involvement with piracy.

Pirates were not the only ones who dined on meat stolen from the barrier islands. Stock was abundant and largely unguarded. Mariners of all ethnicities frequently availed themselves of this source of sustenance. British forces took hundreds of cattle and sheep from Ocracoke and Portsmouth during the War of 1812 alone (Stick, 1958).

The islanders developed a unique dialect with turns of speech distinct from those of mainland North Carolinians (Howren, 1962). On Ocracoke, *high tide* is pronounced \häi 'täid\, usually, inaccurately rendered *hoi toide*. A dragonfly is a *skeeter hawk*. The term *abreast* is used as a preposition to mean "across from," as in "'He lives up here abreast the post office' and 'She went aground abreast the island'" (Howren, 1962, p. 174). A grove of trees is a *hammock* \'həm•ək\. A single cow or bull is a *cattle-beast*. *Airish* means breezy. *Begombed* is soiled, as in "Your shirt's begombed with grease." To *mommuck* is to damage or destroy. Edible shore birds such as curlews and sandpipers are *sea chickens* (Impact Assessment, 2005, vol. 2, pp. 509–511).

Visitors arrived by mail boat, which could carry up to 35 passengers and freight in addition to letters and packages (Goerch, 1956/1995). Around 1915, there were "sailing vessels (sharpies) with auxiliary motors" (Goerch, 1995/1995, p. 55) that left Washington, N.C., Wednesdays and Saturdays at 8 p.m. for the overnight trip to Ocracoke. The Taylor Brothers started a daily ferry service between Atlantic and Ocracoke in the spring of 1960. In 1963, the state took over the ferry and shifted operations to Cedar Island (Goerch, 1956/1995).

Every family on Ocracoke had at least one horse, and each of the free-roaming horses had a nominal owner. But many, if not most, of Ocracoke's horses lived their lives, birth to death, running wild, breeding at will, deliberately handled by human beings only during pony pennings.

Living wild sometimes posed dangers to the horses. Sometimes one would get stuck in a marsh until his herdmates rescued him by pushing him out. During one storm, a band gathered at the tip of a peninsula. Floods cut off their escape, and every one drowned (O'Neal et al., 1976).

A Park Service ethnohistorical research project describes the life of a typical Outer Banker. Before the advent of cars and buses, islanders traveled across land via horse and cart. A villager described "flat carts" that allowed one to sit on the edge with "your feet dangled down between the shaft and the horse," that is, along the front edge of the cart where the horse's harness attached. Bankers also rode horses bareback after catching them on the beach. A villager recalled when she and her brother were sent after one of her father's mares. They approached "Mary" with bridle in hand, but the mare was in heat and a stallion challenged the children:

> We were both going to ride her bareback. But here come this old stallion. He'd rare up on his hind legs and scream. Honey, that was two scared kids. You talking about somebody praying. We'd get on one side and when he'd come at us, we'd go on the opposite side. We kept inching along until he ran back to his brood. We'd

Lawton Howard, a friendly bay pinto gelding, comes to the fence to solicit attention from visitors. Lawton, a son of Santiago, is closely related to many of the horses in the herd and therefore was ineligible for the Park Service breeding program. He was gelded as a colt.

be a-crying telling Dad about it. He said, "Well, next time I'll give you a gig and gig him like that. And, very calmly, mother said, "No, Jim, there won't be a next time." (Impact Assessment, 2005, vol. 1, pp. 243–244)

According to one account, boys of Hatteras Village in the 1920s spent much of their free time

chasing horses on the beach and breaking them. It was easy to break a two-year old in the soft sand; "he'd settle right down and get easy." Horses were typically caught on "Friday or Saturday mornings," ridden on the weekends and "turned loose Mondays—let them graze on the beach the rest of the week." (Impact Assessment, 2005, vol. 2, p. 307)

In 1920, federal agents built concrete dipping vats along the Outer Banks in an effort to eradicate tick-borne disease. The U.S. Department of Agriculture required stockmen to round up free-roaming cattle and herd them through a toxic "creosote mixture" (Impact Assessment, 2005, vol. 1, p. 214) every other week. Many of the solutions used contained compounds of toxic, carcinogenic arsenic (Ellenberger & Chapin, 1919; Thomas, Rhue, & Hornsby, 2009). Stockmen marked the rump of each dipped animal with a splash of green paint and searched for unmarked animals. A local described the dipping process:

They drug them through that dipping vat. They'd swim through [a formula] like Lysol mixed in five-gallon jugs. . . .

And DDT—they'd empty the solution, pull the plug and it would run right out in the sand. So you know it got in the water. That could be the reason the cancer rate is so high. Probably a dozen in Avon right now with it. And that's a lot for a small village like that. (Impact Assessment, 2005, vol. 1, pp. 214–215)

Teenaged boys were happy to participate. Said one resident, "We boys would get a kick out of being cowboys, rounding up the cattle and horses and driving them through this dipping vat" (Impact Assessment, 2005, vol. 2, p. 307).

Apparently many other Bankers resisted the stock-dipping mandate. Lee (2008, pp. 104–105) writes,

Islanders responded first by dynamiting the concrete vats, only to repair them, apparently under threat of prosecution. When agents came to oversee dips in the first years of the program, villagers passively resisted by refusing to participate. After finally compelling villagers to contribute to the federal effort, many of the animals proved too feral to be corralled into pens . . . the county offered villagers five dollars a head to shoot any cattle that could not be dipped, and locals eliminated about two-thirds of the nearly 400 cattle in Dare County before the USDA declared the area tick-free after the 1924 season.

At Cape Hatteras, men gathered free-roaming horses and cattle twice yearly for identification and sale. One elderly resident remembered "'When I was young, there was cattle and horses just roaming around everywhere, like the old west. . . . The old guys used to round the cattle up all the way from Avon. They would bring them to the pound to mark the cows and brand the colts'" (Impact Assessment, 2005, vol. 1, p. 150). Leland Tillett (b. 1913) recalled cattle drives from Oregon Inlet to the Virginia line, and Ermie Bowden (b. 1925) recalled that there were more than 3,000 cattle just from False Cape to Carova (Eaton, 1989). Chater (1926) estimated that there were 5,000–6,000 horses on the Banks. A decade later, all the free-roaming livestock had been removed from the Banks except for small populations on Ocracoke, Currituck Banks, and Shackleford Banks. The last semi-feral cow was shot in 1938.

On Ocracoke, the July 4th pony penning was a long-anticipated celebration, a festival of hard work and hard play. Each Ocracoke family had its own brand, and foals were matched to mothers and emblazoned with an owner's logo. Banker Horses were in demand on the mainland and many were sold off the island during these roundups. Once off the island they had to adapt to confinement and transition from a diet of marsh grass to one of hay and grain, but most were domesticated with relative ease.

Traditionally, the roundup began on July 3, when a handful of Ocracoke's cowboys would ride north, to the periphery of the wild herds that grazed in places with such fanciful names as Tar Hole Plains (Henning, 1985). The horsemen would camp overnight near the sound, and in the morning the roundup would begin. They rode in McClellan saddles, a style devised by General George McClellan, adopted by the U.S. Army in 1859, and used in the Civil War (O'Neal et al., 1976).

Captain James W. Howard, who ran the Hatteras Inlet Lifesaving Station for many years, bought a dapple-gray 2-year-old Arabian colt from the mainland in the late 1800s. An 1888 photograph shows 49-year-old Howard astride the high-headed mount, named "White Dandy" by his son Homer. Homer and White Dandy were renowned for their ability to round up more than 200 wild ponies without assistance (Henning, 1985).

Homer's son Marvin wrote,

On "White Dandy" Homer on many occasions started at the north end of the island in the cool of the morning, driving the herd of wild ponies south. He rode merrily along across Tar-Ho[l]e Plains. There he would come upon a second herd of ponies headed by "Old Wildy," a long, rangy stallion. This herd, too, he would start driving southward. The third herd he encountered at Scraggly Cedars, then the Great Swash. After passing Great Swash he came to Knoll Cedars where the sheep pen used to be, and from there on southward the driving got touchy and more strenuous for the herds from the north were reluctant to go farther south and would try to cut through the thickets or sand hills back northward.

There were about two-hundred wild ponies in those days. They had to be driven over sand hills, through bogs, across creeks, through marshes, and through woodland thickets of myrtle, cedar, oak and yaupon. At about ten o'clock in the morning of pony-penning day, the horses could be seen spread out on the plains around "First Hammock Hills," just north of Ocracoke Village. Each little band was headed by a tough and stringy stallion. They ran hither and thither, their manes and tails flying, heads held high, ears pointed forward, and necks arched to meet a foe. And whenever the stallions met, they did battle—biting, kicking, pawing—until the rider closed in. Then, they veered off from each other, returning to their herds. It was no easy task to drive these wild ponies sixteen miles [sic] southward to the corral in Ocracoke Village. (Howard, 1976, p. 26)

Horse Pen Point was the destination for many years, though other locations were favored throughout the 20th century.

"Interest in the once-wild Banker Ponies is a long tradition in the Howard family," says Philip Howard (2002), grandson of Homer, son of Lawton, and nephew of Marvin.

My father often told me about the time in 1926 when he was 15 years old. It was July and the annual Independence Day pony penning was in jeopardy of not happening because several of the young men were squabbling about something and no one was prepared to round up the horses. My dad and his best friend, Ansley O'Neal, though still teenagers, decided that they were old enough to tackle this responsibility. They mounted their ponies on July 3 and rode all the way to Hatteras Inlet (this was long before there were any paved roads on the island) where they camped out under the stars. Early the next morning the two boys began chasing the first small herd southward, toward the village. As they encountered each succeeding herd they forced them to join the others. Occasionally some of the animals would swim out into Pamlico Sound and make the boys' job much more difficult. Finally, after a grueling day of hard riding in the blazing summertime sun Lawton and Ansley rode proudly into the village behind several hundred stampeding Outer Banks ponies. It was a proud day for them both, and a fond memory for my father until the day he died.

Sometimes the cowboys would break the horses "Wild West style" before shipping them to the mainland. The designated animal was moved into an individual pen and crowded against the rail. The stockman grasped his tail through the fence, tied his head to a post, and placed a blindfold over his eyes before working a saddle onto his back. A cowboy would mount to take the ride. Off came the blindfold, and the horse was released. The frantic pony would run, buck, pivot, twist, rear, and sunfish, but the tenacious cowboy usually stuck to the horse's back like a leech. At last the pony would acquiesce, sweating and blowing, and let the rider direct him around the pen.

Homer Howard was skilled in breaking the willful, powerful wild stallions:

> To catch a wild stallion with nothing but bare hands took wit, agility, strength, and stamina. Homer would walk quietly through the mares, slapping them on the rump, working his way between them slowly, gradually—getting closer and closer to a great stallion—crouching panther-like, ready, alert—and in a flash he was astride the stallion, holding its mane with his left hand, throwing his elbow over the horse's withers, hooking his knee behind the elbow of the horse's front leg, reaching out with his right hand to catch the horse's lower face just above the nostrils, clamping down tight, and sticking there with the tenacity of a bulldog. The stallion would rear, pitch, squeal, snort, paw the air for thirty of forty minutes, but finally, out of wind, tired, and afraid, he stopped his violent struggling. Slowly the horseman eased his grip; immediately, the stallion lunged and reared. Only after several attempts did the horse admit his defeat. "Old Widdie," "Guthrie Sam," and "Rainbow" and others were truly great stallions and had the spunk and grit to put up terrible battles. Their tusks, or cutting teeth, were long from age and could be used to cut and slash, and their forefeet and rear hoofs held a wicked kick. (Howard, 1976, p. 26).

Other horsemen trained Banker Horses with kindness, rewarding them with sweets, petting, and scratching of the itchy spots. Sometimes the first mounting was accomplished in the sound, with the horse belly-deep in water. This way, the horse's movements were restricted, and a thrown rider would have a soft landing. Horses gentled before riding usually put up no resistance.

Shipping the horses off Ocracoke presented a challenge. Ocracoke is 30 mi/48 km from the mainland, and ponies leaving the island were transported on flat barges, freight boats, or fishing boats. In one incident, two horses broke into the engine room of a fishing boat when a storm panicked them. Occasionally horses fell overboard and drowned (O'Neal et al., 1976).

Most Ocracoke natives enjoyed the pesky ponies. Hatteras Island had more villages, and people were less tolerant of livestock making themselves at home in their yards. Ocracoke Island had but one human settlement, at the south end, and Ocracokers comfortably shared their village with the animals.

The ponies wandered into town looking for handouts and would devour vegetable gardens if the gates were left open. One apparently developed a taste for fried fish and would reach his head into open windows to devour the family supper (Henning, 1985). Occasionally a family would be awakened long before dawn by an odd noise only to find an itchy pony scratching himself on the corner of the house. Occasionally a herd would stampede through town. They were small, beautiful animals with long, flowing manes and tails, tractable temperaments, and smooth gaits. The Ocracokers used them to pull carts and for plowing and rode them under saddle or bareback. Overall, the villagers and the horses coexisted in a relationship of mutual toleration and positive regard.

Over the years, horses from the mainland were introduced to improve the island stock. In the early 1900s, locals thought that the fleet, maneuverable Banker Horses would make

A horse's posture communicates volumes to the other horses in the herd. This Cumberland Island mare (center) moved the 2-year-old stallion (left) away from her foal by flattening her ears, raising and tossing her head, moving backwards, and shifting the weight off a hind leg. The juvenile stallion did not question her authority, but yielded ground by moving away until she dropped her threat, head low and tail clamped. If he had not moved fast or far enough, the mare would have followed with a series of forceful kicks or would have whirled to chase or bite the young stallion. Her foal presses close for her protection; his posture and the set of his ears indicate that he is curious and relaxed, but aware of the potential for danger. The foal to the right, who had been playing with the young stallion before the mare moved him away, watched attentively with concern but not fear. He was ready to move if the mare threatened him, but knew she was unlikely to unleash full violence on a young foal.

superior polo ponies if only they were taller. In 1925, the stallion Beeswax was imported to Ocracoke Island for stud. Beeswax was the son of the champion polo pony Christopher Columbus, an English Thoroughbred imported to sire polo ponies and cavalry mounts. Beeswax was not rugged enough to withstand the rigors of life in the wild and was kept confined on a diet of hay and grain. His foals were popular, and most were sold to mainlanders for riding and polo. Elisha Ballance recalled, "'Back in the '20s, I believe, [David] Keppel brought a stud here, and he looked like and I think he was a Thoroughbred. He was a fine horse and couldn't live like the Bankers, had to be kept up and fed'" (Henning, 1985, p. 8). Henning maintains that the blood of Beeswax has been lost to Ocracoke over time, and none of his Thoroughbred genes remain in today's Banker herd.

Ballance also remembered large herds fortified by the introduction of Banker bloodlines from Hatteras Island, probably following the 1935 prohibition against free-roaming livestock. "There were plenty of horses up until the time I went into the Coast Guard in 1938," he said. "Some had been brought from Hatteras and turned out with the ones here, but they were Bankers, too, they got along" (Henning, 1985, p. 8). O'Neal et al. (1976, p. 24) describe a smaller herd: "As late as 1939 there were fifty to one-hundred ponies on Ocracoke, half of them continuing to run wild along the Banks, the other half broken in for riding by the island boys." They went on to say that there were no annual public roundups during the War years of the 1940s, and the "spectacular Fourth-of-July events have never been revived" (1976, p. 25).

Roaming horses were eradicated from Dare County, to the north, beginning in 1935 and forbidden in Currituck County in 1937, though never completely removed. Only the barrier chain between Hatteras Inlet and Beaufort Inlet—Ocracoke, Core Banks, and Shackleford Banks—retained wild herds.

A 1940 news article related,

> When soil conservation forces of the federal government four years ago began their most ambitious endeavor to check sand erosion along the Carolina banks, they soon realized that the banker ponies were rapidly eating the tough beach grasses and shrubs which they were carefully planting in long miles of brush panel fences to hold back the ocean.
>
> The suggestion that the banker ponies ought to be exterminated, in the interest of anchoring the shifting sand dunes, brought forth so many objections from the animals' friends that the proposal was abandoned. But more care was taken to keep the ponies away from the grass fences.
>
> Today there are said to be more ponies along the banks than there were a decade ago; but they still fall far short of the many thousands that roamed the sands years ago. ("Vegetation Scarcity Blamed on N.C. Oceanside Ponies," 1940, p. 10)

The horsemen suspended public roundups during the war years, and afterwards, the pony pennings were never as big or exciting. The number of ponies on the island was declining rapidly as well, from 200–300 in the 1800s (O'Neal et al., 1976) to 70 in 1956 (Brooks, 1956) to 35 in 1957 by order of the Park Service (O'Neal, 2008) to an all-time low of nine in 1976 (Henning, 1985).

Marvin Howard retired from his military career to his home on Ocracoke Island and organized the world's only mounted Boy Scout troop. Howard, who found great satisfaction in working with both children and horses, founded Troop 290 in 1954, and most of the boys on the island enthusiastically joined. For the next 5 years, Howard served as scoutmaster to a total of 25 boys and advised them on the care and training of their mounts (O'Neal, 2009).

For the barefoot boys of Ocracoke, the Scout troop and the ponies were the focus of their lives. Each boy began by selecting a wild pony to catch, train, and ride. Each pony, though living free, had an owner. Some were privately owned, and some were legally the property of the Park Service, which had acquired most of the island for inclusion in Cape Hatteras National Seashore. The price was $50 per pony, a steep sum for a young boy on a poor, remote island in the 1950s and equivalent to more than $400 today (U.S. Bureau of Labor Statistics, n.d.). To earn money, the boys spent countless afternoons mowing lawns and helping fishermen with the day's catch.

The boys knew each of the wild horses well, and after much observation, deliberation, and daydreaming, they would set their sights on particular horses, often fractious young stallions. Usually two boys set out after the chosen pony, which had no desire to be captured. The wild bands evaded the boys at every turn, often venturing out into the marshes or the water to escape. Because castrating a colt makes him more tractable, most male horses are gelded young, but the Scouts largely preferred to ride stallions.

Riding shoulder to shoulder and in groups, the boys galloped bareback down the beach on stallions that had been rivals in the wild. The feisty ponies were used to having their own way and often resisted domestication, especially at first. One time, so the tale goes, a rowdy stallion aptly named "Little Teach" bucked a Scout from his back, kicking him in the head for emphasis. A vacationing doctor, slightly inebriated, successfully sewed up the scalp wound with 44 stitches (Henning, 1985).

The Scouts followed many time-honored Ocracoke techniques of horse breaking, including mounting blindfolded horses as they stood belly-deep in the sound. They experimented with filling an old pair of pants with sand and tying it around the pony so that he could expend his bucking energy on an inanimate object rather than a Scout. Unfortunately, the ponies usually dislodged the pants and trampled them. The Scouts realized that it was best not to reinforce the trampling of what they bucked off in case one of them might be the next victim.

Howard coached the boys in training methods and horsemanship, and they met most of their Scouting requirements on horseback. The boys also had the opportunity to show off their skills at the Pirates' Jamboree, a celebration that featured races and other tests of riding ability. Annually the troop would compete in the horse races held on the beach at Hatteras and at Buxton, which was no small undertaking.

The boys would set out early and ride a total of 26 mi/42 km to reach Buxton. To cross Hatteras Inlet, a dozen boys would lead their blindfolded ponies onto the little ferry and hold them on the open deck of the rocking, groaning boat for the 40-min crossing, a situation that would have panicked other horses. The boys then rode from Hatteras to Buxton, about 12 mi/19 km. After arriving at Buxton, the boys would race in four quarter-mile (400-m) heats, often besting stiff competition that included Arabians and Quarter Horses (Henning, 1985).

Five hundred to 600 cattle still roamed Ocracoke Island during this period, and the Scouts became skilled at sorting and penning them for branding. The fee for filing a brand with the County Register of Deeds was the same as it had been for many years—$0.10. At the July 4th pony penning celebrations, they displayed their superb horsemanship skills for the benefit of the visitors.

The Ocracoke mounted Scouts captured national attention when they were featured in *Boy's Life* magazine (Brooks, 1956) and in a children's novel, *Wild Pony Island* (Meader, 1959). The Scouts also helped around town and served as mounted honor guards for the Coast Guard. During the summer, the boys helped keep Ocracoke's mosquitoes at bay by spraying the marshes with insecticide. Astride sure-footed marsh ponies, they were able to penetrate the wide flats of muck far more easily than anyone else.

The sturdy Banker Ponies cost virtually nothing to maintain. The Scouts could choose to let the horses run free to be captured when needed, because they remained sleek and well-fleshed on a diet of marsh grass. Most of the boys, however, opted to build stalls in their back yards to keep their horses close at hand. Unlike wild ponies, a penned Banker Pony does not have the luxury of grazing 17 hr a day on dense wild grasses. As the ponies got less grazing and more exercise, they needed a dietary supplement such as grain, a concentrated energy source. When the boys offered sweet feed, a tasty grain-and-molasses mixture that most horses relish, the ponies did not know what to do with it. The Scouts initially had to place it in their mouths until they noticed the sweet flavor and realized that it was food. It was not long before they discovered the pleasure of other flavors as well. Many Scout ponies developed a taste for soda pop.

The horses often formed close bonds with their owners and when released to run free would often visit them in the village when they wanted human companionship. Some ponies would come running at the sound of their owner's whistle.

When Cape Hatteras NS took over Ocracoke Island, it was a mixed blessing. The new status would offer some protection against the hotels, cottages, restaurants, miniature golf courses, and tourist attractions that had overtaken many East Coast beaches. But the Park Service did not recognize free-roaming horses as repatriated wildlife and did not want them competing with the native wildlife that it was charged to protect. It removed the cattle, pigs, sheep, and goats that roamed the island by the late 1950s and saw no reason to treat horses any differently.

Ocracokers loved their horses, however. To them, the free-roaming ponies were an important element of the island's character. Lawmakers and regulators argued that not enough was known about the horses' past to support claims that they had historical value. There were high emotions on both sides of the argument.

Marvin Howard, whose family had been involved with the Banker Ponies since the 1700s, worked relentlessly to save the Ocracoke herd. During the late 1950s, at least eight ponies died of Eastern equine encephalitis, a mosquito-borne viral illness. Rudy Austin, an original Scout, lost his two mounts, Diabolo and Blaze, to the disease (O'Neal, 2008).

When Highway 12 was opened on Ocracoke in 1957, the posted speed limit of 50 mph/80 kph posed a new danger to the horses. The Scouts petitioned to fence a large pasture for the horses as a sort of a compromise—the horses would not roam entirely free anymore, but they would remain on the island to be enjoyed as a reminder of Ocracoke's bygone days.

The North Carolina General Assembly finally, but confusingly, ended free range on parts of the Outer Banks not already under the stock law. N.C. General Statute 68-42, which took effect July 1, 1958, made two exceptions: "horses known as marsh ponies or banks ponies" on Ocracoke Island and Shackleford Banks. The next year, the General Assembly directed the Sheriff of Carteret County to kill or remove all other livestock from Core and Shackleford Banks (An Act To Provide for the Removal of Cattle Remaining on Core Banks, 1959), which the state Supreme Court upheld in *Chadwick v. Salter* (1961). The exception to these exceptions was a group of up to 35 horses owned by the Scout troop, which the law implicitly required to be fenced. But G.S. 68-43 made the Scouts' horses subject to removal by the state no matter whether they were fenced. By this time, however, the Park Service owned most of Ocracoke, and as a federal agency it was not bound by state law to let horses live on its property within or without fences.

The Park Service eventually granted the Scouts a special-use permit and provided fence posts. Residents raised money for fencing and for the first year of supplemental feeding. The state of North Carolina also contributed funds toward the new lifestyle of the ponies for the first year. The animals were finally penned in 1959. It was a fairly easy task to put the herd within fences, but keeping them contained proved challenging. When the ponies wearied of confinement, they simply knocked over the posts and broke the wire. The Scouts would leave school to recover the ponies and mend fences. The Scouts enjoyed this task— so much, in fact, that they would often sneak back to the pasture after dark and push over posts, ensuring another holiday from school (Henning, 1985).

Two horses in the Ocracoke Pony Pen communicate without making a sound. Lawton Howard, left, peers around the corner curiously before approaching to investigate a new visitor. Lindeza, right, is uncertain whether she should welcome the stranger. The set of her neck, head, and ears denotes uncertainty, and presenting her hind legs toward the photographer may denote insecurity.

In the mid-1960s, the Boy Scouts of America ruled that if the Ocracokers were to maintain a mounted troop, its members would have to carry insurance. The boys could not afford insurance, and Ocracoke's mounted Scout troop dismounted after only about 10 years.

The Park Service took over management of the ponies in the late 1960s, when the herd was on the verge of extinction. In 1973, Park Ranger Jim Henning was transferred to Ocracoke from Bodie Island. He took an interest in the herd stallion, also named Jim. With his wife, Jeannetta, Henning wholeheartedly devoted himself to the resurrection of the herd.

When the Hennings first arrived, the ponies were malnourished, wormy, and in dire need of veterinary care. Dr. Jasper Needham, a veterinarian on Hatteras Island, came in to vaccinate the animals, trim hooves to resolve gait abnormalities, and deworm the animals.

Internal parasites are more of a problem for domestic horses than for their wild counterparts. Locals say that the marsh grass diet serves as a natural dewormer and is effective also for cattle. More likely, wild horses have fewer parasites because they range over a larger area than their domesticated relatives and avoid areas contaminated by manure. Penned horses feed in areas where others defecate and are likely to pick up worm eggs while grazing. By this time the Ocracoke herd had been confined to a 180-acre pen and exchanging parasites for 14 years.

Chutes and pens made veterinary care less of a rodeo event. Before they were built, dewormer medication was given in a large trough mixed with food. The horses would compete over the offering. Often the dominant animals received too much, subordinates not enough. Using a chute, caretakers could administer dewormer in precise doses via oral syringe.

By the 1970s, fertility decreased until there were no births in one 5-year span. The last remaining Ocracoke stallion was genetically incompatible with three of his mares. These mares could deliver healthy foals, but antibodies in their colostrum, or first milk, would attack and destroy the newborns' red blood cells, a condition called neonatal isoerythrolysis. At the time this condition was attributed to excessive inbreeding, but this assumption was later proved false. Blood-type incompatibility is less likely in an inbred population because closely related horses tend to have the same blood types.

The Hennings rescued three foals by bottle-feeding them, but something more had to be done. To save the herd from extinction, the Park Service brought in an Andalusian stallion named Cubanito in the 1980s. Modern Andalusians exhibit Spanish characteristics similar to those of the ancient Jennet. Cubanito was a handsome example of the breed.

Although bringing in outside blood resulted in live foals, Cubanito has since been criticized as a choice for this role. As it turns out, Cubanito carried both of the genes necessary to perpetuate neonatal isoerythrolysis. When he was bred to the three problem mares, however, the resulting offspring were healthy and robust. The herd was on the increase again, but was no longer true to the original bloodlines.

Andalusians, like Banker Horses, are of Spanish origin, but the modern Andalusian is not closely related to the modern Banker Horse. Conant et al. wrote (2011, p. 61), "North American Colonial Spanish Horses show more similarity to other New World horses of Iberian origin than they do to the modern Lusitano and Andalusian." The matings to Cubanito amounted to crossbreeding, moving the genes of the Ocracoke herd away from their original lines. Several Ocracoke horses had been officially accepted by the Spanish Mustang registry, but Cubanito's foals were ineligible for registration.

The Park Service also tried to introduce a horse named Sailor, who was born on Ocracoke and removed to Hatteras by Dale Burrus when the Park Service was selling off the wild ponies. The mares wanted nothing to do with Sailor. He did manage to sire three or four foals, all males, but the mares never fully accepted him.

Compounding problems of genetic and emotional incompatibility, some of the mares simply stopped conceiving. Hormone therapy was initiated in 1977, and subsequently three of the four problem mares foaled.

In addition, the foals had a run of bad luck. One died at birth. Three died from overeating. One punctured his foot as a newborn, probably on a reed out in the marsh. A tendon contracted until he was walking on his fetlock rather than his hoof. He underwent surgery and was fitted with a cast, and he ultimately survived.

With veterinary care, the Ocracoke herd began to recover. One mare reached the astonishing age of 40, more than 100 in human years. In her last year she gave birth to a foal probably sired by Sailor. (Unlike a woman, whose reproductive capacity ends at about two thirds of her lifespan, a mare is never too old to conceive, though fertility decreases with age.) She was initially unable to rise after the birth, but she lived another 6 months, hand-fed by the Hennings (Henning, 1985).

In 1977, a pinto colt named Mr. Bobby or Mr. Bob was born to a spotted mare, "Old Paint," and the sorrel stallion Jim. Jim Henning broke him to saddle and put him to use in living history programs and parades. In 1981 Park Ranger Howard Bennink used him to patrol the beach and to manage crowds at festivals. Mr. Bob became the first horse in Ocracoke's National Park Service Beach Patrol. He died in the fall of 2010 at the age of 33.

In 2010, there were 21 horses in the Ocracoke herd. Some are larger than many Banker Ponies, apparently because of outcrossing to Cubanito and maybe other outside horses. Most stand between 14 and 15.2 hands at the withers (56–62 in/1.32–1.57 m) and weigh about 1,000 lb/450 kg each.

The Ocracoke pony pen remains a popular attraction. A boardwalk and viewing platform flank the front pastures, allowing even the least adventurous visitor an opportunity to watch the ponies. Park Service educational programs allow tourists and schoolchildren a closer look at these horses. Annual vacationers take pleasure in watching the foals mature into adults.

The Ocracoke horses live in domesticity, accustomed to the usual husbandry routines of feeding, watering, grooming, and mucking stalls. Horse people generally do not find these chores onerous, but approach the physical maintenance of their charges with the same pleasure a gardener finds in working with the earth. A number of volunteers assist with the chores inherent in maintaining the Ocracoke herd, and for them it is a labor of love and a passion.

While many people stop at the Pony Pen just long enough to snap a couple of photographs, others spend peaceful hours observing the ponies' behavior and watching them interact. Wild or domestic, horses communicate mostly through postures, gestures, and facial expressions that can be readily interpreted by human observers. Body language allows them to communicate and maintain social order without extraneous noises that could draw predators.

In the wild, the horse that belongs to a group is less likely to be targeted by predators and will more likely to survive to pass on his or her genes. Equine social behavior minimizes conflict and promotes group stability. The social order within a band is complex and stratified. Mares tend to form long-term social bonds with other mares. These friendships increase foal birth rates and foal survival, probably by reducing harassment by stallions (Cameron, Setsaas, & Linklater, 2009).

Horses are attuned to subtle changes in the body language of their companions. There is meaning in the flick of an ear or the retraction of a nostril. Horses are able to recognize other horses from a distance by appearance and posture as well as by the scent of the body, hoofprints, or manure.

The outline of the horse's body is an important visual signal to herdmates. An excited, alert horse has a high-headed, high-tailed outline; a straight back and low head and tail signal relaxation. The alarm posture—head and tail raised, nostrils flared, eyes wide, usually accompanied by an inspiratory snort—alerts the herd to possible danger.

Many parts of the body can be used independently for signaling. Tails are particularly expressive. An excited horse may carry its tail lifted like a flag, particularly while running. Tail flagging is especially pronounced in Arabian horses. A tail clamped tightly against the rump indicates fear. Horses swish their tails when annoyed, and when the insects are biting, tails in constant motion serve as fly swatters. A mare will raise her tail and carry it to the side to display her vulva when in heat.

Horses have excellent hearing and can move their ears independently. They can turn their ears toward the focus of attention and flatten them to protect them from loud noises. Horses grazing in a group monitor human visitors by keeping one eye and one ear on the newcomers, usually without a pause in feeding. Ears are laid back to show aggression and flattened to the neck in battle to prevent them from being bitten off by a rival. Two ears directed sideways convey uncertainty or insecurity. Horses seem able to use body positioning to amplify sounds by bouncing them off their shoulders (McGreevy, 2004). Dogs and humans are better able to localize sound sources, but horses can hear higher-pitched sounds than humans and can hear a broader range of frequencies than most other animals.

Horses are capable of complex facial expressions and, like people, clearly show thoughts and emotions through their eyes. The white scleras of their eyes may show when the animal is anxious, but the eyes of many Appaloosa and pinto horses show the whites normally.

Because postural dialogs and spotting predators are key activities, horses have evolved an uncommonly large field of vision. With his head down in a grazing position, a horse can see about 350°, depending on the size and shape of the eye and jaw—everywhere but directly under his nose and behind his tail. A horse compensates for the blind spot directly behind him by having powerful hind legs and an instinctive tendency to kick at or run from whoever or whatever surprises him from the rear.

Each eye sees a different field of view, but horses do have 65–80° of binocular vision in the front (Murphy, Hall, & Arkins, 2009). (Humans have about 130°, which is most of our visual field.) Horses have about the same depth perception as a cat or pigeon, and they can accurately judge distance and width using only one eye. The author once owned a mare that lost the sight of her right eye as a young filly; as an adult she had no trouble gauging the height or distance of jumps.

A horse can distinguish details at a distance about as well as a person with 20/33 vision (McGreevy, 2004). This is better than a dog (20/50) or cat (20/100). Kirkpatrick notes that the horse can recognize some objects at a great distance, noticing other horses as far as 2 mi/3.2 km away! A horse has excellent night vision with which to escape nocturnal predators and can navigate challenging terrain in what looks to a rider like total darkness. They cannot, however, see with great clarity in low light.

Horses can see color, but only about as much as a human being with red-green color blindness. A number of studies suggest that horses can readily distinguish orange, yellow, and blue from gray, but have trouble distinguishing red and green (Murphy et al., 2009).

Horses communicate primarily with visual signals—body language and very small movements of the head, ears, and eyes—although they do use their voices as well (Proops & McComb, 2010). When horses meet, they exchange information that may be so subtle as to be inapparent to a human observer. These communications usually establish which horse is dominant. In general, the horse who wins the most conflicts is dominant and is entitled to more and better forage, first access to water, and more opportunities to reproduce.

Aggression usually begins as subtle nose-wrinkling and flattened ears. If these understated threats are ignored, the altercation escalates to head tossing, bared teeth, tail swishing, stamping a foreleg, charging, and biting. A horse on the defensive will use the same signals of nose-wrinkling and ear-flattening, along with tail-flattening and presenting the rump, backing toward the aggressor and swinging a hind hoof in a kick threat, or back-kicking with one or both hind hooves (Goodwin, 2002).

Submissive horses usually escape aggression by turning their heads away or by yielding space, withdrawing with ears half-flattened. If the threat comes from behind, the horse will tuck her tail and drop her croup as she retreats. Many horse trainers consider licking, chewing, and head-lowering submissive signals, but some scientists believe these

Face grizzled and muscles wasting, this horse is well into his 30s. The Ocracoke herd has a number of elderly horses and would soon die out if not for recent emphasis on preserving the strain.

behaviors are a displacement activity that occurs when horses experience conflicting impulses (Goodwin, 2002).

Yawning is also a displacement behavior, serving to release tension when the animal is conflicted. When humans approach too closely, barrier island horses often yawn, showing that they feel uncomfortable with the visitor, but not to the point of fight or flight. Keiper (1985) describes a mare yawning while standing over her dead foal, unsure whether she should stay with him.

Behavioral rules are somewhat lax for foals. Most of the time, elders patiently tolerate a youngster's inquisitive peskiness; but when tempers flare, an impatient kick can be lethal to a foal. To circumvent this risk, foals adopt a posture similar to an exaggerated nursing position—neck and muzzle extended, ears out to the sides, with the corners of the mouth drawn back and jaw clapping, as if trying to chew a thick chunk of bubble gum. Some interpret this snapping or champing as a ritualized version of the posture a horse takes in mutual grooming activity. This gesture was once thought to shut off aggressive tendencies in mature horses, translating as "Don't hurt me, I'm only a little baby!" Crowell-Davis found that snapping did not inhibit aggression and in some cases *caused* it (Crowell-Davis, Houpt, & Burnham, 1985). If champing is a submissive gesture, it is an inefficient one (Goodwin, 2002). It may be that champing is a displacement behavior derived from nursing that has a calming effect mainly on the horse doing the champing (Crowell-Davis & Weeks, 2005).

The set of the mouth communicates emotion. When a horse is tense, a deep triangle becomes visible above the mouth, the lips tighten, and the chin may dimple. Intense dislike or irritation is conveyed by wrinkling the mouth and nostrils. A relaxed horse may have droopy lips, ears, and eyelids. A contented horse may open its mouth slightly to expose the incisors and gums, with nostrils dilated, ears up and forward, and a look of contentment in the eye (Bennett & Hoffman, 1999). When scratching themselves on a branch or a stump, horses show pleasure by elongating the lips with half-closed eyes.

Keeping order on the move requires a great deal of communication. Horses communicate among themselves almost constantly, but the majority of their messages are silent, with occasional sounds that add emphasis or information. Most movies and television programs that feature horses dub inappropriate horse noises into the soundtrack, giving the impression that horses vocalize continually. Spending time with horses, one will quickly notice that they use their voices for only specific communications. Wild horses vocalize far less than domestic horses, particularly in herds where predators are a threat. Horses evolved on grasslands, and herd members usually remain in sight of one another at all times. The flattened ears and head toss of an irate mare is as effective as any vocalization and is less likely to draw the attention of predators.

The precise meaning of a vocalization is often related to context. The whinny is a loud call that carries over long distances. It can be used as a greeting, a call for a lost herd member, an invitation, or a plea for assistance. Stallions have a laugh-like, animated neigh when hopeful of romantic escapades and may trumpet, scream, or grunt when interacting with rival stallions.

The nicker is a low, soft sound that flutters through the nostrils, usually heard when a horse likes or wants something. Stallions and mares nicker in courtship. Mares nicker to foals. Domestic horses nicker when they hear grain rattle into a feed bucket.

A squeal expresses simultaneous fear and aggression. A mare will squeal when rejecting a stallion's advances or when protecting her foal. Horses meeting for the first time often sniff each other's breath intently, then squeal and strike out with a front hoof. Stallions sizing each other up have squealing contests. The longer, louder squeal usually belongs to the dominant animal.

Horses have an excellent sense of smell and frequently sniff objects or other horses. We do not know much about how they process this information. Two horses greeting each other will often softly blow into each other's nostrils. (Horses are unable to breathe through the mouth.) Stallions will greet other stallions and mares by sniffing at the flanks or elbow regions, then sniffing at the genitals—all areas rich with sweat glands and olfactory information, though what it means to the horse is anyone's guess (Feh, 2005).

Soon after birth, a foal will respond to a pungent aroma by curling back the upper lip in a flehmen posture, using another odor-sensing structure, the vomeronasal organ, deep within the nasal passages. Stallions use the flehmen posture to determine whether a mare is in heat, muzzle held high on an extended neck.

A horse compensates for the blind spot under his nose with incredibly sensitive lips and whiskers and an excellent sense of smell and taste. A domestic horse, given a mixture of pelleted dewormer and grain in which all particles are the same size, can eat just the grain, leaving the medicine behind. He can sort extraordinarily small things with his lips. His olfactory acuity allows him to sniff a pile of manure and know immediately which horse deposited it or sniff a trail and determine which way his herdmates went.

Like people, horses are emotionally invested in knowing where they fit into a social hierarchy. After engaging in behaviors that establish dominance—flattened ears, biting, rearing, kicking, striking—the winning horse will often will often lay its head on the loser's rump, and by tolerating this gesture the loser accepts subordinate status (Bennett & Hoffman, 1999). A stallion will also lay his head on a mare's rump before mounting, a final test of receptivity before moving into a more vulnerable position.

A frightened horse may sweat profusely, clamp his tail, raise his head, shift weight onto his hindquarters, and roll his eyes. Horses show annoyance or apprehension by twitching or wringing their tails or by stamping or pawing with their front feet. A horse may extend its upper lip in response to pleasurable sensations such as grooming or painful sensations like a bite on the hindquarters.

A horse's skin is very sensitive—he can respond to the sensation of a single fly walking on his neck or a rider's subtle shift of weight. Horses discourage biting flies by using the "fly-shaker" muscle, which rapidly twitches the skin surface and dislodges insects. When harassed by flies, they also stamp their front feet, shake their heads and necks to mobilize the mane, "cow-kick" at the belly with a hind hoof, or shake like a wet dog, legs apart and neck extended forward. A horse's tail is a supreme fly swatter, and it flicks constantly at the hindquarters and legs when insects are fierce. During fly season, horses often stand nose-to-tail with a friend, tails flicking over faces to keep biting insects at bay. Foals thrust their faces under their mothers' swishing tails for fly relief.

Rubenstein and Hohmann (1989) wrote that on Shackleford Banks, up to 200 flies can be found on a single horse. Flies bite adults in preference to juveniles, stallions in preference to mares, and dominant mares in preference to subordinates (probably because dominant mares stay closer to the stallions). Members of larger bands are bothered by fewer flies than horses in smaller groups, but flies are present in larger numbers on clusters of densely packed horses (Rubenstein & Hohmann, 1989).

Horses dislodge insects and itchy, shedding hair by rubbing against branches and tree trunks, rubbing the face and eyes against the legs, or using the hind foot to scratch the head and ears. Foals self-groom much more often than mature horses, up to 12 times an hour between 5 and 8 weeks of age (McGreevy, 2004). A horse with pinworms (*Oxyuris equi*, different from human pinworms) may rub much of the hair off his tail head in response to anal itching.

Horses also roll on the ground to scratch itches, kill insects, and groom their coats. The horse begins by locating the perfect wallow of fine dirt, sand, or snow with few rocks or roots. He usually lowers his head, sniffs the ground, paws, and circles several times. Then he folds his front legs, followed by the hind legs, tips over onto his side, and rolls up toward his back. He rocks back and forth several times, and if he is fit and coordinated, he may flip himself over and attend to the opposite side. Some horses unable or unwilling to flip will stand, then lie down again on the opposite side to complete the process. When he is done, he will roll onto his chest, thrust his forelegs in front of him, and lurch upwards while getting his hind legs beneath him. Then he adopts a sawhorse posture, extends his nose and neck, and shakes off the loose dirt. Most horses prefer to roll in sand or dry soil, but some prefer mud or even water.

Rolling is a contagious behavior, and other horses often follow suit, using the same wallow to scratch their own hides. The most dominant horse usually rolls last so its scent prevails (McGreevy, 2004). Other behaviors are socially contagious as well. Horses may defecate or urinate at the sight of a herdmate doing the same.

Colonial Spanish Banker Horses in the Ocracoke Pony Pen in the 1990s displaying a variety of colors and sizes.

When one horse lies down to rest, others usually do, too; but at least one horse always remains standing on sentry duty. Most spend around 2 hours out of every 24 on their chests or flat on their sides for deeper sleep. Horses sleep efficiently standing up and doze this way for an hour or more at a time. Horse legs are constructed to lock into place for light sleep, so remaining upright requires minimal muscular effort.

Horses may vocalize or kick and make running movements during REM sleep (Houpt, 2005). Stallions appear to have erotic dreams and may masturbate on waking (Houpt, 2005). Dreaming occurs only while the horse is recumbent because REM (rapid-eye-movement) sleep is accompanied by muscle relaxation so profound, that if he were standing he would fall over. Each dream cycle lasts only 3–10 minutes. Horses deprived of the opportunity to lie down for REM sleep by flooding, harassment by predators, or lack of companions to stand watch can suffer sleep deprivation, even to the point of narcoleptic collapse (Dallaire & Ruckebusch, 1974; Federation of Animal Science Societies, 2010).

Physiologically, it is more stressful for a horse to sleep lying down because of increased pressure on internal organs, so horses are recumbent for only short intervals. Young foals spend almost half their time sleeping and do most of it on the ground. In the first two months of life, they spend 70–80% of their total resting time lying down (Muhonen & Lönn, 2003). Foals spend half their time sleeping, usually sprawled flat on their sides. The average wild horse spends about 5 hr a day standing at rest, with one hind leg cocked, head down, in light sleep (Goodwin, 2002).

Although harem mares sleep 20–27% of the time, dominant stallions spend only 5–6% of their time asleep (Bennett & Hoffman, 1999). Horses sleep mostly at night and spend about

Frederic Remington (1895) depicted Florida Cracker Cowboys and Cracker Horses on canvas. Courtesy of Cornell University Library, *Making of America Digital Collection.*

30–40% of the hours of darkness sleeping (Houpt, 2005). Like new human mothers, mares with newborn foals get significantly less sleep.

A horse's daytime activities include grazing and grooming, seeking shade, and avoiding insects. During the summer, about 71% of daylight hours are spent grazing, which increases to 86% during the winter (Stevens, 1990). Harem stallions become lean and mean during breeding season, spending 25–45% of their time in motion. Other herd members focus on resting or grazing and are active less than 10% of the time.

Keiper and Keenan (1980) observed that Assateague Ponies typically seek water just after sunset, then graze until about 10 p.m. During the evening they migrate from the marsh to preferred resting areas on hummocks supporting stands of pine and oak, on the dunes overlooking the ocean, or on the beach. Ponies are more likely to rest standing between 10 p.m. and midnight and from 3 a.m. to 4 a.m. Ponies are more likely to lie down to rest between midnight and 3 a.m., and about 4 a.m. embark on another round of grazing until after sunrise (Keiper & Keenan, 1980).

A horse typically produces manure every 75–90 min while feeding and every 3 hr when resting. Horses can defecate not only while walking, but also while trotting or even galloping. When one horse defecates, another often follows. Mares and prepubescent foals pass manure at random and urinate every 3–4 hr on average (Houpt, 2005). Horses will often use a latrine area and avoid feeding on manured grass. Horses prefer to pass their water over an absorbent material such as grass or sand, because most dislike the splashing of urine on their legs.

For a stallion during the breeding season, urination and defecation are almost always socially significant. Whenever one of his mares urinates, the herd stallion sniffs it, lifts his lip in the flehmen posture to determine her estrus status, and then urinates over the spot. Marking is done more frequently if the mare is in heat—stallion urine contains pungent chemicals, including *p*-cresol, that disguise the scent of her receptivity from other stallions (Kimura, 2001). Urine from mares in estrus evidently contains about half the *p*-cresol as that of stallions or non-receptive mares. When the stallion marks the urine of a mare in heat, the *p*-cresol in his urine appears to balance the shortage in hers, effectively disguising her status.

Horses have an amazingly acute sense of smell, and scent-marking is one of the few ways they can leave messages for other horses. If another stallion encounters this spot, he would inevitably sniff the mare's manure, and probably conclude "A mare was here, but she is claimed by this particular stallion."

The Ocracoke horses are now sheltered in a sturdy barn, but during the years when they ran wild, they had no protection from the elements. Horses in cool climates grow thick coats that provide insulation beginning in August or September and begin to shed their winter pelage when temperatures exceed 42°F/6°C (Bennett & Hoffman, 1999). Loose hair begins to shed from the head and neck, then proceeds sequentially to the hindquarters, flank, back, front of the legs, and finally to the belly and insides of the legs. Within about 56 days, the horse has a slick, shiny summer coat. Yearlings can take up to 75 days to complete the process, presumably because nature better insulates babies against their first winter.

Coat density is individualized to an animal's needs: old, young, and very thin horses keep their winter coats longer because they have more difficulty staying warm (Gill, 1994). Foals born in late summer or fall have denser coats than those born in mid-spring.

The winter coat has a fine, springy undercoat to trap body heat and an outer layer of guard hairs that channels water off the horse. The mane, forelock, and long hairs under the jaw and down the legs also serve to channel water. When a horse is standing in the rain, his skin usually remains dry as the water runs off and the oily undercoat repels moisture.

Horses' winter coat is so heavy, coarse, and insulating, snow can remain on their backs without melting. Ice storms, however, can foil the animal's natural defenses and cause rapid heat loss. Ponies that have evolved in cold climates tend to have low-set tails, which allow snow to slide off the fan of hair at the top of the tail. During storms, horses turn their tails to the wind to conserve heat. Horses also conserve heat in cold weather by warming air through the large nasal sinuses before it enters the lungs, and they dissipate heat in hot weather by flaring their nostrils and breathing fast. Horses can generate heat by shivering.

The whiskers of the muzzle are easily mistaken for part of the hair coat, but in fact they are very responsive sense organs called vibrissae, with a rich nerve supply that may even detect sound (McGreevy, 2004). In mice, every whisker maps to its own region of sensory cortex in the brain, and it is quite possible the same is true of horses. If a portion of the brain is dedicated to processing information from each of these whiskers, the information they convey is probably more complex and valuable than we realize. At the very least, they communicate such information as the type of plants in front of the muzzle in the horse's blind spot and the distance of the muzzle from a surface. It has been proposed that show horses are more likely to sustain facial trauma during trailer transport if they have had their whiskers removed for cosmetic purposes (McGreevy, 2004).

Equus has thrived for more than 4 million years mainly by virtue of adaptive behaviors that enhance the odds of staying alive long enough to reproduce. To a wild horse, anything unanticipated could threaten survival. Horses, wild and domestic, notice subtle changes in the environment and react to what they do not expect. Contrary to the opinion of many unseated riders, horses' skittishness does not denote stupidity. Horses do not always recognize danger immediately, so the equine strategy is to panic first, then sort out the details later. This could save his life.

Horses instinctively respond to movement. A noisy bird that launches suddenly from the trail ahead may panic a horse, which, for a moment, does not know that the sudden movement was not caused by a predator. Excitability to motion also means that if one horse bolts, the others in the vicinity bolt, too, or risk becoming a predator's next meal.

How does equine intellect compare to what humans call intelligence? Intelligence is hard to quantify or define in people and harder still to measure in other species. We tend to consider an animal intelligent if it behaves as we would in similar circumstances. Horses have evolved to survive in their ecological niche and would not have endured long if they were not well-suited to the task. To the non-horseman, however, equine quirks can seem erratic and irrational.

To survive in the wild, free-ranging horses must learn and remember their social, biological, and physical environments, including circumstances that change predictably at times and not at other times (Hanggi & Ingersoll, 2009). For both wild and domestic horses, almost every activity of daily life involves cognition, learning, and memory. Horses with the greatest capacity to learn, understand, and solve problems are most likely to survive and reproduce.

Horses perform respectably, but not brilliantly, in studies involving mazes and do not rank high in reasoning or problem-solving ability (Nicol, 2005). These skills are better developed in carnivores, which gained an evolutionary advantage by anticipating the movements of prey.

Horses are, however, extremely sharp when it comes to distinguishing differences, associating cues, and remembering associations. One study involving horses choosing symbols on LCD screens—large versus small, solid circle versus hollow, Snuffleupagus versus Cookie Monster—demonstrated that they could remember relatively complex problem-solving strategies for a minimum of 7–10 years and use these experiences to work out new challenges of a comparable nature (Hanggi & Ingersoll, 2009)

Aside from the Shackleford and Corolla horses used to expand the gene pool of the herd, all the horses within Ocracoke's Pony Pen were born in captivity and raised on pasturage supplemented by hay and grain. Breedings are carefully planned to maximize the gene pool. Stallions are kept in separate paddocks, and males not chosen for breeding are gelded. The Park Service has protected Ocracoke from development and has preserved its natural beauty, but the island horses are no longer wild.

The advance of the modern world has pushed numerous strains of the wild Colonial Spanish Horse to the brink of extinction. The wild herds at Corolla and Shackleford and the captive herd at Ocracoke struggle to maintain genetically viable populations. Wild horses of old Spanish bloodlines once lived on islands all along the Southeastern coast, from Florida

Sacajawea, a young Shackleford mare, joined the Ocracoke herd in December 2009. The Park Service hopes that her offspring will breathe new life into the Ocracoke herd.

to Virginia. Some breeds are no longer wild, but conservators have established breeding farms to preserve them.

The Cracker Horse of Florida and the Marsh Tacky of South Carolina and Georgia are breeds of old Spanish blood derived from wild horses that were re-domesticated. The Florida Cracker Horse (known variously as Seminole Pony, Prairie Pony, Florida Horse, Florida Cow Pony, and Grass Gut) was once abundant on the barrier islands of that state, but is no longer wild and is in danger of extinction. It appears that these horses derive not from the original Spanish expeditions, but from the Spanish settlements that came later (Conant et al., 2012). The Spanish brought the ancestors of Cracker Horses to Florida during the colonial period and put them to use working cattle on their ranches. The name of the breed is said to derive from the cracking of the cowboys' bullwhips. When the Dust Bowl desiccated the Western rangelands, stockmen drove their cattle to the lush pastures of Florida, thus introducing more Western horses. The task of roping and holding cattle required a larger, stronger horse, and the Spanish Cracker was replaced by the Quarter Horse, which by this time was a taller and heavier breed (Sponenberg, 2011).

The Marsh Tacky of South Carolina and Georgia was once the most abundant horse in the Low Country. Today the American Livestock Breeds Conservancy believes that there are fewer than 150 pure Marsh Tackies left (Beranger, 2009). Breeders carefully plan matings to preserve the integrity of the breed—some bloodlines have remained pure since the Civil War era. Like the closely related Banker Pony, the Marsh Tacky has been protected from dilution by geographic isolation, so it has retained many strong Spanish characteristics. Tackies are renowned for their intelligence, ruggedness, and kind dispositions.

Tall ladies are not a problem! Luna, a gray mare who was born a pinto, shares a nuzzle with her diminutive stallion Doran over the manger. Luna is uncommonly tall for a Banker horse, standing 15 hands (60 in./1.52 m). Her height and gray coloring apparently trace to Cubanito, an Andalusian stallion introduced into the Ocracoke herd in the 1970s. At 12.2 hands, Doran is about typical in size for a Shackleford.

Wild-horse rehabilitator Steve Edwards (2011) admires the tough little horses:

> I love Marsh Tackies. They are now the state horse of South Carolina, as they should be. Their history is intricately bound with the history of working people from the Low Country to the Mountains. Small, gaited, remarkably calm, trainable and tough—endless endurance. . . .
>
> Like the Bankers, they were the horses of poor, hardworking folks. Indeed, the breed name "Tacky" is rooted in its Colonial meaning, not as being of poor quality but of being common and widespread. Today they are neither common nor widespread. Like the Corollas, their future is very tenuous.

The Florida Cracker, the Marsh Tacky, and the horses of the Outer Banks are considered Colonial Spanish Horses based on conformation and genetic makeup and are worthy of preservation as unique herds with genetic, historic, and cultural value (Conant et al., 2012). They are recognized as among the few remaining vestiges of the ancient Spanish Jennet.

Numerous strains of Colonial Spanish Horses have died out over the years, their genes lost forever. The Ocracoke Ponies were headed down a similar path until people stepped in to save them from extinction. Despite their low numbers, Ocracoke horses show no signs of inbreeding depression, due to introductions from outside horses.

Recently, the Park Service has brought in horses from Shackleford Banks and Corolla to benefit the genetic health of the herd and to breed the Ocracoke horses back toward their historic bloodlines. Cothran (personal communication, April 20, 2011) said,

> They are reasonably good sources of additional variability. Possibly all of the horses on the Outer Banks share some ancestry, [from] the Shackleford Banks and Carrot

Island up to a little population in Virginia at Back Bay. You can maintain variability at a much higher level with as little as one effective migrant per generation. It doesn't take much crossing to restore the variability.

Historically, people have "improved" feral herds by introducing outside stallions, reasoning that one male horse can sire many offspring with many mares throughout his lifetime. Unfortunately, his fecundity can hijack the gene pool and eliminate the characteristics of the original herd. Cothran cautions,

> I usually recommend that you use mares if you are going to make exchanges, if you're trying to preserve the particular characteristics of a population. The reason is that they will have less impact on the population, yet they will, over time, help the variability to recover. Within a feral population in the Theodore Roosevelt National Park [North Dakota], fully 15% of that gene pool was from one introduced stallion, and that percentage is probably going to increase over time. He was a very dominant type of stallion. The local people had introduced big horses in there to create bucking stock for rodeos. This was a big horse—and he was very effective. (Personal communication, April 20, 2011)

In December 2009, Cape Hatteras NS adopted Sacajawea and Jitterbug, young Shackleford mares who will contribute their genes to the Ocracoke herd. Two privately owned Shackleford stallions, Wenzel and Doran, are on loan.

In 2010, Wenzel, the size of a Welsh pony, was cohabiting with Maya, at 15.2 hands (62 in./1.57 m) the tallest horse in the herd. Wenzel had to climb onto a manger to consummate the union. There was a similar height disparity between Doran and his mate Luna. Soon after his arrival on Ocracoke, Wenzel impregnated Spirit, one of the younger mares in the herd. In March 2010 she delivered Paloma. Alonzo, a 3-year-old sorrel Corolla stallion, joined the herd in 2012.

Santiago is an 18-year-old pinto stallion with contrasting irregular markings, long flowing mane and tail, and strikingly Spanish conformation. He is the sire of four mares and one gelding within the herd.

Breedings are carefully planned to increase genetic diversity, but sometimes animal instinct trumps science. Some years ago, Jim, a 25-year-old Banker stallion of the original bloodlines, was diagnosed with cancer. The rangers attempted to breed him with Nevada and Lindessa, but after consorting with him for months, their pregnancy tests came back negative. One day that spring, Ranger Bill Caswell went to the pony pen and saw something small and brown in the pasture. Nevada had delivered a foal unexpectedly during the night. The ponies are largely named by local school children. For this filly, they selected *Bonita Sopresa*, Spanish for "Beautiful Surprise."

One mare, chestnut with a white snip, was conceived after her sire escaped from his corral and had a romantic interlude with her dam. The mare's name? Oops. As a mature mare, Oops went on to cohabit with Santiago for two years, but the rangers never saw them mate.

One morning Park Service volunteer Kimberly Emery arrived to feed the horses and found that Oops had unexpectedly delivered a handsome pinto colt. He was named Lawton Howard in honor of a beloved island native who had died in 2002. Lawton is a healthy, sociable gelding who loves to be the center of attention, persistently soliciting scratching, stroking, and praise from Ranger Laura Michaels, their handler. Michaels, who vacationed on Ocracoke as a child, took a job with the Park Service as a seasonal maintenance worker in 2001. She tended the ponies three times weekly as a volunteer. Seven years later, she was awarded a permanent position as the official Ocracoke wrangler.

Ocracoke's Banker ponies remain on site as memento of what was. Though no longer free to run the beaches in great numbers as their ancestors did for centuries, their confinement does not diminish their importance. Some, however, feel that their importance should diminish their confinement. Restricted to less than 4% of Ocracoke's federal land, this herd is too small to survive without infusions of outside blood. To become self-sustaining, the horses would require greater access to their historic range.

The Banker ponies of Ocracoke represent a rare strain of a rare breed that nearly vanished, and one of the last vestiges of an important aspect of Outer Banks history. Their future remains uncertain. In 2010, there were 21 horses in the Ocracoke herd, and only five mares of breeding age. Most of the reproductively eligible females seem to have fertility issues and remain barren despite access to stallions. When the author visited the herd in 2009, there was a large contingent of elderly horses: Mr. Bob (33) Okies Rainbow (31), Nevada (28) and South Wind (28).

In 2014, the census was 18, and the herd was becoming more genetically diverse, yet truer to its historic lineage. Paloma was foaled in 2010, and two two years later her full sibling was born—a seal brown pinto colt named Rayo, or 'thunderbolt.' In May, 2013, Jitterbug, born on Shackleford Banks, and Alonzo, born on Currituck Banks, became parents to a chestnut colt named Captain Marvin Howard by local school children. This new generation lifts the hearts of countless devotees who fervently hope that this rare breed will survive the challenges ahead.

References

An Act To Provide for the Removal of Cattle Remaining on Core Banks in Carteret County. (1959). NC Sess L 1959 ch 782.

Almendrala, A. (2012, May 16.) Horse named 'Air of Temptation' runs into sea, is rescued a mile from shore. *The Huffington Post*. Retrieved from http://www.huffingtonpost.com/2012/05/16/horse-runs-into-sea-rescu_n_1521798.html

American Livestock Breeds Conservancy. (2009). *Conservation priority equine breeds 2009*. Retrieved from http://www.albc-usa.org/documents/ALBCEquineCPL.pdf

Avitable, J. (2009). *The Atlantic world economy and colonial Connecticut* (Unpublished doctoral dissertation). University of Rochester, NY.

Bank of England. (n.d.). *Inflation calculator*. Retrieved from http://www.bankofengland.co.uk/education/Pages/inflation/calculator/flash/default.aspx

Barnes, J. (2007, November-December). Scattered by the wind: The lost settlement of Diamond City. *Weatherwise*, 60(6), 36–41. doi: 10.3200/WEWI.60.6.36-41

Battling stallions run Banker ponies off Ocracoke Isle. (1951, June 5). *Evening Telegram* (Rocky Mount, NC), p. 3B.

Bennett, D., & Hoffman, R.S. (1999, December). *Equus caballus. Mammalian Species*, 628, 1–14.

Beranger, J. (2009). *The Marsh Tacky Horse—Yesterday and today*. Retrieved from http://blackberryridgehorsefarm.com/history_of_the_marsh_tacky_horse.html

Berenger, R. (1771). *The history and art of horsemanship.* London, United Kingdom: For T. Davies and T. Cadell.

Brooks, B. (1956, March). Riders of the beach. *Boy's Life, 46*(3), 25–26, 69.

Browne, D.J. (1854). Domestic animals. In C. Mason, *Report of the Commissioner of Patents for the year 1853: Agriculture* (S. Ex. Doc. 27, 33rd Congress, 1st Session) (pp. 1–4). Washington, DC: Beverley Tucker.

Cameron, E.Z., Setsaas, T.H., & Linklater, W.L. (2009). Social bonds between unrelated females increase reproductive success in feral horses. *Proceedings of the National Academy of Sciences of the United States of America, 106*(33), 13850–13853. doi: 10.1073/pnas.0900639106

Chadwick v. Salter, 254 N.C. 389, 119 S.E.2d 158 (1961).

Chard, T. (1940). Did the first Spanish horses landed in Florida and Carolina leave progeny? *American Anthropologist, 42*(1), 90–106. doi: 10.1525/aa.1940.42.1.02a00060

Chater, M. (1926). Motor-coaching through North Carolina. *National Geographic, 49* (5), 475–523.

Conant, E.K., Juras, R., & Cothran, E.G. (2012). A microsatellite analysis of five colonial Spanish horse populations of the southeastern United States. *Animal Genetics, 43*(1), 53–62. doi: 10.1111/j.1365-2052.2011.02210.x

Cooper J.F. (1831). *The last of the Mohicans; A narrative of 1757.* London, United Kingdom: Henry Colburn & Richard Bentley.

Crowell-Davis, S.L., & Weeks, J.W. (2005). Maternal behavior and mare-foal interaction. In D.S. Mills & S.M. McDonnell, *The domestic horse: The evolution, development, and management of its behavior* (pp. 126–138). Cambridge, United Kingdom: Cambridge University Press.

Crowell-Davis, S.L., Houpt, K.A., & Burnham, J.S. (1985). Snapping by foals of *Equus caballus. Zeitschrift für Tierpsychologie, 69*(1), 42–54. doi: 10.1111/j.1439-0310.1985.tb00755.x

Culver, F.B. (1922). *Blooded horses of colonial days: Classic horse matches in America before the Revolution.* Baltimore, MD: Author.

"D." (1832). A list of all the stallions that have stood along the Roanoke, in the state of North Carolina, from the Revolution to the present time. *American Turf Register and Sporting Magazine, 3*(6), 272–277.

Dallaire, A., & Ruckebusch, Y. (1974, January). Sleep and wakefulness in the housed pony under different dietary conditions. *Canadian Journal of Comparative Medicine, 38*(1), 65–71. Retrieved from http://www.ncbi.nlm.nih.gov/pmc/articles/PMC1319968/pdf/compmed00045-0071.pdf

Federation of Animal Science Societies. (2010, January). *Guide for the care and use of agricultural animals in research and teaching* (3rd ed.). Champaign, IL: Author. Retrieved from http://www.fass.org/docs/agguide3rd/Chapter08.pdf

Dodge, T. A. (1892). The horse in America. *North American Review, 155*(433), 667–683.

Du Bois, W.E.B. (1896). *The suppression of the African slave trade to the United States of America, 1638–1870.* Harvard Historical Studies. New York, NY: Longmans, Green, and Co.

Dunbar, G.S. (1958). *Historical geography of the North Carolina Outer Banks.* Louisiana State University Studies, Coastal Studies Series 3. Baton Rouge: Louisiana State University Press.

Dunbar, G.S. (1961). Colonial Carolina cowpens. *Agricultural History, 35*(3), 125–131.

Eaton, L. (1989). The last of the Currituck Beach cowboys. *Outer Banks Magazine,* 1989–1990 annual, 22–27, 83–84.

Edwards, S. (2010, March 10). They look smaller up close. *Mill Swamp Indian Horse Views.* Retrieved from http://msindianhorses.blogspot.com/2010/03/they-look-smaller-up-close.html

Edwards, S. (2011, December 16). High on the hog—putting 'Cuz to work. *Mill Swamp Indian Horse Views.* Retrieved from http://msindianhorses.blogspot.com/2011/12/high-on-hog-putting-cuz-to-work.html

Ellenberger, W.P., & Chapin, R.M. (1919). *Cattle-fever ticks and methods of eradication* (Farmers' Bulletin 1057). Washington, DC: U. S. Department of Agriculture. Retrieved from http://books.google.com/books/download/Cattle_fever_ticks_and_methods_of_eradic.pdf?id=JCkbAAAAYAAJ&output=pdf&sig=ACfU3U270HG8hGj4IKNtsQa6ESByhTJt2Q

Engels, W.L. (1942). Vertebrate fauna of North Carolina coastal islands: A study in the dynamics of animal distribution I. Ocracoke Island. *American Midland Naturalist, 28*(2), 273–304.

Feh, C. (2005). Relationships and communication in socially natural horse herds. In D.S. Mills & S.M. McDonnell (Eds.), *The domestic horse: The origins, development and management of its behaviour* (pp. 83–109). Cambridge, United Kingdom: Cambridge University Press.

Gill, E. (1994). *Ponies in the wild.* London, United Kingdom: Whittet Books.

Goerch, C. (1995). *Ocracoke.* Winston-Salem, NC: John F. Blair (Original work published 1956 and revised in later printings).

[Goldsmith, O.] (1768). *The present state of the British Empire in Europe, America, Africa and Asia. . . .* London, United Kingdom: For W. Griffin, J. Johnson, W. Nicoll, and Richardson and Urquhart.

Goodwin, D. (2002). Horse behaviour: Evolution, domestication and feralisation. In N. Waran (Ed.), *The Welfare of Horses* (pp. 1–18). Dordrecht, Netherlands: Kluwer Academic Publishers.

Gregg, A. (1867). *History of the Old Cheraws: Containing an account of the aborigines of the Pedee, the first white settlements . . . extending from about A.D. 1730 to 1810. . . .* New York, NY: Richardson and Company.

Hall, S.J.G. (2005). The horse in human society. In D.S. Mills & S.M. McDonnell (Eds.), *The domestic horse: The origins, development and management of its behaviour* (pp. 23–32). Cambridge, United Kingdom: Cambridge University Press.

Hanggi, E., & Ingersoll, J. (2009). Long-term memory for categories and concepts in horses (*Equus caballus*). *Animal Cognition, 12*(3), 451–462. doi: 10.1007/s10071-008-0205-9

Harriot, T. (1590). *A briefe and true report of the new found land of Virginia.* Frankfurt-am-Main, Germany: Johann Wechel.

Harrison, F. (1929). *The Belair Stud, 1747–1761.* Richmond, VA: Old Dominion Press.

Harrison, F. (1931). *The John's Island Stud (South Carolina) 1750–1788.* Richmond, VA: Old Dominion Press.

Harrison, F. (1934). *Early American turf stock, 1730–1830* (Vol. 1). Richmond, VA: Old Dominion Press.

Henning, J. (1985). *Conquistadores' legacy: The horses of Ocracoke*. Ocracoke, NC: Author.

Herbert. H.W. (1857). *Horse and horsemanship and the United States and British provinces of North America* (Vol. 1). New York, NY: Stringer & Townsend.

Heyward, D., & Allen, H. (1922). *Carolina chansons: Legends of the Low Country*. New York, NY: MacMillan.

Houpt, K. (2005). Maintenance behaviours. In D.S. Mills & S.M. McDonnell (Eds.), *The domestic horse: The origins, development and management of its behaviour* (pp. 94–109). Cambridge, United Kingdom: Cambridge University Press.

Howard, M. (1976). Ocracoke horsemen. In C. O'Neal, A. Rondthaler, & A. Fletcher (Eds.), *The story of Ocracoke Island: A Hyde County bicentennial project* (pp. 25–27). Charlotte, NC: Herb Eaton.

Howard, P. (2002, May 1). *Ocracoke Newsletter*. Retrieved from http://www.village crafts-men.com/news050102.htm#top

Howren, R. (1962). The speech of Ocracoke, North Carolina. *American Speech, 37*(3), 163–175.

Impact Assessment, Inc. (2005). *Ethnohistorical description of the eight villages adjoining Cape Hatteras National Seashore and interpretive themes of history and heritage: Final technical report*. Manteo, NC: Cape Hatteras National Seashore.

Ives, V. (2007). *Corolla and Shackleford Horse of the Americas inspections—February 23–25, 2007*. Retrieved from http://www.corollawildhorses.com/Images/HOA Report/hoa-report.pdf

Keiper, R. (1985). *The Assateague ponies*. Atglen, PA: Schiffer Publishing.

Keiper, R.R., & Keenan, M.A. (1980). Nocturnal activity patterns of feral ponies. *Journal of Mammalogy, 61*(1), 116–118.

Kimura, R. (2001). Volatile substances in feces, urine and urine-marked feces of feral horses. *Canadian Journal of Animal Science, 81*(3), 411–420. doi: 10.4141/A00-068

Lee, F.G. (2008). *Constructing the Outer Banks: Land use, management, and meaning in the creation of an American place* (Unpublished master's thesis). North Carolina State University, Raleigh.

Little, M.R. (2012, April 24). *A comprehensive architectural survey of Carteret County, North Carolina's archipelago: Final report*. Raleigh, NC: Author.

Mallinson, D.J., Culver, S.J., Riggs, S.R., Walsh, J.P., Ames, D., & Smith, C.W. (2008). *Past, present and future inlets of the Outer Banks barrier islands, North Carolina*. Greenville, NC: East Carolina University.

Mallinson, D.J., Riggs, S.R., Culver, S.J., Ames, D., Horton, B.P., & Kemp, A.C. (2009). *The North Carolina Outer Banks barrier islands: A field trip guide to the geology, geomorphology, and processes*. Retrieved from http://core.ecu.edu/geology/mallinsond/IGCP_NC_Field_Trip_Guide_rev1.pdf

McGreevy, P. (2004). *Equine behavior: A guide for veterinarians and equine scientists*. London, United Kingdom: W.B. Saunders.

Meader, S.W. (1959). *Wild pony island*. New York, NY: Harcourt, Brace.

Muhonen, S., & Lönn, M. (2003). *The behaviour of foals before and after weaning in group* (Examensarbete 190). Uppsala: Swedish University of Agricultural Sciences, Department of Animal Nutrition and Management.

Munsell, J.W. (1946, August 22). Hunting and fishing. *Newark Advocate* (Newark, NJ), p. 11.

Murphy, J., Hall, C., & Arkins, S. (2009). What horses and humans see: A comparative review. *International Journal of Zoology, 2009*, Article ID 721798. doi: 10.1155/2009/721798

Newsome, A.R. (1929, October). A miscellany from the Thomas Henderson letter book, 1810–1811. *North Carolina Historical Review, 6*, 398–410.

Nicholls, J. (Artist). (1736). Edward Teach commonly call'd Black Beard. In C. Johnson [Daniel Defoe], *A general history of the lives and adventures of the most famous highwaymen . . . to which is added, a genuine account of the voyages and plunders of the most notorious pyrates . . .* (plate facing p. 86). London, United Kingdom: Oliver Payne.

Nicol, C.J. (2005). Learning abilities in the horse. In D.S. Mills & S.M. McDonnell (Eds.), *The domestic horse: The origins, development and management of its behaviour* (pp. 169–183). Cambridge, United Kingdom: Cambridge University Press.

Norris, D.A. (2006). Shell Castle. *NCpedia.* Retrieved from http://ncpedia.org/shell-castle

Olszewski, G. (1970, September). *Historic resource study for history of Portsmouth Village, Cape Lookout National Seashore, North Carolina.* Retrieved from http://www.nps.gov/history/history/online_books/calo/portsmouth_village.pdf

O'Neal, C., Rondthaler, A., & Fletcher, A. (Eds.). (1976). *The story of Ocracoke Island: A Hyde County bicentennial project.* Charlotte, NC: Herb Eaton.

O'Neal, Jr., E.W. (2008). *Wild Ponies of Ocracoke Island, North Carolina.* Privately published.

Peck, K.J. (2008). *Horse husbandry in colonial Virginia: An analysis of probate inventories in relation to environmental and social changes* (Unpublished honors thesis). College of William and Mary, Williamsburg, VA.

Phillips, D. (1922, May). *Horse raising in colonial New England* (Cornell University Agricultural Experiment Station Memoir 54). Ithaca, NY: Cornell University.

Proops, L., & McComb, K. (2010). Attributing attention: The use of human-given cues by domestic horses (*Equus caballus*). *Animal Cognition, 13*(2), 197–205. doi: 10.1007/s10071-009-0257-5

Purchasing power of British pounds from 1245 to present. (2013). Retrieved from http://www.measuringworth.com/calculators/ppoweruk

Quinn, D.B. (Ed.) (1955). *The Roanoke voyages, 1584–1590: Documents to illustrate the English voyages to North America under the patent granted to Walter Raleigh in 1584.* London, United Kingdom: For the Hakluyt Society.

Ramsay, D. (1809). *The history of South-Carolina, from its first settlement in 1670 to the year 1808* (Vol. 2). Charleston, SC: David Longworth.

Remington, F. (1895, August). [Cracker Cowboys on horseback]. In Remington, F., Cracker Cowboys of Florida. *Harper's New Monthly Magazine, 91*(543), 339.

Rubenstein, D.I., & Hohmann, M.E. (1989). Parasites and social behavior of island feral horses. *Oikos, 55*(3), 312–320.

Ruffin, E. (1861). *Agricultural, geological, and descriptive sketches of lower North Carolina, and the similar adjacent lands.* Raleigh, NC: Institution for the Deaf & Dumb & the Blind.

Ryden, H. (2005). *America's last wild horses* (Revised ed.). New York, NY: Lyons Press.

Smith, J. (1624). *The generall historie of Virginia, New-England, and the Summer Isles.* London, England: Michael Sparkes.

Smyth, J.F.D. (1784). *A tour in the United States of America . . .* (Vol. 1). London, United Kingdom: For G. Robinson, J. Robson, & J. Sewell.

Sponenberg, D.P. (1992). The colonial Spanish horse in the USA: History and current status. *Archivos de Zootecnia, 41*(154/extra), 335–348. Retrieved from http://dialnet.unirioja.es/servlet/articulo?codigo=278710&orden=0& info=link

Sponenberg, D.P. (2011). *North American Colonial Spanish Horse update, July 2011.* Retrieved from http://www.centerforamericasfirsthorse.org/north-american-colonial-spanish-horse.html

Sponenberg, D.P., & Reed, C. (2009). Colonial Spanish type matrix. In U.S. Bureau of Land Management, Billings Field Office, *Pryor Mountain wild horse range herd management area plan and environmental assessment* (EA #MT-010-08-24) (pp. 140–143). Billings, MT: Bureau of Land Management, Billings Field Office.

Stevens, E.F. (1990). Instability of harems of feral horses in relation to season and presence of subordinate stallions. *Behaviour, 112*(3-4), 149–161. doi: 10.1163/156853990X00167

Stick, D. (1952). *Graveyard of the Atlantic: Shipwrecks of the North Carolina Coast.* Chapel Hill: University of North Carolina Press.

Stick, D. (1958). *The Outer Banks of North Carolina, 1584–1958.* Chapel Hill: University of North Carolina Press.

Thomas, J.E., Rhue, R.D., & Hornsby, A.G. (2009). *Arsenic contamination from cattle-dipping vats* (Publication SL 152). Gainesville: University of Florida Institute of Food and Agricultural Sciences. Retrieved from http://edis.ifas.ufl.edu/ss205

Turner, F. (n.d.). *Money and exchange rates in 1632.* Retrieved from http://www.google.com/search?q=pound+peso+exchange+rate+16th+century&rls=com.microsoft:en-us:IE-Address&ie=UTF-8&oe=UTF-8&sourceid=ie7&rlz=1I7ACEW_enUS482

U.S. Bureau of Labor Statistics. (n.d.). *CPI inflation calculator.* Retrieved from http://www.bls.gov/data/inflation_calculator.htm

Vegetation scarcity blamed on N.C. oceanside ponies. (1940, March 12). *Portsmouth Herald* (Portsmouth, VA), p. 10.

Wallace, J.H. (1897). *The horse of America in his derivation, history, and development.* New York, NY: Author.

Westvang, D. (2008). *Finding the foundation.* Retrieved from http://www.foxtrotterfoundation.com/FINDINGTHEFOUNDATION.doc

Welch, W. L. (1885). *An account of the cutting through of Hatteras Inlet, North Carolina, September 7, 1846. Also through which inlet did the English adventurers of 1584, enter the sounds of North Carolina and Some changes in the coast line since their time.* Salem, MA: Salem Press.

Winsor, J. (1886). *Narrative and critical history of North America* (Vol. 2, Spanish explorations and settlements in America from the fifteenth to the seventeenth century). Boston, MA: Houghton, Mifflin.

Wiss, Janney, Elstner Associates & John Milner Associates. (2007). *Portsmouth Village cultural landscape report.* Atlanta, GA: U.S. National Park Service, Southeast Regional Office. Retrieved from http://www.nps.gov/calo/parkmgmt/upload/CALO Portsmouth Village CLR_Site History.pdf

Chapter 7

Down East

Shackleford Banks, Carrot Island, and
Cedar Island, North Carolina

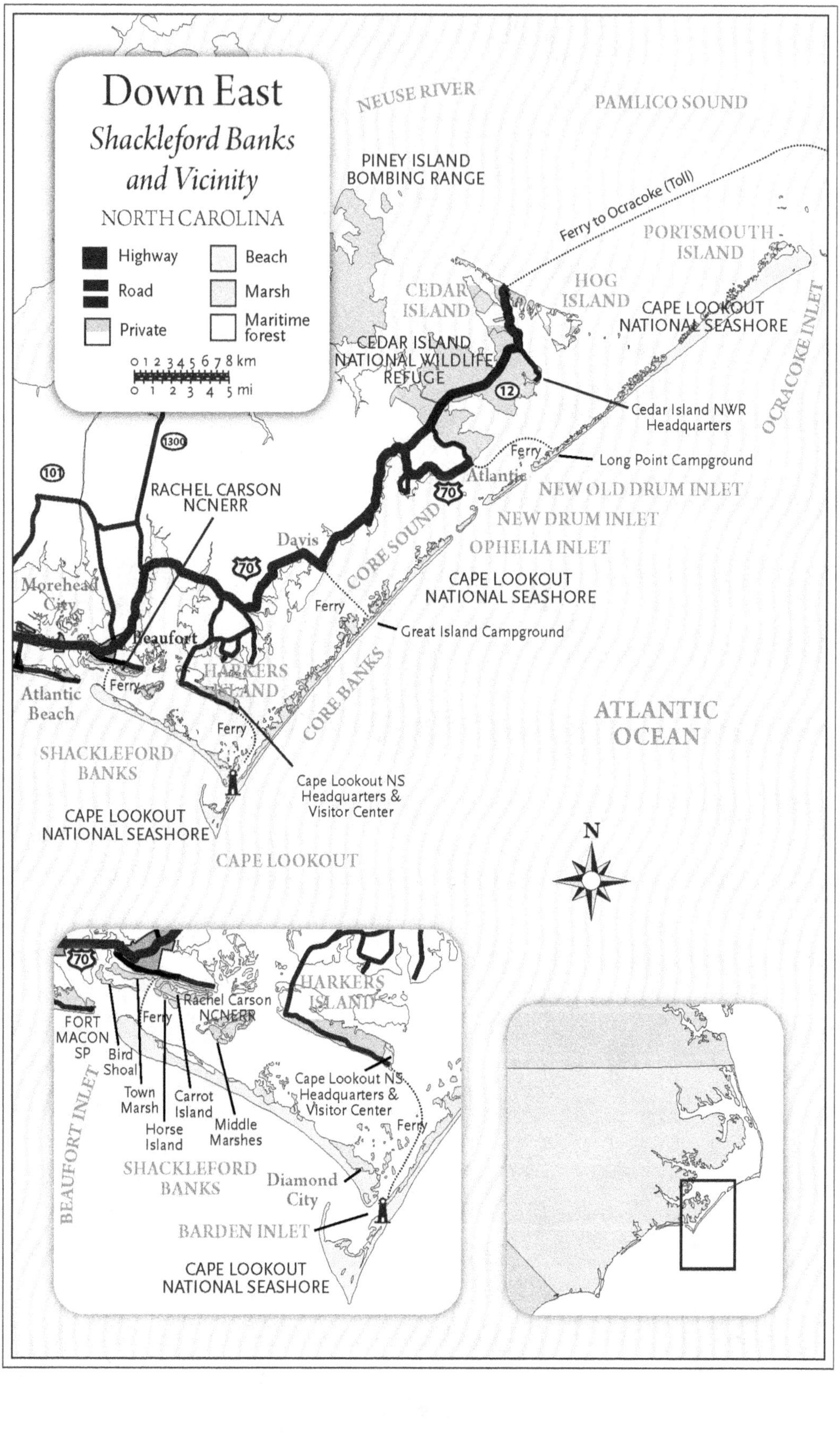

Down East
Shackleford Banks and Vicinity
NORTH CAROLINA

Legend:
- Highway
- Road
- Private
- Beach
- Marsh
- Maritime forest

0 1 2 3 4 5 6 7 8 km
0 1 2 3 4 5 mi

NEUSE RIVER

PAMLICO SOUND

PINEY ISLAND BOMBING RANGE

Ferry to Ocracoke (Toll)

PORTSMOUTH ISLAND

CEDAR ISLAND

HOG ISLAND

CAPE LOOKOUT NATIONAL SEASHORE

CEDAR ISLAND NATIONAL WILDLIFE REFUGE

OCRACOKE INLET

12

Cedar Island NWR Headquarters

1300

101

RACHEL CARSON NCNERR

Ferry

Atlantic

70

Long Point Campground

Davis

NEW OLD DRUM INLET

NEW DRUM INLET

CORE SOUND

OPHELIA INLET

CAPE LOOKOUT NATIONAL SEASHORE

Morehead City

70

Ferry

Great Island Campground

Beaufort

HARKERS ISLAND

CORE BANKS

Atlantic Beach

Ferry

Ferry

ATLANTIC OCEAN

SHACKLEFORD BANKS

Cape Lookout NS Headquarters & Visitor Center

CAPE LOOKOUT NATIONAL SEASHORE

CAPE LOOKOUT

N

70

HARKERS ISLAND

FORT MACON SP

Ferry

Rachel Carson NCNERR

Bird Shoal

Town Marsh

Carrot Island

Cape Lookout NS Headquarters & Visitor Center

Horse Island

Middle Marshes

Ferry

BEAUFORT INLET

SHACKLEFORD BANKS

Diamond City

BARDEN INLET

CAPE LOOKOUT NATIONAL SEASHORE

The seal-brown mare known to the Cape Lookout National Seashore staff as 13R was thirsty. The day was warm, and the only available water came at the price of hard labor. When rainfall is sparse, the horses of Shackleford Banks dig holes in the sand, which fill with freshwater when they reach the water table. After grazing in the relentless sun for most of the morning, the band led by the rugged bay stallion 14K arrived at the drinking spot. The alpha mare, the 9-year-old sorrel 14L, quickened her step at the scent of water. As they rounded the dunes and stepped into the clearing, the stallion moved in front of the sorrel with animated steps, head and tail high, prepared to defend his mares from competing stallions. But the clearing was empty except for five crows that took to air, complaining in raspy voices.

As the dominant animal in the group, the muscular stallion drank first, drawing easily at the water that had accumulated in the depression. Next in line was the sorrel, distinctive for her shining red-gold coat and a milky blaze that coursed down her face and pooled over her left nostril. She took a few sips of standing water, then dug deeper to let more flow into the hollow.

The seal-brown mare's mouth felt as dry as the sand that blew around her fetlocks, but she was 4 years old and low in rank. Protocol demanded that she wait her turn.

Two-year-old 15U pushed around the sorrel to steal a sip out of turn, but the lead mare flattened her ears and threatened to bite. The filly retreated reluctantly. The past winter had left the adolescent looking somewhat ragged—an undershot jaw and angular limb deformity had made her life challenging, as did her position at the very bottom of the pecking order.

As the alpha mare signaled her satiation and stepped away, she was replaced by the filly's dam, 22-year-old Number 16, who was leaner and less muscular than the lead mare. Despite abundant forage, the mare remained thinner than other adults in the herd, probably because of her advanced age, repeated reproduction, and recent lactation. The elderly mare pawed vigorously to excavate the depression, then lowered her muzzle to sip while the brown mare and the filly stood by, obvious distress in their eyes.

When the seal-brown mare finally took her turn, she had barely lowered her lips to the murky fluid when the stallion decided that it was time to leave. He aggressively pushed his harem into motion with flattened ears and snaking head. The mares trotted beyond his reach and obediently ascended the dune, but when the stallion looked away, the brown doubled back and rushed toward the drinking hole. Apparently, the young mare had resolved to slake her thirst regardless of the consequences.

The stallion was incensed at her insolence. This was not the first time the mare had flouted his authority. He spun and flew at her with teeth bared. The brown mare nimbly evaded him, escaping to the safety of the dunes. Satisfied that he had driven her from the water source, the bay stallion shifted his attention to the other mares and pressed the herd north. The seal-brown saw opportunity and spun away from her herdmates in another unauthorized rush to the water hole. The stallion appeared dumbfounded—this low-ranking upstart was persisting in challenging his decree?

Ears flat, head low, and thoroughly enraged, the muscular bay flew at her like an arrow, and she poured on the speed, raising a cloud of fine sand. With the stallion interposed between the water hole and the band, the seal-brown had no choice but to rejoin the herd and allow the bay to drive her over the dune to the grasslands.

An hour later, a younger battle-scarred bay stallion with the freeze-brand 2L was taking his turn at the water hole, accompanied by his two mares. They lifted their heads to the unexpected sound of galloping hoofbeats and saw the seal-brown mare round the dune at full tilt and rush straight to the water hole in total disregard of social protocol. Recklessly, the mare thrust her muzzle into the hole.

Her stallion arrived seconds later, infuriated. But as he followed her into the clearing, he found himself face to face with a surprised and indignant 2L. The older stallion hesitated for an instant, and then launched himself at 2L, catching him off guard. The younger stallion fell back, off balance, then rallied and descended on his rival, biting him savagely and raking him with his front hooves. The older bay, to this point brash and confident, suddenly realized that he was outmatched and withdrew. He tucked his tail in submission and retreated across the dunes.

The seal-brown mare gulped greedily at the water hole, uninterested in the drama around her. When her mate finally returned, exhausted and humiliated, the little rogue mare was waiting with a self-satisfied expression, her thirst now slaked. She allowed him to herd her back to the band without a protest.

As the reader will note, the preceding tale refers to these horses only by color, number, age or other characteristics. It was a challenge to create an engaging account of this water-hole drama without using names, especially when two look-alike bay stallions squared off for battle. Moreover, all these horses do indeed have formal names, but the Cape Lookout NS asked that the author omit them from this book.

Each year, graduate students of Dr. Daniel Rubenstein, a professor of ecology and evolutionary biology at Princeton University who has been conducting research on Shackleford Banks since the 1970s, name the new foals of the Cape Lookout herd. They give each new filly or colt a name that begins with the first letter of its dam's name and conforms to an annual theme—Shakespeare, music, cartoon characters, countries, Biblical figures, and so on (Prioli, 2007).

These names are used by the Foundation for Shackleford Horses on public Web sites, by reporters in newspaper articles, and by Carmine Prioli in his comprehensive work, *The Wild Horses of Shackleford Banks* (2007). They are even used among biological professionals at the Cape Lookout NS and in official documents accessible by the public (such as Bardenhagen, Rogers, & Borrelli, 2011). But in conversations with lay people, the agency now refers to horses by number only "to help the public see them as more wild" (S. Stuska, personal communication, February 19, 2013)—an unexpected stance for an agency with a long history of denying horses status as wildlife.

Surprisingly, choosing to name horses in a wild herd or publicize these names can become a point of contention. Names empower us to organize and understand a forbidding universe

The sorrel alpha mare, 14L (far left), admonishes 2-year-old 15U, an impatient adolescent, to wait her turn. The bay harem stallion looks on, thirst slaked. On the far right, the seal-brown mare 13R gazes longingly at the sorrel, the set of her ears communicating her discomfiture.

With ears flat and hind leg cocked, mare number 16, age 22, warns her filly to back off while she drinks. The filly scrambles away in a fearful, submissive posture and bumps into the seal-brown mare, who similarly warns her out of her personal space with a flat-eared head toss.

Finally it was the brown mare's turn, and she pushed the adolescent filly aside to claim her rightful place at the water hole.

The stallion decided that it was time to move on and drove his band from the water hole.

Head low and ears flat, the stallion drove the brown mare from the water and pressed her toward the other mares. The adolescent filly saw an opportunity and began sucking down the contents of the muddy pool.

But the brown mare was not finished drinking. She dodged the stallion and cantered back to the water hole.

Her insolence was not to be tolerated. The stallion flattened his ears and galloped after her, but she was faster than the stallion and nimbly kept out of reach.

of impersonal information, but also affect our emotions and reason. Upon naming wild animals, some people feel a sense of ownership that creates a perceptual shift, which can cause them to mistake the natural attributes of the creature for those that they assign (Borkfelt, 2011). Often it takes a determined effort to allow wild creatures to be their authentic selves without remaking them through our own projections.

Even naming inanimate objects creates emotional attachment. The city of Boston, Massachusetts has successfully encouraged residents to take responsibility for keeping fire hydrants clear of snow by allowing businesses, organizations, and private citizens to adopt and name them (Code for America, n.d.; Mayor's Press Office, 2012).

Naming wild animals stimulates empathy. Some recreational boaters change their behavior around wild orcas in Puget Sound, Washington, when they have learned the identities of

Eventually, the bay stallion succeeded in driving the mare from the water hole, but she refused to give him the respect his station demanded, and he was frustrated. Determined to exert his power over someone, he changed course and charged straight at the author, who was standing nearly 2 bus lengths away documenting the events through a telephoto lens. She clutched her camera and ran! Fortunately this was the desired response—the stallion turned and smugly returned to his mares.

The stallion succeeded in driving his harem away from the water hole, but the seal-brown mare remained thirsty and determined. Sometime later, the author heard the hammering of hoofbeats and looked up to see the mare rushing back to the water hole with the bay stallion in tight pursuit.

individual whales. In one case, a private boater was speeding dangerously, pursuing whales and disregarding safety guidelines. The driver was surly and uncooperative until the naturalist explained that the whale in question was Granny, a 99-year-old pod leader. Milstein (2011, p. 11) wrote, "Granny's name was . . . easy for visitors to remember, and it enfolded her matriarchal position into a culturally comprehensible package, eliciting similarities with humans, in this case familial relations, and likely mediating points of connection and even empathy." Upon recognizing the whale as an individual, the boater willingly yielded space to the elderly Orca.

Similarly, at Assateague Island NS in Maryland, officials use horse names to engage and educate the public. When Dr. Ronald Keiper began his research with the herd in the 1970s, he assigned each band a letter, and each pony in the band a number (Keiper, 1985). Each foal was labeled with its mother's identifier, plus a letter indicating its birth year—A for 1976, B for 1977. To the initiated, the birth year, mother, grandmother, and great-grandmother are immediately evident in the designation of foal N2B-E-H.

This system allows managers and researchers to follow the maternal pedigree of the entire herd easily and indefinitely, and to trace maternal lineage back for nearly four generations, without time-consuming use of reference lists. The designations alone can help a scientist keep entire genetic lines in mind as they surge and ebb in often surprising ways, as prolific lines stop reproducing and vanish while poorly represented lines unexpectedly increase.

However useful, these designations are not quite names. Though Keiper's work was strictly scientific, it is evident from his writing that he warmed to the animals he studied. Eventually he saw a need to use real, meaningful names. He wrote (1985, p. 19), "As in other behavioral studies, some ponies earned conventional names related to their appearance, personality, or family life." He named Comma for the white crescent on his forehead, Nasty for her aggressive behavior, and Park Service Pony for a marking resembling the agency's uniform patches. He named an elderly mare Irene because, like his grandmother of the same name, she produced three female offspring, followed by a male.

Today, most horses in the Assateague NS herd have names, and the seashore keeps an identification book in its visitor center to help people investigate horses or nominally adopt their favorites. The nonprofit Assateague Island Alliance (2010) maintains a Web site where visitors can identify and "adopt" these horses or bid on naming rights for new foals, with proceeds benefitting the seashore.

Mustangs in many of the Western herds have names. Wild horse advocates and researchers name Pryor Mountain horses, and even the U.S. Bureau of Land Management (2012, October 15) refers to these horses by name on its official Web site. The herd gained a high public profile after Ginger Kathrens featured them in three extraordinary documentaries, following the life of the wild stallion Cloud, for the Public Broadcasting System series *Nature* (Kathrens, 2001, 2003, 2009). Kathrens balances scientific objectivity with genuine concern and interest and paints honest, yet emotionally stirring portraits of mustangs and the challenges of their mountainous Montana range. These films, along with numerous books, articles, and plastic Breyer® horse models, have promoted understanding of wild horses in general by helping people appreciate them as individuals. Cloud lent his name to a nonprofit support group, the Cloud Foundation.

Raising awareness of these horses has not spoiled their wild character in their natural range, but it has transformed the hearts and minds of the public. Whereas the average 2012

As they approached, the stallion abruptly found himself face to face with another alpha stallion using the water hole. They scuffled, and the bold bay quickly realized that he was outmatched.

The younger stallion 2L drove the harem stallion 14K into a humiliated retreat.

adoption rate of BLM horses and burros was 31.37% (USBLM, 2013), 100% of the gathered Pryor Mountain horses were quickly adopted or placed in private care—as in every previous year (USBLM, 2012, April). The BLM infrequently offers Pryor Mountain horses for adoption, and few horses are gathered from the herd.

Most often, people choose names for animals when they feel affinity and numbers when they are, or wish to be, disengaged, as in the case of laboratory rabbits or feedlot steers (Borkfelt, 2011). In Alberta, Canada, horses bred specifically for meat production and fattened on feedlots are somehow deemed more suitable for slaughter than companion horses, because "Purpose bred meat horses do not have names, nor have they served humans in another role" (Alberta Equine Welfare Group, 2008, p. 31). Before modern medicine increased the survival rate of human infants, parents often delayed naming their babies until it became clear that they would live.

In most of the Atlantic coast herds, managers struggle with the problem of visitors who approach and interact with wild horses as if they were already tame. As Dwyer (2007, p. 86) wrote, "we tend to generalize the successes we have with our pets to all animals. If we can have a loving relationship with a dog, why not one with a fox or a wolf? If we can get a horse to love us, why not a zebra?" Close contact with wild creatures, however, often ends badly, resulting in injuries to humans and bold, aggressive behavior in wildlife. In our culture, animals that hurt people are usually removed or destroyed.

At Cape Lookout NS, administrators say officially that numerical identifiers will emphasize the enigmatic aura of wildness that keeps visitors at a distance. Not everyone accepts this explanation, however. Years of controversy over drastic state and federal management practices, such as mass euthanasia of horses, pressing for unsustainably small herd size, and removing other livestock, have created suspicion on both sides of the public-private divide. Naming can lead to attachment. Whereas few may have noticed or cared about the sudden disappearance of some horse with a name like 12M, the death of a white foal named Spirit of Shackleford generated widespread sympathy and support for the herd (Foundation for Shackleford Horses, 2005). In this climate, some imply that the seashore numbers its horses to prevent public sentiment from complicating the next unpopular decision or simple mishap.

For centuries, free-roaming horses have lived and died on the 60-odd miles (97 km) of North Carolina islands that now constitute Cape Lookout NS. Now they remain only on 3,000-acre/1,200-ha Shackleford Banks, a narrow ribbon of sand roughly 9 mi/14.5 km long that runs perpendicular to the southernmost end of Core Banks.

North and South Core Banks extend for 44 mi/71 km from southwest to northeast, a long, slender stretch of low dunes, grasses, shrub thickets, maritime forest, and spreading salt marshes. Periodically, inlets divide the island, opening during storms and closing again in a matter of months or years. At this writing, three are open: the sporadically maintained New Drum Inlet; New Old Drum Inlet, reopened by Hurricane Irene in 2011; and Ophelia Inlet, opened by Hurricane Ophelia in 2005. The natural migration of the chain is unimpeded by human structures, and they are free to respond to the pressures of waves and overwash. From north to south, the landscape changes from wide tidal sand flats at Portsmouth Island to continuous dunes at Cape Lookout. Small freshwater marshes form in the depressions between dunes or where sandbars or spits close the mouth of a small bay. A jetty built in the early 1900s led to the formation of a sandy hook at the western end of Cape Lookout.

Compared to Core Banks, Shackleford displays more environmental diversity, with dunes as high as 44 ft/13 m at the western end, a dense maritime forest, marshes, and grassland. About 3.5 mi/5.6 km east of Beaufort Inlet, the island's profile shifts abruptly from high dunes to low overwash flats.

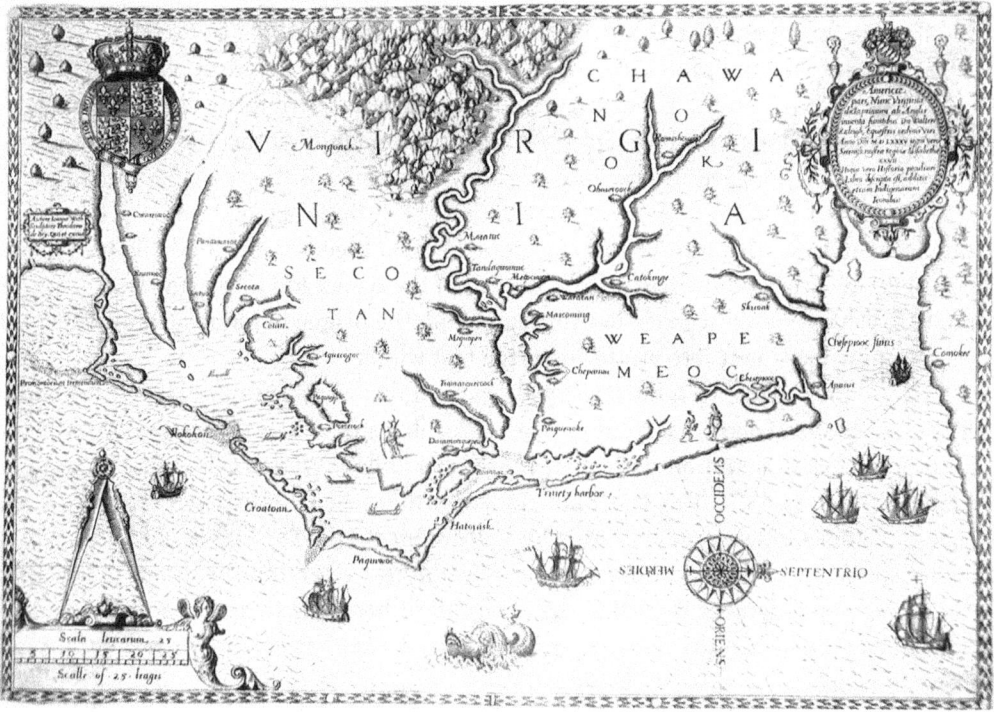

Theodor De Bry's 1590 map of Virginia, based on manuscript maps by John White, was the first accurate representation of the Outer Banks and vicinity. Courtesy of the Library of Congress, Geography and Map Division.

Here, as on Assateague, Currituck Banks, and Ocracoke, many believe that the original horses swam ashore from Spanish shipwrecks long before the English colonized the New World. The shoals off Cape Lookout and along the banks are certainly treacherous. One of the region's earliest cartographers named Cape Lookout *Promontorium tremendum*, or "horrible headland" (Jones, 2004). Ships, mostly bound to and from the English colonies, often passed this hazard carrying horses and other livestock, especially in the mid-to-late 1600s and 1700s. Again, shipwrecks might have brought horses to the island, but there is no proof, and perhaps no possibility of proof.

Sue Stuska, the Park Service biologist who manages the Shackleford herd, says that she does not know of any convincing evidence that the original horses arrived by shipwreck:

> As a scientist, I can't say that they swam to the island from sinking ships. The horses could have come by way of early explorers, colonists, or overland trade. Any shipwrecked horses that reached the island would have been the same type of horses that the colonists already had. The islands were like a common grazing ground. Anyone could take an animal over and drop him off. (S. Stuska, personal communication, May 26, 2010)

Few records exist, and the horses have been there for so long, many generations grew up believing that they were native to the islands. Edmund Ruffin, an agriculture authority who visited the area around 1858 and published his observations three years later, concluded that these horses were at that time a genetically unique, closed population, and that outside horses had not contributed their genes to the herd in recent history:

The race, of course, was originally derived from a superior kind or breed of stock; but long acclimation, and subjection for many generations to this peculiar mode of living, has fixed on the breed the peculiar characteristics of form, size, and qualities, which distinguish the "banks' ponies." It is thought that the present stock has suffered deterioration by the long continued breeding without change of blood. (1861, pp. 132–133)

Ruffin went on to say that introducing the blood of other breeds to the herd might decrease their ability to withstand the harsh environment of the Banks and that mainland horses "if turned loose here, would scarcely live through either the plague of bloodsucking insects of the first summer, or the severe privations of the first winter" (1861, p. 133).

Blood samples analyzed in the late 1980s showed that Shackleford horses were genetically similar to Ocracoke horses, "from which breeding stock reportedly was derived," but have other genes in common with draft breeds (Goodloe, Warren, Cothran, Bratton, & Trembicki, 1991, p. 417). This study concluded further "the Cumberland, Ocracoke, and Assateague horse populations are not genetically unique" (p. 418) in relation to the blood markers studied and that the horses did not qualify for retention on federal lands. The authors reserved judgment regarding the Shackleford herd due to small sample size: two of four horses died when the researchers immobilized them for phlebotomy, and the team halted sampling.

The findings of Goodloe et al. (1991) correspond with the historical evidence. Throughout the 20th century and probably earlier, the herds at Cumberland, Ocracoke, and Assateague Islands saw numerous infusions of outside bloodlines, mostly from American breeds. The Shackleford and Corolla herds are unrelated members of the same rare breed, probably derived from the same foundation population, but isolated from each other for hundreds of years. As discussed previously, the horses of North Carolina's Outer Banks probably descend from Spanish Horses from the West Indies, Chickasaws from the mainland, and perhaps Narragansett Pacers, racing horses, and saddle horses from New England and elsewhere. The Shackleford herd probably saw few introductions in the 20th century aside from genetically similar horses from nearby islands and the mainland.

On the other hand, Conant, Juras, and Cothran (2012, p. 60), write that

The Marsh Tackies, Florida Crackers and Shackleford Banks populations show similarity to each other ... and all show affinity to the Puerto Rican Paso Fino and other South American Iberian breeds, which is consistent with popular history and the Colonial Spanish designation for these populations. Given that early European explorers and settlers often obtained livestock from Puerto Rico and other Caribbean islands before heading to the eastern coast of the United States, this signature may truly show shared ancestry, which is rather remarkable after 400 years of separation.

Certain DNA markers are very characteristic of Spanish bloodlines, and these occur at relatively high frequency within the population (U.S. Bureau of Land Management, 2001). Sponenberg (2011) writes, "Due to the inheritance pattern of these markers it is easily possible for an absolutely pure Colonial Spanish Horse to have missed inheriting any of the Iberian markers. It is likewise possible for a crossbred horse to have inherited several." DNA typing provides useful insights and information, but leaves much open to interpretation. Sponenberg (2011) says,

After hundreds of years of relative isolation on barrier islands, the North Carolina Banker Horses still resemble the Spanish Jennet that carried the conquistadors.

These techniques do offer great help in verifying the initial results of historic and phenotypic analysis, but are by themselves insufficient to arrive at a final conclusion. . . .

A conservation program based heavily on blood types without considering other factors could then easily exclude the very horses whose conservation is important, and could include some that should have been excluded. Therefore, conformational type is a more important factor than blood type or DNA type, and will always remain so. It is impossible to determine the relative percentage of Spanish breeding in a horse through blood typing or DNA typing, at least currently.

Several variants (including Pi-W and Q-ac) clearly denote old Spanish origin. The only other breeds known to carry the Q-ac variant are the Puerto Rican Paso Fino and the Pryor Mountain wild horses of Wyoming and Montana.

Cothran says that although the marker cannot prove where or how this ancestry came about, he believes that the North Carolina Banker Horses have descended from a very old lineage of Spanish horses:

The Q-ac is a very uncommon variant that we see in what is called the Q blood group. Horses have blood groups just like people do. When I was doing some of my early work on mustangs, I was noticing this uncommon variant in some populations. And when I say uncommon, I mean that in a herd where it is relatively common, you'll see it in 2–5% of the population. When I see that variant in a population, I consider that to be evidence of old Spanish bloodlines. . . . I observed

Several horses in the Shackleford Banks herd carry the rare Q-ac blood type variant, which strongly suggests Colonial Spanish ancestry.

the highest frequency in the Puerto Rican Paso Fino, which is a very old Spanish breed. That population had a frequency of that variant at 5 percent—the highest of anywhere. I did not see the variant in any of the North American breeds such as the Morgan, the Saddlebred or the Quarter Horse. . . . On Shackleford Banks, only 3 or 4 horses in the herd had the Q-ac variant. At the population level, it is saying something. It is likely that there is old Spanish blood in that population . . . They show a high similarity to some of the English pony breeds like the Dales. (Personal communication, April 20, 2011)

Shackleford horses and certain Western herds of Spanish mustangs contain individuals with the rare, archaic Q-ac blood type variant. The most likely explanation is that both Western and Eastern Spanish horses descend from the same source population—Spanish ranches established in the 1500s in the Caribbean. It is well documented that Native American traders used horses from Western populations of pure Spanish horses to transport goods and sold the animals to lowland farmers in the Carolinas (Ramsay, 1809, p 403, Harrison, 1929, p. 23).

Another possible, but less likely, explanation relates to the use of horses to patrol the American coast during WWII. In June 1942, a U-boat landed four German saboteurs equipped with explosives, ammunition, cash, and forged documents on a beach near Amagansett, Long Island, New York. Another contingent landed on Ponte Vedra Beach, near Jacksonville, Florida. Federal authorities quickly caught them, but the episode pointed out the weakness of coastal defenses. Within months, the Coast Guard created an armed Beach Patrol with motorized, canine, and mounted elements to cover the Atlantic, Pacific, and

The Q-ac blood type variant is also present in the Pryor Mountain mustang herd of Wyoming and Montana.

Gulf coasts. Mounted units eventually served everywhere in the East except parts of New England (Noble, 1992). The U.S. Army Remount Service provided saddles and other gear. Because it also had procurement depots in Wyoming, Montana, and Nebraska, it provided more than 3,200 horses, some of them unbroken mustangs (Durham, 2009). There are no accessible records of where each horse originated; where each was sent; or what percentage were mares, stallions, or geldings. After the war, the horses were no longer needed and expensive to maintain, and most were immediately removed from service. Few records reveal the fate of these horses. Auctions in Ocean City, Maryland, Virginia Beach, Virginia, and elsewhere that began in 1944 do not account for all the horses (Bishop, 1989). The Coast Guard seems to have returned many to the Army, but might have abandoned some of them or sold them to coastal stockmen who added them to their free-roaming herds. Could horses that carried the Q-ac allele—military horses of old Spanish lineage or Beach Patrol horses from the Pryor Mountains or elsewhere—have added their genes to the Shackleford herd? There is no hard evidence that they did, and there may be no way to find out; but it remains an interesting possibility.

The first known residents of these islands were the Coree Indians, who had lived on and around them seasonally since prehistoric times. In fact, the name Core Banks may be derived from *Coree*. A settlement was established in the Core Sound area by 1710. The Tuscarora and allied tribes staged a coordinated attack on the colony in 1711, igniting years of violence before a peace treaty ended the Tuscarora War in 1715. Subsequently, Eastern North Carolina opened to further colonization after many Tuscarora migrated to western New York, and other native peoples succumbed to disease or were displaced onto reservations (Wiss Janney Elstner Associates & John Milner Associates, 2007).

In 1713 John Porter acquired Core Banks and Shackleford Banks, then parts of a single island connected by a tidal flat known as "The Drain." Porter sold the 7,000-acre/2,800-ha island to Enoch Ward and John Shackleford. When they divided the properly in 1723, Shackleford took the western part and gave it its current name (Prioli, 2007).

Colonists of predominantly English descent settled the island in the 1760s, and by the mid-1800s, Shackleford Banks was home to more than 600 people in several communities who called themselves "Ca'e Bankers" (dialect for "Cape Bankers"). The U.S. Coast and Geodetic Survey topographic map of 1853 shows about 20 residences along the sound side of Lookout Woods in the area that became known as Diamond City (Engels, 1952). Although most of the Cape Bankers lived in this area, there were homes all along the sound side of the Banks.

Diamond City, named after the distinctive diamond pattern of the nearby Cape Lookout Lighthouse, was the largest town ever established on Shackleford Banks. It was situated on the east end before Barden Inlet divided the island from Core Banks, and it boasted 500 residents. People had lived on the east end of Shackleford since the Banks were settled, but the community was not named Diamond City until around 1885 (Stick, 1958).

Other Shackleford villages included Wade's Shore, Mullet Pond, Bell's Island, and Windsor's Lump. The island supported an oyster house, a factory that extracted oil from porpoises, a crab-packing plant, schools, businesses, and churches. Feral sheep, goats, hogs, cattle, and horses freely roamed the island, while the Bankers lived in fenced homesteads built from maritime-forest timber and salvage from shipwrecks (Prioli, 2007).

A Reconstruction-era report of the North Carolina Geological Survey noted,

> These islands are inhabited by a hardy race of people, called Bankers, who subsist by fishing and whaling, occasionally by wrecking, and by raising for market a small, wiry, tough-sinewed, splay-hoofed variety of horse, called the bank pony, or marsh pony, which subsists on the coarse salt grasses of the wide marshes which margin the sound. These animals receive no care, save at the annual "penning" frolic, when the banks and marshes are "driven," as in a deer hunt, and the horses collected in hundreds in order to be claimed and branded, or sold. (Kerr, 1875, p. 15)

By the mid-1700s, Banker watermen and whalers from New England regularly hunted pods of right whales around Cape Lookout as they migrated north from their calving grounds (Stick, 1958). It appears that these short-range whaling operations were successful for more than 150 years.

As recently as 1875, the North Carolina state geologist described a thriving whaling industry around Cape Lookout:

> Whaling is carried on chiefly along the Shackleford Banks, between Cape Lookout and Fort Macon. The whales are taken in April and May, sometimes 5 or 6 in the course of one or two weeks. They are the common right whales, 40 to 60 feet [12–18 m] long; and a single animal frequently yields, in oil and bone, $1,200 to $1,900 [roughly $25,000–39,000 today (U.S. Bureau of Labor Statistics, n.d.)]. On one occasion, two sperm whales were taken, one of which measured 62 feet [19 m] in length. (Kerr, 1875, p. 15)

After the American Revolution, shipping increased along the Atlantic Seaboard, and consequently, so did shipwrecks. A 95-ft-tall (30-m) light station was built at Cape

Members of the U.S. Coast Guard Beach Patrol at Hilton Head Training Station, South Carolina ("Beach Pounders," *circa* 1943). Mounted Guardsmen patrolled the beaches along the Atlantic coast south of New England during World War II. Many of their horses, provided by the U.S. Army Remount Service, were mustangs from the West. Did their genes enter the barrier island herds? Photograph courtesy of the U.S. Coast Guard Historian's Office.

Lookout in 1812, but the structure was too low and dim to warn ships off the perilous shoals that extended 10 mi/16 km into the ocean. In 1859, the structure was replaced by the present 163-ft/50-m lighthouse, which was outfitted with a Fresnel lens that directed its beam far across the water. In 1873 the lighthouse was painted with a bold pattern of black and white diamonds to enhance visibility. Even so, shipwrecks remained commonplace.

Shackleford is generally higher than Core Banks and has a more varied landscape, including high dunes at the western end. The dunes provide a windbreak that allowed the myrtle (*Myrica* spp.), cedar (*Juniperus virginiana*), and live oaks (*Quercus* spp.) of the maritime forest to take hold almost 2,000 years ago, which in turn offered storm protection for people and animals alike.

Bratton and Davison (1987) wrote that the larger trees were cut aggressively for shipbuilding and homesteads, and by 1819 the supply of live oak and cedar was noticeably diminished. But coastal charts drawn between 1850 and 1870 and published in 1888 show extensive maritime forest on Shackleford all the way to its connection with Core Banks. A chart issued in 1966, however, showed forest cover on Shackleford greatly reduced and confined to the western half (Godfrey & Godfrey, 1976). The 1906 biennial report of the state geologist described Shackleford Banks as once covered with dense forest within the memory of the island's oldest inhabitants (Senter, 2003). Engels, however, paints a different picture. He wrote in 1952 (p. 714)

> There is no evidence that extensive logging ever was practiced; the "cutting of timber" was limited to the immediate needs of the inhabitants. The latter were primarily fishermen; they lived on the soundward side of the island, not on the ocean beach. . . . It is unlikely that at that time, or for some time afterward, the number of sheep on the island was great enough to seriously decrease plant reproduction.

The oldest tree on Shackleford identified in a 1974 study was a cedar about 100 years old, but old trees on the Outer Banks are notoriously hard to date. Their centers are often rotted away, and during hot, dry periods they form false growth rings, sometimes five or six in a season, that can make the tree appear much older than it is (Godfrey & Godfrey, 1976).

Evidence of the dynamic nature of the barrier islands is everywhere. Layers of peat and broken stumps along the beach indicate the presence of a swamp forest about 200 years before, when sea level was lower and the beach much farther to the south (Godfrey & Godfrey, 1976). Stumps are also found in the salt marshes along the north border of Shackleford, where historic maps showed forests, indicating sea level rise.

Around 1840, dune fields on Shackleford Banks began to migrate and engulf the woods on the eastern end of the island. A photograph from 1917 shows a "ghost forest," the gnarled trunks of dead cedars protruding from sandy hillocks like fingers from a giant grave.

Livestock took the blame for creating migratory dunes refractory to stabilization attempts—yet by the 1970s, the island showed "extensive vegetation cover over most of the island" (Godfrey & Godfrey, 1976, p. 115) despite the continued pressure from horses, sheep, goats, and cattle. Eventually, geologists came to realize that it is a natural characteristic for some dune systems to mobilize spontaneously and swallow everything in their path, independent of the activities of man or beast.

Because Barden Inlet did not separate Shackleford from Core Banks until the Hurricane of 1933, horses and other animals were free to move from one island to the next. Horses typically forage within a preferred home range and do not often travel great distances, but over time there was a genetic interchange between the Core Banks and Shackleford herds as various bands interbred and expanded their ranges.

Ruffin described them unflatteringly: "These horses are all of small size, with rough and shaggy coats, and long manes. They are generally ugly. Their hoofs, in many cases, grow to unusual lengths. They are capable of great endurance of labor and hardship" (1861, p. 132). He also spoke of horses introduced to the herd "from abroad" that were not as hardy as the Banker horses, commenting "they seldom live a year on such food and under such great exposure" (1861, p. 132). He added,

> In applying the term *wild* to these horses, it is not meant that they are as much so as deer or wolves, or as the herds of horses, wild for many generations on the great grassy plains of South America or Texas. A man may approach these, within gunshot distance without difficulty. But he could not get much nearer, without alarming the herd, and causing them to flee for safety to the marshes, or across water, (to which they take very freely,) or to more remote distance on the sands. Twice a year, for all the horses on each united portion of the reef, (or so much as is unbroken by inlets too wide for the horses to swim across,) there is a general "horse-penning,"

Diamond City derived its name from the distinctive diamond pattern of the nearby Cape Lookout Lighthouse. Today wild horses roam where a village once prospered.

to secure, and brand with the owner's marks, all the young colts. The first of these operations is in May, and the second in July, late enough for the previous birth of all the colts that come after the penning in May. If there was only one penning, and that one late enough for the latest births to have occurred, the earliest colts would be weaned, or otherwise could not be distinguished, as when much younger, by their being always close to their respective mothers, and so to have their ownership readily determined.

The "horse-pennings" are much attended, and are very interesting festivals for all the residents of the neighboring main-land. There are few adults, residing within a day's sailing of the horse-pen, that have not attended one or more of these exciting scenes. A strong enclosure, called the horse-pen, is made at a narrow part of the reef, and suitable in other respects for the purpose—with a connected strong fence, stretching quite across the reef. All of the many proprietors of the horses, and with many assistants, drive (in deer-hunters' phrase,) from the remote extremities of the reef, and easily bring, and then encircle, all the horses to the fence and near to the pen.

There the drivers are reinforced by hundreds of volunteers from among the visitors and amateurs, and the circle is narrowed until all the horses are forced into the pen, where any of them may be caught and confined. Then the young colts, distinguished by being with their mothers, are marked by their owner's brand. All of the many persons who came to buy horses, and the proprietors who wish to capture and remove any for use, or subsequent sale, then make their selections. After the price is fixed, each selected animal is caught and haltered, and immediately

The wild horses of Shackleford Banks in this 1948 photograph appear very similar to their modern-day descendants and look well-nourished. Photograph by Aycock Brown, courtesy of the Outer Banks History Center.

subjected to a rider. This is not generally very difficult—or the difficulties and the consequent accidents and mishaps to the riders are only sufficient to increase the interest and fun of the scene, and the pleasure and triumph of the actors. After the captured horse has been thrown, and sufficiently choked by the halter, he is suffered to rise, mounted by some bold and experienced rider and breaker, and forced into a neighboring creek, with a bottom of mud, stiff and deep enough to fatigue the horse, and to render him incapable of making more use of his feet than to struggle to avoid sinking too deep into the mire. Under these circumstances, he soon yields to his rider—and rarely afterwards does one resist. But there are other subsequent and greater difficulties in the domesticating [sic] these animals. They have previously fed entirely on the coarse salt grasses of the marshes, and always afterwards prefer that food, if attainable. When removed to the main land, away from the salt marshes, many die before learning to eat grain, or other strange provender. Others injure, and some kill themselves, in struggling, and in vain efforts to break through the stables or enclosures in which they are subsequently confined. All the horses in use on the reef, and on many of the nearest farms on the main-land, are of these previously wild "banks' ponies." And when having access to their former food on the salt marshes, they seek and prefer it, and will eat very little of any other and better food. (Ruffin, 1861, pp. 131–132)

Shackleford is situated where hurricanes and other large storms frequently strike. Looking at a map, one will note that the coastline is oriented from southwest to northeast from northern Florida to Cape Hatteras. On reaching the Beaufort area, the shoreline runs in almost an east-west direction. This means that a broad expanse of ocean lies to the south and southwest of Shackleford and Cape Lookout. This positions the Outer Banks for direct hits from hurricanes that have strengthened over the warm Gulf Stream (Engels, 1952). These

Horses graze on *Spartina alterniflora* in a large, persistent saltwater marsh on eastern Shackleford Banks.

storms rotate counterclockwise, exposing Core Banks to the unmitigated onslaught of frenzied wind and waves. Situated perpendicular to Core Banks and advantageously positioned in the lee of the longer island, Shackleford is exposed to a direct strike from the south, but it is protected to a degree from storms from the east.

The year 1893 was particularly bad for storms—at least five major tropical storms or hurricanes swept Shackleford Banks, including one in August that killed 18 people and another in October that killed 22. In 1896, two major hurricanes brushed Shackleford Banks, convincing a number of residents to relocate to the mainland.

On August 17, 1899, the great San Ciriaco hurricane flattened Shackleford. Named for the saint's day on which the storm made landfall on Puerto Rico, the storm struck San Juan as a Category 4 hurricane, flogging the island with estimated sustained winds of 135–140 mph/217–225 kph (Barnes, 2007). In the days before satellites and radar, most residents were unaware that the storm was coming, and 3,369 drowned or were killed by flying debris and mudslides.

After passing north of the Dominican Republic and Haiti, the mighty storm grazed Florida, then strengthened and hit North Carolina as a Category 3 hurricane. At least 25 died in North Carolina, and Buxton clocked sustained winds at greater than 100 mph/161 kph, with gusts up to 140 mph/225 kph.

On Shackleford, seven ships wrecked. The storm washed homes off their foundations, and destroyed crops and gardens. Carcasses of dead horses and sheep lodged in trees, and families watched in horror as the waves unearthed deceased family members and scattered their bones.

On Ocracoke, 100 horses and cattle drowned. The storm surge grabbed two porpoises and lodged them in an oak tree. A U.S. Weather Bureau employee assessed the damage soon after the storm:

Beneath the sand of barrier islands, rainwater collects to form a freshwater lens. Thirsty horses need only locate where the aquifer lies closest to the surface and excavate a drinking hole, which slowly fills with freshwater when they reach the water table.

The chief force of the storm was spent on the Banks between Cape Lookout and Cape Hatteras, and the waves sweeping into Ocracoke Inlet nearly flooded the towns of Portsmouth and Ocracoke. Some houses were destroyed, blown over, or moved from their foundations, but no loss of life occurred in either town. Many cattle and other domestic animals were drowned undoubtedly. Many of the "Banker Ponies" roaming wild on Core Bank, opposite Carteret county, were also drowned, but not all; in fact the relief committee that visited this section observed numbers of marsh ponies all along the Banks, as well as cows, sheep, and some flocks of geese, which seem to have found places of safety. (Moore, 1899, p. 4)

The *Atlanta Constitution* reported,

It is now thought that the total drowned will run close to 100 if it does not overreach it . . . fully sixty to seventy houses four or five churches and numerous stores, barns and warehouses were either washed away or damaged beyond repair, and as a result numbers are homeless and destitute and others have lost crops and flocks. . . . There were several thousand wild ponies on the banks which divide the ocean from the sounds and nearly all these were drowned. ("Terrible Record of Recent Storm," p. 2)

The devastation was too much to bear. After the storm, even the most tenacious residents of Shackleford Banks decided to pack up what was left of their belongings and move to stable ground. Many settled in the Promised Land section of Morehead City or on Harkers

A stallion and his mare watch another stallion drive his mares through a corner of their home range. Home ranges often overlap at resources such as water sources, and these stallions worked out a compromise and saw no reason to come to blows.

Island. Those lucky enough to have intact homes—some 50 to 70 families—floated them on barges to places such as Beaufort and Broad Creek (Barnes, 2007).

By 1902, Shackleford was deserted. The island was once again left to wildlife and free-roaming livestock. At the turn of last century, Shackleford reportedly supported a vast number of wild horses.

> On Shackelford's banks alone are the little ponies referred to. It is strange but true, that they are found in their wild state nowhere else. There are said to be about twelve hundred of them on the banks. Inquiry made of residents as to whether the number of ponies had decreased during the past fifty years brought the response that it had, and that until about 1850 the ponies increased. . . .
>
> The colts are covered with hair several inches in length, a nature's protection against the weather. This is called colt hair, and looks like felt. It falls off in large flakes. Most of the colts are of a faded brown color, but are sometimes black. . . . Their instinct is remarkable. They know by means of it the way to get to the mainland or to islands with the minimum amount of swimming, and the writer has seen them wade great distances without getting out of their depth, making various turns and changes of direction to conform to the shoals, yet they are fearless swimmers.

The social dynamics of a wild horse herd are complex. Some animals form close friendships, and others are perpetually at odds. Horses within a band are usually patient and indulgent with young foals, and a mare's status rises temporarily after giving birth.

> Though an inlet only about two miles in width separates Shackelford's banks from Bogue banks, yet the ponies never go to the latter banks, nor do they cross the Ocracoke inlet. (Olds, 1902, p. 384)

In 1933, Barden Inlet severed Shackleford from Core Banks. William Engels, a zoologist who camped on Shackleford for a month, wrote that in 1939 "there were only two buildings on the island, each a one-room shed, the one just west of The Mullet Pond, the other on Wade Shore" (1952, p. 705). When he returned in 1952, he mentioned a summer cottage built on Wade Shore, and a hunter's camp east of Whale Creek Bay. He also encountered horses:

> Horses, cattle, sheep, goats, pigs and house-cats live on the island in a semi-wild state .... The horses have been famed since colonial times as the "bank-ponies"; they are popularly supposed to have reached the outer banks through the wreck of an early colonial Spanish vessel. The ponies do not enter the woodland, but roam the grassland eastward of Whale Creek Bay, and frequently visit the numerous marshy islands (1952, p. 721).

Cape Lookout somehow escaped the commercialism and population growth of the Cape Hatteras area. At one time, developers considered building a bridge to Shackleford and developing the island as a tourist destination, but the state began acquiring land in the 1960s, and ultimately it became part of Cape Lookout NS. Under the management of the Park Service, the island is evolving toward a more natural state.

Horses form bonds that can last for a lifetime. This Cedar Island pair is inseparable. They lived together on Shackleford Banks as mare and stallion, and she bore him a foal. They were moved to Cedar Island to replace horses lost to an epidemic of EIA, and in time herd managers gelded the stallion. Even with reproduction no longer a possibility, the bond endures.

Congress created Cape Lookout NS in March 1966, extending from Ocracoke Inlet in the north to Beaufort Inlet in the south, but it would be another 20 years before the seashore was fully established. Mostly undeveloped and accessible only by boat, the seashore comprises four barrier islands that buffer 56 mi/90 km of the central coast of North Carolina. Unlike most other parts of the Outer Banks, Cape Lookout is poised to cope with sea-level rise in the most natural of ways. Because there are no buildings, jetties, seawalls, or other man-made structures to block sand, the islands will naturally migrate and adapt.

Cape Lookout NS is situated on the Atlantic Flyway, and is home or seasonal stopover to at least 275 bird species, including the bald eagle, peregrine falcon, and piping plover. Wildlife is everywhere. Raccoons hunt for invertebrates in the shallows, probing the tidal creeks with sensitive fingers as if reading Braille. Nonvenomous snakes, coiled like pretzels, bake on sunny logs. In the summer months, newborn loggerhead turtles emerge on moonlit nights and flipper their way to the sea while other sea turtle species forage in the adjacent waters. Feeding black skimmers unzip the glassy water with their wakes. The island also supports terns, mergansers, herons, egrets, snapping turtles, rabbits, nutria, river otters, and many other varieties of wildlife.

Three grass communities run along the long axis of the island, comprising marsh, dune, and swale. Small tangles of thick woods, branches interlaced like Velcro®, grow randomly in the hollows. The landscape to the west consists of tall dunes, dense maritime forest, small freshwater marshes, and, on the western tip, a large saltwater marsh (Rubenstein, 1989).

Horses are social animals, but sometimes build community in surprising ways. One Cedar Island mare does not typically interact with other horses, but instead prefers the company and protection of three large bulls that share the marshlands. In heat she consorted with a stallion, but after conceiving her foal, she retreated to the security of her bovine guardians.

Sue Stuska led a research team that studied Shackleford equine grazing behavior and discovered that horses primarily consume four plant species (Stuska, Pratt, Beveridge, & Yoder, 2009). In the fall, 78.0% of the horses' diet consists of sea oats, centipede grass, and saltmarsh cordgrass (*Spartina alterniflora*). In the winter, they eat more centipede grass, less sea oats, and much less saltmarsh cordgrass, and consume small quantities of several other plants. In the spring, horses prefer sea oats, saltmarsh cordgrass, and pennywort and consume considerably less centipede grass. In summer, almost 65% of equine foraging included sea oats, centipede grass, and smooth cordgrass. Horses eat saltmeadow cordgrass (*S. patens*) consistently throughout the year without seasonal fluctuations.

The average horse consumes about 2% of its body weight in vegetation each day. One would think that the selection of fodder is related to the nutrition and digestible energy content of the plants season by season, but this does not appear to be true (Stuska et al., 2009). Horses appear to graze preferentially on plants that taste good to them. They prefer to eat short new growth rather than old, fibrous, relatively dry mature plants.

The horses of Shackleford flourish with the same vigor that kept their predecessors alive here for centuries. They demonstrate distinctive Colonial Spanish conformation, and average a diminutive 12 hands (48 in./1.22 m) high, though individuals raised on the mainland sometimes grow taller. Shacklefords have long, thick manes that unfurl contrarily in multiple directions. Bays, chestnuts, sorrels, and blacks are common. Dun, buckskin, and palomino colors have been uncommon or absent since the Park Service euthanized carriers of equine infectious anemia in 1996. Many have white markings, but none is pinto. The relentless sunshine bleaches highlights into light-colored coats, and gives blacks a rusty burnish. Their muzzles are well-furred, especially in winter, when their jaws are bearded with guard hairs. Their eyes are intelligent and unguarded, emotions sparkling within them.

The water in the holes dug by horses is muddy and fills the excavation slowly, but its salt content is minimal.

Bands of feral horses vary widely in size, usually 2–20 individuals (Ransom & Cade, 2000). Just as there is no "average" human family, scientists disagree on what constitutes normal horse behavior, and the horses themselves sometimes disagree with the scientists. Behavioral and social norms may differ widely from herd to herd and from band to band.

In general, a herd stallion will not permit another male to remain past puberty, but sometimes makes exceptions. The typical wild horse band consists of an alpha male and up to five or six unrelated mares with their foals. Herd composition reflects habitat, food availability, and sometimes sex ratio within the population.

Harem stallions sometimes tolerate a young subordinate stallion unrelated to the mares in the harem as long as he remembers his place and does not attempt to mate with the mares. Multiple stallions are generally present only in bands of more than 9 horses (Ransom & Cade, 2009). Usually, this arrangement begins when a subordinate stallion follows at the fringes of the band and defends it from intruding stallions (McDonnell, 2005). The patriarch initially tries to chase the interloper away, but over time skirmishes become less frequent.

In time, the harem stallion realizes that the subordinate can lighten the workload, giving him more time to graze, mate, and rest. The lower-ranked stallion assumes the task of scent-marking the urine and feces of the mares to discourage rivals. He defends the group from competing males and shepherds wayward youngsters back to the band.

The beta stallion may not mate with mature harem mares, but sometimes mounts the patriarch's adolescent daughters or young mares from other bands who visit during their pubescent heat cycles. Bands with more than one stallion are more stable, and mares are less likely to leave. Uncommonly, three or more stallions may share a band.

Although these arrangements are temporary, seldom lasting more than one breeding season, there are benefits for the subordinate stallion, too. When a stallion who has worked in this sort of "internship" moves on to join a bachelor band, he ranks higher in the in the social hierarchy (Rubenstein, 1982).

In some cases, a pair of half-brothers will leave their natal herd and form an alliance, driving off rivals. When they acquire mares, sometimes the more dominant of the two accomplishes most or all of the matings in the band; at other times, the stallions divide up the mares in estrus—"you take that one, I'll take this one"; and sometimes it is a free-for-all in which either stallion mates with whoever is convenient whenever he wants.

Two-male harems were uncommon on Shackleford in the early 1980s, when the sex ratio was typically two females to every male. During the late 1990s, the ratio for the entire population was closer to 1:1. Perhaps this explains the increase in two-stallion bands. In 2005–2006, 19% of the Shackleford bands contained more than one adult stallion (Nuñez, Adelman, Mason, & Rubenstein, 2009).

The band generally adheres to a set pattern of activity, moving along well-worn trails to locations within the home range that provide food, water, and relief from insects. The alpha mare will usually lead the herd in daily activities, and the stallion will bring up the rear; but he will move to the front if there is danger ahead. When a threat appears, the stallion will snort an alarm and the herd will mobilize, foals toward the center, stallion placing himself between his band and the menace (Bennett & Hoffman, 1999).

Stallions gather the band using a driving posture—head low, ears flat, menacing look in the eye. Moving the lowered head from side to side, "snaking," implies extreme threat that, unheeded, is followed by a nasty bite. Rank has privileges. The most dominant horse is the first to drink and has first choice of the available food, followed by the beta, and then the subordinates in descending order of status.

Horses are accomplished at both flight and fight. Mares usually back-kick for defense or offense or to discipline an unruly foal, though they may strike with front hooves when rejecting the advances of a stallion. Stallions may kick, bite, lash out with a hoof, or rear and strike.

Ordinarily, wild horses do not defend a territory. Instead, they maintain a movable "sphere of intolerance" within a home range that overlaps the spheres of other bands. The stallion grazes his band of mares within a preferred range, attacking any rival males that violate his invisible boundaries. In the 1970s, Shackleford Banks had the distinction of being one of the only places in the world where biologists observed horses defending territories rather than simply guarding harems. The island is so narrow that home ranges spanned from sea to sound in a band across the island, and stallions could easily spot intruders from a distance over the low dunes and grassland. (A single Assateague stallion, Comma, was observed displaying territorial behaviors on a similarly narrow, treeless stretch of island.)

The territorial stallions on Shackleford outcompeted weaker rivals and controlled home ranges with prime forage and easy access to water. Mares associated with dominant males grazed for an average of two extra hours per day and enjoyed increased fertility. Rubenstein was surprised to find these territorial stallions maintaining very large harems of mares. While his studies were in progress, around 1980, an increase of bachelor males upset the social order by overthrowing the older harem studs. These younger stallions divided the mares, and the territorial system dissolved (Rubenstein, 1981).

A study involving Przewalski Horses found that the ratio between adults and juveniles in the herd strongly influenced the rate of aggression between individuals (Bourjade, de Boyer

The rainy season brings vernal pools, freshwater that collects in depressions. Horses use this water source until it evaporates or sinks down to replenish the freshwater lens that forms below the sand.

des Roches, & Hausberger, 2009). When there are many foals in a herd, they tend to bond like gangs of naughty children and show aggressive behavior. This is probably because their mothers and other adults have less opportunity to regulate their behavior when they are off romping with playmates.

The dominance hierarchy is crucial to the social structure of the herd. When a new horse enters the band, fights ensue until the newcomer establishes a position in the hierarchy. After this initial trial, he or she generally maintains status by nonviolent threats such as pinning back the ears whenever another horse intrudes. These threats are well heeded by lower-ranking individuals, so actual kicking and biting are usually unnecessary. A more subtle manifestation of rank is the avoidance order. Lower-ranking horses will yield to dominant herd members by staying out of their way, dropping eye contact, and trying not to attract their attention. As long as a horse's position is well defined, he or she apparently feels secure, even if very low in rank.

The effects of a dominance system are to reduce overall aggression in the group and to control personal space. Status is a learned relationship between individuals, and a horse can learn to be more dominant by developing effective manipulation strategies such as bluffing (McGreevy, 2004).

One privilege of rank is that a dominant stallion spends more time grazing and less time defending his mares against other stallions. Dominant mares spend more time eating and have access to the highest-quality forage, have faster-growing foals, and engage in more sexual behavior. Dominant mares sometimes prevent subordinate mares in heat from mating with the stallion (Boyd & Keiper, 2005).

At a Cape Lookout pony penning in 1946, men and boys crowd the fence to get a closer look at the restless horses corralled within. The horses appear to have Spanish characteristics, and they bear a strong resemblance to their modern relatives on Shackleford Banks, as well as the horses pictured in the 1910 roundup near Oriental, N.C. (see chapter 5). Photograph by Aycock Brown, courtesy of the Outer Banks History Center.

The established hierarchy persists for as long as the group remains a stable, closed unit (Goodwin, 2002). Once defined, position in the social hierarchy remains relatively static until injury, old age, departure, or death of herd members changes it (Boyd & Keiper, 2005). An individual's place in the hierarchy is influenced by age and order of arrival in the harem, but not necessarily by size and weight. In domestic pastures, a Shetland Pony can be dominant over a Belgian draft horse.

The first mares to arrive in a harem tend to be of higher rank, and belligerent horses also rank higher. Older mares tend to rise in social standing. The sons of dominant mares tend to be higher in rank as well, but not their daughters. A new mare that joins a band is usually dominant only over the female offspring of the resident mares—unless the mares in the new herd are young and already acquainted with the new mare. Unsettled dominance relationships are mostly found between young horses and horses new to the band (van Dierendonck & Goodwin, 2005).

Social dynamics are usually more complicated than simple ranking. One horse may be dominant over another horse, who is dominant over a third, while the third horse is dominant over the first (Goodwin, 2002). As confusing as these relationships can be to human researchers, every horse knows his or her status.

Horses form cliques and friendships, keeping company throughout the day's activities. The strongest alliances are between mare and foal and within "best friend" pair bonds. Most

These men seem mesmerized by the strength and beauty of a defiant black stallion as he tests the rope that restrains him at a 1946 penning at Cape Lookout. Photograph by Aycock Brown, courtesy of the Outer Banks History Center.

horses have one or more friends with whom they graze, rest, engage in mutual grooming, and conduct most of their daily activities. Pair bonds can withstand periods of long separation, documented as long as 5 years in Icelandic mares (McGreevy, 2004). Horses of pair bonds often show jealousy when their companion mutually grooms or is courted by another horse.

When horses initiate mutual grooming, one horse will approach another with an invitation readable on its face. If the other horse agrees, each horse, beginning at the neck, gently nibbles and scratches her companion with her teeth, sometimes working her way to the root of the tail. Studies have shown that this grooming is more than pleasant scratching. It cleans the coat and tends to strengthen the social bonds between horses. Zoologists Claudia Feh and Joanne de Mazières (1993) discovered that when a horse's withers are nibbled in this fashion, her heart rate slows by 10% on average. Mares typically engage in mutual grooming with their foals, the stallion grooms with his favorite mares, foals of similar ages groom one another, and friends groom friends.

The stallion is not always the most dominant animal in the herd, especially if he is young and inexperienced. Houpt and Keiper (1982) found that all the stallions they observed were subordinate to some of the mares in their bands. Other studies including one on Shackleford showed stallions were usually the highest-ranked horse in the herd, but Keiper's study in 1986 showed that the stallions were not dominant in any of the Assateague Island bands (Keiper,1986).

The personalities of horses are as varied as those of dogs or people. Each forms friendships within the group and displays unique preferences and quirks. Like people, horses have personality conflicts. A stallion may be affectionate with one mare and bicker constantly with another. Even mothering skills differ from mare to mare. Foals from previous years, especially fillies, may remain close to their mothers. Understanding a horse's basic need for

By the mid-1990s, the Carrot Island herd had apparently recovered from its population crash, and the horses grazing across from the Beaufort waterfront appeared well-nourished.

companionship, a role or status within the herd, and physical contact with other horses can make one uncomfortable to realize just how lonely and sterile are the lives of many domestic horses kept solitary in a box stall or paddock much of the time.

Just as some gentlemen prefer blondes, a stud may be attracted only to mares of a certain color, usually the color of his dam, and will even go so far as to collect a band of identically marked mares. Some stallions, wild and domestic, will refuse to mate with mares of the "wrong" coat color. Most stallions prefer high-ranking mares, show little interest in fillies under the age of 3, and avoid mating with their own daughters.

The Shackleford Banks and Carrot Island horses make their own water holes by digging in the sand with their hooves until they reach the water table. A study by Blythe (1983) compared the salt content of water from various sources on the Rachel Carson National Estuarine Research Reserve and found these water holes to hold surprisingly fresh water. A tidal flat on the ocean side had 483 mEq of salt per liter of water, Taylors Creek (the narrow passage between Carrot Island and the mainland) had 477 mEq /L, and a tidal lake that connected with the sound had 339 mEq /L (Blythe). But water holes dug by horses at various locations on the island had a salt content ranging from 68 mEq /L to less than 10 mEq /L. Aside from temporary rainwater pools, the holes dug by horses are the only freshwater sources on Carrot Island, and they probably provide drinking water for other wildlife.

How is it possible for freshwater to lie below the surface of an island with no permanent freshwater source? The Ghyben-Herzberg principle explains this phenomenon, which concerns the relationship between freshwater and salt water on islands and peninsulas in proximity to the ocean (Blythe, 1983). The barrier islands of the East Coast and many nearby areas, such as Carrot Island, are made mostly of sand surrounded and permeated by salt or brackish water. Rainfall seeps into the sand, and a lens-shaped body of freshwater

Horses congregate on the grassy arms of Horse Island, where the sea breeze keeps flies at bay. The author waded from Carrot Island through a quarter mile (400 m) of shallows to approach close enough for photographs.

develops. This lens floats atop the denser salt water, "much as an iceberg floats in the ocean with most of its mass submerged" (Blythe, p. 70).

For every foot (30 cm) that this freshwater stands above sea level, it extends 40 ft/12.2 m below sea level. During periods of frequent soaking rains, the water table rises and the bottom of the lens sinks; the opposite happens during periods of drought.

To access the water table, the horses sometimes dig so deep that only their rumps are visible as they drink from the pool. Freshwater can be slow to seep into these depressions, and pony herds may spend long periods near them to ensure that each herd member gets enough. Horses return to these same holes for as long as they continue to produce, enlarging them with every visit. When water is particularly scarce, low-ranking horses may become dehydrated (Taggart, 2008).

On Shackleford Banks, a large, shallow freshwater pool named Mullet Pond provides water to horses on the west end of the island. Spring rains also create seasonal freshwater pools—hoofprints along the edges advertise that the horses use this source as well.

Edmund Ruffin wrote,

On the whole reef, there are no springs; but there are many small tide-water creeks, passing through and having their heads in marshes, from which their sources ooze out. Their supply must be from the over-flowing sea-water. I could not learn, and do not suppose, that these waters, even at their highest sources, are ever fresh. Water that is fresh, but badly flavored, may be found anywhere (even on the sea-beach), by digging from two to six feet deep. The wild horses supply their want of fresh water by pawing away the sand deep enough to reach the fresh-water, which

Today the horses on the Rachel Carson Reserve appear healthy. Horses in this small, isolated population are similar in size, color, and conformation because of a shrinking gene pool.

oozes into the excavation, and which reservoir serves for this use while it remains open. (1861, p. 133)

One 1900 account in a Scottish magazine describes Banker Horses catching and eating fish in these holes! The author explained, "they catch [fish] for themselves at low-tide, using their hoofs to dig deep holes in the sand below high-water mark; and they greedily devour the fish so left stranded, often fighting fiercely over an especially tempting one" ("Fishing Horses," 1900, p. 493). Because it is unlikely that horses would purposely catch live fish, the observer—who may have been separated by several degrees from the final publication—must have been watching horses drink from their water holes, perhaps so close to the tideline that fish were trapped within them. There is also an account of a couple of marsh ponies devouring an angler's sea bass, but this too may be a "fish story" (Clarke, 1892). On reconsideration, however, filching a fish might not be any more aberrant than raiding coolers for unlikely comestibles such as tuna sandwiches and soda pop, as do the Assateague horses.

Historically, horses grazed other islands in the area, including Hog Island, Browns Island, Harbor Island, Chain Shot Island, and Cedar Island. Steve Yeomans, a Cedar Island horseman, remembers when many of the locals would turn unneeded domestic horses out to join the herds in the marsh. Roundups were held in the summer. About 30 people would spread out to form a chain, and "walk them in." Stockmen would brand the horses, provide veterinary care, and trim overgrown hooves. They freed most back to the marshes (S. Yeomans, personal communication, October 15, 1995).

The same technique of "walking them in" was used on Shackleford Banks. An account from 1946 relates,

Whereas overgrazing can damage the environment, light grazing creates disturbances favorable to many native animals and plants.

Early in the morning, the beaters begin driving the ponies down the banks toward the trap, with a line of men which thinly stretches from the ocean to the sound. . . .

When the ponies approach the funnel leading into the pens, the volunteers armed with leafy twigs help prevent them passing around the pen or dashing off into the sound. Once in the pen, owners apply their brands to the colts, identifying them by watching whose mares they are following. Brands are registered in Carteret County courthouse. If any bidders are present, a few may be sold, at from $50.00 to $100.00, and taken to the mainland, where they are used as pets and riders. . . .

The mares and foals follow their stallion-leader unhesitatingly, and even the horse-hunters cannot restrain a cheer when a wise old banker stallion outwits [h]is adversaries. Sometimes he will climb to the top of a sea-oat-tufted dune, and sniff at the danger ahead and behind, while his herd huddles quivvering [sic] below, the frightened foals nuzzling close to their mares[.] At his moment of quick decision, the leader sometimes plunges headlong toward the pen, his crowd galloping blindly behind him[.] But once in awhile he will find a break in the line, and wheeling swiftly, plunge into the sound, reach deep water, and swim around behind the discouraged beaters who have walked miles in the sun, only to have their quarry outwit them. (Munsell, 1946, p. 11)

Margaret Willis (1999) had similar impressions of the horses when she attended the Park Service gathers of the late 1990s. She wrote,

These smart little horses know well how to hide just on the other side of a dune or in a little valley and within the thickets and forest. One can walk right into them without being aware of their presence until near contact or right by them then turn to see alert, soft brown eyes quietly watching.

Once out in the open, these horses can disappear before your eyes then reappear some distance away, speeding in excess of 32 miles an hour, noses in the air and showing you their heels. Most seem to enjoy playing these hide and seek games, almost smiling as they slip away.

Banker Ponies had long been valued for riding and draft throughout the area. "Not all of them make good riding horses," Yeomans explained. "Some are unridable. They're just too smart. Most of them are easier to train than other breeds. Once you get them used to a saddle and bridle, they'll do anything in the world for you" (personal communication, October 15, 1995). Some were terrific swimmers and could even cross wide, swift currents in the channel separating Harkers Island from the mainland with a child astride. He told of a mare that would not flinch when a large rifle was shot from between her ears. Yeomans' grandmother lived on Shackleford Banks when livestock still roamed freely. For generations, Banker people and Banker Horses worked as partners. The horses were transportation, muscle, and recreation. Children grew up riding barefooted on bareback horses, running the beaches and swimming the sounds. For generations, for as far back as anyone could remember, the horses were there.

A 1902 article in the *New York Times* reads,

> East of Greensboro . . . it is the ambition of every child to own a Banker, and each town is likely to boast three or four of the pretty creatures. Hitched to diminutive buggies and wagons, they trot around the streets, taking steps about as long as a dog's. They are then as gentle as lambs for once tamed their tameness is absolute. ("Wild Horses of North Carolina," 1902, p. 27)

At one time, Shackleford roundups were conducted by men and boys, but during the 1950s, the pennings became part of July 4th celebrations and involved most of the community. Horses were removed to Harkers Island or the mainland for riding. Sometimes Bankers were brought over to serve as summer mounts for children. Come September, the horses were regretfully returned to the island to resume their wild lives while the children resumed their domesticated ones.

Shackleford horses are rugged and hardy, curious and smart. Carolyn Mason of the Foundation for Shackleford Horses commented, "When most horses who haven't been handled see something unusual, they often will spook and run away. A Shackleford will walk right up and investigate" (personal communication, May 25, 2009).

In 1960, an edict by the 1959 General Assembly of North Carolina required stockmen to remove all livestock from Shackleford Banks. A news account reported,

> The legislation instructed the Sheriff of Carteret County to shoot them, unless their owners herded and removed them to the mainland. The "head" count is estimated to be 500 sheep and goats and 25 head of cattle on the Island.
>
> Owners had constructed a corral to complete evacuation, but along came violent hurricane Donna to blow it away. The Sheriff has not commenced firing. But there is little time left. The move is aimed at conservation, to permit grass to grow and anchor the sand blown inland from the Atlantic ocean [*sic*].
>
> Still, the natives and the tourists who look for the "Banker Ponies" in the area more than goats and cows, and forward to the roundup, will miss this attraction. ("Take 'Em Alive," 1960, p. 4)

This magnificent black stallion, Dionysius, was gathered in the roundup of 1996 and released back to the island as a healthy horse. Other herd members were not so fortunate. In 1996, 76 Shackleford Banks horses tested positive for EIA and were subsequently euthanized.

Stockmen did remove large numbers of livestock in the early 1960s, but apparently not all. The remainder multiplied, and between 1978 and 1981 the livestock census on Shackleford ranged from 81 to 108 horses, 64 to 89 cattle, 104 to 144 sheep, and 100 to 150 goats (Wood, Mengak, & Murphy, 1987). By that time there were no year-'round residents on Shackleford, but landowners maintained fishing cabins and other structures for seasonal use.

In 1986 the Park Service razed the remaining buildings and allowed natural processes to resume. It requested that stock owners remove sheep, goats, and cattle from the islands because the grazing of feral ungulates appeared detrimental to saltmarsh and grass-shrub areas and to dune-stabilizing grasses. Then the agency slaughtered any remaining livestock "to prevent the spread of disease" (Saffron, 1987).

Local residents, however, strenuously resisted the removal of Banker Horses from Shackleford Banks. Park Service resource management specialist Michael Rikard said in an interview in the New Bern *Sun Journal,* "For some reason, there's very strong public sentiment for the horses, as opposed to goats and pigs" (Gengenbach, 1994, p. C1).

Cape Lookout NS Superintendent Preston Riddel planned to decrease the herd to a "representative number," professedly for its own good (Saffron, 1987). "The horses, for their own sakes, would be better off somewhere else," he said. "It's a very, very hard life. They belong in an environment where they can be protected." He went on to assert that if left to themselves, herds of wild horses would inevitably increase until the vegetation could no longer support them—disregarding the fact that they had been in residence for centuries.

In the 1980s, biologists determined that horses primarily consumed *Spartina alterniflora,* saltmarsh cordgrass, which readily recovered. Goats, on the other hand, browsed relentlessly in the maritime forest and had a negative impact on the ecosystem. Ultimately, in 1987, the Park Service allowed a representative herd of feral horses to remain on Shackleford "because of their potentially historic origin" (Prioli, 2007).

Lacking sheep, goats, and cattle to compete with them for resources, horses multiplied rapidly, from a relatively stable count of roughly 100 from the 1970s to 1986 to an estimated high of more than 221 in 1994 (Prioli, 2007). A management crisis had developed. The Park Service declared that the equine population overgrazed the island and strained the ecosystem. Rubenstein (1982) wrote, "Grazing competition on Shackleford is very intense. Despite the fact that horses spend over 75% of each hour grazing, bodily condition remains poor, and juvenile death rates remain high" (p. 484).

Mares who bear foals at the extremes of reproductive age are particularly vulnerable to malnutrition, as are foals and yearlings of either sex. Nature's way of keeping the balance is to allow the most vulnerable horses to die of starvation when food sources are exhausted. It was feared that if the Park Service did not intervene, large numbers of horses would suffer preventable deaths, and in the meantime certain native plants could also be grazed out of existence. Rikard said "We are concerned about a major die-off of horses. . . . Some out there are real thin and looking bad" (Park service [sic] study," p. 3).

The Park service had reason for concern. In the mid-1980s, a wild horse die-off struck Carrot Island, a small, marshy island lying to the west of Shackleford. Carrot, Town Marsh, Bird Shoal, Horse Island, and Middle Marshes make up the 2,315-acre/938-ha Rachel Carson NCNERR, one of 27 reserves around the country designated for research, education, and stewardship. The four reserves in North Carolina represent the diversity of habitats in the state's estuarine ecosystems. Of these, Rachel Carson and Currituck both reluctantly tolerate horses.

Carrot Island appears on the Moseley map (1733), and it was the site of a fishery in the early 1800s. In 1782, during the Revolutionary War, a small British party landed near the mouth of Taylors Creek, exchanged fire with locals, and then withdrew to Carrot Island. The next morning, the British overcame the local troops in Beaufort and briefly occupied the town (Fear, 2008).

A fishery existed on Carrot Island as early as 1806 (Fear, 2008). In the 1920s, the U.S. Army Corps of Engineers dredged Taylors Creek and deposited the spoil on Carrot, building it higher and increasing its stability. Carrot Island, Town Marsh, Bird Shoal, and Horse Island, acquired for the reserve in 1985, total more than 3 mi/4.8 km in length and less than

1 mi/1.6 km in width. Middle Marsh, acquired in 1989, is roughly 2 mi/3.2 km long and less than 1 mi/1.6 km wide. The Reserve was named in honor of Rachel Carson, who conducted research at the site in the 1940s.

In the late 1940s, a Beaufort physician named Luther Fulcher, who also owned horses on Shackleford Banks, released six of his horses to graze Carrot Island and its associated salt marshes, intertidal flats, and tidal creeks. These animals were probably not the first equids to live there. Paula Gillikin, Rachel Carson site manager, said, "Horses were likely on the property long before the 1940s; although at this time it cannot be proved as there isn't enough documentation. . . . Horse Island has been on the maps since the 1800s. It probably got that name because at some point there were horses there" (personal communication, March 11, 2013).

Amy Muse (1941, p. 69) wrote of the early 1900s,

> At some elusive period early in this century, Beaufort changed considerably. Banker ponies were prohibited on the Town Marsh and Bird Shoal, so they were no longer able to swim across the channel at low tide to graze along the sidewalks or run up and down the streets at night. . . . Dr. Maxwell came out with his Maxwell automobile in 1911, and from then on the familiar two-wheeled carts drawn by banker ponies began to disappear from the streets.

In July 1976, about 40 acres/16 ha of the island were almost auctioned off for development. Beaufort residents who enjoyed the wild beauty of the island and its horses took action. After a legal battle, the Nature Conservancy, aided by funds raised by concerned local residents, paid $250,000 for Carrot Island and Bird Shoal.

With nothing to curb their fertility, the free-roaming horses proliferated and overgrazed the marsh. By 1986, the horse population on these small islands and marshes had reached 68. There simply was not enough food for all of them, and palatable plant species began to disappear until minimal resources were available to other species.

Unchecked, populations of large herbivores can proliferate quickly until they reach carrying capacity, the maximum number that the environment can support, whereupon the numbers typically stabilize at or below that level and the herd stops growing (De Roos, Galic, & Heesterbeek, 2009). A population at carrying capacity is unhealthy for both the herbivore and the vegetative community. At maximum population density, plant diversity suffers, and certain plant species may become locally rare or extinct. Likewise, an ungulate population maintained at carrying capacity has a greater incidence of disease and malnutrition, reproductive challenges, and a higher death rate compared to a herd in balance with its habitat.

A population overshoot occurs when the census temporarily exceeds the carrying capacity, resulting in overpopulation until more deaths or fewer births restore the balance. A population overshoot may precede a large-scale die-off, which occurs when the demand for food so greatly exceeds supply that the animals fail to reproduce and die in great numbers (Wood, Mengak, & Murphy, 1987). On St. Matthew Island, Alaska, a group of 29 reindeer introduced in 1944 multiplied to 6,000 within 19 years—many more animals than the island could support. Weakened by starvation and disease, the population plummeted to 50 during one especially severe winter. All survivors were female, and the herd became extinct. Similar explosions and die-offs have occurred with reindeer on the Pribilof Islands of Alaska; sika deer in Japan and on James Island, Md. (Diefenbach & Christensen, 2009);

In the 1990s, the Park Service freeze-branded large numerals on the left rump of the Shackleford horses that were re-released to the island.

cottontail rabbits on Fishers Island, New York; and moose on Isle Royale, Michigan, in Lake Superior (Klein, 1963).

The 68-horse herd on Carrot Island experienced a similar die-off during the winter of 1986–1987. Abundant rainfall in the late 1970s caused lush vegetative growth on Carrot, allowing the well-nourished herd to expand through increased fertility and greater survival of the oldest and youngest members of the herd. The severe drought that followed this abundance killed or stunted plants, and suddenly there was not enough food for the enlarged equine population.

Moreover, the Army Corps of Engineers dredged more than 150,000 tons/136 million kg of silt and sand from Beaufort Channel and deposited it atop the island, forming an 18-foot-high (5.5 m) dike that prevented the horses from accessing a freshwater pond (Saffron, 1987). Barry Holliday, an Army Corps of Engineers official, explained that dumping sand was necessary to eliminate ground cover used by small mammals that prey on marsh birds, acknowledging that brush and grass vital to the horses was interred. "'By pumping dredge material on the island from time to time, it basically purges small bushes and shrubs where predators hide,' he said. The island's ground cover, he said, was never meant to sustain a non-indigenous species like the horses" (Saffron, 1987).

To shore up his position, Holliday reportedly added that "dumping of dredged sand created the island in the first place" (Saffron, 1987). One wonders how this dumping affected species of concern. One also wonders how horses could possibly be more damaging to the island than burying native plants and animals with dredge spoil. And it remains a mystery how dredging begun in the early 20th century, in part to reverse the natural growth of Town Marsh (Warshaw, 2010), could have created a recognizable Carrot Island, bearing that very name, nearly 200 years earlier (Warshaw, 2012; Moseley, 1733).

The Foundation for Shackleford Horses saved the lives of five horses that tested positive for EIA by finding them a home where they could live in quarantine.

Famine, disease, and parasites killed 29 horses within a few months. Once concerned locals realized what was happening, they brought in hay as supplementary feed; but the starving horses, accustomed to native grasses, were reluctant to eat it. Spring brought numerous foals, and by August 1988 the herd numbered 51. The North Carolina Division of Marine Fisheries had helped them through the 1987–1988 winter by providing 20 bales of hay each week. A point well ensured adequate freshwater. No horses died that winter, which was significant; typically, during an equine population collapse, about 75% of foals die, and almost all the yearlings (De Roos et al., 2009). But clearly this level of human assistance could not continue. If the horses were to remain as on the island, managers must ensure that their numbers remain in balance with their environment.

Biologists determined that Carrot Island could comfortably sustain between 15 and 25 horses, and in 1988 the state removed 33 of 52. Nine of the 33 removed tested positive for EIA and were euthanized. Private individuals adopted the remainder. "To avoid another population collapse, protect herd health, and minimize environmental damage, the herd size is currently managed through a birth-control program similar to those administered by the National Park Service at Cape Lookout and Assateague Island National Seashores," says Gillikin (personal communication, April 18, 2011). Additionally, the Army Corps of Engineers currently deposits dredge spoils *alongside* Carrot Island rather than on top of it.

In 1998, the Secretary of the North Carolina Department of Environment and Natural Resources permitted a representative herd of 30 horses to remain on the Rachel Carson Reserve. The reserve keeps a record book that tracks each horse, noting parentage, appearance, reproductive record, contraceptive doses, general health, social habits and, eventually, death. Occasionally, horses have wandered to places where their presence is even less

Cedar Island wild horses graze under a leaden sky. The horse at the far left is "Shack," otherwise known as Deliops (*spoiled* in reverse), the 1998 son of the elderly mare #16 of Shackleford, pictured elsewhere in this chapter. He has sired numerous foals in the privately owned horse sanctuary on Cedar Island.

welcome In 1994, three bachelor stallions swam from Carrot to Radio Island—off limits for horses (Gengenbach, 1994).

Public opinion has influenced the policies that have allowed the herds to remain, but people still voice concerns. Personnel at the reserve have received requests to plant forage for the horses to eat, offers from local businesses to buy supplemental feed, appeals to evacuate horses to the mainland before storms, and opposition to immunocontraception and Coggins testing (Taggart, 2008).

Some horse advocates argue for a hands-off approach. They support leaving these horses to live entirely as wild animals with minimal interference from people, despite challenges such as limited freshwater and exposure to weather extremes including heat waves, ice storms, nor'easters, hurricanes, and floods. Others demand that the horses be managed with all the accoutrements of domesticity. But if they are to be sheltered, vaccinated, fed, wormed, and given farrier care, the reserve might as well take it one step farther and offer them for adoption on the mainland.

A herd large enough to maintain genetic diversity without periodic introductions of unrelated horses is too large for the food resources in the Rachel Carson Reserve. The Reserve comprises several islands separated by shallows, creeks and mud flats that frustrate access by personnel and make immunocontraception difficult to implement. In the 1990s, veterinarians repeatedly attempted to test each Carrot Island horse for EIA within the same week and were unsuccessful because of the inaccessible terrain (Taggart, 2008).

Horses remain on Carrot Island, but relations with their caretakers are strained. The state of North Carolina does not view wild horses as repatriated native wildlife, but as invasive exotics. The managers of the state lands on which they graze see the horses as incompatible

Bucky, a buckskin mare whose mane and tail bleach to blonde in the sun, was the last survivor of the original Cedar Island herd. When horses were brought in from Shackleford to resurrect the herd, Bucky found a mate, and 2010 in she produced a filly named Gay. The original bloodlines will continue.

with the management goals of the reserve and fear long-term ecological consequences (Taggart, 2008). They believe that grazing and trampling of marsh grasses will accelerate erosion, and sea-level rise could further limit resources (Taggart, 2008). Yet, ironically, Rachel Carson herself, namesake of the reserve wherein the horses roam, did not consider horses harmful to coastal wildlife refuges. She wrote—on behalf of the U.S. Fish and Wildlife Service—that the 300-odd cattle and ponies grazing the roughly 9,000 acres/3,642 ha of Chincoteague National Wildlife Refuge were not detrimental to the waterfowl for which the refuge was established (1947, p. 17).

Reserve policies mandate maintenance of these herds, but their presence conflicts with the core mission of the reserve. Taggart, an assistant professor of environmental studies at the University of North Carolina at Wilmington wrote (2008, p. 187),

> Among the Atlantic Coast herds, conditions at the Rachel Carson site are least accommodating for the animals. With a combination of pertinent research results plus 20 years of site-specific management experience as a basis, I argue that feral horses of the Rachel Carson site should be removed for programmatic, ecological and humane reasons.

One concern is the plight of the crystal skipper (*Atrytonopsis* new species 1), a federally protected butterfly that ranges only along Bogue Banks from Fort Macon to Emerald Isle, Radio Island, and the eastern end of the Rachel Carson Reserve. This unique insect prefers untouched open-beach dune habitat and revegetated dredge spoil areas with a variety of plants (Taggart, 2008). The larvae of the crystal skipper eat only seaside little bluestem (*Schizachyrium littorale*), a plant that appears unable to establish itself on the reserve, ostensibly due to pressures from grazing horses. Horses do consume small amounts of little bluestem, but mostly during the winter months, when butterflies are not reproducing (Stuska et al., 2009).

In 2009, the reserve established four 8 ft x 8 ft/2.4 m x. 2.4 m plots and planted them with seaside little bluestem to encourage breeding butterflies. Three of the four plots were edged with pony-proof fencing, and the fourth was left open to monitor the effects of equine disturbance. As it turned out, the horses browsed the plot only once, and the plants recovered fully. No skippers used the plants, but their avoidance had nothing to do with grazing.

Nonetheless, the presence of horses on the Reserve creates a management conflict for its managers. "The Rachel Carson component of the North Carolina Coastal Reserve and National Estuarine Research Reserve is protected to provide opportunities for long-term research, education, and interpretation," says Gillikin:

> Essentially, this means that the site is managed to remain as close to a natural state as possible. Horses remain on the Rachel Carson Reserve due to the strong public sentiment attached to them. Humane management of herd size is essential to protect herd health and the natural environment. (P. Gillikin, personal communication, April 18, 2011)

At first, the North Carolina State University College of Veterinary Medicine administered a hormone-blocking vaccine to limit herd growth (Taggart, 2008). This treatment was ineffective, so reserve officials began to dart mares with porcine zona pellucida vaccine in 1999. The birth-control program is the only regular intervention in their lives.

> Gillikin closely monitors the equine census. In corresponding with the author, she wrote, As of April 18, 2011, there are 31 adult horses and 1 foal present in the herd. At this time, there are 7 harems and no bachelor bands; our bachelors are generally loners. It is not known if any mares are pregnant, as we do not perform pregnancy tests. Visual observations are used to assess pregnancy; no mares appear to be pregnant at this time. (Personal communication, April 18, 2011)

While the Rachel Carson Reserve horse population crested, crashed, and restabilized, the National Park Service took a hard look at the equine population dynamics at Cape Lookout National Seashore. By the late 1990s, Park Service personnel voiced concern that the Shackleford Banks herd would suffer a die-off similar to what had occurred on Carrot Island if they did not take action to decrease the population. The agency used research by Gene Wood and Daniel Rubenstein to better understand the relationship of the horses to the native wildlife and vegetation. Data in hand, the Park Service pondered how best to handle the equine population dilemma (Wood, Mengak, & Murphy, 1987).

Barrier island ecosystems are intricate, naturally maintaining a dynamic equilibrium of biology, geology, and physical processes that have remained in balance for tens of thousands of years (Zedler & Callaway, 2001). Researchers who attempt to study a single component of this ever-changing web have difficulty extracting meaningful data from the lively interplay. The scientific method can usually prove that change has taken place, but can only speculate on causation and key influences. When managers manipulate habitat, wild species often respond in an unexpected manner for reasons that are not immediately obvious. Our conclusions are invariably generalizations, because ecologists cannot yet accurately model the relationships of organisms and environment in detail. Moreover, we often filter conclusions through our expectations and biases.

Sporting comically fuzzy ears and an undisciplined Shackleford mane, a young filly named Sprite dozes in the shade near her dam. Immunocontraception allows herd managers to restrict herd growth with minimal disruption to the animals. Very few foals are born on the island each year, most of them to mares with uncommon bloodlines.

Land managers around the world use managed grazing to maintain a healthy environmental balance. Godfrey and Godfrey (1976, p. 115) wrote that Shackleford's horses graze the *Spartina* in the salt mashes "down to within an inch of the mud," but concede that it is difficult to accurately assess the overall impact of grazing livestock on the Outer Banks. "Some localized areas were undoubtedly overgrazed and thus livestock were blamed for the 'deteriorated' condition of the entire Outer Banks" (Godfrey & Godfrey, 1976, p. 115).

Whether horse grazing is ultimately helpful or harmful to the environment depends largely upon herd size. Overpopulation is clearly detrimental to horses and habitat alike. By the mid-1990s many of the horses in the Shackleford herd were underweight, particularly the mares, and researchers were seeing signs of increasing environmental damage. It appeared that Shackleford Banks had more horses than it could comfortably support.

Whenever the management of wild horses is up for discussion, the enigmatic pull horses have on our emotions tends to polarize people for or against their continued residency. Heated arguments erupt between advocates and detractors, many of them armed to the teeth with hard facts, half-truths, conjecture, and strong feelings. Scientific data does little to resolve issues rooted primarily in culture, economics, and politics (Linklater, Stafford, Minot, & Cameron, 2002).

In the 1980s and 1990s, the Park Service saw free-roaming horses as uninvited guests at a dinner table set to nourish native wildlife. The agency proposed several possible management plans, including simply removing the horses from Shackleford Banks to allow the island to regain its former character. According to Park Service policy, "management—up to and including eradication" is warranted if a species deemed exotic "disrupts the accurate presentation of a cultural landscape, or damages cultural resources" (USNPS, 2006, p. 48). Paradoxically, the Park Service concedes that the horses are themselves a cultural resource. Thus the agency can neither deport them nor keep them without neglecting part of its mission.

Many Down-Easters—residents of the lowlands and villages of eastern Carteret County—considered the horses an important part of part of their cultural heritage and wanted them to remain. Many had grown up on the backs of the rugged little Bankers, and had fond memories of participating in annual pennings. Comments on the Seashore's 1982 *Draft General Management Plan* revealed strong public support for the horses, and a commitment to maintaining the herd on Shackleford Banks. The *Final General Management Plan* specified that a representative herd would remain on Shackleford Banks after federal acquisition was complete (Smith, 2003).

The Park Service considered removing a number of horses from the island and "managing" the rest. The agency proposed that a herd of roughly 60 horses could be self-sustaining if outside genes were periodically added to revitalize the gene pool. A roundup could be held and the surplus horses adopted by the public. Stallions could be castrated. Mares could be maintained on contraceptives. Another workable alternative would have been to restore part of the herd to its historic range on Core Banks and limit numbers though contraception, but the Park service vetoed this option.

A contingent of local people favored controlling the population through annual roundups. Horse pennings had been a tradition on these islands for centuries. As on Chincoteague to the north, an annual roundup could stimulate tourism, and proceeds from the sales of young stock could support horse management on the island. The Park Service opposed this plan and argued that removal of foals alters the social and reproductive dynamics of natural horse behavior. Horse penning would trample the vegetation disrupt the environment in the vicinity of the pens, and place humans and horses at risk of injury. This last concern was well-founded: wild horse gathers in the American West often result in equine injury and death. In 1998, the Park service gathered the Shackleford herd with a helicopter and hunting dogs to test for disease and euthanized two injured horses (Willis, 1999).

In 1994 officials from the Cape Lookout NS held a series of open meetings with the public and interested groups at the North Carolina Maritime Museum in Beaufort. Surprisingly, more than half of those present felt that the horses should be removed or, if permitted to remain, prevented from breeding and simply left to live out their natural lifespans. After 20 or 25 years, the horses would be gone.

Ultimately the Park Service decided to gather all the horses, test them for EIA with the assistance of the North Carolina Department of Agriculture, destroy any positive reactors, and offer the remainder for adoption. The Park service would return 50–60 horses to the island and use contraception vaccines to limit fertility. With the birth rate in check, the population would never again exceed what the island could comfortably sustain. Both the horses and their environment would be healthier.

EIA, also known as swamp fever, a disease that affects only equids, is caused by a lentivirus—a retrovirus with a long incubation period—similar to the ones that cause AIDS in humans and feline leukemia. The disease was identified in France in 1843, and it was probably first diagnosed in the United States in 1888. It occurs worldwide, but predominantly in warmer climates, and there is no vaccine or cure. It is not contagious, or spread

Mare number 16, age 22 at the time of this 2010 photograph, shows loss of muscle on the neck and prominent spine and hips. These age-related changes are normal, exacerbated by the nutritional stress of nursing her most recent foal, 15U, and the eight offspring that preceded her. All in all, she appeared in good health. She died in 2012 at the age of 24, a good long life for a multiparous wild mare. Her son Deliops was moved to Cedar Island, where his healthy genes breathed new life into the dwindling Cedar Island herd.

directly from one animal to another; but it is infectious, transmitted by vectors. Horseflies can transfer blood from infected hosts into healthy animals. So can people when they use the same syringe to draw blood from multiple horses. After an incubation period of 2–4 weeks, infected horses may show signs such as fever, weakness, weight loss, jaundice, lack of coordination, and swelling of the legs and underbelly. Some may die.

Once infected with EIA, horses remain infected for life. Some positive testers are highly infectious, others are less so. Horses that survive the initial attack or contract a mild case become asymptomatic carriers that can infect others (Cheevers & McGuire, 1985). The viral content of blood samples from positive-testing horses may differ by a factor of 1 million (Cordes & Issel, 1996). An inapparent carrier may carry an infinitesimal amount of virus, but under stress may have a flare with a viral load high enough to infect a whole herd with a few drops of blood. Cordes and Issel (1996) state that 1 ml/0.03 oz of blood from a horse with chronic EIA having a feverish episode contains enough virus to infect 10,000 horses.

In 1970 Dr. Leroy Coggins developed a serologic test for antibodies specific to the virus. The spread of EIA has been slowed by regulations that require testing of domestic horses before crossing borders, breeding, racing, entering a horse show, or any other formal activity that will bring horses from one location into contact with horses from another (Cordes & Issel, 1996). Most adult horses that have a positive Coggins test are inapparent carriers. They show no obvious signs and have very small amounts of the virus in their blood, but they remain reservoirs of the disease.

Horseflies feed by slashing open the skin and lapping the blood. When a horse feels the bite, he attempts to dislodge the fly by twitching his fly-shaker muscle, stamping, biting,

Henneke Body Condition Scoring System

1—Poor (Extremely emaciated; no fatty tissue)	Neck	Bone structure easily noticeable
	Withers	Bone structure easily noticeable
	Shoulder	Bone structure easily noticeable
	Ribs	Ribs protruding prominently
	Loin	Spinous processes projecting prominently
	Tailhead	Tailhead, pinbones, and hook bones projecting prominently
2—Very Thin	Neck	Bone structure faintly discernible
	Withers	Bone structure faintly discernible
	Shoulder	Bone structure faintly discernible
	Ribs	Ribs prominent
	Loin	Slight fat covering over base of spinous processes. Transverse processes of lumbar vertebrae feel rounded. Spinous processes prominent
	Tailhead	Tailhead prominent
3—Thin	Neck	Neck accentuated
	Withers	Withers accentuated
	Shoulder	Shoulder accentuated
	Ribs	Slight fat over ribs. Ribs easily discernible
	Loin	Fat buildup halfway on spinous processes, but easily discernible. Transverse processes cannot be felt
	Tailhead	Tailhead prominent but individual vertebrae cannot be visually identified. Hook bones appear rounded, but are still easily discernible. Pin bones not distinguishable
4—Moderately Thin	Neck	Neck not obviously thin
	Withers	Withers not obviously thin
	Shoulder	Shoulder not obviously thin
	Ribs	Faint outline of ribs discernible
	Loin	Negative crease (peaked appearance) along back
	Tailhead	Prominence depends on conformation. Fat can be felt. Hook bones not discernible
5—Moderate (Ideal weight)	Neck	Neck blends smoothly into body
	Withers	Withers rounded over spinous processes
	Shoulder	Shoulder blends smoothly into body
	Ribs	Ribs cannot be visually distinguished, but can be easily felt
	Loin	Back is level
	Tailhead	Fat around tailhead beginning to feel soft
6—Moderately Fleshy	Neck	Fat beginning to be deposited
	Withers	Fat beginning to be deposited
	Shoulder	Fat beginning to be deposited
	Ribs	Fat over ribs feels spongy
	Loin	May have a slight positive crease (a groove) down back
	Tailhead	Fat around tailhead feels soft
7—Fleshy	Neck	Fat deposited along neck
	Withers	Fat deposited along withers
	Shoulder	Fat deposited behind shoulder
	Ribs	Individual ribs can be felt with pressure, but noticeable fat filling between ribs
	Loin	May have a positive crease down the back
	Tailhead	Fat around tailhead is soft
8—Fat (Fat deposited along inner buttocks)	Neck	Noticeable thickening of neck
	Withers	Area along withers filled with fat
	Shoulder	Area behind shoulder filled in flush with body
	Ribs	Difficult to feel ribs
	Loin	Positive crease down the back
	Tailhead	Fat around tailhead very soft
9—Extremely Fat (Fat along inner buttocks may rub together. Flank filled in flush)	Neck	Bulging fat
	Withers	Bulging fat
	Shoulder	Bulging fat
	Ribs	Patchy fat appearing over ribs
	Loin	Obvious crease down the back
	Tailhead	Bulging fat around tailhead

Score each element separately, add the scores, divide by 6, then round to the nearest ¼ point to allow for uneven fat deposit. Adapted from USBLM, Billings Field Office (2009, pp. 132–133).

The author took this photograph on the east end of Shackleford Banks in October 1995. There was little forage available—the tall grasses around the horses appear to be species unpalatable to horses. Many mares in this band had a Henneke score of 1.5–2. Note this mare's prominent spine, hips, and ribs beneath her winter coat, and the wasting of her neck and hip musculature. Many of these horses tested positive for EIA the year after this photograph was taken. What combination of factors led to the emaciated state of the lactating mare? Chronic disease? Repeated reproduction? Advanced age? Inadequate forage? Heavy parasite load? Regardless of cause, the lactating mares as a group had the lowest Henneke scores of all horses observed. Although thin mares produce less of milk than well-nourished mares, this mare's foal appears in reasonably good condition.

shaking his mane, or switching his tail. Meal interrupted, the fly alights again and makes another slash or zooms off to the next victim with wet blood on her mouthparts. More typically, the fly feeds on a single horse and remains nearby after its meal; the disease does not spread easily (Willis, 1999). The virus does not live long on the horsefly carrier, and the disease can be passed only between animals that are near each other. Whereas a single fly may spread the disease from an infected horse with a high viral count, transmission from an inapparent carrier may require 25 or more blood transfers. In the summer, there is no shortage of vectors: a wild horse may receive more than 1,000 horsefly bites per hour (Cordes & Issel, 1996). Mosquitoes do not spread the disease.

If mares test positive, their foals will initially test positive as well. These foals may be infected or may simply carry passive antibodies to the disease obtained through the mare's colostrum. When mares are stable, inapparent carriers of EIA virus, the vast majority of their foals will eventually test negative, even when weaned at 5–8 months of age in areas with high populations of insect vectors. A three-year study of Choctaw and Cherokee horses in Oklahoma revealed that 97% of the foals born to infected mares eventually tested negative for the disease (U.S. Department of Agriculture, 2006).

The Humane Society of the United States, U.S. Representative Walter B. Jones (NC 3), scientists and horse advocates vigorously opposed EIA testing of the Shackleford Banks herd. Local residents petitioned to block Park Service interference with the island horses. If EIA existed on Shackleford, they argued, it had been there for a very long time, and the herd had developed an ability to live with the disease. Clearly, if overpopulation was an issue, the presence of the disease was not limiting herd growth, and most of the horses appeared to be in good flesh and lived long lives. Horses in the asymptomatic stage are usually healthy overall, and most carriers are entirely asymptomatic.

Many of the Down-Easters resented the Park Service takeover of their ancestral lands. Their families had used these barrier islands for generations, and they felt they had a right to continue using them as they always had, gathering livestock and fishing from seasonal fishing shacks. They argued that the horses had been there "forever" and had the right to continue living on Shackleford. They accused the Park Service of manipulating data to be rid of the horses. Many believed that the push for EIA testing was a ploy to justify destroying the animals.

The people rose up and pushed back. Jerry Hyatt of, Newport, N.C., collected 1,700 signatures on a petition in favor of greater protection for the horses. He organized a rally at the Carteret County Courthouse in Beaufort, and hundreds of Down-Easters turned out in support of the horses. Said Hyatt in a letter to the editor of the New Bern *Sun Journal* (1996, p. A11),

> In a conversation with Assistant Superintendent Chuck Harris and Dr. Michael Rickard [sic] at the Cape Lookout National Seashore office on Harkers Island, I was told that, "The National Park Service does not care about the people of Carteret County, or their heritage, only the people of the United States....
>
> They say they have held public meetings. Yes, they have. But at a time when working people could not attend, in a place with no parking and advertised only locally. Did the Park Service want public opinion, or did they "go through the motions" as required by the law?

Rikard maintained that the Park Service had no intention of eliminating the herd. "We're not trying to get rid of the horses. We're just trying to control the population," he said (Gernert, 1996, p. A3). Protests and pleas did not deter the Park Service. On November 12, 1996, the agency corralled all 184 horses—considerably fewer than the original estimate of 221. A Coggins test was performed on every horse. Seventy-six horses tested positive for EIA. Sixteen of the 18 dominant herd stallions on Shackleford Banks—89%—tested positive (USDA, 2006). None of these horses showed obvious signs of the disease.

Because no one can predict the risk posed by any given infected horse over time, veterinarians take the conservative position and assume that each infected horse poses the same threat at all times (Cordes & Issel, 1996). North Carolina state law mandates the destruction or quarantine of any horses with positive Coggins tests to prevent them from infecting others. Because the disease can be spread only by fresh blood, and blood dries rapidly on the mouthparts of biting flies, horses are considered quarantined if they are stabled 200 yards/183 m from other horses.

The Park Service made plans to euthanize all the horses with positive Coggins tests. Down-Easters were horrified. Most positive testers showed no outward sign of illness and appeared in robust good health. The horses who were inapparent carriers could pass the virus only to herdmates. And the risk of transmission is exceedingly low: only one horsefly out of 6 million is likely to pick up and transmit EIA virus from an inapparent carrier (Cordes & Issel, 1996). In other situations, the Park Service practices non-intervention in the natural processes of wildlife, even in the presence of disease. These horses were on a barrier island and posed no threat to mainland horses. Were they not *already* quarantined?

Coggins testing of feral horses was not a federal mandate. On Assateague Island NS, horse advocates pointed out, feral horses were *not* tested, despite being in fairly close proximity to domestic horses that campers bring to the island for beach riding (though not during fly season). Because many of the Chincoteague National Wildlife Refuge horses tested positive in the 1970s, it is quite possible that the untested Maryland herd harbors positive reactors. The Park Service recognizes that EIA may exist in that herd, but views it as a natural disease and does not intervene.

Down-Easters tried to find an alternative to euthanasia, protesting that the Park service disregarded their input and concerns. Carolyn Mason, a retired librarian committed to protecting the culture and heritage of the area, organized local residents to form the Foundation for Shackleford Horses, a 501(c)(3) nonprofit corporation.

The foundation proposed isolating the infected animals on Davis Ridge, a remote island-like hammock of 1,200 acres/486 ha in Core Sound and Jarrett Bay, connected to the mainland by marsh. Representatives from the Park Service, the N.C. Department of Agriculture, and the North Carolina Horse Council investigated the site, but eventually rejected it because of lack of security for the horses and inaccessibility to state veterinarians who would need to examine them periodically. No other practicable options surfaced, but officials did not allow much time.

North Carolina state law stipulates that any equid testing positive for EIA be quarantined for 60 days before retesting; a second positive result "likely means a death sentence" (Prioli, 2007, p. 62). Yet on November 20, 1996, only 8 days after the horses were corralled, the N.C. Department of Agriculture and the Park Service killed the 76 positive testers "in a clandestine middle of the night debacle" in Clinton, N.C. (Willis, 1999). Also euthanized was an uninfected foal who slipped into the quarantined herd to remain with its dam. Their bodies were unceremoniously buried in a landfill.

The 108 horses with negative Coggins tests were released back to the island after the Park Service freeze-branded large numerals on their rumps to make identification easier. Freeze-branding involves applying a supercooled instrument to the horse's skin, which damages the pigment-producing cells (melanocytes) without causing the greater tissue injury of hot branding. The shape of the brand eventually grows in as white hair.

Several horses evaded captors in the 1996 roundup and were not tested (Willis, 1999). A second roundup in March 1997 captured 103 horses. Five of these were positive for EIA. The foundation was ready this time and had obtained prior approval for an isolation site. The Park Service gave these animals to the foundation, sparing them from destruction. In 1998, three of 106 horses tested positive for EIA and were quarantined (Willis, 1999). Since 1997, these inapparent carriers have been maintained at an approved 192-acre quarantine site managed by the foundation and regulated by the N.C. Dept. of Agriculture (Willis, 1999). Carrying a dormant disease has robbed them of freedom, but they found a comfortable life in quarantine, courting attention from the volunteers who bring hay and water twice daily and companionably grooming one another in the shade.

The herd on Cumberland Island is unmanaged, and the equine census had apparently stabilized at the carrying capacity of the island many years prior to this 2011 photograph. The population could no longer grow because mortality balanced birth rate, and the horse count was roughly the same from year to year. When years of drought reduced the available forage, horses began to starve. Many of the lactating mares on Cumberland Island had Henneke scores of 1.5–2.5, though the author noted some individuals scoring as high as 3. Muscle wasting was evident on the gaunt mares. Where horses grazed in lawn areas, grasses were continuously cropped short with bare patches, while marsh areas showed varying degrees of impact. In 2011, the horse count was 148; by 2013, the population had crashed to 108.

At that point it was still possible that annual roundups would continue to eliminate EIA carriers until only a few horses remained. The quarantined horses might have been the only survivors of a rare strain of an ancient breed. Since almost no offspring of positive testers carry the disease, the foundation considered the possibility of breeding the quarantined horses to preserve bloodlines should the wild herd become extinct (Willis, 1999)—but the state required the castration of quarantined males. The 114-member wild herd was gathered again in 1999. Margaret Willis wrote of this roundup, "all but a few, mostly the old, are in excellent condition . . . AND they all tested negative for EIA" (Willis, 1999).

In September 2008, 11 years after their capture and quarantine, two horses developed clinical signs of EIA, one so severe that euthanasia became necessary. The surviving horse developed a second exacerbation the following month. In May 2009, a third horse in quarantine experienced a particularly severe acute episode and was euthanized (Capomaccio et al., 2012).

In February 1996, shortly before undertaking the scheduled gathers for EIA testing, the Park Service arbitrarily decided that the Shackleford Banks component would maintain a herd of 50–60 horses through immunocontraception (H.R. Rep. No. 105-179, 1997; Prioli, 2007). This census goal had no science to support it, and it represented a compromise only between allowing the horses to set their own maximum population

On the western end of Shackleford Banks in May 1994, many lactating mares had Henneke scores of 2.5-3.5 but some scored as high as 4. The grass was very short throughout most of the areas favored by the horses, indicating that they were consuming the forage as fast as it could grow, yet many fell short nutritionally.

This 9-year-old Shackleford mare, 16K, photographed in May of 2009 while nursing her 4-month-old filly, scored a 4 on the Henneke scale. While her spine and ribs are discernible, her hip, neck, and shoulder muscles are well developed.

and removing them entirely. In February 1997, Rep. Jones, introduced H.R. 765, the Shackleford Banks Wild Horses Protection Act, and on August 13, 1998, President Clinton signed An Act To Ensure Maintenance of a Herd of Wild Horses in Cape Lookout National Seashore (16 U.S.C. §459g–4, 2010), which ordered the Interior Department to "allow a herd of 100 free roaming horses" and forbade removal unless the population exceeded 110.

It also mandated a partnership between the Park Service and the Foundation for Shackleford Horses, Inc. (or another qualified nonprofit entity), for the management of free-roaming horses in the seashore (S. Rep. No. 109-154, 2005). Equine geneticist Dr. E. Gus Cothran evaluated the markers present in the DNA of the horses and concluded that the herd had excellent genetic variability—but for how long?

In the early days of co-management, the relationship between the Park Service and the foundation was often turbulent. Rubenstein wrote,

> Debate often turned rancorous over disagreements on how best to manage this historical horse population. In particular, too much effort was expended in trying to interpret at what threshold management should begin and end, as well as on balancing the mechanisms of selective removal and fertility control. (H.R. 1521, H.R. 1658 and H.R. 2055, 2003a)

In October of 2002, a new superintendent, Bob Vogel, took the reins of Cape Lookout NS and organized a meeting that included scientists, representatives from the foundation, and other stakeholders to rethink the existing management plan. The team reached a consensus that the horse population should never fall below 110 horses and that occasional expansions to 130 animals would allow successful genes to increase in frequency and benefit the population. In May 2003, Rep. Jones introduced H.R. 2055 to give this determination the force of law. Although legislation never reached the full Senate in the 108th Congress, it was a defining moment because an agency of the Department of the Interior publicly endorsed increasing the size of a wild horse herd. P. Daniel Smith of the Park Service spoke to the House Committee on Resources in support of increasing the number of horses in the herd. He said, "The Department is strongly committed to conserving, protecting, and maintaining a representative number of horses on the Shackleford Banks portion of the Seashore, as we have done in other units of the National Park System which contain horses, and believes that the number of horses on Shackleford Banks should be determined by the ecology of the island and by means which protect the genetic viability of the Shackleford Banks horses" (H.R. 1521, H.R. 1658 and H.R. 2055, 2003b).

Jones tried again in January 2005 with H.R. 126, sometimes referred to as the Cape Lookout National Seashore Free-Roaming Horse Law Amendment (S. Rep. No. 109-154, 2005). In October 2005, Pete Domenici (NM), chairman of the Senate Committee on Energy and Natural Resources, reported, "The range of 110 to 130 horses is based on sound science and provides the population changes, which are necessary for maintaining the genetic viability of the herd" (S. Rep. No. 109-154, 2005). The recommendation became Public Law 109-117 in 2005 (An Act . . . To Allow for an Adjustment in the Number of Free Roaming Horses Permitted in Cape Lookout National Seashore, 2005).

The author was surprised to learn that through this legislative process, the Department of the Interior had unequivocally supported, and Congress had eventually approved, a proposal virtually identical to one for improving the genetic health of the dangerously inbred Corolla herd, which both entities have repeatedly rejected. Says Karen McCalpin of the Corolla Wild Horse Fund,

> H.R. 306 (formerly H.R. 5482), the Corolla Wild Horse Protection Act, mirrors the Shackleford Banks Act with one important exception—it allows for the introduction of mares from Shackleford Banks. This would immediately breathe new genes into our dying gene pool. . . .
>
> The Shackleford horses live on 3,000 acres, have been managed at a target of 120–130 (with never less than 110) for the last 12 years, and with no documented negative impact to the National Park. The Corolla horses have access to nearly 8,000 acres. Only a third of that is owned by the Department of Interior—the rest is private land. It is not an issue of lack of carrying capacity to support 120–130 horses. (McCalpin, 2011)

Cedar Island, across the sound from Core Banks, is perhaps best known as the southern terminus of a state ferry to Ocracoke. It is also home to a small, little-known herd of wild horses. Although Cedar Island NWR (a low-profile operation with no resident staff) takes up more than half the island, the horses live on private land. A series of low islands and marshes owned by longtime residents has functioned for years as a wild horse sanctuary. The range lies east of the ferry dock and extends for about 8 mi/13 km, in a swath about 2 mi/3.2 km wide. Sponenberg (2011) writes that the original Cedar Island herd comprised horses taken from Core Banks, "which were supplemented by a later addition of Ocracoke horses." For more than a century, perhaps much longer, herds of 100–200 bays, chestnuts, buckskins, and blacks lived wild in these marshes, sharing the abundant forage with other wildlife and with cagey feral cattle that charged when surprised. After helping the Park Service euthanize nearly half the Shackleford herd in unseemly haste, the Veterinary Division of the N.C. Department of Agriculture turned its attention to the few wild horses left on Cedar Island and nearly eradicated them as well.

"Local people used to round up the horses every 4th of July," says Nena Hancock, who manages the Cedar Island herd with her husband, Woody. "Many people came to help and to watch. Some of the horses were branded by their owners during the roundups and others were sold" (personal communication, March 9, 2011). Locals attest that horses have roamed Cedar Island for more than a century, their numbers augmented in 1958 by a group of horses removed from Portsmouth Island. An Associated Press article documented the translocation:

> Those shaggy wild things called Banker ponies which roam this off-shore North Carolina island, will get their annual penning-up next Friday.
>
> The island animals are a herd of the famed outer banks ponies. In previous years some of them have been sold to private owners after the annual roundup, but few of them will be sold this year. The Cedar Island Banker Pony Assn. which looks after them, said plans are to build up the herd.
>
> The pony herd was moved to Cedar Island, 12 miles east of Atlantic from Portsmouth Island several months ago. The transfer was ordered by the 1957 General Assembly in an effort to halt beach erosion. ("Round-Up Slated for Banker Ponies," 1958, p. 10B)

White Sands Stable, owned by Wayland Cato, offered boarding and trail riding along the beach near the Cedar Island ferry dock. Equestrians from all over brought their horses to

White Sands for the opportunity to ride on the beach. In 1996, one of these visiting horses apparently brought EIA to the stable, where it spread rapidly to infect all 11 of Cato's domestic horses. Seven of these horses were quarantined in Virginia, and the rest were destroyed.

In June 1997 veterinarians from the N.C. Department of Agriculture gathered the wild herd from the Cedar Island marshes and tested each horse for EIA. Of the 15 wild horses, 13 tested positive for EIA and were euthanized. "When the original herd became sick, Woody and Clyde helped the state vets round up the horses and test them," says Hancock. "After the testing was done there were only two remaining mares."

The wild horses of Cedar Island were a beloved part of local history and culture. Island residents were determined to save the herd and introduced genetically similar horses from Shackleford Banks to replace those that had been euthanized. "Woody and Clyde asked the local land owners, who agreed to allow us to reintroduce the horses," says Hancock (personal communication, March 9, 2011). "The land is still owned by several different people who graciously allow us to keep the wild herd of horses around, as they have been there for as long as the local people remember. Because we help with the Shackleford horses, we were able to get a stallion to reintroduce with the two remaining mares. Over the next couple of years, the Shackleford foundation gave us several more mares to restart the herd."

New foals are named for elderly or deceased Cedar island residents—Becky, Kassie, Ronald, Ulva, Ina May. Bucky, the lone buckskin mare, is the sole survivor of the original herd. She has produced a number of foals, and in 2010, she delivered Gay, a lovely buckskin filly very much like herself.

By 2010, there were 39 horses in the Cedar Island marshes, 6 of them gelded, most between the ages of 8 and 12. Thirty to 40 wild cattle also roam these marshes, as they have for more than a century. Steve Edwards writes "They are managed by people that care greatly about them. This is no BLM "round 'em up and lock 'em up story. The briefest conversation with Woody Hancock is all that it takes to see where his heart is. He wants only the best for these horses" (2011, February 23).

Despite strong local support, argument continues about whether the grazing of wild horses on Cedar Island and Shackleford Banks damages the environment (Barber & Pilkey, 2001). At the northeast end of Cedar Island, two actively migrating barrier spits with grassy dunes extending into southern Pamlico Sound mirror the migration mechanism of barrier islands. The northwestern and southeastern spits share similar tidal dynamics and sand supply, but have been divided by a sturdy fence for 45 years. The free-roaming horses and cattle of the island use the southeastern spit. The grazed spit has shorter growth of *Ammophila* and *Spartina* grasses, low or absent dunes, and broad overwash fans, while the ungrazed area is lush, with tall, grassy dunes and narrow, localized overwash channels. Could the light to moderate grazing occurring now cause changes such as these, or were they a result of historic overgrazing when horse and cattle populations were higher? More research is necessary to answer this question.

Private citizens have managed the Cedar Island herds humanely, cooperatively, and at no cost to the public for many years. In contrast, federal law required Cape Lookout NS and the Foundation for Shackleford Horses to manage the Shackleford herd jointly beginning in 1998, but real cooperation took years to develop.

The Park Service now makes an apparently sincere effort to integrate public opinion and information into its management plans. From a distance, Sue Stuska delivers contraceptives by dart gun, documents new foals, and evaluates the health status of each individual, but does not feed, touch, or interact with the horses except in extraordinary circumstances (King, 2007). Cape Lookout NS treats them as wildlife and grants them the space in which they can be wild horses. It does not provide veterinary care, farrier care, or vaccines. If a horse is suffering from a terminal condition, they decide whether to euthanize on a case-by-case basis.

The PZP contraceptive vaccine has been in use on Shackleford since January 2000. In January, Stuska's team performs pregnancy tests on each mare's manure. They administer PZP vaccines between late February and April, beginning when a mare is 2 years old. The vaccine reduces pregnancy by 97% the first year and 76% following an annual booster in the second year of treatment (Nuñez et al., 2009). The side effects of immunocontraception include an occasional small abscess at the injection site, a higher rate of late season births, and changes in the genetics or social order of the herd (Kirkpatrick, 2010). The advantages are better health and longer lives for contracepted mares, and increased foal survival in a herd kept in balance with its environment. By vaccinating individual mares, the Seashore saves the entire herd from the stress of repeated gathers.

Horses are selected for removal or contraception based on matrilineage (how many horses represent the mare's line) and genetics. The pedigrees comprise at most 4 generations and trace back to any of 47 horses identified at the onset of record keeping for whom the Park knows neither parent. In reality, all the horses on the island are probably related to one another in some way.

An article in the Cape Lookout NS publication *Preserve and Protect* asks, "If you were managing a barrier island population for the future, and wanted to be sure it had the best chance to adapt to changing conditions, what attributes would you choose?" ("Managing wildlife for a changing ecosystem," 2008, p. 7). On Shackleford, herd managers have made preserving the herd's diverse gene pool a priority so that when conditions change, some animals within the population should have whatever adaptive genes will be needed.

The seashore staff bases its decisions about contraception and removal on the number of representatives of a line, individual factors such as whether a horse is socially and physically able to reproduce, and a concept called "mean kinship" ("Managing wildlife," 2008, p. 7). Mean kinship is a number assigned to a horse that shows how closely it is related to other members of the herd. A foal born to well-represented parents would have many aunts, uncles, cousins, or siblings and would be assigned a higher mean kinship number than one born from rare lineages. A foal with a high mean kinship number is a likely target for contraception or for removal from the island, especially if it has a number of full siblings.

Managers limit the reproduction of these horses so that the fertile family will not eclipse the rarer lineages over time. As of 2008, seven lines were represented by a single horse, five by two horses, and nine by three horses ("Managing wildlife," 2008, p. 7). "We don't have a policy to let every mare reproduce," says Stuska. "We look at each horse as an individual" (personal communication, May 26, 2010). She pointed to a wall chart depicting the lineage of each horse. One family line had more than 30 representatives.

Number 79, a sorrel stallion, is the only representative of his line. His branch of the family tree will probably end with him because he has been unable to acquire and keep a mate and is unlikely to do so. Because records have been kept for only a short time, however, and because paternity can be uncertain, he may well have unidentified aunts, uncles, or cousins on the island, and they may perpetuate some of the same genes.

Poco Latte, the smallest horse on Chincoteague NWR, scores a 5 for condition despite the fact that she gestates and nurses a foal almost every year. On the Chincoteague Refuge, lactating mares have Henneke scores of 4–5, although elderly mares can score as low as 3 or 3.5 after a severe winter. This photograph was taken in a corral during the April roundup in 2010.

On Cedar Island in May of 2010, forage remained lush despite grazing by free-roaming horses and cattle. Favored grazing areas regrew rapidly when horses moved on. All horses had Henneke scores of 5–7, including lactating mares (center) and older multiparous mares such as the grandmother of the foal (right).

On Currituck Banks, lactating mares typically have Henneke scores of 5–6, and the herd appears in balance with its environment. This photograph was taken in May 2012. Research indicates that the effects of grazing are temporary, and grasses recover from grazing by early summer (Rheinhardt and Rheinhardt, 2004).

On the north end of Assateague, this mare scored a 5.5 while nursing a foal in October 2009. In April, the onset of foaling season, lactating mares tend to score lower, about 3.5–4. Most mares on the Maryland end receive contraceptive vaccines annually and do not experience the stress of serial pregnancies and lactation. Though forage is abundant, favored spots sometimes show evidence of overgrazing and are slow to regrow to optimal density.

Assateague Island NS uses a different approach, allowing the horses themselves to decide which lines prevail and which extinguish (J. Kirkpatrick, personal communication, May 29, 3014). On Assateague, each mare is given the opportunity to reproduce and make a genetic contribution to the herd. Opportunity does not always equal success, and the drama unfolds in often surprising ways. For example, the offspring of the unusually prolific mare M4 failed to reproduce, and her once well-represented line ended. On the other hand, the N4 and N6 lines became dominant despite limiting each mare to a single foal. Kirkpatrick explains, "Certain lines will be more successful than others—that's natures's way and the stuff of evolution. By trying to manipulate genetics through contraception you may very well be shorting the most successful lines in the herd and trying to inflate the least successful" (personal communication, May 29, 3014). Eggert et al. (2010) studied kinship and pedigree using DNA in fecal samples of Assateague horses and found that although mitochondrial DNA diversity (inherited through the mother only) is low, nuclear DNA remains as diverse as that in established breeds. After considering various strategies to maximize genetic health, they concluded that the current method of allowing each mare to reproduce once optimally preserves the long-term viability of the herd.

Research has repeatedly shown that immunocontraception does not alter daily activity patterns, social relationships, or harem fidelity (Kirkpatrick, Rutberg & Coates-Markle, 2012, June 6). Other studies seem to contradict some of these findings and conclude that contracepted mares are more likely to switch harems, undermining the cohesion among females (Nuñez et al., 2009; Madosky, Rubenstein, Howard, & Stuska, 2010). "When a mare has a foal," says Stuska, "her whole focus is to eat, eat, eat to meet the tremendous energy requirements of pregnancy and lactation. She tends to stay within a harem because she has no energy to spare straying to another" (personal communication, May 26, 2010). A vaccinated mare, on the other hand, spends a considerable amount of time soliciting and receiving sexual interest. "Perhaps she instinctively realizes that if this stallion isn't impregnating her, she should follow her biological mandate to find another mate who will," she added (personal communication, May 26, 2010). A stallion is unlikely to impregnate a contracepted mare, so in theory she may move from band to band, unsatisfied.

Earlier research (Rutberg, 1990) demonstrated that the age and experience of the harem stallion is the primary influence on band stability. Assessments by Gray (2009) and Turner (2011), both cited in NRC (2013), showed no change in band fidelity among PZP-vaccinated mares in Nevada and on Assateague Island, respectively. It is difficult to compare the results of studies involving different designs, methods, objectives, environments, time frames, and variables. Disagreement among researchers and confusion among readers is therefore understandable. After looking at available research, the National Research Council (2013, p. 110) concluded,

> The importance of harem stability to mare well-being is not clear, but considering the relatively large number of free-ranging mares that have been treated with liquid PZP in a variety of ecological settings, the likelihood of serious adverse effects seems low.

Some social behavior patterns hold true for most wild horses; others are unique to individuals or their herds. Although some researchers have reported that wild equine harems show little change in composition from year to year, others have found that band fidelity differs widely from population to population (Cameron, Setsaas, & Linklater, 2009). Before immunocontraception was initiated in the Carrot Island herd, about 30% of the mares changed bands in late winter (Stevens, 1990). During a 5-year study on Cumberland Island, about 38% of the mares remained in their original harems, and the majority of mares changed bands from one to four times (Goodloe, Warren, Osborn, & Hall, 2000).

On the one hand, mares who live in unstable harems weigh less, carry more parasites, and have less time for grooming or grazing, and their foals are more likely to die (Linklater, Cameron, Minot, Stafford, 1999). On the other hand, immunocontraception decreases foal mortality in the herd as a whole and extends the life of mares, which probably balances any negative effects of changing harems (Kirkpatrick, & Turner. 2007).

Maintaining a smaller herd has benefitted the horses greatly. Rubenstein (1982) wrote that Shackleford foals had only a 48% chance of surviving the first two years. Since 2000, about 82% of Shackleford foals have survived the first 2 years (S. Stuska, personal communication, March 8, 2013). Stuska et al. (2009) indicate that the diet of the Shackleford horses is adequate to maintain relatively good health.

Some Shackleford horses are ribby, with angular hips and narrow necks. Most of the horses that appear excessively thin are elderly, and, like many people of advanced years, often have trouble maintaining weight because of poor dentition and digestive issues. Body fat reserves are used as an energy stash in case the animal needs to burn it for fuel in times of physical stress. If a horse does not have these reserves and energy is needed, the animal will break down muscles and burn protein for fuel.

The Henneke Body Condition Scoring System is one scale scientists use to assess body condition in horses. This tool yields an accurate assessment of condition regardless of breed, body type, sex or age (Evans, 2005). Assessors press on the horse's body to obtain accurate information—or in a wild setting, use visual inspection only. Each body area is assessed and scored, and the numbers are totaled and divided by 6.

Any given horse's ideal score varies with breeding role and workload. For endurance and polo horses, the ideal Henneke score is 4–5, whereas dressage horses and show jumpers should score 5–7. A breeding stallion should score 4–6, and a pregnant mare should score 6–8 (Kentucky Horse Council, n.d.). A mare with a score of less than 4.5 at foaling is considered thin and is less likely to successfully produce a foal the following year (Evans, 2005). Pregnant and recently delivered mares often become thinner over the topline because the weight of the foal pulls the flesh tighter over the back and ribs. The presence of fat deposits and well developed muscle along the neck, shoulder, withers, and tail head more accurately indicate good condition, while muscle wasting indicates the horse has run out of carbohydrate and fat resources and is breaking down muscular protein. Aged horses will have lower scores as their muscle structure softens.

Dr. Alberto Scorolli used body condition scoring to assess the general nutrition, health, and growth rate of the wild equine population in Tornquist Park, Argentina (Scorolli, 2012). His team rated the condition of each horse using an alternative method, the visual body condition score, with a scale from 0=very thin to 5=obese. He found that the condition of the horses showed an annual cycle, peaking in late summer and early fall. He found that stallions had higher values than adult females—average score 3—in every assessment—a finding consistent with studies involving other populations. Tornquist Park mares, including yearlings and 2-year-olds, had condition scores significantly lower than those of stallions;

With a dirty look at the author, a wild Pryor Mountain mare escorts her newborn son away from human intrusion. This photograph was taken in May 2011, early in the growing season. After a tempestuous Montana winter, this mare scored a 3.5, like most of the lactating mares observed by the author on that trip.

45% of females had a BCS equal to or less than 1.5 at some point during the study, with scores highest in March and May (late summer/early autumn in the Southern Hemisphere) and lowest in September (late winter–early spring). Scorolli submits that the lower values of adult females probably reflect the caloric drain of pregnancy and lactation (Scorolli, 2012), while the weight of growing juvenile horses is influenced by both nutritional reserves and forage availability.

Body condition scoring becomes a measure of whether an individual meets its energy requirements for maintenance, growth, and reproduction, and lactation. In 1977, Eberhardt observed that as population density increases (1) the age at first reproduction increases, (2) fertility of females in their reproductive prime decreases, and (3) mortality of juveniles and vulnerable adults increases, especially through the winter (Eberhardt, 1977, 2002). Although Eberhardt's original hypothesis applied to marine mammals, subsequent research has demonstrated that it generally holds true for ungulates.

De Roos, Galic, and Heesterbeek (2009) found that nutrition and growth *in utero* and in youth are pivotal to population dynamics in free-roaming horses. Foals of well-nourished mares are born with nutritional reserves equal to that of their dams, and for the first months of life profit from additional caloric intake through suckling (De Roos et al., 2009). If forage is abundant, they remain in good flesh through their first two winters. Conversely, a foal born to and suckled by a malnourished mare starts life at a nutritional disadvantage. When challenged to meet the demands of maintenance and growth with inadequate winter forage, these foals exhaust their bodily reserves and begin their second summers depleted. These undernourished yearlings often die over their second winter. Those that survive have a later onset of puberty and first reproduction than optimally nourished horses and face higher mortality throughout life, especially during pregnancy and lactation (De Roos et al., 2009).

Horses grow larger and heavier when climatic conditions favor growth of forage and when the population is diffuse enough to reduce competition for resources. When a

The wild horse management team keeps track of each individual in the herd and its habits, home ranges, and preferred companions. This black mare of the "D" matrilineage, famed for her distinctive low-set ears, lives on the far west end of the island.

Athletic wild stallions can grow ribby during the breeding season. Whereas a malnourished horse would show muscle wasting, this stallion has a well-developed neck and chest and is lean and fit for successful combat.

population of free-roaming horses reaches the carrying capacity of its range, the condition of both habitat and herbivores decline.

Although environmental damage is not always quantifiable in the absence of long-term studies, the body condition of horses is easy to determine, even from a distance, and it appears to give a rough benchmark of environmental conditions. Scorolli proposes using the average body condition scores of adult females in a herd as a tool to determine a herd's "health, its potential growth rate and the proximity of the population size to carrying capacity" (2012, p. 92). A more sensitive indicator, however, would be an assessment of the body condition scores of *only* the herd members under the greatest caloric stress—lactating mares.

When forage is abundant, pregnant mares typically store body fat for the first 270 days of gestation (Harper, 2010). As gestation advances, a wild mare is challenged to consume sufficient calories, and body condition will usually decline slightly. This decline continues during the first 120 days of lactation, because milk production can nearly double a mare's daily energy requirements. On average, mares produce about 3 gallons of milk daily—approximately 450 gallons/1,700 L in the first 5 months of her foal's life. Foals nurse about five times an hour for the first 5 weeks and will usually double in weight their first month of life. Burdened by this relentless need for calories, lactating mares may fail to meet energy demands when resources are scarce.

As Scorolli and others have observed, lactating mares are usually among the first herd members to lose condition when food resources are scarce. A biologist might consider a Henneke score of 4 "good" for a lactating wild mare younger than 4 or older than 15 years of age at the end of winter (Kane, 2011). By early spring, before the burgeoning of new vegetative growth, a BCS of 3 for these same mares—who could very well be pregnant again—"might not be alarming" (Kane, 2011, p. 444). If a lactating mare has a BCS of 3 at the end of autumn, her poor condition indicates that either forage has been inadequate during the growing season, or she has some other health problem. A mare that heads into winter with a BCS of 3 or less has virtually no fat reserves (Wood, 2002), and when faced with climatic challenges or famine, may break down muscle to meet energy demands.

Kane proposed that if fewer than 5% of mares are in poor condition, it is likely that those individuals have health issues, such as advanced age, dental abnormalities, chronic illness, or parasite overload (Kane, 2011). An imbalance between grazers and forage may exist if 20–30% of the lactating mares have poor BCS scores. When greater than half the nursing mares are extremely thin, the rangeland is probably inadequate to support the herd, and the potential for disastrous consequences is great. Because conception occurs in the previous growing season, some underweight mares will continue to foal until they are elderly, or their Henneke scores fall below 2. Overall, however, underweight mares have lower conception rates and decreased fetal and foal survival as compared to mares at a healthy weight (Kentucky Equine Research, 2003).

While researching this book, the author made informal visual surveys of the condition of barrier island herds and observed a relationship between Henneke body condition scores of nursing mares and the health of herds as a whole and their environment. She proposes using the Lactational Condition Index described in the accompanying table to assess the health of herds and their proximity to the carrying capacity of their ranges. An LCI score of 1 indicates that the majority of lactating mares have Henneke scores of 1–2. Their poor condition

Lactational Condition Index

LCI 1

Majority of lactating mares with Henneke scores of 1–2.

Herd less likely to survive short-term insults (for example disease, drought, harsh climatic events).

Lower mare fertility and poorer pregnancy outcomes. Pregnant and lactating mares, juveniles, and elderly horses risk higher mortality.

Range at high risk for overgrazing.

LCI 2

Majority of lactating mares with Henneke scores of 3–4.

Herd more likely to survive short-term stresses.

Higher fertility, better pregnancy outcomes, greater survival.

Range at overall lower risk for overgrazing, but may show signs of overgrazing in places favored by horses.

LCI 3

Majority of lactating mares with Henneke scores of 5–6.

Herd most likely to survive short-term stresses.

Best pregnancy outcomes, high fertility, best survival.

Range at low risk for overgrazing.

LCI 4

Majority of lactating mares with Henneke scores greater than 7.

Herd most likely to survive short-term stresses.

Best pregnancy outcomes (Kentucky Equine Research, 2003), highest fertility, best survival. Horses with very high Henneke scores are at risk for obesity-related health problems.

Range at lowest risk for overgrazing.

If lactating mares, the horses under the greatest nutritional stress, remain in good condition, one can reliably expect to find the rest of the herd and the herd's environment in good condition. A low Lactational Condition Index at the end of the growing season is a telling indicator of individual and herd health. Lactational Index tool © Bonnie Gruenberg 2013, may be used freely with attribution.

suggests that the herd is probably at or near carrying capacity, reproductively compromised, and at risk of high mortality due to malnutrition, parasites, and disease. Herds in which the majority of lactating mares have Henneke scores of 3–4 receive an LCI score of 2. These herds are generally healthier, more fecund, and have lower mortality than those with an LCI score of 1. Their ranges are at lower risk of depletion than at LCI-1, though some areas may be overused. Herds in which most lactating mares have Henneke scores greater than 5 receive LCI scores of 3 and 4. These well-nourished herds are generally healthier and optimally fecund, and their ranges are at low risk of grazing stress. Overweight mares are the most fertile, but may develop obesity-related health problems.

Although federal legislation supports a target population of 110–130 horses on Shackleford Banks as long as they remain in balance with their environment, that number may change as the dynamic island shrinks, grows, or experiences changes in vegetation. Stuska keeps close watch on herd dynamics, utilizing a GPS system to follow the movements of bands within the herd (Prioli, 2007). Key goals of management are to minimize human interference with the herd and to promote the perpetuation of genes to keep the herd optimally healthy.

In 2005, 18 horses were removed to the mainland; two joined the Cedar Island herd, and the rest were adopted (Pippin, 2005). If more than 12 horses required removal, the Park Service would consider conducting a large-scale roundup of most of the herd, but no such roundups are planned (Cape Lookout NS & Foundation for Shackleford Horses, 2010).

Stuska is authorized to remove any horse from the island if necessary, such as if a horse sustains a severe injury or its life is endangered by the actions of people; if it shows "consistent,

Grinning with utter delight, the author sets out for a trail ride on Steve Edwards's Shackleford gelding Holland. Holland floated down the trail with easy, tireless gaits reminiscent of those of an Icelandic. He was remarkably surefooted, trotting and cantering over exposed roots and rocks without missing a step. He was also bomb-proof, unconcerned when deer exploded from the underbrush.

repetitious, unprovoked aggressive behavior" toward people; or if a foal is orphaned before its first birthday (Cape Lookout NS & Foundation for Shackleford Horses, 2010, p 2).

If the population "blooms" beyond the target 130-horse maximum, the 2010 management plan provides for the removal of about 2–4 foals per year, mostly males. These animals are taken young, so that they might adapt easily to life in domestication, and upon leaving the island become the property of the Foundation for Shackleford Horses. These removals are a low-key affair; Stuska and her assistants typically approach the periphery of the band, quietly sedate the animal, load it into a boat, and transport it to the mainland. The band usually continues grazing, unconcerned.

The Seashore conducts an equine census each December, and that year's foals are numbered in order of birth. Stuska closely monitors births, deaths, and band composition. When the author visited in 2010, there were 114 horses on Shackleford Banks, divided into roughly 25 harems and about 7 bachelor bands (S. Stuska, personal communication, May 26, 2010). The oldest living horse was 28, and the second-oldest was 27. There were five births in 2009, although one foal died soon after birth, and the herd saw the deaths of one 8-year-old horse,

Adagio, a diminutive Shackleford gelding removed from the island as a foal, was adopted in 2012. Pictured here as a yearling, Adagio loves people and followed visitors around Carolyn Mason's farm, soliciting attention like an oversized dog.

two 17-year-olds, one 21-year-old and two 24-year-olds. Five foals were born in 2010. Stuska suspended contraception and removal to allow the population to expand. As of March 30, 2012, there were 109 horses on Shackleford Banks (*Shackleford Banks Horses 2011 Findings Report*, 2012) because of an unexpectedly low birthrate and higher mortality in 2011—one juvenile, five horses in their teens and two in their 20s. Six foals were born in 2011, "Y" year, and 11 in 2012, "Z" year. Two of these "Z" foals were born out of season, one in January and one December (*Shackleford Banks Horse Herd Update*, 2013).

In 2009, four youngsters were removed from the island to be adopted. To preserve rare genes, Stuska targets a young horse for removal only if its bloodlines are well represented on the island. "If there are aunts, uncles and siblings out there, the horse is more likely to be removed. We consider whether the dam has had other offspring, and whether the grand dam has had other offspring," she explained. "Nobody knows the pedigrees of these horses for more than four or five generations back" (personal communication, May 26, 2010).

In 2012, Stuska recorded the first known horse in the herd to reach the age of 30, a mare that died the following winter. In 2013, another mare reached the 30-year milestone—#68, mother of nine. Her last colt, black stallion 4W, was born to a stallion of uncommon lineage, after her birth control vaccine failed to prevent pregnancy. She was 26 at the time. Since the Park Service started using contraception, mares typically outlive stallions on the island.

Stuska sponsors four horse-watching field trips annually for the public. She explained,

> It is important for the public to get information directly from park rangers who are knowledgeable within the subject area. On these trips, we take a boat to Shackleford Banks, walk around, look for horses, identify which ones we see, talk about what they are doing, and talk about how they are related to each other. It's a whole day of horses, walking through the marsh, braving the bugs. They always have waiting lists but you can get in if you reserve months ahead of time. (Personal communication, May 26, 2010)

Shackleford Banks is accessible to visitors by boat; passenger ferry service is available from Beaufort and Harkers Island. The Park Service requires that visitors remain at least 50 ft/15 m from the horses—about the length of a large bus.

The Shackleford horses are relatively tolerant of people, but like all wild horses bite, kick, and charge unpredictably. The author took all the photographs in this book with a telephoto lens from a distance greater than a bus length away, but came to realize that she was still too close when she was charged by an angry stallion who suddenly spun and came after her. (Fortunately, he did not pursue when she ran away.) The prospect of injury is especially daunting when one considers the inaccessibility of the island, the significant distance from a hospital, and patchy cell-phone reception.

Banker Horses are astonishingly rugged, durable, and unflappable. Their endurance is legendary. Steve Edwards describes an informal race between Holland, his 14-hand (56-in./1.42-m) Shackleford, and a Spanish Mustang stallion descended from the legendary Choctaw Sundance. Carrying a 160-lb/73-kg rider, Holland was allowed to choose his own speed and gait. He ran 5 mi/8 km in 20:54, finishing a half mile (0.8 km) ahead of the well-bred stallion (Edwards, 2009, July 16). On a 50-mi/80-km long-distance ride, Edwards (2009, September 14) described Holland as "absolutely impeccable. He carried 220 lbs forty miles with about 35–38 of those miles at a trot and averaged 5.5 mph. I have no idea how many more miles he could have done." In a blog post Edwards wrote of Holland:

When I ask him to go, he goes. Where I ask him to go, he goes. When we get to the briers, he goes. When we reach deep water, he goes. As far as I ask him to go, he goes. As smoothly as I asked him to go, he goes. With the ground frozen rock hard, he goes. With the ground parched and baked rock hard, he goes. With the sun glaring down on us, he goes and in the pitch darkness of the night, he goes (2011, July 17).

Shackleford horses are sometimes available for adoption. The Foundation charges a $600 adoption fee and seeks homes where the horses will be well-treated. They are highly intelligent and quickly learn to enjoy the company of people. On Carolyn Mason's farm, a Shackleford named Adagio, removed from the island as a foal, followed visitors around the yard, nuzzling and expecting to be hugged and scratched. When the inspectors for the Horse of the Americas registry came to evaluate the herd, they commented, "The domesticated ones were 'pocket ponies' that wanted to please, and were quite willing to follow one around just to get more attention" (Ives, 2007, p. 10). Shackleford Banks is loved by countless people who hold memories of the island close to their hearts. Shackleford, the Thoroughbred racehorse who won the 2011 Preakness Stakes was named for the island—his owners evidently visit Shackleford Banks frequently and find great peace and pleasure there.

The congressionally mandated partnership between the Park Service and the Foundation for Shackleford Horses has been in place since 1999. The agencies work well together, combining resources and reaching goals more effectively than either could alone. In the face of climate change and other future challenges, this management team should act to preserve both the island ecology and the herd, keeping the horses of Shackleford in a healthy balance with the island that has been their home for hundreds of years.

References

An Act To Amend Public Law 89-366 To Allow for an Adjustment in the Number of Free Roaming Horses Permitted in Cape Lookout National Seashore, Pub. L. No. 109-117 (2005).

An Act To Ensure Maintenance of a Herd of Wild Horses in Cape Lookout National Seashore, 16 U.S.C. § 459g-4 (2010).

Assateague Island Alliance. (2010). *eBay foal-naming auction*. Retrieved from http://www. assateagueislandalliance.org/name.html

Barber, & Pilkey, O.H. (2001). *Influence of grazing on barrier island vegetation and geomorphology, coastal North Carolina*. Paper No. 68-0 given at the Geological Society of America Annual Meeting, November 6, 2001. Retrieved from https://gsa.confex.com/gsa/2001AM/finalprogram/abstract_28327.htm

Bardenhagen, E., Rogers, G., & Borrelli, M. (2011). *Cape Lookout National Seashore storm recovery plan 2011: Final draft*. Washington, DC: U.S. National Park Service.

Barlow, C. (2000). *The ghosts of evolution: Nonsensical fruit, missing partners, and other ecological anachronisms*. New York, NY: Basic Books.

Barnes, J. (2007, November-December). Scattered by the wind: The lost settlement of Diamond City. *Weatherwise, 60*(6), 36–41. doi: 10.3200/WEWI.60.6.36-41

"Beach Pounders" [U.S. Coast Guard Beach Patrol personnel at Hilton Head Training Center, SC] (Photograph). (*circa* 1943).

Bennett, D., & Hoffman, R.S. (1999, December). *Equus caballus. Mammalian Species, 628,* 1–14.

Bishop, E.C. (1989). *Prints in the sand: The U.S. Coast Guard Beach Patrol during World War II.* Missoula, MT: Pictorial Histories.

Blythe, W.B. (1983). The Banker ponies of North Carolina and the Ghyben-Herzberg principle. *Transactions of the American Clinical and Climatological Association, 94*(6): 63–72.

Borkfelt, S. (2011). What's in a name?—Consequences of naming non-human animals. *Animals 2011, 1*(1), 116–125; doi: 10.3390/ani1010116

Bourjade, M., de Boyer des Roches, A., & Hausberger, M. (2009). Adult-young ratio, a major factor regulating social behaviour of young: A horse study. *PLoS ONE, 4*(3), e4888. doi: 10.1371/journal.pone.0004888

Boyd, L., & Keiper, R. (2005). Behavioural ecology of feral horses. In D.S. Mills & S.M. McDonnell (Eds.), *The domestic horse: The origins, development and management of its behaviour* (pp. 55–82). Cambridge, United Kingdom: Cambridge University Press.

Bratton, S.P., & Davison, K. (1987). Disturbance and succession in Buxton Woods, Cape Hatteras, North Carolina. *Castanea, 52*(3), 166–179.

Cameron, E.Z, Setsaas, T.H., & Linklater, W.L. (2009). Social bonds between unrelated females increase reproductive success in feral horses. *Proceedings of the National Academy of Sciences of the United States of America, 106*(33), 13850–13853. doi: 10.1073/pnas.0900639106

Cape Lookout National Seashore & Foundation for Shackleford Horses. (2010). Management plan for the Shackleford Banks horse herd. Harkers Island, NC: Cape Lookout National Seashore.

Capomaccio, S., Willand, Z.A., Cook, S.J., Issel, C.J., Santos, E.M., Reis, J.K.P, & Cook, R.F. (2012). Detection, molecular characterization and phylogenetic analysis of full-length equine infectious anemia (EIAV) gag genes isolated from Shackleford Banks wild horses. *Veterinary Microbiology, 157*(3–4), 320–332. doi: 10.1016/j.vetmic.2012.01.015

Carson, R. (1947). *Chincoteague: A National Wildlife Refuge* (Conservation in Action 1). Washington, DC: U.S. Fish and Wildlife Service. Retrieved from http://digitalcommons.unl.edu/usfwspubs/1

Cheevers, W.P., & McGuire, T.C. (1985). Equine infectious anemia virus: Immunopathogenesis and persistence. *Reviews of Infectious Diseases, 7*(1), 83–88.

Clarke, S.C. (1892). Sea-bass and other fishes. In G.O. Shields (Ed.), *American game fishes: Their habits, habitat, and peculiarities; how, when and where to angle for them* (pp. 287–343). New York, NY: Rand, McNally.

Code for America Brigade. (n.d.). *Work on Adopt-a-Hydrant.* Retrieved from http://brigade.codeforamerica.org/applications/8

Conant, E.K., Juras, R., & Cothran, E.G. (2012). A microsatellite analysis of five colonial Spanish horse populations of the southeastern United States. *Animal Genetics, 43*(1), 53–62. doi: 10.1111/j.1365-2052.2011.02210.x

Cordes, T., & Issel, C. (1996, June). *Equine infectious anemia: A status report on its control, 1996* (APHIS 91-55-032). Washington, DC: U.S. Department of Agriculture, Animal and Plant Health Inspection Service.

De Bry, T. (1590) (Engraver). *Americae pars, nunc Virginia dicta . . .* [Part of America, now called Virginia . . .]. In T. Harriot, *A briefe and true report of the new found land of Virginia*. Frankfurt-am-Main, Germany: Theodor de Bry. Retrieved from http://memory.loc.gov/gmd/gmd388/g3880/g3880/ct000777.jp2

De Roos, A.M., Galic, N., & Heesterbeek, H. (2009). How resource competition shapes individual life history for nonplastic growth: Ungulates in seasonal food environments. *Ecology, 90*(4), 945-960.

Diefenbach, D.R., & Christensen, S.A. (2009, August). *Movement and habitat use of sika and white-tailed deer on Assateague Island National Seashore, Maryland* (Technical Report NPS/NER/NRTR—2009/140). Philadelphia, PA: National Park Service, Northeast Region.

Durham, R.S. (2009, July 6). *The mounted beach patrol*. Retrieved from http://www.army.mil/article/23935/The_mounted_beach_patrol/

Donlan, C.J., Berger, J., Bock, C.E., Bock, J.H., Burney, D.A., Estes, J.A., . . . Greene, H.W. (2006). Pleistocene rewilding: An optimistic agenda for twenty-first century conservation. *American Naturalist, 168*(5), 660–681. doi: 10.1086/508027

Dwyer, J. (2007). A non-companion species manifesto: Humans, wild animals, and "the pain of anthropomorphism." *South Atlantic Review, 72*(3), 73–89.

Eberhardt, L.L. (1977). Optimal policies for the conservation of large mammals, with special reference to marine ecosystems. *Environmental Conservation, 4*, 205–212.

Eberhardt, L.L. (2002). A paradigm for population analysis of long-lived vertebrates. *Ecology, 83*(10), 2841–2854.

Edwards, S. (2009, July 16). I could not have hit him with a shotgun. *Mill Swamp Indian Horse Views*. Retrieved from http://msindianhorses.blogspot.com/2009/07/i-could-not-have-hit-him-with-shotgun.html

Edwards, S. (2009, September 14). Lab results are in. *Mill Swamp Indian Horse Views*. Retrieved from http://msindianhorses.blogspot.com/2009/09/lab-results-are-in.html

Edwards, S. (2011, February 23). The Shacklefords of Cedar Island. *Mill Swamp Indian Horse Views*. Retrieved from http://msindianhorses.blogspot.com/2011/02/shacklefords-of-cedar-island.html

Edwards, S. (2011, July 17). Yesterday I rode a horse that was not beautiful. *Mill Swamp Indian Horse Views*. Retrieved from http://msindianhorses.blogspot.com/search?q=holland

Eggert, L., Powell, D., Ballou, J., Malo, A., Turner, A., Kumer, J., . . . Maldonado, J.E. (2010). Pedigrees and the study of the wild horse population of Assateague Island National Seashore. *Journal of Wildlife Management, 74*(5), 963–973. doi: 10.2193/2009-231

Engels, W.L. (1952). Vertebrate fauna of North Carolina coastal islands II. Shackleford Banks. *American Midland Naturalist, 47*(3), 702–742.

Evans, P. (2005, April). *Body condition scoring: A management tool for evaluating all horses* (AG/Equine/2005-01). Logan, UT: Utah State University Cooperative Extension. Retrieved from http://extension.usu.edu/files/publications/publication/AG_Equine_2005-01.pdf

Fear, J., et al. (2008, August). *A comprehensive site profile for the North Carolina National Estuarine Research Reserve*. Retrieved from http://www.nerrs.noaa.gov/Doc/PDF/Reserve/NOC_SiteProfile.pdf

Feh, C, & de Mazières, J. (1993). Grooming at a preferred site reduces heart rate in horses. *Animal Behaviour, 46*(6): 1191–1194. doi: 10.1006/anbe.1993.1309

Fishing horses. (1900). *Chambers's Journal, 6th Series* (3), 493.

Foundation for Shackleford Horses. (2005). *Spirit's page.* Retrieved from http://www.shacklefordhorses.org/stories/spirit.htm

Gengenbach, L. (1994, March 27). Reined in? Human population starting to threaten banks horses. *Sun Journal* (New Bern, NC), pp. C1, C5.

Gernert, T. (1996, March 15). People to protest pony control. *Sun Journal* (New Bern, NC), p. A 3.

Godfrey, P.J., & Godfrey, M.M. (1976). *Barrier island ecology of Cape Lookout National Seashore and vicinity, North Carolina* (National Park Service Scientific Monograph 9). Washington, DC: Government Printing Office.

Goodloe, R.B., Warren, R.J., Cothran, E.G., Bratton, S.P., & Trembicki, K.A. (1991). Genetic variation and its management applications in eastern U.S. feral horses. *Journal of Wildlife Management, 55*(3), 412–421.

Goodloe, R.B., Warren, R.J., Osborn, D.A., & Hall, C. (2000). Population characteristics of feral horses on Cumberland Island, Georgia and their management implications. *Journal of Wildlife Management, 64*(1), 114–121. doi: 10.2307/3802980

Goodwin, D. (2002). Horse behaviour: Evolution, domestication and feralisation. In N. Waran (Ed.), *The Welfare of Horses* (pp. 1–18). Dordrecht, Netherlands: Kluwer Academic Publishers.

H.R. 1521, H.R. 1658 and H.R. 2055: Legislative hearing before the Subcommittee on National Parks, Recreation, and Public Lands of the Committee on Resources, U.S. House of Representatives, 108th Cong. 6 (2003a) (letter of Daniel I. Rubenstein).

H.R. 1521, H.R. 1658 and H.R. 2055: Legislative hearing before the Subcommittee on National Parks, Recreation, and Public Lands of the Committee on Resources, U.S. House of Representatives, 108th Cong. 16 (2003b) (statement of P. Daniel Smith).

H.R. 2055, 108th Cong. (2003).

H.R. Rep. No. 105-179 (1997).

Harper, F. (2010, September 15). *Broodmare management in fall.* Retrieved from http://www.extension.org/pages/29125/broodmare-management-in-fall

Houpt, K.A., & Keiper, R. (1982). The position of the stallion in the equine dominance hierarchy of feral and domestic ponies. *Journal of Animal Science, 54*(5), 945–950.

Hyatt, J. (1996, March 17). After being around 400 years, Shackleford Banks herd deserves better treatment [Letter to the editor]. *Sun Journal* (New Bern, NC), p. A 11.

Ives, V. (2007). *Corolla and Shackleford Horse of the Americas inspections—February 23–25, 2007.* Retrieved from http://www.corollawildhorses.com/Images/HOA Report/hoa-report.pdf

Jones, T.H. (2004). *Cape Lookout Life Saving Station historic structure report.* Atlanta, GA: U.S. National Park Service, Southeast Regional Office, Historical Architecture, Cultural Resources Division.

Kane, A.J. (2011). The welfare of wild horses in the western USA. In W. McIlwraith &. B.E. Rollin (Eds.), *Equine Welfare* (pp. 442–462). Chichester, United Kingdom: Wiley-Blackwell.

Kathrens, G. (Writer & Cinematographer). (2001). Cloud: Wild Stallion of the Rockies [Television series episode]. In THIRTEEN (Producer), *Nature.* New York, NY: WNET.

Kathrens, G. (Director, Writer, & Cinematographer). (2003). Cloud's Legacy: The Wild

Stallion Returns [Television series episode]. In THIRTEEN (Producer), *Nature*. New York, NY: WNET.

Kathrens, G. (Writer & Cinematographer). (2009). Cloud: Challenge of the Stallions [Television series episode]. In THIRTEEN (Producer), *Nature*. New York, NY: WNET.

Keiper, R. (1985). *The Assateague ponies*. Atglen, PA: Schiffer Publishing.

Keiper, R. (1986). Social structure. *Veterinary Clinics of North America: Equine Practice, 2*(3), 465–484.

Kentucky Equine Research, Inc. (2003). *Optimal body condition scores for breeding mares* (Equine Review N21). Retrieved from http://www.ker.com/library/EquineReview/2003/Nutrition/N21.pdf

Kentucky Horse Council. (n.d.). *The Henneke system of body condition scoring*. Retrieved from http://www.kentuckyhorse.org/henneke-body-condition-scoring

Kerr, W. (1875). *Report of the geological survey of North Carolina*. Raleigh, NC: Josiah Turner.

King, M. (2007). *The wild horses of Shackleford Banks: An interview with Dr. Sue Stuska*. Retrieved from http://network.bestfriends.org/3448/news.aspx

Kirkpatrick, J., & Fazio, P. (2010, January). *Wild horses as native North American wildlife*. Retrieved from http://awionline.org/content/wild-horses-native-north-american-wildlife

Kirkpatrick, J. F. and A. Turner (2007). Immunocontraception and increased longevity in equids. Zoo Biology 26:237-244.

Klein, D. (1963). The introduction, increase, and crash of reindeer on St. Matthew Island. *Journal of Wildlife Management, 32*(2), 350–367.

Levin, P., Ellis, J., Petrik, R., & Hay, M. (2002). Indirect effects of feral horses on estuarine communities. *Conservation Biology, 16*(5), 1364–1371. doi: 10.1046/j.1523-1739.2002.01167.x

Linklater, W.L., Cameron, E.Z., Minot, E.O., Stafford, K.J., (1999). Stallion harassment and the mating system of horses. Anim. Behav. 58, 295–306.

Linklater, W.L., Stafford, K.J., Minot, E.O., & Cameron, E.Z. (2002). Researching feral horse ecology and behaviour: Turning political debate into opportunity. *Wildlife Society Bulletin, 30*(2), 644–650.

Madosky, J.M., Rubenstein, D.I., Howard, J.J., & Stuska, S. (2010). The effects of immuno-contraception on harem fidelity in a feral horse (Equus caballus) population. *Applied Animal Behaviour Science, 128*(1), 50–56. doi: 10.1016/j.applanim.2010.09.013

Managing wildlife for a changing ecosystem. (2008). *Preserve & Protect*, 2008–2009, 7. Retrieved from http://www.nps.gov/calo/naturescience/loader.cfm?csModule=security/getfile&PageID=317930

Mayor's Press Office. (2012, January 19). *Mayor Menino invites residents to "Adopt-A-Hydrant" this winter*. Retrieved from http://www.cityofboston.gov/news/default.aspx?id=5444

McCalpin, K. (2011, January 31). Why is the federal legislation critical? *Wild and Free Weekly*. Retrieved from http://corollawildhorses.blogspot.com/2011_01_01_archive.html

McDonnell, S.M. (2005). Sexual behaviour. In D.S. Mills & S.M. McDonnell (Eds.), *The domestic horse: The origins, development and management of its behaviour* (pp. 110–125). Cambridge, United Kingdom: Cambridge University Press.

McGreevy, P. (2004). *Equine behavior: A guide for veterinarians and equine scientists*. London, United Kingdom: W.B. Saunders.

Milstein, T. (2011). Nature identification: The power of pointing and naming. *Environmental Communication, 5*(1), 3–24.

Moore, W. (1899). *Climate and crop service of the National Weather Bureau*. Raleigh, NC: U.S. Department of Agriculture.

Moseley, E. (Cartographer). (1733). *A new and correct map of the province of North Carolina*. Retrieved from http://digital.lib.ecu.edu/1028

Munsell, J.W. (1946, August 22). Hunting and fishing. *Newark Advocate* (Newark, OH), p. 11.

Muse, A. (1941). *The story of the Methodists in the port of Beaufort*. New Bern, NC: Owen G. Dunn.

National Research Council. (2013). *Using Science to Improve the BLM Wild Horse and Burro Program: A way forward*. Washington, DC: The National Academies Press.

Noble, D.L. (1992, March). *The Beach Patrol and Corsair Fleet: The U.S. Coast Guard in World War II*. Washington, DC: Coast Guard Historian's Office.

Nuñez, C.M.V., Adelman, J.S., Mason, C., & Rubenstein, D.I. (2009). Immunocontraception decreases group fidelity in a feral horse population during the non-breeding season. *Applied Animal Behaviour Science, 117*(1), 74–83. doi: 10.1016/j.applanim.2008.12.001

Olds, F.A. (1902). The wild horse of the Banks. *Forest and Stream, 59*(20), 384. Retrieved from https://play.google.com/books/reader?id=9kMhAQAAMAAJ& printsec=front cover&output=reader&hl=en_US

Park service [*sic*] study says wild horse herd should be thinned. (1995, September 2). *Sun Journal* (New Bern, NC), p. A3.

Pippin, J. (2005, January 11). Several wild horses to be removed from Shackleford Banks, N.C., to thin herd. *Knight Ridder/Tribune Business News*. Retrieved from http://www.accessmylibrary.com/coms2/summary_0286-7781584_ITM

Prioli, C. (2007). *The wild horses of Shackleford Banks*. Winston-Salem, NC: John F. Blair.

Ransom, J.I., & Cade, B.S. (2009). *Quantifying equid behavior—A research ethogram for free-roaming feral horses* (Techniques and Methods 2-A9). Reston, VA: U.S. Geological Survey.

Round-up slated for Banker ponies on Cedar Island. (1958, July 1). *Evening Telegram* (Rocky Mount, NC), p. 10B.

Rubenstein, D.I. (1981). Behavioural ecology of island feral horses. *Equine Veterinary Journal, 13*(1) 27–34.

Rubenstein, D.I. (1982). Reproductive value and behavioral strategies: Coming of age in monkeys and horses. In P.P.G. Bateson & P.H. Klopfer (Eds.), *Perspectives in Ethology: Vol. 5. Ontogeny* (pp. 469–487). Princeton University Press.

Rubenstein, D.R., Rubenstein, D.I., Sherman, P.W., & Gavin, T.A. (2006). Pleistocene Park: Does re-wilding North America represent sound conservation for the 21st century? *Biological Conservation, 132*(2), 232–238. doi: 10.1016/j.biocon.2006.04.003

Ruffin, E. (1861). *Agricultural, geological, and descriptive sketches of lower North Carolina, and the similar adjacent lands*. Raleigh, NC: Institution for the Deaf & Dumb & the Blind.

Rutberg, A.T. (1990). Intergroup transfer in Assateague pony mares. *Animal Behaviour, 40*(5), 945–952. doi: 10.1016/S0003-3472(05)80996-0.

S. Rep. No. 109-154, at 6 (2005).

Saffron, I. (1987, April 20). *As ponies die, an entire town feels the pain.* Retrieved from http://articles.philly.com/1987-04-20/news/26195580_1_wild-horses-dead-horses-carrot-island

Scorolli, A.L. (2012). Feral horse body condition: A useful tool for population management? In *International Wild Equid Conference, Vienna 2012: Book of abstracts* (p. 92). Vienna, Austria: Research Institute of Wildlife Ecology, University of Veterinary Medicine. Retrieved from http://www.vetmeduni.ac.at/fileadmin/v/fiwi/Konferenzen/Wild_Equid_Conference/IWEC_book_of_abstracts_final.pdf

Senter, J. (2003). Live dunes and ghost forests: Stability and change in the history of North Carolina's maritime forests. *North Carolina Historical Review, 80*(3), 334–361.

Shackleford Banks Wild Horses Protection Act, H.R. 765, 105th Cong. (1997).

Shackleford Banks horse herd update. (2013, January 7). Retrieved from http://static-horse-journal.s3.amazonaws.com/wp-content/uploads/2013/02/Shackleford-Banks-Horse-Herd-Update-2013-01-07-final.pdf

Shackleford Banks horses 2011 findings report. (2012, April 5). Retrieved from http://www.nps.gov/calo/parknews/2012-04-05.htm

Smith, P.D. (2003, June 24). *Statement of P. Daniel Smith . . . before the Subcommittee on National Parks, Recreation, and Public Lands, of the House Committee on Resources, concerning H. R. 2055, to amend Public Law 89-366 to allow for an adjustment in the number of free roaming horses permitted in Cape Lookout National Seashore.* Retrieved from http://www.nps.gov/legal/testimony/108th/capelook.htm

Sponenberg, D.P. (2011). *North American Colonial Spanish Horse update, July 2011.* Retrieved from http://www.centerforamericasfirsthorse.org/north-american-colonial-spanish-horse.html

Stevens, E.F. (1990). Instability of harems of feral horses in relation to season and presence of subordinate stallions. *Behaviour, 112*(3-4), 149–161. doi: 10.1163/156853990X00167

Stick, D. (1958). *The Outer Banks of North Carolina, 1584–1958.* Chapel Hill: University of North Carolina Press.

Stuska, S., Pratt, S.E., Beveridge, H.L., & Yoder, M. (2009). *Nutrient composition and selection preferences of forages by feral horses: The horses of Shackleford Banks, North Carolina.* Retrieved from http://www.nps.gov/calo/parkmgmt/upload/Nutrient Composition and Selection Preferences.pdf

Taggart, J.B. (2008). Management of feral horses at the North Carolina Estuarine Research Reserve. *Natural Areas Journal, 28*(2), 187–195. doi: 10.3375/0885-8608(2008)28[187:MOFHAT]2.0.CO;2

Take 'em alive. (1960, November 16). *Burlington Daily Times-News* (Burlington, NC), p. 4.

Terrible record of recent storm. (1899, August 25). *Atlanta Constitution* (Atlanta, GA), p. 2.

U.S. Bureau of Labor Statistics. (n.d.). *CPI inflation calculator.* Retrieved from http://www.bls.gov/data/inflation_calculator.htm

U.S. Bureau of Land Management. (2012, April). *Pryor Mountain Wild Horse Range quick facts.* Retrieved from http://www.blm.gov/pgdata/etc/medialib/blm/mt/field_ offices/billings/wild_horses/popular_pages.Par.94314.File.dat/FinalPMWHR_Quick_Facts.pdf

U.S. Bureau of Land Management. (2012, October 15). *Pryor Mountain Wild Horse Range.* Retrieved from http://www.blm.gov/mt/st/en/fo/billings_field_office/ wildhorses.

html

U.S. Bureau of Land Management. (2013, March 8). *Wild horse and burro quick facts.* Retrieved from http://www.blm.gov/wo/st/en/prog/whbprogram/history_and_ facts/quick_facts.html

U.S. Bureau of Land Management, Billings Field Office. (2001). *Environmental assessment and gather plan, Pryor Mountain Wild Horse Range, FY2001 wild horse gather and selective removal* (EA #MT-010-1-44). Billings, MT: Bureau of Land Management, Billings Field Office.

U.S. Bureau of Land Management, Billings Field Office. (2003, April). *Environmental assessment, Pryor Mountain Wild Horse Range, FY03 fertility control on select young wild horse mares; selective removal of young wild horse stallions* (EA #MT-010-03-14). Billings, MT: Bureau of Land Management, Billings Field Office.

U.S. Bureau of Land Management, Billings Field Office. (2009, May). *Pryor Mountain Wild Horse Range/Territory environmental assessment MT-010-08-24 and herd management area plan.* Billings, MT: Bureau of Land Management, Billings Field Office. Retrieved from http://www.blm.gov/pgdata/etc/medialib/blm/mt/field_offices/billings/wild_ horses.Par.30079.File.dat/pmwhrFINAL.pdf

U.S. Department of Agriculture, Centers for Epidemiology and Animal Health. (2006, September). *Equine infectious anemia (EIA)* (APHIS Info Sheet). Washington, DC: U.S. Department of Agriculture, Animal and Plant Health Inspection Service. Retrieved from http://www.aphis.usda.gov/vs/nahss/equine/eia/eia_info_sheet.pdf

U.S. National Park Service. (2006). *Management policies 2006.* Washington, DC: Government Printing Office.

van Dierendonck, M., & Goodwin, D. (2005). Social contact in horses: Implications for human-horse interactions. In F. de Jonge & R. van den Bos (Eds.), *The human-animal relationship: Forever and a day* (pp. 65–82). Assen, Netherlands: Royal Van Gorcum BV.

Vavra, M. (2005). Livestock grazing and wildlife: Developing compatibilities. *Rangeland Ecology & Management, 58*(2), 128–134. doi: 10.2111/1551-5028(2005)58<128:LGA WDC>2.0.CO;2

Warshaw, M. (2010, August). Dredge-spoil islands—Town Marsh and Carrot Island. *Beaufort, North Carolina History.* Retrieved from http://beaufortartist.blogspot. com/2010/08/dredging-created-islands.htmlWild horses of North Carolina. (1902, December 21). *New York Times,* p. 27.

Warshaw, M. (2012, April). Carrot Island. *Beaufort Harbor* [Web log]. Retrieved from http://beaufortinlet.blogspot.com/2012/04/1854-map-carrot-island.html

Willis, M. (1999). Shackleford Banks, NC, wild horses free of EIA: Roundup on Shackleford Banks, January 16–22, 1999. *Caution: Horses, 4*(3). Retrieved from http://asci.uvm.edu/ equine/law/articles/shackle.htm

Wiss, Janney, Elstner Associates & John Milner Associates. (2007). *Portsmouth Village cultural landscape report.* Atlanta, GA: U.S. National Park Service, Southeast Regional Office. Retrieved from http://www.nps.gov/calo/parkmgmt/upload/CALO Portsmouth Village CLR_Site History.pdf

Wood, C.H. (2002). *Body condition scoring for your horse* (University of Maine Cooperative Extension Publications, Bulletin 1010). Retrieved from http://umaine.edu/ publications/1010e/

Wood, G.W., Mengak, M.T., & Murphy, M. (1987). Ecological importance of feral ungulates at Shackleford Banks, North Carolina. *American Midland Naturalist, 118*(2), 236–244.

Zedler, J.B., & Callaway, J.C. (2001). Tidal wetland functioning. *Journal of Coastal Research,* Special Issue 27, 38–64.

Chapter 8

Cumberland Island, Georgia

Island of Contradictions

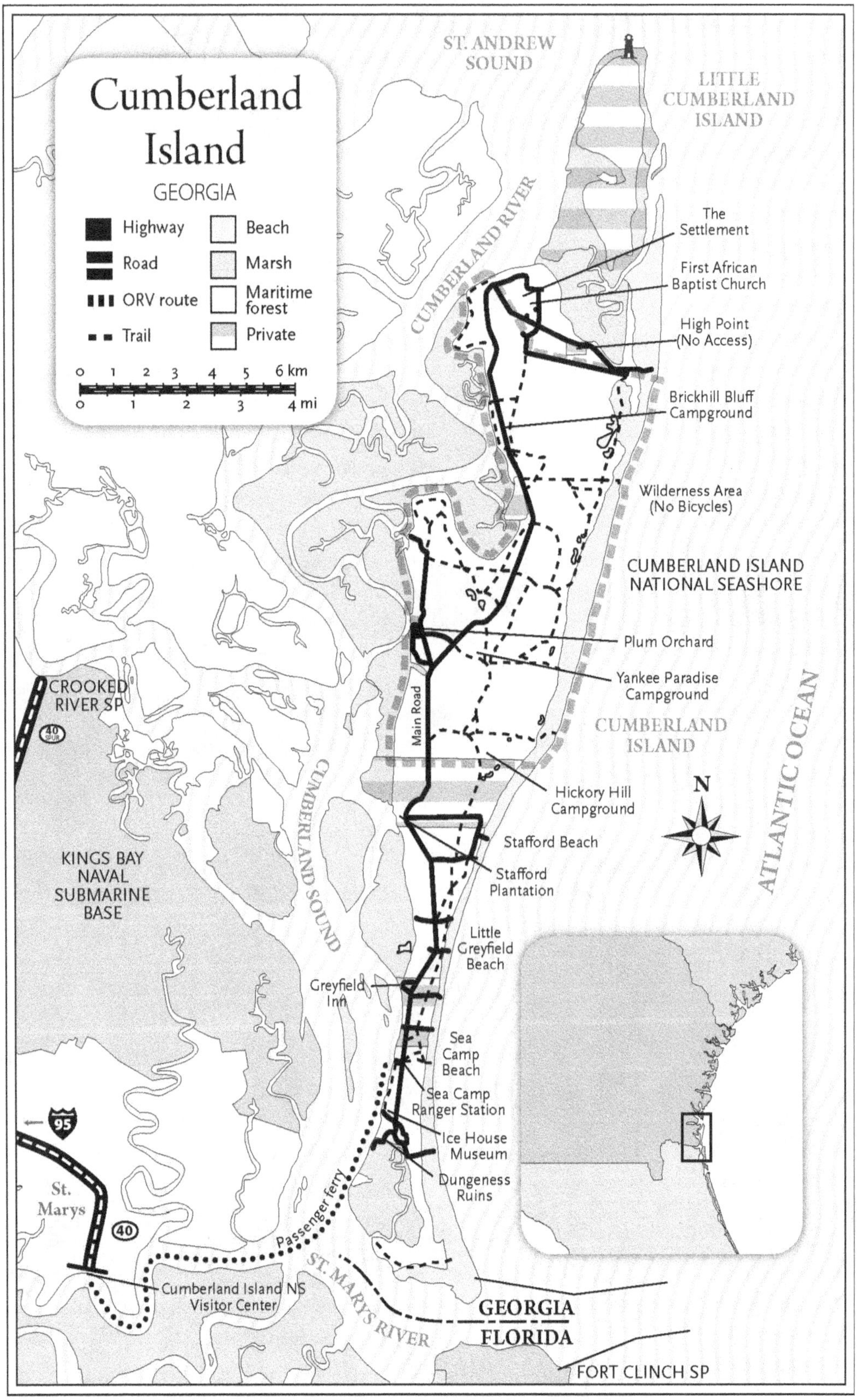

Cumberland Island

GEORGIA

Legend:
- Highway
- Road
- ORV route
- Trail
- Beach
- Marsh
- Maritime forest
- Private

Scale: 0 1 2 3 4 5 6 km / 0 1 2 3 4 mi

ST. ANDREW SOUND

LITTLE CUMBERLAND ISLAND

CUMBERLAND RIVER

The Settlement

First African Baptist Church

High Point (No Access)

Brickhill Bluff Campground

Wilderness Area (No Bicycles)

CUMBERLAND ISLAND NATIONAL SEASHORE

Plum Orchard

Yankee Paradise Campground

CUMBERLAND ISLAND

ATLANTIC OCEAN

N

Hickory Hill Campground

Stafford Beach

Stafford Plantation

CROOKED RIVER SP

40

KINGS BAY NAVAL SUBMARINE BASE

CUMBERLAND SOUND

Main Road

Little Greyfield Beach

Greyfield Inn

Sea Camp Beach

Sea Camp Ranger Station

Ice House Museum

Dungeness Ruins

95

St. Marys

40

Passenger ferry

Cumberland Island NS Visitor Center

ST. MARYS RIVER

GEORGIA
FLORIDA

FORT CLINCH SP

At first glance, the bay mare with the crooked blaze appeared to be a free spirit. When all of the other horses in the band were content to graze beneath the moss-laden live oak trees in the heat of the day, the mare was off in search of greener pastures. It was not difficult to see what drove her. She was hungry. Her ribs were prominent, her hip bones jutted, and individual vertebrae were discernible across her topline.

Her young foal, an Appaloosa with a white blanket across his hips, learned from an early age that his mother lived life on her own terms, and if he wished to be part of her activities she required him to keep up. If he wanted to nurse, he had to convince her to stop for a moment. This generally meant moving suddenly into her path to cut her off, forcing her to stand for him, or walking backwards to keep his hold on the teat while she pressed ahead. Often he kept company with other mares in the group who kindly looked after him while his mother foraged elsewhere. Between basic body maintenance and the enormous caloric drain of lactation, the bay mare had to eat almost constantly, going out of her way to select forage with the greatest nutritional benefit. The harem stallion, a powerful steel gray with bright dapples, was powerless to keep her in line, despite his large size and bossy attitude. Two years before, he claimed her as an adolescent filly and spent that summer driving her around and monitoring her every move. Since then, he had acquired three additional mates, cooperative mares who stayed together in a manageable group. None of them had foals, so they remained in reasonably good condition despite competing with other bands to graze on the lawn of the Dungeness ferry dock complex.

When rainfall is abundant, Cumberland Island blooms. Healthy-looking horses graze lush lawns with evident contentment. Foal survival improves, fertility increases, and the herd expands. When rainfall is scarce for long periods, upland grass withers, and ravenous horses crop it short. Marsh grasses, stressed by rising salinity in drought-stricken estuaries, are further pressured by overgrazing. Unable to meet their caloric requirements, horses burn fat and muscle for fuel, and become weakened by infections and parasite infestations. The most vulnerable members of the herd—foals, the elderly, and lactating mares—suffer most. If drought persists, many will die from starvation and illness.

For the bay mare, the need for social interaction was secondary to her relentless appetite. She knew when her favorite food, saltmarsh cordgrass (*Spartina alterniflora*), was exposed by the ebbing tide, and she made sure she was there to devour it, regardless of the activities of her herdmates. It was a daily challenge for her to consume enough calories to sustain herself and her foal.

Unlike the other East Coast islands roamed by wild horses, Cumberland is mostly dense subtropical forest of saw palmetto, magnolia, cedar, holly, pine, and myrtle. Majestic 300-year-old live oaks stand like wizened sentries along the pathways, Spanish moss trailing from their limbs.

In 2011, this 5-year-old mare was starving. She grazed almost constantly to meet the caloric requirements of lactation, yet remained bone thin. The wild herd of Cumberland Island had reached the carrying capacity of the island's forage resources while rainfall was adequate. When years of drought reduced the vegetation, lactating mares had trouble finding enough to eat.

One of the largest undeveloped barrier islands in the world, Cumberland is almost 3 mi/4.8 km wide at one point and nearly 18 mi/30 km long. The island is a mosaic of diverse ecosystems including primary and secondary dunes, interdune meadows, wax myrtle thickets, mowed lawns, fresh and salt marshes, and pristine beaches, as well as one of the largest maritime forests in the United States. More than 15 mi²/40 km² are federally designated wilderness.

John F. Kennedy, Jr., brought Cumberland Island to national attention in the fall of 1996 when he married Carolyn Bessette in a quiet ceremony in the humble First African Baptist Church, then retreated to the privacy of Greyfield Inn. Cumberland Island was one of the few places where such a wedding could be conducted in utter seclusion. The event was coordinated by Carnegie descendant Janet "Gogo" Ferguson, part-owner of Greyfield, who developed a friendship with Kennedy during his regular visits to the island.

A few tracts remain privately owned, but most of the island is now a national seashore. Greyfield is situated in a private compound including more than 200 acres/81 ha of mostly undeveloped oceanfront property. The elegant mansion was built in 1900 for Margaret Ricketson, the daughter of Lucy and Thomas Carnegie and niece of Andrew Carnegie. Since 1962, the property has operated as an inn to generate income for its own upkeep.

A century ago, Cumberland Island was well known as a playground for the wealthy and powerful, but it lapsed into obscurity. Aside from nature lovers and birders, few people outside Georgia had heard of it before the Kennedy wedding thrust it back into the headlines.

Like the rope in a tug-of-war, this idyllic island is wrenched by conflicting laws, policies, and philosophies. The Wilderness Act (1964) demands that the National Park Service preserve much of the island without roads, structures, motorized equipment, or mechanical

In 2009, this same mare was a healthy 3-year-old, newly claimed by the gray stallion as the start of his harem. It was an exceptionally wet year, and grass was abundant. Note the stallion's "snaking" posture as he drives her across the lush lawn.

transport; yet within this wilderness, some can legally drive cars and live in houses. The National Historic Preservation Act (1966) calls for the protection of historic structures and objects in or near the wilderness area, but the Park Service was been obliged to maintain them without the benefit of trucks or power tools. Various laws protect legal contracts for real estate and reserved rights of use, but the rights upheld by these contracts frequently conflict with federal policies. The Park Service struggles to balance these competing interests, but it seems that every objective is at odds with another.

Each of Cumberland's resources has its own advocacy groups, and the agendas of these organizations frequently collide. Impassioned environmentalists oppose current and former landowners, federal agencies contend with politicians, wilderness activists argue with historic preservationists, and horse-lovers clash with birders. Their only common ground is a love for the island and a desire to act in its best interest.

Popular legend holds that the wild horses of Cumberland Island descend from Spanish Jennets imported and maintained there between 1566 and 1675, when Spain had a fort and missions on the island. It is possible that Spanish horses were allowed to range freely to multiply and forage for themselves. Some speculate that the Spanish abandoned the island and left livestock behind, perhaps leaving horses too old, lame, or intractable to be useful to the Spanish elsewhere. Though this origin is possible, there is no evidence.

English imports probably came to Cumberland Island with James Oglethorpe in the late summer of 1739. He wrote, "I left a stud of the Trust's horses and mares when I went last for England, and the colts bred out of them are very good" (Bullard, 2003, p. 29). Two years later, Benjamin Martyn wrote that on Cumberland and Amelia islands "a stud of Horses and Mares . . . are bred without any expense" (Bullard, 2003 p. 29).

After being twice rebuffed for trying to nurse, the foal approaches a third time, hesitancy showing in his expression and the set of his ears. This time the mare allowed him to feed, but she continued to tear at the tender *Spartina* revealed by the ebbing tide near Dungeness Dock. Without the pressures of overgrazing, these wispy sprouts could grow into a dense and healthy salt marsh.

As on other islands, stockmen frequently released horses, cattle and other livestock into "feeding marshes" and onto islands to make use of otherwise unusable land and to contain the animals without use of fences These animals, sometimes subject to periodic roundups and at other times ignored for decades, often became wild.

Free-roaming horses were first documented on Cumberland Island in a 1788 letter to Edward Rutledge from Phineas Miller, but we do not know whether any descendants of the early lines remain (Goodloe, Warren, Osborn, & Hall, 2000). In a 1789 letter to Captain John McIntosh, a Cumberland wrangler arranged shipment on a flat boat for 13 young horses gathered from the island, "much better ones than the last" (Bullard, 2003, p. 70). Some of the remainder were probably eaten by 900 starving soldiers and freedmen during this time (Bullard, 2003; Hoffman, 2011).

After the Civil War, Cumberland residents used most of the island for grazing free-range livestock (Bullard, 2003). Records indicate that semi-gentled horses were swum from Cumberland Island to be sold on the mainland in 1866 much as the Chincoteague Ponies are penned today (Bullard, 2003). This gather and swim might have been an annual activity, the drive probably starting in the marsh south of the Plum Orchard dock. At slack low tide, wranglers would press the horses to swim the Brick Hill River, then to the eastern shore of the shell-bottomed Cumberland River, to land at a point about 300 yd/274 m south of Cabin Bluff (Bullard, 2003).

When the Carnegies came to the island in the 1880s, they stabled a number of well-bred domestic horses in their compounds. Plum Orchard had an impressive 15-stall stable filled with high-quality animals, and, farther south, the stable at Dungeness accommodated 60.

While the bay mare forages alone at the dock, the well-nourished gray stallion (right) waters at the artesian well with the rest of his band. Cumberland mares that are neither pregnant nor lactating are frequently underweight, but it is the nursing mothers that walk the narrow edge between survival and starvation.

The Georgia State Archives documents many horses that the Carnegies bought or sold on Cumberland Island and the presence of feral herds that ran at liberty, like the cattle and hogs, all over the island. The residents held wild horse roundups and "rodeos" to break them to saddle (Seabrook, 2002).

In the 1800s, free-roaming Cumberland Horses were popularly classified as Marsh Tackies. They were small in stature, and had the reputation of being exceptionally strong and rugged. At one point the horses might have carried a high proportion of colonial Spanish blood, but over time outcrossing changed the character of the herd.

Many breeds have contributed genes to the free-roaming herd. Through much of the 19th century, visitors to the Robert Stafford plantation captured free-roaming "marsh tackies" and bought them for $5 each (Wright & Lawrence, 2010). It appears that Stafford imported foreign horses in the 1850s and maintained about 30–40 horses on the plantation through 1877 (Bullard, 2003; Wright & Lawrence, 2010). Thomas Carnegie purchased the Stafford and Dungeness plantations in 1881 (Bullard, 2003) and proceeded to improve the free-roaming herds by adding purebred horses of various breeds to the mix. From the 1880s through the 1960s, the Carnegie family released many horses—Appaloosas, Tennessee Walking Horses, Thoroughbreds, even retired circus horses (Goodloe, Warren, Cothran, Bratton, & Trembicki, 1991; Seabrook, 2002). In 1896, the Carnegies bought a white stallion from the stud farm of Czar Nicholas II (Bullard, 2003). In 1921, a train-car load of beautiful Arizona mustang mares probably brought an infusion of Spanish blood (Bullard, 2003). It remains unclear which, if any, of these horses contributed to the modern free-roaming herd. In the 1950s, Lucy Ferguson and her staff castrated every wild male foal that they could capture, then released their own fine stallions to breed with the wild mares (Seabrook, 2002). As recently as 1992, a resident added four registered Arabians to the wild herd to improve conformation (Seabrook, 2002).

A thin black mare with the conformation of a Tennessee Walking Horse grazes under the moss-draped live oaks at Plum Orchard. After the author took this photograph, the mare's yearling son emerged from a copse of cedars, shedding light on the reason for her emaciation. When resources are scarce, lactating mares become vulnerable to malnutrition. A well-nourished gray stallion with Appaloosa characteristics forages in the distance.

In 1991, Goodloe et al. reported that the Cumberland Horses are genetically similar to other East Coast island populations, especially the Assateague Ponies. Assateague Ponies carry mostly old Spanish, mustang, and Arabian bloodlines, with Shetland Ponies introduced in the 1920s to add lively markings to the herd. The horses of the North Carolina Outer Banks are a unique, mostly homogeneous strain of Colonial Spanish horse, apparently descended from old Spanish bloodlines.

The horses of Cumberland also carry genetic markers from American breeds such as the American Thoroughbred, the Tennessee Walking Horse, and the American Quarter Horse. These breeds descended largely from a handful of Spanish, English, Dutch, French, Arabian, and Barb horses imported into the colonies in the 17th and 18th centuries. Consequently, each of the American breeds share so many genetic markers with one another that pinpointing lineage can be difficult or impossible. Arabian and Paso Fino horses have also made their mark on the genome of the Cumberland herd. It is probable that the horses currently on the island descended primarily from stock associated with the Carnegie estates, which probably carried these bloodlines.

Compared to other East Coast island horses, members of the Cumberland herd are generally taller, longer-legged, and sleeker. Whereas Assateague Ponies are mostly pinto or sorrel, Cumberland Horses range from bays and chestnuts to grays, blacks, browns, and roans. Many horses have white markings such as blazes, forehead stars, and socks. Some horses have Appaloosa patterns—such as roan with white blankets, dark with white "snowflake"

The horses of Cumberland Island descend from a number of breeds, including the Tennessee Walking Horse, Thoroughbred, Paso Fino, Quarter Horse, Appaloosa, Spanish Colonial Horse, and Arabian.

markings, or spotted blankets across the hindquarters—and show other Appaloosa traits such as a sparse mane and tail, striped hooves, mottled skin, and white scleras.

Horses have been in continuous residence on Cumberland Island for hundreds of years, but exactly how long is anyone's guess. Over the last 500 years, the island has been repeatedly developed and exploited, then abandoned, and many of these contingents brought horses with them. Native Americans made minimal impact, but the Spanish, English, and American residents transformed the island with forts, missions, logging, cotton plantations, estates, a hotel complex, and numerous other projects. Between waves of human habitation, mansions crumbled and native vegetation and wildlife encroached. Since the Cumberland Island National Seashore was established in 1972, and especially since Congress designated 9,886 acres/4,001 ha (more than half the seashore) wilderness in 1982, the island's primitive persona has been encouraged to reassert itself in many places. Up to 10,731 acres/4,343 ha may be added to the wilderness area in the future.

Paleo-Indians arrived in the Florida region as early as 13,000 years ago, and it is reasonable to think that these early people inhabited the area that would become Georgia at least as early. It is likely that eastern North America was peopled long before this date, but evidence of this habitation is scarce. Between 18,000 and 12,000 years ago, the shoreline hugged the edge of the continental shelf, about 70 mi/113 km east of the present beachfront (Dilsaver, 2004). When people first made their homes in the area, it was far from the ocean, and they began foraging on the present island more than 4,000 years ago.

On a visit to Cumberland Island in the 1990s, the author photographed an exquisite white Arabian mare running wild in a group of roans and Appaloosas near the Dungeness ruins.

At the time of European contact, the indigenous Timucua Indians burned land for agriculture and to create a favorable environment for the white-tailed deer (Dilsaver 2004). They subsisted largely on seafood and wild game and left behind ceramic shards and numerous shell middens throughout Cumberland Island. Many roads and trails on the island today follow their ancient footpaths. They spoke a language apparently unrelated to any other. Although the Timucua survived until the early 19th century, and Spanish missionaries preserved some of their language in writing, little else can be said with confidence. Those who lived on Cumberland Island and the nearby mainland are sometimes called, and may have called themselves or their village, Tacatacuru. But Milanich points out,

> Just how the natives did refer to themselves is something of a puzzle. Evidence exists that when asked by a Frenchman or a Spaniard what a place was and who its occupants were, villagers responded, "this is our land" and "we are us." (1996, p. 45)

The first European to visit Cumberland was probably Jean Ribault, a Frenchman, in 1562. The Spanish named the island San Pedro, and in 1566 the Spanish governor of Florida directed construction of a fort on the island, also named San Pedro. The governor then established the Franciscan missions of San Pedro de Mocama and San Pedro y San Pablo de Puritiba (Puturiba, Potohiriba . . .) to convert the natives to Christianity. Researchers believe that missionaries brought livestock to the island. The first mission seems to have lasted until about 1660. The second was apparently active only in the mid-1590s. A third, San Felipe, lasted till around 1684. After attacks by pirates in 1683 and 1684, missionaries and natives alike seem to have fled to St. Augustine. The Yamasee from farther north had settled the island by the time Oglethorpe arrived.

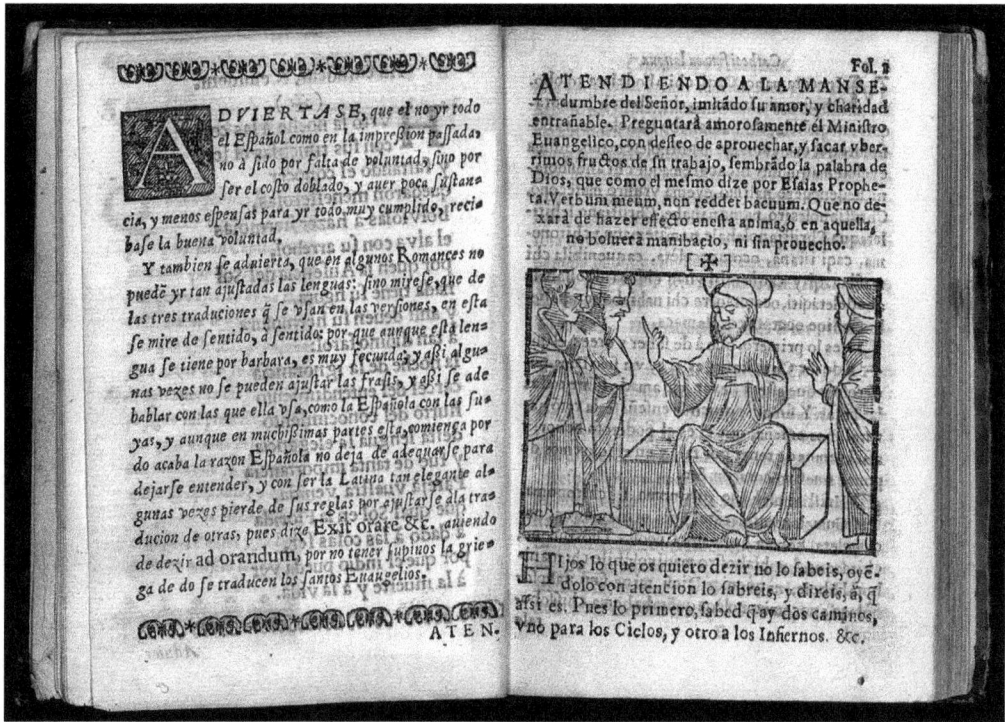

Spanish missionaries spread the Catholic faith to native tribes in what is now North Florida and Georgia by teaching from a catechism printed in both Spanish and Timucua (Pareja, 1627). When Europeans first made contact in the 16th century, the Timucua may have numbered as many as 200,000. Three hundred years later, they were extinct, victims of warfare, bondage, and European diseases. Courtesy of the Jay I. Kislak Collection, Rare Book and Special Collections Division, Library of Congress.

In 1736, Oglethorpe, founder of the English colony of Georgia, established two forts on the island—Fort St. Andrew at the north end and Fort Prince William to the south. Cumberland's present name came about through the suggestion of an Indian boy named Toonahowie, who visited London with his uncle Chief Tomochichi and General Oglethorpe. Toonahowie struck up a friendship with the 13-year-old son of King George II, Prince William Augustus, Duke of Cumberland. To seal their friendship, on parting William gave Toonahowie a gold watch. Toonahowie, in turn, asked General Oglethorpe to name this island after the Duke of Cumberland. Fort Prince William, a substantial structure built on South Point to accommodate 200 men (Thorn, 1977), also honored the young duke. Dungeness, originally constructed as Oglethorpe's hunting lodge, was named after a headland in Kent.

Spanish forces clashed unsuccessfully with British soldiers, many of them Scottish Highlanders, at the southern fort in 1742, and their defeat ended Spanish incursions into English Georgia. After the battle, however, the English believed the southern coast of Georgia to be indefensible against Spanish, French, and Indian raids. The forts, constructed of timber and sand, were reclaimed by the elements; no trace remains today. From 1748 through the 1750s, the island was inhabited mostly by "bandits, pirates and other lawbreakers" from the English and Spanish colonies (Dilsaver, 2004, p. 23).

In 1763, the Treaty of Paris ended the French and Indian War (Seven Years' War) and granted the English rights to Florida, removing Cumberland Island from the frontier and

A collection of skulls, many of them from horses, decorates the tabby wall surrounding the Stafford mansion.

making land grants there desirable. Before the American Revolution, there were few settlers on Cumberland, but thereafter planters moved in to grow rice, indigo, and corn and raise horses, cattle, and hogs. Colonists harvested Cumberland's live oak and cedar for ships' timber, lumber, and shingles.

After the war, General Nathanael Greene, commander of American forces in the south, acquired Mulberry Grove plantation near Savannah. He later bought extensive property on Cumberland Island, intending to harvest live oak to sell for shipbuilding and to build a home for his family on the site where Oglethorpe had maintained his hunting lodge, Dungeness. Unfortunately, Greene died before he could begin to build or turn a profit. Greene wrote that in 1785, shortly after the war, at least 200 horses and mules roamed the island. Goodloe et al. (1991) offer corroboration, a 1788 letter from island resident Phineas Miller to Edward Rutledge that mentioned free-roaming horses.

In 1796, Greene's widow, Catharine, married Phineas Miller, the general's former secretary and her children's tutor. They built a four-story mansion, also named Dungeness, surrounded by 12 acres/5 ha of terraced gardens that grew many exotic foods. The Millers frequently held elegant parties, and Cumberland became a social hub for the affluent. Phineas Miller also died, leaving Catharine once again alone. In the years to come, the pattern would repeat—men would die young, leaving Cumberland Island molded and controlled by strong, intelligent widows (Dilsaver, 2004).

When British Admiral George Cockburn briefly occupied Cumberland, he and his men used Dungeness as their headquarters and hospital. To produce social turbulence, Cockburn freed the slaves there and so attracted many runaways from around the region, who were ultimately transplanted to Trinidad.

The third Dungeness, still impressive despite decades of neglect, as photographed for the Historic American Buildings Survey the year before vandals burned it down. Photograph by P.E. Gardner (1958), courtesy of the Prints and Photographs Division, Library of Congress.

Another Revolutionary hero, General Henry "Light Horse Harry" Lee, III, suffered internal injuries while defending a friend and freedom of the press in Baltimore, went to the West Indies to recover, and retreated to Cumberland Island as his health declined. He died on Cumberland in 1818 in the home of Louisa Shaw, the daughter of his former commander, Nathanael Greene. He was later exhumed and reinterred beside his son, Robert E. Lee, in Lexington, Va.

From the end of the Revolution to the Civil War, Sea Island cotton—a variety that commanded higher prices by virtue of its unusually long fibers, was the major cash crop of Cumberland Island. More than half of the island was cleared for farming. Slaves lived in small cabins. Today a cluster of stone chimneys marks their locations. During the heyday of cotton plantations, Cumberland supported a population of 65 whites and 455 slaves (Dilsaver, 2004). Other products cultivated included timber, citrus fruit, figs, dates, olives, and pomegranates. Horses were necessary for working these plantations and for traversing the 18-mi/30-km length of the island.

When the Civil War began, the planters of Cumberland retreated to the mainland for safety. They returned to find Dungeness burned to the ground except its stone walls and chimneys.

Robert Stafford, Jr., son of Thomas and nephew of Robert Stafford, acquired land over time until his holdings totaled 8,125 acres/3,288 ha, located around the present-day Sea

Camp area (Dilsaver, 2004). (At the time, the terms *junior* and *senior* were often used to distinguish between two men with the same name, lineal relatives or not.) Census reports show that Stafford kept 30–40 horses on his plantation between 1850 and 1870 despite the fact that most Cumberland Island horses were sold to help the Confederate effort during the Civil War. From 1800 to 1880, the Robert Stafford plantation sold Marsh Tackies, ponies captured from free-roaming island stock, for $5 each.

Stafford never married, but fathered 6 children with a slave named Elizabeth. During the 1850s Stafford owned up to 348 slaves, who were emancipated when he was in his 70s. The profitable island plantations owed their success to slave labor, and after the Civil War, agriculture was no longer sustainable. Before the Civil War, there were 10–12 large, productive plantations on Cumberland, but by 1876, not one acre was in cultivation (Dilsaver, 2004).

Rather than leave his land and see it divided among his freed slaves as abandoned property, Stafford remained until he died in 1877, destitute and bitter about his change of fortune. After his death, his nephews inherited his holdings. The six biracial children he fathered with Elizabeth did not receive any land on Cumberland.

Stafford's story was not unique. After 1865, there was a reshuffling of the social order, and many prominent landowners faced poverty. Planters could not afford to hire laborers or resisted employing former slaves. Some former masters evicted the freedmen and burned their cabins.

By 1878, two hotels were operating on the northern end of Cumberland Island in an area known as High Point. They reached a peak in operations in the 1890s and 1900s and shut down by 1920. This property later passed to the Candler family of Atlanta, whose wealth came from the astonishingly popular beverage Coca-Cola.

Thomas Morrison Carnegie, brother and partner of steel magnate Andrew Carnegie, bought the plantation at Dungeness in 1881 for his wife, Lucy, and their family. They built another Dungeness mansion on the site of the original, graced by verandas, turrets, and gables and boasting 50 rooms. Dungeness was evidently built on a prehistoric shell midden, which undoubtedly provided the material for the 6-inch-thick walls of the tabby structure.

Like the earlier Dungeness, this great house soon became a social center for the wealthy. Thomas and Lucy entertained guests with activities such as hunting, fishing, golfing, and cruising aboard their yacht, also named *Dungeness*.

Thomas Carnegie died in 1886 at the age of 43, leaving Lucy with nine children and a large inheritance. Lucy went on to acquire 90% of Cumberland Island, turning it into a vast, self-sufficient family preserve staffed by about 200 employees. She set about expanding the Dungeness complex to include more than 20 buildings, as well as walls, decorations, and a pergola (colonnaded walkway that supports climbing plants). A recreation and guest house east of the mansion included a heated pool, a steam room, a recreation room, a squash court, and guest bedrooms. Other houses, docks, and structures were built all over the island. Seven of Lucy's nine children married, and she presented four of them with mansions on Cumberland Island.

The Carnegie estate maintained an ornamental upper garden while the lower garden became a subsistence garden of vegetables and fruits. The complex included greenhouses to grow cut flowers, an olive orchard, crops such as prize-winning Sea Island cotton, and livestock such as beef and dairy cattle, pigs, and poultry. Carriage and pleasure horses roamed free on the island, and about 60 were kept stabled at Dungeness.

This pronghorn (top) and elk (bottom) are native North American wildlife, but when brought to Cumberland Island in the late 19th century as a novelty, they became unequivocally exotic species in this environment. More enigmatic is the status of horses as native or exotic wildlife within the barrier island environment. If native is defined as present at the arrival of the first Europeans, horses are clearly exotic; but if defined as present at the arrival of people, horses are a native species, and our Pleistocene ancestors probably contributed to their extinction. Though horses are native to East Coast salt marshes, there is no evidence that they lived on barrier islands before reintroduction. Ocean levels were about 400 ft/122 m lower than they are at present, and what is now submerged continental shelf was exposed as much as 60–90 mi/97–145 km east of the present shoreline (Carroll, Kapeluck, Harper, & Van Lear, 2002). Cumberland Island Ricketson Family Photograh Album *ca.* 1898–1900. Courtesy of the Lucy Coleman and Thomas M. Carnegie Family Papers, ac 1994-0003M, Georgia Archives.

A wild horse grazes beneath the wing of a private airplane on the grounds of the Stafford plantation in late May, 2011. Even on the expansive lawn area, grass is dry and sparse.

Lucy bestowed the Stafford mansion on her eldest son, William, along with cash for renovations; but in 1900 the old homestead burned. On the site, William built a second Stafford house, which was similar to the previous structure, but with a fire-resistant stucco exterior.

Lucy's wedding gift to her son Thomas Carnegie II was a 29-room house known as The Cottage, built on the grounds of Dungeness. It burned in the 1940s. Lucy's daughter Margaret (nicknamed Retta) used her husband's wealth to build Greyfield, a two-story house 2 mi/3.2 km north of Dungeness. Greyfield now operates as an elite and expensive inn. In 2013, the most economical room was $425 a night off-season, single occupancy, for a minimum of two nights (three on holidays). Air conditioning, meals, and access by private ferry were included; telephones, television, and Internet access are not.

Lucy Carnegie built Plum Orchard for her son George when he married Margaret Thaw in 1898. Eight years later, Margaret expanded the mansion to include 30 principal rooms, 12 bathrooms, and many smaller rooms.

Lucy set up a trust that protected the family's holdings until her last child died in 1962. She ensured that she would maintain control of Cumberland Island and that she would not lose her fortune to any irresponsible actions of her children. Her six sons lived a life of leisure, and none of them ever held a job.

When Lucy died in 1916, Andrew II, Thomas II, and Margaret remained trustees of the estate. Initially, the Carnegies maintained their affluent lifestyle, but finances grew tight in the mid-1920s and tighter with the stock market crash of 1929. Although the Carnegies valued the primitive nature of the island, financial difficulties forced the family to generate income by closing Dungeness and capitalizing on Cumberland's resources. They considered raising cattle and hogs, lumbering, sawmilling, planting, shrimping, operating freight and passenger boats, and investing in developments such as hotels and amusements. In

The multi-tiered lawns surrounding the ruins of Dungeness are the favored grazing areas for several bands of horses.

the 1920s, A.A. Ainsworth of New York almost purchased Cumberland to found a major development similar to Coral Gables, Florida, but the project never materialized.

In the 1950s, several companies bid to strip-mine titanium-bearing ilmenite from 7,000 acres/2,800 ha of Cumberland Island, an undertaking that would have destroyed vegetation and other surface features of the island (Dilsaver, 2004). Luckily, legal difficulties and a sudden drop in the price of titanium saved Cumberland from disembowelment.

By the 1950s, the gardens at Dungeness were no longer maintained. In 1952 Robert Ferguson sought bids to demolish Dungeness, but did not follow through on razing it.

Around that time, Cumberland Island was subject to poaching and vandalism, and Lucy Ferguson and the other landowners resorted to "hunting" the poachers to repel them from the island. Employee J.B. Peebles shot at one such poacher in 1959, and shortly afterward, Dungeness was destroyed by a fire that could be seen for miles along the mainland coast. Several Florida poachers were the likely culprits, but nobody was ever arrested for the crime. The ruins of Dungeness remain a popular feature of Cumberland to this day, and wild horses favor grazing on the grassy grounds.

In 1951, Lucy Ferguson and her husband turned to cattle ranching. Their cattle opened the forests, destabilized sand dunes, trampled marsh, and altered the population dynamics of deer and horses so that it is difficult to assess the true carrying capacity of the island today (Dilsaver, 2004).

The Carnegie heirs considered options that would maintain the island's wild beauty while allowing them to maintain a presence. In 1955, the Park Service published a report titled *Our Vanishing Shoreline*, which expressed concern that the wild coast of the United

Sea turtles have been making a comeback on Cumberland Island, largely because of aggressive conservation efforts. Wildlife biologist Doug Hoffman demonstrates the effectiveness of a barrier that prevents raccoons and hogs from preying on nests. The year 2010 saw a record 486 successful nests.

States was being swallowed by developers (USNPS, 1955). Of the 3,700 mi/6,000 km of coast, only 6.5% was preserved for public recreation, and most of that was around Cape Hatteras. The Park Service sought to acquire 15 additional sites and considered Cumberland Island ideal because it contained "practically all the desirable features for public enjoyment" (Dilsaver, 2004, p. 80).

Carnegie landowners invited the Park Service to consider Cumberland as a potential national park, and it eagerly agreed. Secretary of the Interior Stewart Udall promised that if the Park Service acquired Cumberland Island, it would maintain the property largely undeveloped, preserving the unspoiled beauty that the Carnegies had cherished. The Park Service offered to obtain land over time, paying the residents substantial sums but allowing them to live and drive vehicles on the island.

At about the same time, the state of Georgia was eyeing Cumberland Island as a state park with the intention of developing it into a resort "for the average man" (Dilsaver, 2004, p. 81). A bridge would connect Cumberland to the mainland, roads would be built to and on the island, and the state would "propagate it as a vacationland" (Dilsaver, p. 83).

At the time, land owners had differing visions for the future of the island. In 1968, while the Park Service negotiated for a national seashore, three of the Carnegie landholders sold large tracts totaling 3,000 acres/1,200 ha, or one fifth of the total Carnegie holdings, to Charles Fraser of the Sea Pines Development Company. Fraser was the developer and major owner of a large resort complex on Hilton Head Island, South Carolina.

Fraser envisioned Cumberland as a residential retreat for the affluent with about 150 homes; golf courses, natural areas, and recreational facilities and a fire department, airport,

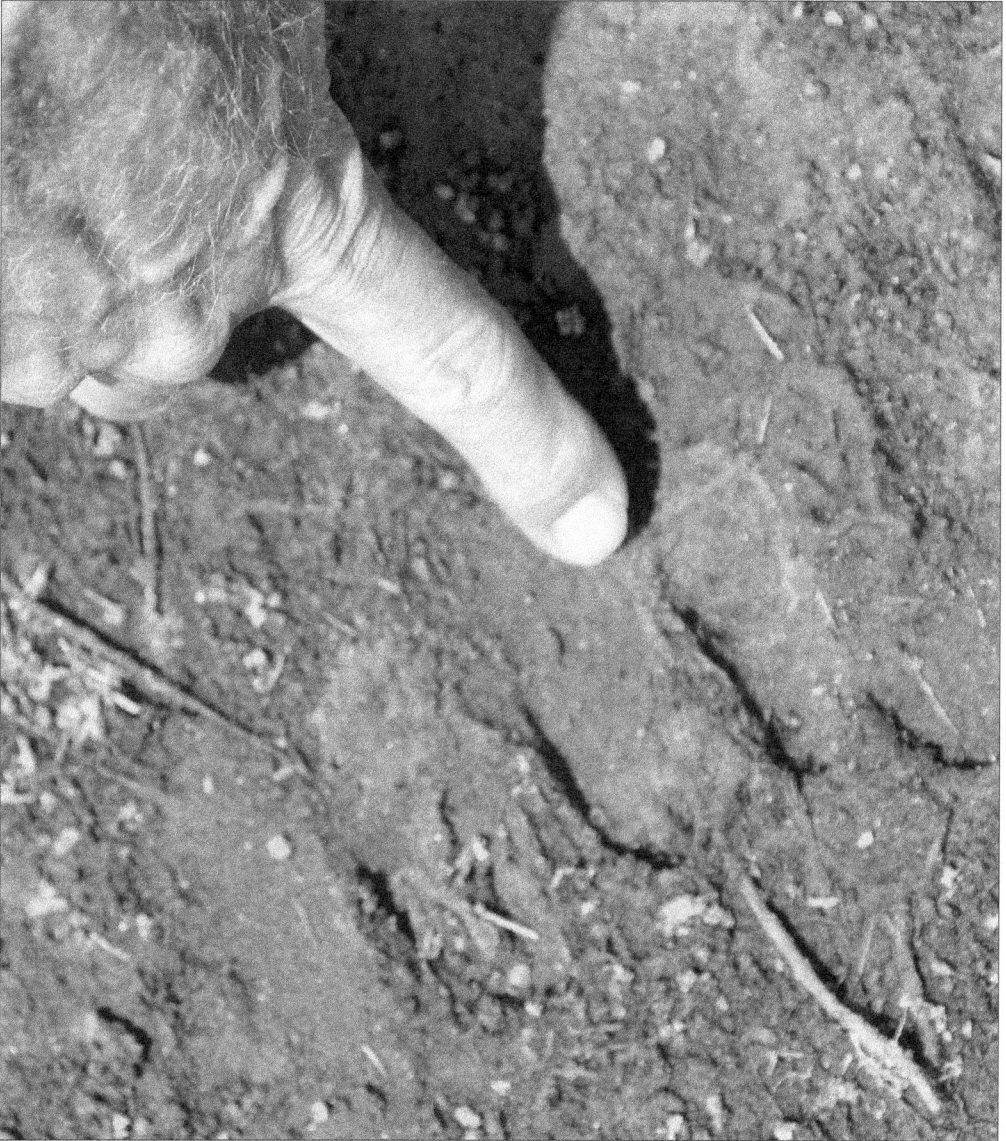

Hoffman explains that the track of a wild hog can be distinguished from that of a deer by the rounded tips of its cloven digits. As summer takes hold, Hoffman spends long nights on the beach hunting pigs that have the potential to consume sea turtle nests.

and medical center. He even incorporated the Park Service into his plans, supporting a small national park within his resort community. As relations soured, Fraser set out to gain control of the island by any means, intending to push the landowners off their property and block their influence. Fortunately, he failed, and Cumberland escaped development.

In 1961, the U.S. Air Force and the National Aeronautics and Space Administration sought an East Coast launch site for the developing space program. They winnowed the list of potential properties to two, Cumberland Island and Cape Canaveral. NASA actually preferred Cumberland for this purpose, but Cape Canaveral was chosen because of its proximity to Patrick Air Force Base. If Cumberland had been acquired by NASA, neither the Carnegies nor the Park Service would have had use of it.

The remaining Carnegie and Candler descendants, who together held 85% of the island, were disturbed by the prospect of outsiders transforming an island that had been in their families for decades. Together with a group of conservationists, they succeeded in blocking development and establishing a national seashore on Cumberland Island while retaining their homes.

In the early 1960s, Little Cumberland Island, at just north of the main island, changed hands twice and was bought by conservation-minded investors who planned to develop 200–300 residences, leaving the rest of the island in a natural state.

In 1971 the Carnegie heirs sold or donated most of their property to the Park Service. On October 23, 1972, President Nixon signed legislation to establish Cumberland Island as a 40,500-acre/16,400-ha national seashore, making it one of the largest mostly undeveloped barrier island preserves in the world (Sharp & Miller, 2008). Several parcels of private property or retained-rights lands escaped Park Service control. There were 18 retained-rights lots within the seashore boundaries deeded to allow residents to live on them for periods ranging from decades to life, with the Park service to assume ownership of the parcels when the agreements expired (Laliberté, 2007). Some residents rent out their properties, some use them seasonally, and at least one lives on the island year-'round.

When the Park Service acquired Cumberland Island, its original plan was to create a recreational park that would attract 10,000 visitors a day with numerous interpretive sites and activities. Environmental groups, residents, and the public argued that the park should preserve the island's unique wild character, and the Park Service agreed to a primitive park with a 300-person daily limit and minimal public facilities (Sharp & Miller, 2008). The *Cumberland Queen*, operated by Lang's Seafood, the concessionaire for the park, ferries people to and from the island, taking up to 150 per trip.

Cumberland Island is not far offshore. At the Cumberland Dividings, between Crooked River and Black Point, the distance from high land on the mainland to high land on the island is only about 300 yards/274 m. Marshes block direct east-west passage for boats, however. To venture from the port of Saint Marys to a landing on the island, they must travel up to 6 mi/10 km.

The first Europeans to claim Cumberland Island found large stands of live oak and pine with less understory of saw palmettos (*Serenoa repens*) and other ground cover than exists today because of centuries of intentional burning by the native people. The forest remains dense despite extensive British and American harvesting of live oak for shipbuilding.

The marshes, beaches, meadows, forest, and ocean habitats teem with wildlife. Tree frogs trill loudly for much of the night, their throats ballooning like bubble gum, at daybreak sleeping off the evening's debauchery camouflaged on palmetto leaves. Log-like alligators (*Alligator mississippiensis*) lurk half-submerged in the freshwater shallows, and river otters (*Lontra canadensis*) cavort in their quest for fish. Throughout the moist, steamy woods, armadillos (*Dasypus novemcinctus*) root for grubs in rich humus. White-tailed deer (*Odocoileus virginianus*), smaller than in northern forests, venture onto the meadows as the sun drops behind the horizon. Venomous snakes such as diamondback rattlesnakes (*Crotalus adamanteus*), timber rattlesnakes (*Crotalus horridus*), and cottonmouths (*Agkistrodon piscivorus*) hunt scampering rodents.

The waters off Cumberland are critical breeding and calving grounds for endangered right whales (*Eubalaena glacialis*). Manatees (*Trichechus manatus*), up to 15 ft/4.6 m long

The lawns of the Dungeness ruins are favored by the horses as a grazing area, and the home ranges of several bands overlap at this resource. Stallions make concessions to avoid coming to blows. In this picture, the mare in the middle, a mate of the stallion on the right, had strayed to flirt with the harem stallion on the left. The stallion on the left was lower ranking and apparently thought it prudent to return her to her mate. Unwilling to risk the ire of the dominant male, he drove the mare toward the alpha stallion's band. When the alpha claimed her, stepping high with neck arched, the beta stallion defecated, signaling subordinate status. The alpha defecated atop the pile, signifying dominant status. Both stallions are similar in size, shape and markings—they are probably closely related.

and weighing as much as 3,000 lb/1,361 kg, graze on *Spartina* and algae in salt-marsh creeks. Groups of as many as seven manatees commonly loiter in the salty water around the ferry docks at low tide, drinking the freshwater runoff while people wash their boats.

Sea turtles nest on the wild beaches. These reptiles were hunted on the island through the early 1900s. At one point, the Carnegies preferred turtle eggs to hen's eggs and enjoyed green turtle soup. Eggs were found by thrusting a sharp stick into the sand of a potential site. If it was a nest, the stick would withdraw covered in sticky raw egg. Over the past century, once-abundant marine turtles became rare on Cumberland and elsewhere.

They are, however, apparently making a comeback, with Hoffman counting 336 nests in 2006, 252 in 2009, and 486 in 2010, 700 nests in 2012, and 561 in 2013. "As a whole, the state of Georgia has recorded its highest nest totals during the four nesting seasons in years 2010–2013," says Hoffman (personal communication, July 20, 2014). Interns from the Student Conservation Association—volunteers that provide manpower for Park projects— document the location of the nests, secure them against predators, and give educational talks to visitors.

Migratory birds use the island as a stopover. Cumberland Island is designated a United Nations International Biosphere Reserve because it offers critical habitat for endangered

This leggy newborn on the Stafford Plantation has roughly a 68% chance of surviving her first year. High mortality rates are common among wild-born foals. The sparse, short-cropped grass foretells the difficulty his mother will have consuming enough calories to sustain him.

wildlife such as least terns (*Sternula antillarum*) and loggerhead sea turtles (*Caretta caretta*) (Harlan, 2007).

More than 323 species of birds are residents or seasonal visitors, including threatened and endangered species such as bald eagles, least terns, Wilson's and piping plovers, and wood storks. "We have some of the highest shorebird numbers in the state, both during nesting season and the winter migratory period," says Hoffman.

> Primary beach/dune nesters in the spring/summer include Wilson's Plovers, American Oystercatchers, Least Terns, and Willets. Our wading-type birds—wood storks, herons, and egrets—do not nest on the island, and we presume that the nesting habitat has been lost or severely decreased due to prolonged droughts and the absence of wildfires that would have kept freshwater wetlands free from woody vegetation encroachment (D. Hoffman, personal communication, March 18, 2011).

Black bears (*Ursus americanus*) and endangered Florida panthers (*Puma concolor*) were native to the island, but were hunted to extinction over the last two centuries. Bobcats (*Lynx rufus*), a species that vanished from Cumberland in 1907, either from human activities or from natural causes, have been reintroduced.

Researchers from the University of Georgia released fourteen adult bobcats in the fall of 1988 and 18 more in fall of 1989 (Diefenbach, Hansen, Warren, Conroy, & Nelms, 2009). They thrived on the island, preying primarily on marsh rabbits but rounding out their diets with cotton rats, grey squirrels, raccoons, birds, cotton mice, white-tailed deer and young hogs. To monitor the genetic health of the felines, in the winter of 2011–2012 researchers sampled DNA from their scat and compared it to genetic samples from the founding population (Hoffman, 2011, October–November-b).

Following the band to the lower lawns, a young colt hesitates at the top of a flight of stairs on the Dungeness grounds. His mother gave him an encouraging nudge to start him on his way, then led him down the incline.

Destructive, elusive, and prolific feral hogs are perpetually vexing to the Park Service. They are adept at evading capture and multiply rapidly, with large litters, short gestation periods, and early puberty. Hogs can breed as early as 6 months of age, but more typically do not start to reproduce until 8–10 months. The usual litter size is 4–6 piglets, and they can produce up to two litters annually, though one a year is typical. They live in family groups called sounders, which usually include two sows and their offspring. Adult boars live alone and join a group only to mate.

Their preferred habitats are palmetto forest and salt marshes. Feral hogs will eat anything from grasses, fruit, nuts, tubers, and seeds to rodents, frogs, fawns, and carrion. The hogs stand accused of consuming the eggs of ground-nesting birds and sea turtles, uprooting and destroying endangered plants, and competing with native animals for resources.

Lactation is nutritionally demanding and can nearly double a mare's daily energy requirements. On average, mares produce about 3 U.S. gal./11.4 L of milk daily—approximately 450 U.S. gal./1,700 L of milk in the first five months of her foal's life. During a dry year like 2011 on Cumberland Island, it can be extremely challenging for a mare to support both herself and her foal.

The Park Service instituted a program of live-trapping and removal in 1975, and by 1979, 1,100 feral hogs had been removed from Cumberland (Mayer, Brisbin, & Brisbin, 2008). The designation of nearly 10,000 acres of Wilderness in 1983 significantly impacted the NPS' ability to manage the feral hog population by closing roads that were used for trapping. While hog control efforts continued at a reduced level, large scale hog management efforts did not resume again until 2000. Drought and hunting pressures have reduced the hog population considerably. In 2009, 465 hogs were harvested; in 2010, 356; in 2011, 233; in 2012, 221; and in 2013, 328 (D. Hoffman, personal communication, July 20, 2014). Hoffman estimates the 2014 hog census at fewer than 200 animals.

Not as charismatic as horses and usually more numerous, feral swine are seen universally as a nuisance animal at Cumberland and on many other federal lands. With all of the fervently negative sentiment, it comes as a surprise that some research portrays the hogs (in reasonable numbers) as a relatively benign presence. Jill Baron (1982) found that the feral hogs of Horn Island, Mississippi, ignored the sea oats and other vulnerable plants to scavenge tideline trash, dead fish and crabs, and other carrion. Disturbances created by the rooting of hogs recovered within months, sometimes with greater plant cover. After 150 years of hog habitation, the vegetation of Horn Island was indistinguishable from that of a swine-free island nearby.

The feral hogs of the Sea Islands may be longer-established than the horses. A study of mitochondrial DNA from feral hogs on Ossabaw Island (near Savannah) suggests origins

During a wet year like 2009, the island is lush and verdant, and forage is abundant. This mare has remained in good flesh despite months of nursing her robust colt.

in the Canary Islands from Spanish stock. The American Livestock Breeds Conservancy (2009) puts them on its critical list. On Cumberland, this Spanish ancestry is more dilute. The first swine were probably brought to the island by Spanish colonists and missionaries in 1562, and they probably have been free-roaming since that time. Their bloodlines have blended with the genes of other domestic stock that escaped or was released over the centuries (Mayer et al., 2008). In the 1940s and 1950s, lifetime resident Lucy Ferguson released a number of domestic swine to enlarge the gene pool of the feral population, and European wild boars may have been added to make the swine more desirable to hunters.

When the author toured Cumberland Island with Hoffman in 2011, more than 150 horses had maintained a relatively stable population, their numbers neither increasing nor decreasing significantly over the last few years. Hoffman said the lack of growth indicated that the herd had reached the carrying capacity of the island (personal communication, March 18, 2011).

As discussed previously, carrying capacity concerns the number of animals that can be supported by available habitat. Over the long term, horse herds managed at carrying capacity overgraze certain species of plants and inflict damage that reduces the future carrying capacity of the habitat. The system becomes less productive and less diverse, in effect taking a step backwards in evolution (Flueck, 2000).

Moreover, carrying capacity naturally fluctuates with environmental conditions. For a few years prior to 2009, rainfall was intermittently abundant, allowing forage to grow robustly, and the herd to expand. Then, from 2010 into 2013, severe drought took hold, and forage grew sparse. Then the rains came with a vengeance, and the island became lush once again. "We have had so much rain since the late spring of 2013 that all freshwater wetlands

Foals, like human infants, sleep often and spend much of their naptime in REM sleep.

and dune fields have essentially been flooded for the last 15 months," said Hoffman. "Truly a boom or bust situation" (personal communication, July 21, 2014).

This waxing and waning of resources had little effect on the herd: the census was 143 in 2003, 137 in 2004, 139 in 2005, 143 in 2006, 120 in 2007, 124 in 2008, 131 in 2009, 120 in 2010, 148 in 2011, and 136 in 2012. Observers only counted 108 horses in 2013, apparently due to a series of severe thunderstorms which disrupted the census. In 2014, the count was consistant with that of previous years—143 horses were identified (D. Hoffman, personal communication, July20, 2014).

Park Service policy is to remove exotic animals that threaten the environment. But as we have seen, it also is mandated to preserve and interpret cultural resources. "The horses are a non-native species just like the free-ranging cows that were removed in the 1980s and the feral pigs that we manage 365 days a year," says Hoffman. "Yet public sentiment differs significantly for each species" (personal communication, March 18, 2011).

Free-roaming horses have become a symbol of Cumberland Island to many people. Hoffman admits,

> This is a tough one because of the emotions and wide range of opinions regarding horses on this island . . . and emotion tends to override fact when you're trying to get a message across. Opinions on this issue vary even among the park's staff. The park realizes that while the horses are not native to the island, they are a big part of the visitor attraction. Ideally, a plan should be developed that allows for management of the herd and protection of the island's natural resources, while allowing for continued public enjoyment. (Personal communication, March 18, 2011)

On Cumberland Island, each band claims a home range of about 650–1,062 acres/263–430 ha. As in other locations, dominant stallions control harems, and dominant mares lead their bands to the best forage, the best water sources, and the best places to avoid insects.

Young foals engage in coprophagy, or feces-eating. It seems that consuming the manure of adult horses, especially their mothers, establishes the proper balance of helpful intestinal bacteria and helps them to learn which plants to eat.

Goodloe et al. (2000) found that most of the bands on Cumberland contained 2–10 horses, with an average number of 4–5. Almost 85% of bands were harems containing mares, youngsters, and one dominant stallion. About 10% of bands had more than one adult stallion, and mare and foal groupings comprised about 3.4% of bands. Dominant stallions typically maintained control of their harems for short intervals ranging from a few months to less than 3 years. Only 37% of dominant stallions retained harems over the duration of the 5-year study.

Foals are usually fathered by the dominant stallion within a mare's band. Stallions avoid mating with their own adolescent daughters, but in multi-stallion herds, the beta male sometimes mates with these fillies. If a number of mares come into heat simultaneously while two bands are in close proximity, sometimes a mare will mate with a rival stallion (Ashley, 2004).

Mares return to estrus 1–3 weeks after foaling, and most will conceive in this "foal heat." With an 11-month gestation (range of 320–365 days), it is possible for a mature mare to be pregnant nearly all the time, year after year, even while nursing foals (Crowell-Davis & Weeks, 2005). More commonly, mares will foal in alternate years or two years out of three (Boyd & Keiper, 2005).

Studies conducted on domestic horses demonstrate that about 80% of sexually active mares will achieve pregnancy over the course of a year, which, incidentally, is close to the 85% conception rate for women having unprotected intercourse for a year (Trussell, 2011).

Dominance plays a prominent role in nourishment. This mare is a high-ranking member of a high-ranking band. She and her offspring have increased access to higher-quality forage. Consequently, this 2011 photograph shows that she remained in good flesh while nursing her offspring even though she had borne and suckled a colt little more than 2 years earlier. This photograph was taken during a drought of 2 years' duration. Other mares grazing the Dungeness grounds in her vicinity were significantly thinner.

Of these equine pregnancies, more than 12% miscarry, most commonly between days 25 and 31 (Kirkpatrick & Turner, 1991). Pregnancy loss is more common in lactating mares who conceive immediately postpartum and among adolescent mares.

Interestingly, a well-nourished, healthy mare is more likely to conceive a male foal, while a mare in poor condition is more likely to conceive a female, though at mid-gestation mare condition was not a significant predictor of offspring gender (Cameron, Linklater, Stafford, & Veltman, 1999). Male foals are great caloric investment for the mare. Colts gestate longer than fillies and nurse 40% more often during their first two months, presumably to give them a competitive edge in their lives as stallions (Berger & Cunningham, 1987; Kirkpatrick, 1994). In nutritionally poor years, however, newborn colts are more likely to die than newborn fillies (Monard, Duncan, Fritz, & Feh, 1997).

Goodloe et al. (2000) studied the population dynamics of the Cumberland horses from 1986 to 1990 and saw the herd increase from about 186 to about 220 individuals. An estimated 61% of mares foaled each year. Of the 178 foals born during the study, 123 (69%) survived to their first annual census—61.1% of the fillies and 58.8% of the colts "during 1987 to 1989 (the 3 years of most intensive observations)" (Goodloe et al., 2000, p. 117). Among wild herds, high mortality for young foals is a natural mechanism that keeps the population in check.

The Cumberland herd does have a foal-loss rate considerably higher than the 20–25% average for North American wild horses, but it is not by any means the highest (Kirkpatrick, 1994). The Kaimanawa wild horses of New Zealand average a 50% foal-loss rate, and

Weather has a tremendous influence on the health of the horses and how many the range can support. In 2009, frequent rainfall made the forage lush and abundant, and even lactating mares remained in good flesh.

the Nevada Great Basin herd saw foal mortality as high as 88% due to predation by mountain lions (Greger & Romney, 1999; Linklater, Stafford, Minot, & Cameron, 2002). By comparison, among humans the mortality rate for children less than 1 year of age was up to 20–30% as recently as the 1800s, and it remains as high as 16% in Afghanistan (Johannson & Lindgren, 2008).

From 2003 through 2011, about 14% of foals survived each year. Hoffman explained,

Stories from island residents and park personnel suggest that foal mortality is common. The earliest foals are born during a time when there is normally not much in the way of grass from spring green-up. One would suspect that mares nursing during that time would not be in optimal condition to produce quantity/quality milk versus mares that began nursing in April after the green-up. Also, this island goes through long periods of drought conditions on occasion, further affecting the availability of food and mare condition/milk production. (Personal communication, March 18, 2011)

During Goodloe's five-year study period, the Cumberland herd increased by only 4.3% annually (Goodloe et al., 2000). Hoffman says that the mortality rate is very close to the birth rate, as evidenced by the fact that the population has remained stable over the last few years. Causes of mortality include high parasite loads, drought-related stress, age, coyote predation, accidents, and perhaps Eastern equine encephalitis and West Nile virus (D. Hoffman, personal communication, 2011).

Unlike sheep, goats, or even cattle, horses gestate only one offspring at a time. Twin conceptions are uncommon in horses and usually abort well before term. If the pregnancy continues until viability, the foals usually die at or soon after birth, especially in a wild herd. Among domestic horses, only 14% of twin foals survive to 2 weeks of age (Bennett & Hoffman, 1999). Typically, one or both are undersized and malnourished because of fetal competition for nutrients.

"The first foals are usually born in early to mid-March and most are born by the middle of April," says Hoffman (personal communication, March 18, 2011). Foals are active *in utero*, often moving 55 times an hour from the 5th to the 9th month of gestation, and up to 84 times an hour in the last few days before delivery (McGreevy, 2004). As the mare nears term, her abdomen visibly ripples with the movements of her growing fetus.

Days or weeks before the onset of labor, the mare's udder will swell and her teats will develop a waxy coating of honey-colored dried colostrum. The muscles around her tail head soften and sink. The lips of her vulva elongate.

As parturition approaches, the stallion may show increased sexual interest in the mare, courting her and herding her around (McDonnell, 2005). She rebuffs his advances, but he continues to pester her, pushing his nose under her tail and growing excited by the odor of the fluids. Sometimes he will take advantage of her helplessness by mounting her. Any other males in the harem that are beyond puberty may have a similar response.

Mares have some control over when they deliver their foals. Eighty percent of mares foal at night, usually shortly before dawn, when darkness makes them less noticeable and predators are sated from a night of hunting (Goodwin, 2002). If unusual activity or potential danger disturbs the mare, she may delay the birth until she is alone. One study of domestic mares showed that 10% have difficulty foaling, usually due to a malpositioned fetus (McCue & Ferris, 2012). In the wild, when a foal is undeliverable, both mare and foal often die.

When birth is imminent, barrier island mares typically leave the herd and sequester themselves in the marshes or in dense thickets. Sometimes a group of her closest mare-friends will accompany her and appear to offer moral support (Crowell-Davis & Weeks, 2005).

During the first part of labor, the cervix dilates. The mare becomes restless and sweaty, looking at her sides with a concerned look and pacing for about 1–4 hr before delivery.

Delivery typically begins with the breaking of the water sac. During the expulsive process, the mare cannot stop delivery and is consequently vulnerable. The mare lies on her side and pushes repeatedly by contracting her abdominal muscles to assist the expulsive efforts of her uterus. She may rise and then lie down again.

Glistening white membranes appear at the vulvar lips. The foal's front hooves usually present first, one slightly ahead of the other, padded with rubbery keratin to prevent injury to the mare before and during birth. The nose of the foal is positioned atop the forelegs. Further contractions bring the shoulders, body, hips, and hind legs over the course of 15–20 min. With a final strain, the mare deposits the foal on the ground. Mother and newborn rest for a few minutes, breathing heavily.

The mare cranes her neck to glimpse the newborn lying behind her, and bonding begins. She nickers softly to the new arrival, fills her nostrils with his scent and begins to lick him ambitiously. Maternal licking helps to dry the coat, and the vigorous stimulation encourages the newborn to breathe, move, and stand. Licking also acquaints the mare with the odor of her foal and facilitates bonding. Chemicals in the mare's saliva transfer to the fuzzy wet coat and help her to identify her offspring by smell (Crowell-Davis & Weeks, 2005). It is

In this 2011 photograph, horses graze the sparse dry lawn surrounding Plum Orchard. The horse in the foreground, the adolescent son of the mare in background, is in excellent condition, having been sustained by her milk for more than a year. The mare, however, is extremely thin and appears to have difficulty recovering her body weight with the forage available to her.

crucially important for the mare to understand that the foal is hers and accept responsibility for his care. If she rejects him, he dies. A mare will not nurse a foal that is not her own.

Some of the foal's behaviors are instinctive, and some are learned. Soon after reaching the outside world, the foal's imperative is to stand. The longer he is immobile, the more likely he is to be devoured by some marauder. He comes into the world with the overwhelming instinct to lurch to his feet, unfolding great lengths of spidery legs and assembling them beneath himself at reasonably appropriate angles.

The umbilical cord breaks when the mother or foal stands after birth, and the placenta is expelled soon after. Most mares will not eat the placenta.

The next objective for the unsteady, spraddle-legged foal is to find sustenance. His instinctive programming prompts him to thrust his muzzle into any dark corner until he finds a teat. He randomly probes the shadows on the underside of his dam, using smell and touch to find his target. The mare stands unmoving to facilitate his search, hind leg pulled back to expose the target, gently guiding him as he noses into her elbow and flank before reaching his goal. At birth, foals sport a bouquet of long, silky whiskers that function as sense organs to help them find an udder. Eventually, he makes contact with one of her two teats and suckles enthusiastically.

Within 2 hours the foal has fed, and by dawn of his first day he can keep up with his mother as she rejoins the herd. At first, he follows his mother with great clumsiness, but this is short-lived. Having mastered walking, he soon attempts an awkward run, often punctuated by playful hops and bucks. This is quite a feat, considering that hours before he was unacquainted with balance, sequential leg movement, and the inexorable pull of gravity.

Immunocontraception would be difficult to implement on Cumberland Island for the same reason it is difficult to take a census: the horses in the wilderness area tend to avoid human contact and escape into dense woodland when disturbed.

Colostrum, thick yellow early milk, is liquid gold. Its immunologic benefits are so profound that the foal is likely to die of infection if he fails to receive it. Immunoglobulins, or antibodies, are proteins of the immune system, produced in the bone marrow of an adult horse and circulated in the bloodstream, specifically in the plasma. There are many kinds, each designed to bind to specific bacteria, viruses, or foreign proteins to target them for destruction. Where human infants acquire some defense against disease from maternal immunoglobulins transferred during pregnancy, foals are born with minimal circulating antibodies in their blood and must suckle immediately after birth to protect them until their immune systems mature. This antibody cocktail is customized for that particular foal in that particular environment. The mare passes antibodies against diseases that she has been exposed to, which are the diseases that the foal is most likely to encounter.

The mare produces about 5 L of colostrum in the first 12–18 hr. When the newborn nurses, the large antibodies in colostrum are not digested in the foal's small intestine, but are absorbed directly into the bloodstream by specialized epithelial cells. The capacity to absorb antibodies is greatest during the first hours after birth and quickly declines. After the first day, "gut closure" occurs and this mechanism is no longer functional.

Foals are born with the instinct to follow any large moving object. The mare's task is to make sure that she is that object and to keep other horses at a distance for the first few days. For the first 2–3 months, the mare looks after the foal, waits for him to catch up, guides him, and protects him from danger.

When he is a little older, she is less patient with his transgressions, and it is the foal's responsibility to keep up with the herd and find his mother. Likewise, a mare is very tolerant

Horses use the beach for insect relief, but their activities there do not cause significant damage to the natural dune system on the island. This young stallion is a bachelor, unaffiliated with a band.

of a newborn's awkward attempts at nursing, but if an older foal is rough or clumsy she will discipline him or walk away.

Newborn foals try a number of nursing positions before settling on the one that yields the best results: side by side with the mare, facing the rear, in close contact. Newborns nurse about five times an hour for the first five weeks. Foals older than 10 weeks nurse only about once an hour. A hungry foal will sometimes walk alongside his dam, then move directly in front of her, slipping under her neck so she is forced to stop and let him nurse.

As long as a foal continues to nurse, a mare will produce milk. In contrast to domesticated horses, which are usually weaned before 6 months of age, wild horses nurse until they are yearlings if the mare foals again the following year and often until age 2 or even 3 years if they do not have a new sibling (Kirkpatrick & Turner, 1991). In the wild, horses have been known to nurse for as long as 5 years as a comfort behavior, not for nutritive benefit (Crowell-Davis & Weeks, 2005). Occasionally a mare may allow both the yearling and the new foal to nurse. The author observed one Assateague mare was still occasionally suckling her dam while she had a young foal of her own!

Typically, the other horses in the band are extremely curious about the new arrival. A mare with a new foal is very protective and will temporarily rise in the pecking order as she assertively keeps even dominant horses from her newborn. Maternal protectiveness persists for the first few days and serves to make the foal absolutely sure which mare is his dam while insulating the foal from potentially lethal kicks and bites. The stallion may circle the mare and foal and keep the other horses from approaching too closely.

A newborn foal spends much of his time sleeping or nursing. Unlike mature horses, foals take frequent long naps flat on their sides in deep slumber. When a young foal sleeps, the

A white doe comes to attention atop a dune overlooking a sheltered pocket of maritime forest near the Dungeness dune crossing. In other places on Cumberland Island, dunes migrate like waves of sand, engulfing and killing trees as they progress. Deer and horses are not to blame—dune migration is a natural process and also occurs on islands with undisturbed dunes. Dune movement is primarily the function of sand supply, wind, waves, rain, and the vegetation over which the dune is migrating.

mare positions herself close to him in a behavior called the "recumbency response." She may graze nearby, mowing the grass short in a circle around her foal, or doze herself while standing beside or over him (Crowell-Davis & Weeks, 2005). Foals will often sleep beside one another in a sort of equine kindergarten naptime, limbs or necks overlapping.

Foals are born with a fleecy coat made of long, soft, curly hairs, which begin to shed by about 7 weeks of age. The molting of the baby fuzz takes about 30–40 days, during which patches of mature coat show in odd places around the eyes and muzzle, often in a different color. The fuzzy upright newborn mane gives way to a more mature growth that stands up comically before it begins to lie down on one side of the neck.

Newborn foals taste grasses with toothless mouths and chew without ingesting, initially not discriminating against poisonous plants. They do the same with driftwood, sand, and stones. Over the next months, foals learn which foods are palatable and spend more time grazing. The diet of the Cumberland horses centers on five kinds of grass. *Spartina alterniflora*, saltmarsh cordgrass, is the primary item on the menu.

Foals also ingest manure for the first three or four months of life, especially the fresh dung of the mother (Crowell-Davis & Weeks, 2005). This appears to be necessary for establishing optimal intestinal flora and helping foals identify desirable dietary plants.

Two pairs of incisors begin to emerge in 6–8 days, allowing the foal to nibble at grass. Two more incisors emerge by 2–4 months, and by 6–8 months the foal has a full set (six pairs) of deciduous "baby" teeth.

It is possible to determine the age of a horse, especially a younger one, by his teeth. Like children, horses have deciduous teeth followed by permanent teeth. At age 3, the central four incisors are shed; at age 4, the four teeth lateral to these; and at the age of 5 she has all

Some areas of marsh are barren mud flats, while others nearby are densely vegetated. Why is the foreground bare and trampled, while thick cordgrass grows in the marsh beyond? Horses have unlimited access to both areas. Although some scientists blame horse grazing for these changes, similar diebacks of *Spartina* have occurred in ungrazed marshes from Cape Cod to Louisiana. New evidence suggests that the culprits are not horses, but snails—namely marsh periwinkles, *Littoraria irrorata* (Bertness, Silliman, & Jefferies, 2004).

her adult incisors. After the age of 5, other features indicate age, such as the presence and depth of cups on the incisors, dental stars or exposed pulp, and the shape and angle of the tooth. Canine teeth, also known as tusks or tushes, appear in the gap between the incisors and the molars of male horses at the age of 4–5 years, but only occasionally develop in mares.

Foals, like all young mammals, have an innate curiosity and playful spirit. Scientists define play behavior as activities having no immediate use or function and involving a sense of pleasure (van Dierendonck & Goodwin, 2005). Play develops strong bodies, provides practice for adult behaviors like mating and fighting, and builds social and communication skills.

Solitary and object play develop in the first weeks of life while foals remain close to their mothers. By the age of 4 months, foals become more independent. As maternal protectiveness diminishes, foals spend more time in social play, seeking out playmates close to their own age and size. Through their play, they develop coordination and balance, become socialized, and refine behaviors necessary to survival. They gallop madly in circles around the band, not always gracefully, and mimic adult behavior.

Colts play at sexual activities, climbing astride their playmates, mothers, and herdmates with erections but no penetration. Fillies have less interest in sexual play.

Colts with same-age peers develop a dominance hierarchy among themselves based mostly on birth order. Older colts are dominant, as are the sons of high-ranking mares (McGreevy, 2004). These buddies play-fight boisterously and spend much more time

In the marsh that lies to the south of Dungeness, there is no sign of dieback.

running than do colts raised singly. This roughhousing gives colts with same-age friends a biological advantage, allowing them to develop greater skills, motor abilities, and physical conditioning than colts raised alone. This aptitude leads to higher rankings in bachelor herds and probably increases the likelihood of acquiring and maintaining a harem.

Fillies romp, too, but do not usually engage in mock battles with the same enthusiasm as colts. They play running games, but not to the extent their brothers do. Because of these differences, fillies prefer to play with fillies, colts with colts.

In males, playful behavior peaks between the ages of 2 and 4. Adult stallions will play with their adult sons and with other stallions outside the breeding season (Feh, 2005). Young mares are just as playful as colts living in a bachelor band, but lose interest in most play once they have had their first foals.

Dr. Robin Goodloe focused on the Cumberland Horses as part of her doctoral research and became intimately familiar with their habits and behavior (Goodloe et al., 1991). What she saw convinced her that the horses do not belong on Cumberland Island, and based on ecological and genetic considerations, she recommended that the Seashore reduce the size of the herd or remove it entirely. Laliberté (2007) similarly concluded "feral horses and hogs are having a disastrous impact on the island's ecosystems, particularly the salt marsh."

Dr. Mark Lenarz conducted the first Cumberland Island free-roaming horse census in 1981, recorded 144, and estimated 10–20 animals were uncounted ("Feral horses," 1983). From 1986 to 1990, a research team headed by Goodloe performed annual horse counts and found the population steadily rose, Increasing from 144 in 1981 to 186 in 1986 at an annual rate of 5.3%, then dropped to 103 by 1999. Between 2000 and 2011, the Park Service horse census counted 120–154 horses on the island, 148 in 2011 (D. Hoffman, personal communication, March 18, 2011). The census for 2014 was 143.

Vincent J. Bellis, aquatic biologist and professor emeritus at East Carolina University, explained the dynamic:

This barbed-wire exclosure in the Abraham Point marsh seems to show horse damage on the outside of the fence and healthy vegetation within. This photograph, however, does not tell the whole story. Along the opposite fenceline, grass is equally tall on both sides. Marsh dieback is evident both within and without the exclosure. The exclosure was installed circa 2001 as part of a graduate research project. When the perimeter was intact, horses somehow jumped into the exclosure and were unable to escape. They died there—the white objects visible in the mud are bones. Researchers cut a section of wire to prevent other animals from getting caught inside, and horses have had free access to the exclosure for years.

The number of individuals in a population is thus largely dependent on the available food supply, so that in the absence of human management, such populations can be expected to increase beyond the carrying capacity of the environment. Because migration to better range is seldom an option on a small barrier island, feral populations typically experience cycles in abundance and vigor. A bad year of food production when carrying capacity has already been exceeded can result in drastic losses from to starvation and disease. The survivors may operate with a severely reduced gene pool. Local extirpation may result after several cycles if genetic vigor is not restored by introduction of new individuals. (1995, p. 58)

While breeders of domestic animals keep genetic loss at <2% per generation to avoid inbreeding depression, Goodloe's team, which included renowned equine geneticist E. Gus Cothran, recommended "a more conservative rate of 1%" for most other species, using a generation length of 10 years for stallions and 9 for mares (Goodloe et al., 1991, p. 419). To prevent genetic loss in excess of this threshold, 23 males and 27 females in the Cumberland Island herd must breed each year. When Goodloe (2000) studied the herd, about 32% of the herd were juveniles not yet reproducing. The team recommended that the current population be halved to 122 individuals, a number that would minimize genetic erosion while reducing pressure on island vegetation. Turner (1988) found that feral horse grazing reduced plant biomass up to 55% and salt marsh vegetation up to 98%. She developed simulation models that indicated an equine population greater than 49–73 would cause serious damage to the salt marshes. A population that size would need outside horses introduced every 9–10 years to minimize inbreeding.

Since 1991, Cumberland Island NS has been monitoring the sex, age, coat color, habitat, body condition, and location of horses and the number of horses in each band through an annual census. Initially, technique differed from year to year (D. Hoffman, personal communication, March 18, 2011). Since 2000, seashore staff has standardized annual counts for the greatest possible accuracy, using the same group of about 30 volunteers during the same month and same tidal conditions along the same survey routes. The staff then evaluates census data alongside research into horse-related impacts on the island vegetation to develop a long-range plan for horse management. The census does not record every horse—Hoffman estimates as many as 50 horses may be uncounted every year—but the tabulation serves as an "index to abundance" (D. Hoffman, personal communication, 2011).

As explained in chapter 2, a population's ability to grow maximally under optimal conditions is its *biotic potential*, represented by the letter *r* in formulas. Usually environmental conditions are not optimal, and increase is limited by environmental resistance—limitations in food, water, climate, habitat, and other conditions. The carrying capacity, the number of individuals that a habitat can support indefinitely if food, water, space, and other resources remain constant, is represented by the letter *K*.

For a maximally healthy herd and habitat, managers of wild horse herds should strive to balance a genetically healthy herd size with the carrying capacity of the environment. Carrying capacity (K) is the maximum population threshold that the ecosystem can support, and may vary with climatic fluctuations, i.e., higher with abundant rainfall, lower in a drought. When populations of grazers reach carrying capacity, both herbivores and habitat become stressed and unbalanced. The vegetative environment becomes less diverse, which potentially unbalances the habitat for other species. Oftentimes, dense grasslands are chewed to stubble, trampling becomes widespread, the abundance and distribution of plant species changes, and palatable species are grazed out of existence. When population grows too large, mares begin to produce their first foals later in life, fewer foals and yearlings survive their first winters, and mortality of older animals increases (Eberhardt, 1977, 2002).

A stable horse population, in which births balance deaths and there is no net growth over time, indicates that the herd has reached the carrying capacity of its range. A herd's proximity to carrying capacity can be assessed by annual census—is it stable, growing, or shrinking?

As the herd reaches carrying capacity, lactating mares, which require more calories than other members, grow uniformly thin; conversely, as the population density decreases, the body condition of lactating mares improves (Scorolli, 2012). The photographs of the lactating mares in this chapter taken during abundant rainfall indicate that when well-watered, Cumberland can support a herd near the top of its historical census without environmental compromise; but when drought conditions prevail, as often they do, there is fierce competition for food resources, and mares become emaciated even if the population is comparatively small. See Chapter 7 for fuller treatment of this topic.

In studies of domestic stock, a typical horse on pasture has been shown to eat, trample, and damage about 1,000 lb/454 kg of forage per month (Blum, 2013; Warren, 2004), and an acre (0.4 ha) may support two average-size horses, depending on season, climate, soil, rainfall, percentage of edible grasses and forbs, and a number of other variables.

Hoffman explained:

> Even if one adjusted this figure to more accurately fit the range of sizes and age classes of the CUIS horse herd, there is still a significant amount of dry matter removed daily that would otherwise be subject to only minimal grazing/browsing pressure by native wildlife. Such pressure exerted over a long period of time can permanently change vegetative communities. (Personal communication, March 18, 2011)

Any species can cause environmental damage through overpopulation. Horses can live on barrier islands without destroying habitat, but they must be maintained below the carrying capacity of the island. If grazing pressures are reduced, balance will eventually return.

In 1995, the Park service assembled twenty-six agency, state, and university specialists to discuss options for managing the herds at both Cumberland Island and Cape Lookout NS (Dilsaver, 2004). The experts agreed that the horses were causing environmental damage on Cumberland and that the herd should be reduced or removed, a sentiment echoed by several environmentalist groups. One option involved a one-time removal that would reduce the herd to 50–70 animals (Laliberté, 2007; Turner, 1988)—the same arbitrary, nonscientific 60-horse population that the government advocated at Shackleford Banks and Corolla. The seashore would use immunocontraception to limit growth and restrict the diminished herd to the part of the island south of Sea Camp, where grass is most abundant and the horses are most observable. A herd of that size, however, is not genetically sustainable, and it would require periodic additions to decrease inbreeding over the long term. Another option was to limit the herd to 120 and allow continued access to the entire island, with immunocontraception to control fertility.

Hoffman explained:

> The same obstacles that make the census challenging would make any of these management options difficult to implement on Cumberland Island. The forests are dense with saw palmetto understory that is difficult to penetrate, complicating roundups and immunocontraceptive administration. Some of the horses living at the south end of the island routinely travel distances of greater than two miles over the course of a day or two. A contraception program would necessitate finding, vaccinating, and recording data on every mare on a yearly basis. (Personal communication, March 18, 2011)

In April 1996, the seashore chose the 120-horse option and sought authorization to implement it. At a series of public hearings, a Park Service resource management specialist presented evidence that horses were damaging the marsh and dunes and proposed limiting the population. Her words were a match in a powder keg. Island residents and other locals with strong emotional ties to the herd fervently opposed removing horses from the island (Dilsaver, 2004). Many speakers questioned the motives of the Park Service, the validity of the studies, and the biases of the researchers. They maintained that the horses had been in residence for centuries and had not caused environmental collapse. They pointed out that wild horses were an integral part of the history, character, and beauty of the island and that their presence increased visitor satisfaction.

Like an emotional magnet, the conflict inexorably drew people in and moved them to one pole or the other. People were passionately for or against the horses. There was no middle ground. Opinions fueled agendas, and impassioned arguments broke out in editorials, at home, in the workplace, and of course in the political arena. In fall of 1996, U.S. Representative Jack Kingston (GA 1) toured the island with one of the residents and concluded that the equine census had decreased and there was little evidence of horse-related damage.

Dilsaver recounts (2004, p. 242),

> Without discussing the issue with the Park Service, he added a rider to the fiscal year 1997 budget bill banning all horse management at Cumberland Island National Seashore. The rider suggested that the Park Service use the money allocated for the horse program to further sea turtle protection.

The rider expired the following year, but hampered by budget cuts, the National Park Service has not authored a new management plan. Instead, they have focused on gathering data about the impact of free-roaming horses.

"There are multiple hurdles to creating a management plan for the Cumberland horse herd," says Hoffman:

> We have a very small staff and currently cannot commit the time needed to develop a management plan. Considering the high visibility of this park's horse herd, a concrete and scientifically sound management plan will be needed. This is a very emotional and political issue... there is the possibility that a great deal of effort could go into the development of a new, more current plan, only to be derailed by public pressure and political interests (personal communication, March 18, 2011).

A scientifically sound management plan will be based on accurate environmental assessment, but the Park Service says that it does not have the funds for such an undertaking. Scientists are currently studying the impact of horses on the environment in hopes that their findings will clarify the best options for management. A management plan involving immunocontraception is in place for the horses that range on privately-owned Little Cumberland Island, a 2,400-acre tract separated from the main island by Christmas and Brockington creeks.

"If and when the time comes that NPS develops a management plan," said Hoffman, "I envision that we would draw from other management experiences along the Atlantic Coast and likely establish a committee including personnel from various federal and state agencies and private groups that have been involved in the development of plans for the other feral horse herds. Representatives from the local public would be included in the committee as well" (personal communication, March 18, 2011).

The Cumberland Horse Project, a special-interest group organized by Debbie Middleton, a Camden county resident, proposed to collect data as is done on Assateague, Shackleford, and elsewhere. Cumberland Horse Project members volunteered to monitor the condition and social status of each horse; document births, deaths, and social interchange; and assess the health of each horse from a distance. The Park Service cannot afford to undertake this task independently, and these data would lay the groundwork for an evidence-based management plan.

Middleton met with Fred Boyles, superintendent of Cumberland Island NS, and offered the services of her group free of charge. "He stated emphatically that the NPS preference would be to do that themselves," said Middleton (personal communication, January 30, 2011). She says that she left that meeting disheartened, feeling that if the Cumberland Horse Project drafted a formal proposal to monitor the health and welfare of the herd for the sake of the horses, it would hit a wall of government pushback.

Nonprofit entities affiliated with the Park Service monitor the welfare of Cumberland Island NS and try to assess its problems. A prominent example is the National Parks Conservation Association, a quasiofficial support group that the first Park Service director and others formed in 1919, just three years after the agency itself came into being. A branch of the NPCA, the Center for Park Research, assesses the strengths and weaknesses of National Parks. The agency's 2009 report (Saxton, 2009), based largely on the research of Jennifer Laliberté and Molli Songco, alleges that equine trampling leaves the marshes highly vulnerable to storm damage and erosion, alters the food web, and negatively affects species such as loggerhead sea turtles.

Yet 2010 was a record year for loggerheads—225 nests a year is the average tally for the island, but in 2010 that number doubled, and hatching success increased (Georgia Department of Natural Resources, 2010). It appears that the horses are not affecting the loggerhead turtles after all. Hoffman offers corroboration:

> Horses do occasionally step on our screens used to protect nests from raccoons, but this is a random occurrence resulting from them walking along dunes and the beach. We have never had a nest predated due to a screen damaged from horse hooves, as we replace or repair the screen soon after it is damaged. (Personal communication, March 18, 2011)

Laliberté's environmental assessment blames the horses for other disruptions as well, stating "The main stressor to the park is the presence of feral horses that graze in the salt marsh, trample the dunes, interrupt sea turtle and shore bird nesting and contribute to erosion" (Laliberté, 2007, p. 7). But other wildlife is doing well on Cumberland Island. Recent avian censuses show high counts and excellent species diversity each year (Hoffman, 2008). Bald eagles (*Haliaeetus leucocephalus*) are on the increase. Species uncommon on the island are sighted with increasing frequency, including Franklin's gull (*Leucophaeus pipixcan*), the Nashville warbler (*Vermivora ruficapilla*), and the black-throated blue warbler (*Dendroica caerulescens*). The reddish egret (*Egretta rufescens*), a rare and range-restricted species, has been counted on the island 8 times, all within the last 15 years (*Christmas Bird Count*, 2011).

Hoffman, who is intimately familiar with the entire island, sees no indication that horses are damaging dunes. "While the horses do graze sea oats and other plants on the dunes, we have a healthy dune system that is impacted by severe storm, wind, and tide events more than anything else," says Hoffman (personal communication, March 18, 2011).

Horses are implicated in disturbing shorebird nests, but when researchers installed video monitors to determine the cause of American oystercatcher (*Haematopus palliatus*) nest failure, the true culprits were unmasked (Schulte, Brown, & Reynolds, 2007). Raccoons were devouring the eggs, and people walking on the beach near the nests disrupted incubation by frightening birds off the nests. The cameras also caught a young child handling and breaking eggs. "It is likely that horses occasionally step on a nest while walking the beach and dunes," says Hoffman, "But I suspect this is rare" (personal communication, March 18, 2011).

Some reports, including the Cumberland Island NS resource assessment, blame horses for increasing island erosion, which is removing about 2.6 ft/0.8 m of the western shoreline a year (Saxton, 2009). But closer investigation suggests that the unnatural movement of sand around Cumberland Island is the consequence of two jetties, built to maintain a navigational channel through St. Marys Inlet (Byrnes & Hiland, 1995). The jetties disrupt the longshore and channel flow of sediments, resulting in an unnatural accumulation on the

Although Cumberland Island is relatively large, most of it is woodland. Horses forage primarily in the marsh and lawn areas, crowding one another and promoting erosion and overgrazing. Drought reduces the amount of forage available to the horses.

southern tip of Cumberland Island, which has added more than 500 acres/200 ha of new land on the ocean side (Larson, Rosati, & Kraus, 2002).

The short-term rate of shoreline change calculated over an approximate 27-year period shows a similar pattern of erosion and accretion—from 396 ft/121 m of accretion in one spot to 346 ft/105 m of shoreline retreat in another. Cumberland Island gained 6.5 ft/2 m a year over a period of about 6 years, a rate consistent with observations over the past 150 years, and Cumberland and Little Cumberland "showed accretion over 80%" of their combined ocean beach (Stockdon, Thompson, & Fauver, 2007). Hoffman says:

> There are certainly some spots where horse activity is impacting shoreline but it is likely not a factor in the major erosion areas. We view erosion as a major issue on Cumberland, with the most significant losses along the western boundary. There are several areas that experience almost constant erosion, primarily as a result of high winds, water currents, and boat wakes. We are planning on conducting an oyster reef restoration project in the main problem areas to attempt to slow the losses.
> (D. Hoffman, personal communication, March 18, 2011)

Large sand dunes on the island are migrating and losing height. The back dune ridge at Dungeness dune crossing has decreased in height 13 ft/4 m since the 1960s, and the dune has advanced to engulf trees and spill over into Beach Field at a rate of 3.3–6.6 ft/1–2 m a year. Many blame horse grazing for these changes.

Although overgrazing can destabilize sand dunes, dune migration is a natural process that occurs on barrier islands with or without horses. The movement of the dune at Dungeness dune crossing is the function of sand supply, wind, waves, rain, and the vegetation over which the dune is migrating. The landform is dynamic, and its morphology and rate of migration are ever-changing (Cofer-Shabica, 2010).

Most of Cumberland Island is dense maritime forest with a nearly impenetrable understory of saw palmetto and little forage for horses.

In 2011, most of the lactating mares on Cumberland Island were in poor condition. Was their gauntness caused by insufficient food supply, internal parasites, chronic disease, or something else? The root cause was probably overpopulation, which increases the likelihood of the other problems.

Hoffman offered his perspective:

Most of our foredunes are well covered in sea oats, suggesting that they can withstand some level of grazing pressure. We have large expanses of dune fields in some places between the foredunes and the forest edge, and there are also large dunes along the forest edge that show signs of migration over a long period of time. I do not think that these dune fields and large inner dunes would have extensive vegetation growing on them naturally if the horses were not here; I suspect this is indeed a natural process. The sea oats are not clipped off and overgrazed, but appear to be healthy with decent leaf growth and seed production. (Personal communication, March 18, 2011)

As on Assateague, Cumberland horses are blamed for degrading water quality (Saxton, 2009). But other sources point to pollution from nearby watersheds, effects of acid rain, agricultural runoff, and aging septic systems on the island. The levels of mercury in the sound waters are elevated, perhaps by runoff from mercury-based fungicides once used in nearby pine plantations (Alber, Flory, & Payne, 2005). Other possible sources include the defunct Gilman/Durango-Georgia paper mill near St. Marys (a prospective Superfund site) and active paper mills in Fernandina Beach, Fla., and Brunswick, Ga. (Alber et al., 2005; *Georgia Toxics Release Inventory Report, 2001*, 2003; Goodson, 2009). Mercury accumulates in the flesh of predatory fish, and there is a regional advisory regarding consumption of local seafood.

Cumberland Island is also threatened by a population explosion on the mainland spurring new housing developments on the coast, together with waterfront mansions and marinas directly across Cumberland Sound from the seashore. In 1980, only 13,371 people lived in Camden County, Ga. By 2000, that number had mushroomed to 43,664 (Saxton, 2009).

Global climate change will pose challenges to the Cumberland Island ecosystem. Over the next 100 years, rising sea levels may drown the marsh and push seawater into the aquifers that provide freshwater on the island (Saxton, 2009).

Some of the marshes of Cumberland Island are verdant and healthy, while others are muddy deserts devoid of vegetation. Some studies blame overgrazing for damaging these sensitive wetlands. These mysterious diebacks, however, are not confined to Cumberland Island, and they afflict ungrazed marshes in Louisiana, North and South Carolina, Massachusetts, Connecticut, and Virginia (Ogburn & Alber, 2006).

As it turns out, the culprit responsible for marsh dieback may be marsh periwinkles (*Littoraria irrorata*), which are most abundant where *Spartina* is dead or dying (Bertness et al., 2004). The snails feed primarily on fungus. The snails score the stalks of live cordgrass to facilitate fungal infection of the plants, fertilize the wounds with their own dung, and harvest crops of nutritious fungi. Grazing snails and fungal infection kill cordgrass and cause marsh die-offs (Bertness et al., 2004). Horse grazing actually helps protect marshes against die-off by creating conditions favorable to crabs and birds that prey on *Littoraria* snails (Levin et al., 2002; Turner 1987).

Wildlife ecologist Craig Downer (2011) says,

For years, it has been the all-too-easy custom for established interests to "scapegoat," or lay the blame, for overgrazing, erosion, threats to native species, and other ecological abuses upon wild horses or burros wherever they occur, by magnifying their impacts out of proportion—all the while conveniently ignoring mankind's own enormously destructive traditions.

Some question the Park Service assertion that there are too many horses on Cumberland. Shackleford Banks, North Carolina, easily supports 110–130 horses though it is much smaller—only 9 mi/14.5 km long and barely 1 mile/1.6 km wide—and not nearly as lushly vegetated as Cumberland. They argue that a population of 150–200 horses should easily maintain a balance with the Cumberland ecosystem.

Most of Cumberland, however, is covered by dense forest that contains little horse forage. Hoffman says that although the island is relatively large, the horses can only use certain areas, and large numbers of horses crowd onto them:

> The majority of the island is not suitable habitat for horses, so the areas that have attractive habitat are used by multiple animals. For example, we have approximately 300 total acres of pasture grass-type habitat and most of it is in two places. These see constant grazing pressure. In addition, the salt marsh accounts for approximately 15,000 acres. . . . Not all of the total marsh acreage is accessible to horses due to tidal creeks, deep mud, etc. The areas that are used by horses are used routinely and show grazing pressure and trampling damage. Considering that the island is 36,000 acres, there is really minimal horse-friendly habitat. (Personal communication, March 18, 2011)

The thin lactating mare at the Dungeness dock introduced in the beginning of this chapter suggests that this overcrowding is detrimental to at least some of the horses. In dry years, the majority of lactating mares appear lean and voraciously hungry over most of the island.

Perhaps the least disruptive method for thinning the herd would be to target young horses from well-represented lines for remote sedation, removal, and adoption, as on Shackleford Banks. If horses are removed to the mainland, however, it is likely that another ugly issue may surge to the forefront—the management of equine infectious anemia, or swamp fever.

EIA, discussed at length in Chapter 7, is a retrovirus transmitted by flies from horse to horse. An infected horse has the disease for life and is subject to exacerbations that can make the animal seriously ill and much more infectious than when in the asymptomatic carrier state. According to Georgia law, all horses testing positive for EIA "are to be branded and quarantined for life or disposed of by approved slaughter or euthanasia with burial" (Equine Act of 1969).

When state veterinarians administered the Coggins test for EIA to other free-roaming East Coast herds, the results were devastating. The herd on Chincoteague National Wildlife Refuge lost 53 of 113 horses; the herd in Rachel Carson Estuarine Reserve (N.C.) lost 9 of 33; Shackleford Banks, N.C., lost 76 of 184, with five more identified and quarantined the following year; and the privately owned herd on Cedar Island, N.C., lost 13 of 15.

The Cumberland herd has never been screened for EIA. It is entirely possible that a number of horses on Cumberland Island carry the disease. If officials made a concentrated effort to test the herd, many animals might be marked for euthanasia. Testing this herd would present a tremendous challenge. On thickly forested Cumberland Island, it is likely that some elusive horses would evade a large-scale gather; and if they carry EIA, they could perpetuate the disease.

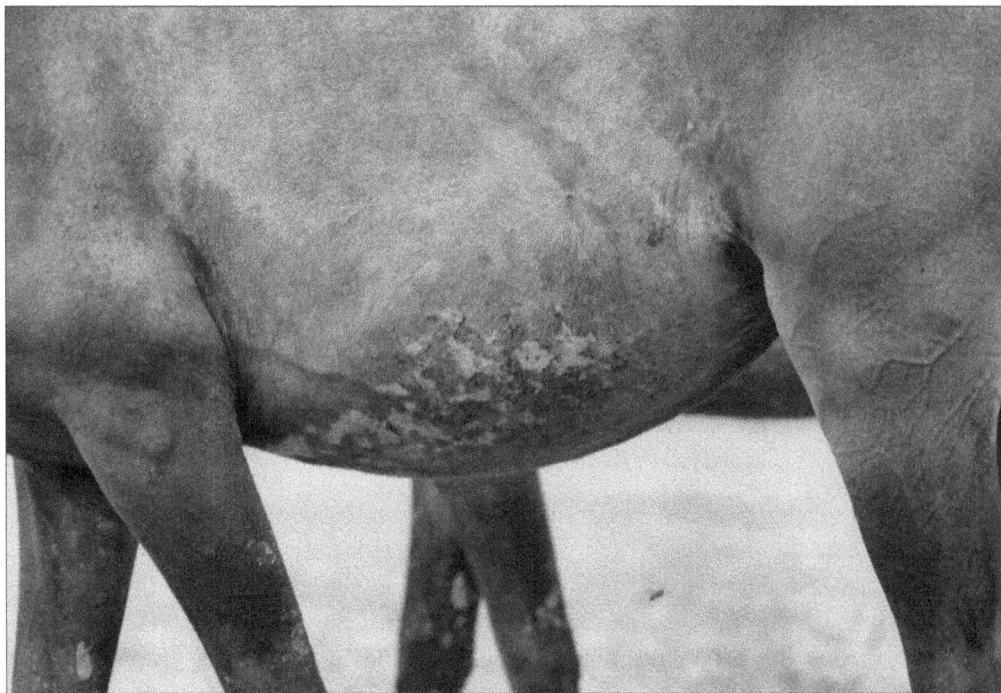

Many Cumberland horses show evidence of what appears to be chronic fungal infection, mostly on the abdomen and across the withers. Widespread fungal infections typically take hold in old, young, debilitated, or immunocompromised animals. They are highly contagious and can spread quickly through a herd.

The issue could also become socially and politically explosive. The destruction of healthy-looking chronic carriers in great numbers always sparks public outcry and resistance.

Environmentalists hoped that as a national seashore, Cumberland would remain safe from exploitation and would become progressively more primitive. Although the wilderness area designated in 1982 was legally closed to vehicular traffic, most private property owners on the island were permitted to use the main road until their rights ended in 2010 (Sharp & Miller, 2008). Without vehicular access through the wilderness area, the historic sites at the north end of the island were no longer accessible to visitors unable or unwilling to hike 17 mi/27 km to see them. The Greyfield Inn provided vehicle tours despite the constraints of the Wilderness Act, Park Service regulations, and even a 2004 ruling by the 11th Circuit Court of Appeals (*Wilderness Watch v. Mainella*, 2004) that they are illegal and must cease (Harlan, 2007).

In 2004, island residents lobbied U.S. Rep. Jack Kingston (GA 1) to attach a back-door amendment to the Consolidated Appropriations Act of 2005 (2004), often cited as the Cumberland Island Wilderness Boundary Adjustment Act of 2004. Without any public input, the legislation removed the federal wilderness designation from 25-ft/7.6-m corridors around three dirt roads (Main Road, North Cut Road, and Plum Orchard Spur) and requires the Park Service to maintain them for vehicle access. This was the first time that Congress rolled back a wilderness designation (Harlan, 2007). Conservation and environmental groups were incensed, but unable to overturn the legislation.

The new law required the Park Service to develop a transportation-management plan that would provide *five to eight* bus tours daily to the historic sites on the northern part of Cumberland (Sharp & Miller, 2008). To accommodate the tours, the road required grading,

The Park Service and a group of young volunteers removed the pews from the First African Baptist Church during a restoration and painting project.

widening, and defoliation. Conservationists argued that the prevailing vision for Cumberland Island NS was for wilderness, not tour buses and vehicular traffic, and that these incursions would detract from the native beauty of the island (Harlan, 2007). The Park Service, however, has a statutory and regulatory duty to make its publicly funded sites accessible to mobility-challenged visitors.

By 2013, the Park Service was running the 30-mi/48-km, 5–6 hour Lands and Legacies tours twice daily, acquainting visitors with the remains of Robert Stafford's plantation and cemetery, Plum Orchard Mansion, Cumberland Wharf, and The Settlement, remnants of a village established in the late 19th century for black employees of the High Point Hotel. This area, rich in African American history, includes the Alberty House and the First African Baptist Church (1893), which also served as a schoolhouse and community center.

The long-term goal of Cumberland Island NS is "to maintain and restore the natural, functioning ecology of the Island" (Brumbelow, 2007). Natural processes such as wild fires will be allowed to resume, and "alien influences" such as feral hogs and exotic plants will be managed until their influence on the ecosystem is insignificant (Brumbelow, 2007). The Park Service hopes to reverse the environmental damage done by farming, logging, and other human endeavors while preserving the flavor of Cumberland's unique cultural and historic heritage.

The seashore plans to make improvements to historic sites as funds become available. One project is to restore and maintain the grounds, gardens, viewpoints, roads, buildings and structures of the Dungeness Historic District (Brumbelow, 2007). Another is to extend the boardwalk over the Dungeness dune crossing to make it wheelchair accessible and build a wheelchair-accessible trail into the wilderness area.

Cumberland Island NS has experienced over 40 years of controversy and compromise, pressures from the public, conflict between interest groups, disagreement over management priorities, and contradictions of policy. There is every hope that in time, those who so passionately love the island will find common ground and work collaboratively toward a more harmonious future in which wild horses have a secure place.

References

Alber, M., Flory, J., & Payne, K. (2005). *Water quality conditions near Cumberland Island National Seashore*. Retrieved from http://www.gcrc.uga.edu/PDFs/2005_gwrc_CUIS.pdf

American Livestock Breeds Conservancy. (2009). *Ossabaw Island hog*. Retrieved from http://albc-usa.org/cpl/Ossabaw.html

Ashley, M.C. (2004). Population genetics of feral horses: Implications of behavioral isolation. *Journal of Mammalogy, 85*(4), 611–617.

Baker, L.A., Warren, R.J., Diefenbach, D.R, James, W.E., & Conroy, M.J. (2001). Prey selection by reintroduced bobcats (*Lynx rufus*) on Cumberland Island, Georgia. *American Midland Naturalist, 145*(1) 80–93. doi: 10.1674/0003-0031(2001)145[0080:PSBRBL]2.0.CO;2

Baron, J. (1982). Effects of feral hogs (*Sus scrofa*) on the vegetation of Horn Island, Mississippi. *American Midland Naturalist, 107*(1), 202–205.

Bellis, V.J. (1995, May). *Ecology of maritime forests of the southern Atlantic coast: A community profile* (Biological Report 30). Washington, DC: U.S. National Biological Service.

Bennett, D., & Hoffman, R.S. (1999, December). *Equus caballus. Mammalian Species, 628*, 1–14.

Berger, J., & Cunningham, C. (1987). Influence of familiarity on frequency of inbreeding in wild horses. *Evolution, 41*(1), 229–231.

Bertness, M., Silliman, B.R., & Jefferies, R. (2004). Salt marshes under siege. *American Scientist, 92*(1), 54–61.

Blum, M. (2013, April 22). Complete forage diet ideal way to feed horses. *AgriNews*. Retrieved from http://agrinews-pubs.com/Content/News/Latest-News/Article/Complete-forage-diet-ideal-way-to-feed-horses/8/6/6832

Boyd, L., & Keiper, R. (2005). Behavioural ecology of feral horses. In D.S. Mills & S.M. McDonnell (Eds.), *The domestic horse: The origins, development and management of its behaviour* (pp. 55–82). Cambridge, United Kingdom: Cambridge University Press.

Brumbelow, J. (2007, August). *The future of America's national parks: First annual centennial strategy for Cumberland Island National Seashore*. Washington, DC: U.S. National Park Service. Retrieved from http://nps01.origin.cdn.level3.net/cuis/parkmgmt/upload/cuis_centennial_strategy.pdf

Bullard, M. (2003). *Cumberland Island: A history*. Athens: University of Georgia Press.

Byrnes, M., & Hiland, M. (1995). Large-scale sediment transport patterns on the continental shelf and influence on shoreline response: St. Andrew Sound, Georgia to Nassau Sound, Florida, USA. *Marine Geology, 126*(1–4), 19–43. doi: 10.1016/0025-3227(95)00064-6

Cameron, E.Z., Linklater, W.L., Stafford, K.J., & Veltman, C.J. (1999). Birth sex ratios relate to mare condition at conception in Kaimanawa horses. *Behavioral Ecology, 10*(5), 472–475.

Carroll, W.D., Kapeluck, P.R., Harper, R.A., & Van Lear, D.H. (2002). Background paper: Historical overview of the Southern forest landscape and associated resources. In D.N. Wear & J.G. Greis (Eds.), *Southern forest resource assessment* (General Technical Report SRS-53) (pp. 583–605). Asheville, NC: U.S. Forest Service, Southern Research Station.

Christmas bird count, Cumberland Island. (2011). Retrieved from http://cbc.audubon.org/cbccurrent/current_table.html

Cofer-Shabica, S. (2010, March). The back dune ridge complex at the Dungeness dune crossing. In O.C. Reedy, F.C. Watts, & F.L. Pirkle, *Quaternary barrier island: Cumberland Island National Seashore*. Guide for the National Society of Consulting Soil Scientists annual meeting scholarship field trip.

Consolidated Appropriations Act, 2005, Pub. L. No. 108-447 (2004).

Crowell-Davis, S.L., & Weeks, J.W. (2005). Maternal behavior and mare-foal interaction. In D.S. Mills & S.M. McDonnell, *The domestic horse: The evolution, development, and management of its behavior* (pp. 126–138). Cambridge, United Kingdom: Cambridge University Press.

Diefenbach, D.R., Hansen, L.A., Warren, R.J., Conroy, M.J., & Nelms, M.G. (2009). Restoration of bobcats to Cumberland Island, Georgia, USA: Lessons learned and evidence for the role of bobcats as keystone predators. In A. Vargas, C. Breitenmoser, & U. Breitenmoser (Eds.), *Iberian lynx ex situ conservation: An interdisciplinary approach* (pp. 423–435). Madrid, Spain: Fundacion Biouniversidad and IUCN Cat Specialist Group.

Dilsaver, L.M. (2004). *Cumberland Island National Seashore: A history of conservation conflict*. Charlottesville: University of Virginia Press.

Downer, C. (2011, March 3). *[Letter to National Wild Horse and Burro Advisory Board][Web log post]*. Retrieved from http://tuesdayshorse.wordpress.com/2011/03/06/craig-downer-appeals-to-national-wild-horse-and-burro-advisory-board/

Equine Act of 1969, Ga. Code Ann. § 4-4-110 (2013).

Feh, C. (2005). Relationships and communication in socially natural horse herds. In D.S. Mills & S.M. McDonnell (Eds.), *The domestic horse: The origins, development and management of its behaviour* (pp. 83–109). Cambridge, United Kingdom: Cambridge University Press.

Feral horses. (1983). *Park Science, 3*(4), 19–20.

Flueck, W. (2000). Population regulation in large northern herbivores: Evolution, thermodynamics, and large predators. *Zeitschrift für Jagdwissenschaft, 46*(3), 139–166. doi: 10.1007/BF02241353

Gardner, P.E. (Photographer). (1958, April). Front elevation from approach road (northeast)—Dungeness, Cumberland Island, Saint Marys, Camden County, GA. Retrieved from http://lcweb2.loc.gov/pnp/habshaer/ga/ga0200/ga0298/photos/055111pu.tif

Georgia Department of Natural Resources, Wildlife Resources Division. (2010, November 1). *Turtle watchers cap 'good year' for loggerheads*. Retrieved from http://www.georgiawildlife.com/node/2391

Georgia toxics release inventory report, 2001. (2003, February). Atlanta: Georgia Dept. of Natural Resources, Environmental Protection Division.

Goodloe, R.B., Warren, R.J., Cothran, E.G., Bratton, S.P., & Trembicki, K.A. (1991). Genetic variation and its management applications in eastern U.S. feral horses. *Journal of Wildlife Management, 55*(3), 412–421.

Goodloe, R.B., Warren, R.J., Osborn, D.A., & Hall, C. (2000). Population characteristics of feral horses on Cumberland Island, Georgia and their management implications. *Journal of Wildlife Management, 64*(1), 114–121. doi: 10.2307/3802980

Goodson, E. (2009, January 14). Is another paper mill auction in the future? Financial defaults lead to higher demands from attorneys. *Tribune & Georgian Online* (St. Marys, GA). Retrieved from http://www.tribune-georgian.com/articles/2009/01/16/news/top_stories/2topstory1.14.txt

Goodwin, D. (2002). Horse behaviour: Evolution, domestication and feralisation. In N. Waran (Ed.), *The Welfare of Horses* (pp. 1–18). Dordrecht, Netherlands: Kluwer Academic Publishers.

Greger, P.D., & Romney, E.M. (1999). High foal mortality limits growth of a desert feral horse population in Nevada. *Western North American Naturalist, 59*(4), 374–379.

Harlan, W. (2007, August). Line in the sand. *Blue Ridge Outdoors, 29–31,* 34.

Harvey, D., Atteberry, K., Hill, G.T., Fry, J., & Kinzer, M. (2009, April). *Final north end access and transportation management plan and environmental assessment, Cumberland Island National Seashore.* Retrieved from http://www.nps.gov/cuis/parkmgmt/loader.cfm?csModule=security/getfile&PageID=214864

Hoffman, D. (2008, Winter). Mid-winter waterbird survey conducted. *Mullet Wrapper,* p. 4.

Hoffman, D. (2011, October–November-a). Current status of the CUIS hog population. *Mullet Wrapper,* p. 4.

Hoffman, D. (2011, October–November-b). Researchers to study CUIS bobcat population. *Mullet Wrapper,* p. 4.

Johannson, K., & Lindgren, M. (2008). *Infant mortality rate (per 1,000 live birth* [sic]*).* Retrieved from https://spreadsheets.google.com/spreadsheet/pub?key=0ArfEDsV3bBwCcGhBd2NOQVZ1eWowTnBGMlBUb3YyQ3c&gid=1

Keiper, R., & Houpt, K. (1984). Reproduction in feral horses: An eight-year study. *American Journal of Veterinary Research, 45*(5), 991–995.

Kirkpatrick, J. (1994). *Into the wind: Wild horses of North America.* Minocqua, WI: Northword Press.

Kirkpatrick, J.F., & Turner, J.W. (1991). Compensatory reproduction in feral horses. *Journal of Wildlife Management, 5*(4), 649–652.

Laliberté, J. (2007). *National Parks Conservation Association state of the parks report: Natural resource assessment, Cumberland Island National Seashore, Camden County, Georgia* (Unpublished master's project). Nicholas School of the Environment and Earth Sciences, Duke University, Durham, NC.

Larson, M., Rosati, J., & Kraus, N. (2002, February). *Overview of regional coastal sediment processes and controls* (Coastal and Hydraulics Engineering Technical Note ERDC/CHL CHETN-XIV-4). Vicksburg, MS: U.S. Army Corps of Engineers Research and Development Center. Retrieved from http://chl.erdc.usace.army.mil/library/publications/chetn/pdf/chetn-xiv-4.pdf

Lewis, J.M. (2009, February 1). Problems mounting for wild-horse management. *DVM Newsmagazine, 40*(2). Retrieved from http://veterinarynews.dvm360.com/ dvm/Veterinary+Equine/Problems-mounting-for-wild-horse-management/ ArticleStandard/Article/detail/581081

Linklater, W.L., Stafford, K.J., Minot, E.O., & Cameron, E.Z. (2002). Researching feral horse ecology and behaviour: Turning political debate into opportunity. *Wildlife Society Bulletin, 30*(2), 644–650.

Mayer, J.J., Brisbin, I.L., & Brisbin, I.L., Jr. (2008). *Wild pigs in the United States: Their history, comparative morphology, and current status.* Athens: University of Georgia Press.

McCue, P.M., & Ferris, R.A. (2012). Parturition, dystocia and foal survival: A retrospective study of 1047 births. *Equine Veterinary Journal, 44,* Issue Supplement s41, 22–25. doi: 10.1111/j.2042-3306.2011.00476.x

McDonnell, S.M. (2005). Sexual behaviour. In D.S. Mills & S.M. McDonnell (Eds.), *The domestic horse: The origins, development and management of its behaviour* (pp. 110–125). Cambridge, United Kingdom: Cambridge University Press.

McGreevy, P. (2004). *Equine behavior: A guide for veterinarians and equine scientists.* London, United Kingdom: W.B. Saunders.

Milanich, J.T. (1996). *The Timucua.* Oxford, United Kingdom, and Cambridge, MA: Blackwell Publishers.

Monard, A.-M., Duncan, P., Fritz, H., & Feh, C. (1997). Variations in the birth sex ratio and neonatal mortality in a natural herd of horses. *Behavioral Ecology and Sociobiology, 41*(4), 243–249. doi: 10.1007/s002650050385

National Historic Preservation Act, 16 U.S.C. § 470 (1966).

Ogburn, M.B., & Alber, M. (2006). An investigation of salt marsh dieback in Georgia using field transplants. *Estuaries and Coasts, 29*(1), 54–62.

Pareja, F. (1627).*Catecismo en lengua Timuquana y Castellana* [Catechism in the Timucuan and Castilian languages]. Mexico, New Spain: Juan Ruiz. Retrieved from http://mem-ory.loc.gov/master/ipo/qcdata/qcdata5/early_americas/tifs/ea0104_02.tif

Saxton, D. (2009). *State of the parks: Cumberland Island National Seashore, a resource assessment.* Washington, DC: National Parks Conservation Association.

Seabrook, C. (2002). *Cumberland Island: Strong women, wild horses.* Winston-Salem, NC: John F. Blair.

Schulte, S., Brown, S., Reynolds, D., & the American Oystercatcher Working Group. (2007, June). *A conservation action plan for the American oystercatcher* (Haematopus palliatus) *for the Atlantic and Gulf coasts of the United States* (Version 2.0). Retrieved from http:// www.fws.gov/migratorybirds/CurrentBirdIssues/Management/FocalSpecies/Plans/ AMOY.pdf

Sharp, R.L., & Miller, C.A. (2008). Wilderness, wildness, and visitor access to Cumberland Island National Seashore, Georgia. In D.B. Klenosky & C.L. Fisher (Eds.), *Proceedings of the 2008 Northeastern Recreation Research Symposium* (Gen. Tech. Rep. NRS-P-42) (pp. 216–222). Newtown Square, PA: U.S. Department of Agriculture, Forest Service, Northern Research Station.

Stockdon, H., Thompson, D., & Fauver, L. (2007). *Vulnerability of National Park Service beaches to inundation during a direct hurricane landfall: Cumberland Island National Seashore* (U.S. Geological Survey, Open-File Report 2007-1387). Retrieved from http://

pubs.usgs.gov/of/2007/1387/ofr2007-1387-CUIS.pdf

Thorn, V. (1977). *Cumberland Island: A place apart.* Washington, DC: U.S. National Park Service, Division of Publications.

Trussell, J. (2011). Contraceptive efficacy. In Hatcher, R.A., Trussell, J., Nelson, A.L., Cates, W., Kowal, D., & Policar, M. (Eds.), *Contraceptive technology* (20th ed.) (pp. 19–47). New York, NY: Ardent Media.

Turner, M. (1988). Simulation and management implications of feral horse grazing on Cumberland Island, Georgia. *Journal of Range Management, 41*(5), 441–447.

U.S. National Park Service. (1955). *A report on our vanishing shoreline: The shoreline, the survey, the areas.* Washington, DC: author.

van Dierendonck, M., & Goodwin, D. (2005). Social contact in horses: Implications for human-horse interactions. In F. de Jonge & R. van den Bos (Eds.), *The human-animal relationship: Forever and a day* (pp. 65–82). Assen, Netherlands: Royal Van Gorcum BV.

Warren, L.K. (2004). *Determining the grazing capacity of your horse pasture.* Retrieved from http://www.coopext.colostate.edu/boulder/sam/pdf/PastureGrazingCapacity.pdf

Wilderness Act, 16 U.S.C. §§ 1131–1136 (1964).

Wilderness Watch v. Mainella, 375 F.3d 1085 (11th Cir., 2004).

Worth, J.E. (1998). *Timucuan chiefdoms of Spanish Florida: Vol. I. Assimilation.* Gainesville: University Press of Florida.

Wright, H., & Lawrence, R. (2012). *Feral animals on Cumberland Island.* Retrieved from http://wildcumberland.org/?page_id=670

Conclusion *and*
Author's Commentary

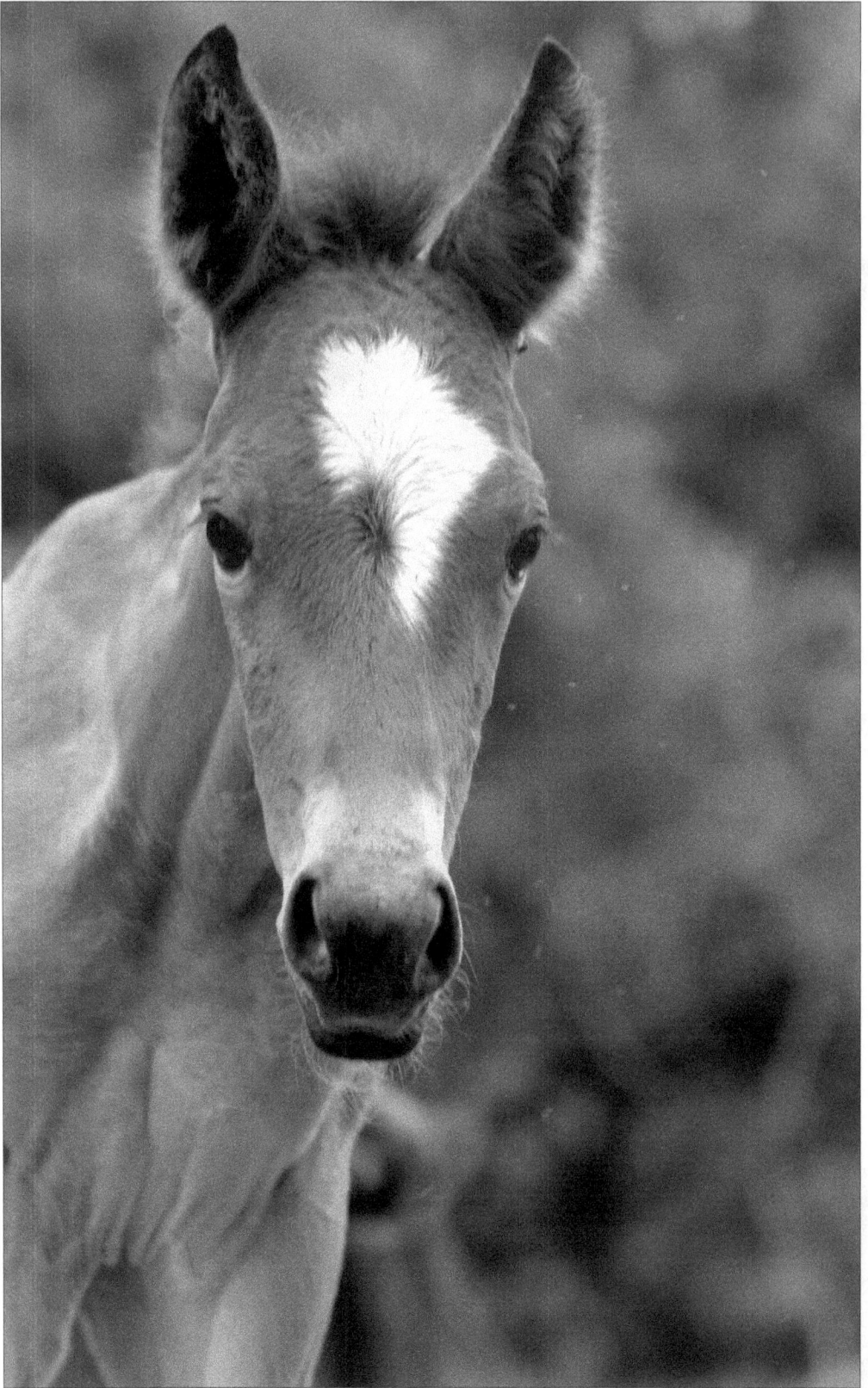

In all likelihood, the origins of the wild island horses of the Atlantic coast will remain a mystery to us. They staked their claim on the shifting sands of their island homelands, and they have remained a constant presence for centuries despite storms, floods, diseases, starvation, and concerted efforts to remove them. Most of them live their lives independent of human contact, icons of freedom and wildness against the backdrop of a roaring sea. We watch them from a distance, yet this unilateral connection kindles an indelible, enigmatic pull on our hearts, souls, and imaginations. Whether or not we want them to remain wild, most of us have strong feelings about what happens to them.

The wild horses themselves, if they had a penchant for the philosophical, probably would not regard themselves as enigmatic at all. They take each day at face value, living by their wits in unforgiving environments where only the strong survive. Their mustang mystique derives from human projections that transform them from living animals to divisive abstractions. We embrace or resist them. We respect or revile them. Whichever direction our passions run, we are incapable of indifference towards wild horses.

With our regard for wild horses so firmly rooted in our emotional matrix, it is unsurprising that the question of how to manage them on public lands has become highly polarized and emotionally charged. "Wild horses generate strong feelings, running on a continuum from strong hate on one end to strong love on the other," writes Dr. Ronald Keiper in the foreword of Jay Kirkpatrick's book *Into the Wind* (1994, p. 8). "Survival of wild horse populations, however, depends less on emotions than on learning as much as possible about how wild horses live."

While some federal officials have supported the horses as free-roaming wildlife, others have tried to discredit them through intense disinformation campaigns. They label them non-native, feral, and exotic to justify removing them as pests and intruders. But Dr. Ross D.E. MacPhee, curator in the Division of Vertebrate Zoology at the American Museum of Natural History, said in a forum at the New York University School of Law, "Scientifically, the idea that horses are an invasive species is utterly wrong. . . . A native species cannot be considered invasive in its native ecosystem" (American Wild Horse Preservation Campaign, 2011). Wildlife biologists Jay Kirkpatrick and Patricia Fazio are also proponents of recognizing the horse as a reintroduced native species, a designation that would place these animals within a new category for management (Kirkpatrick & Fazio, 2010).

"The future of the wild horse is dependent on a number of factors," writes Jay Kirkpatrick (1994, p. 154), "and the largest single factor controlling the fate of wild horses anywhere will be public opinion." The American public cares passionately about the fate of wild horses, and the majority of us want to see them remain wild and independent. "When we capture a free-roaming horse, we destroy an aspect of the very freedom we treasure," says Don Höglund, DVM (2010), author of *Nobody's Horses*. Paradoxically, we must manage wild horses to keep them free. "Human population growth, urban sprawl, and natural resource development have made it necessary to manage everything—even wild things," says Höglund (2010).

We are guardians of the herds. The quality of their continued existence depends on the choices we make on their behalf—and anything we do or do not do will profoundly affect their destiny.

We must apply scientific management principles to ensure that our wild horses do not become genetically compromised, overwhelm their food sources, or do irreparable damage to their environment. To many, Assateague Island National Seashore is a model of successful compromise. Like every virtually other wild horse range, Assateague NS must balance competing objectives and interests in developing a management plan. The seashore must safeguard the health and viability of the herd while granting visitors opportunities to observe free-roaming horses, yet it is required to protect the natural barrier island ecosystem and its sensitive flora and fauna. In 2006, the seashore held a series of Feral Horse Population and Habitat Viability Assessment workshops in which resource experts, stakeholders, and advocacy groups studied data and explored management strategies (Zimmerman, Sturm, Ballou, & Traylor-Holzer, 2006).

After much discussion, data review, and computer modeling, the seashore set the target horse population at 80–100 and authorized an immunocontraceptive program that gives each mare the opportunity to reproduce once. This strategy represented an effective compromise that significantly reduced the negative impacts by horses on the environment and protected the health and viability of the herd going forward (Assateague Island National Seashore, 2008).

"The success at ASIS was as much—or more—the result of non-scientific issues as scientific ones," says Kirkpatrick" (personal communication, May 29, 2014). A sound management plan provided a study backbone, and ongoing research provides periodic reassessments to monitor herd viability and environmental impact. Administrators committed to the plan over the long term and stayed the course, allocating resources to stay on task despite changes in upper-level management. Capable field personnel with no job turnover provide expert assessment and implementation. And an effective education program encourages public involvement. "ASIS is a remarkable place, not so much because we succeeded there, but because of extraordinary administrative forethought and support" (J. Kirkpatrick, personal communication, May 29, 2014).

The horses at Assateague NS form a unique herd. They have seen no introductions and few removals since 1968, and every horse is closely monitored from birth to death by Park Service personnel and intermittently by independent researchers. The horses of Assateague have no fear of people—in fact, many are overly fond of attention—making vaccination and census-taking easier than among the skittish horses on densely wooded Cumberland Island or an expanse of Western desert.

After a very rough start, Cape Lookout NS also eventually developed a management plan that embraces a balance between a healthy herd and a healthy barrier island environment. Managers of other national parks and seashores with wild horses have been unable or unwilling to compromise, to say nothing of local managers in the Fish and Wildlife Service or the Bureau of Land Management, other agencies of the U.S. Department of the Interior. "The overwhelming question is, why does one NS solve its problem in such an effective and publicly acceptable manner and others do not," says Kirkpatrick (personal communication, May

29, 2014). "The tools are at hand, but the political will to develop a unified approach seems to be missing." He went on to recount a meeting with a director of the National Park Service: "He told me that each park was like a fiefdom and that it was a pure reflection of the superintendent. ... unless it is illegal, a superintendent can do pretty much what he/she wants. That's not all bad, but it helps explain why one NS will solve its problem and another wrings its hands." How can a diverse country of 316 million people work out a program for dealing with wild horses if the field units of one federal agency do not agree and cannot be budged?

On the other hand, one-size-fits-all plans are seldom appropriate to every situation, and some could turn out disastrous. To balance environmental concerns, politics, economics, public sentiment, and the needs of the horses themselves, wild-horse management policy should be determined by multidisciplinary panels comprising land managers, scientists, horse advocates, ranchers, politicians, and other stakeholders working in the open, where biases and agendas may be identified.

To balance environmental concerns, politics, public sentiment, and the needs of the horses themselves, wild-horse management policy should be determined by multidisciplinary panels comprised of land managers, scientists, horse advocates, ranchers, politicians, and other stakeholders. "Most public servants and land managers tell me that they want to manage the free-roaming horse in the wild," says Höglund (2010). "They need the proper laws, proper policy, dedicated staff, time, and mandated funding to do it right and to study the effects of the management tools. ... After all, the horses belong to all of us, the land belongs to all of us, and the elected representatives who control our money work for all of us."

Such sentiments are not new. Indeed, they were very old in 1913, when William Hornaday wrote, "The wild things of this earth are *not* ours, to do with as we please. They have been given to us *in trust,* and we must account for them to the generations which will come after us and audit our accounts" (1913, p. 7).

Wild horses can continue their wild existence within boundaries defined by unbiased research towards a goal of finding harmony between equid and environment. We can largely let nature take its course unless we need to intervene to solve a problem or avert a crisis. Small adjustments in the early stages of trouble that reestablish a natural balance are preferable to heroic measures at the brink of disaster. Options include curbing their fertility to restrict herd growth or restoring balance with the natural death rate with adoption or removal, a balance once kept by natural predators.

When considering the role of wild horses, we cannot overlook the fact that these animals touch our souls in ways that few other species can. Even people who have never been close to a horse marvel at its grace and beauty. Horses are symbols of power, elegance, and freedom. Books like *The Horse Whisperer* resonate in a broad array of human hearts. Classics like *Black Beauty, National Velvet, Misty of Chincoteague,* and *The Black Stallion* remain popular with children generation after generation, and series such as the *Saddle Club, Horse Diaries,* and *Winnie the Horse Gentler* speak to the most recent crop of young hippophiles who regularly weave horses into their fantasies. Wild horses mean something to us, and to remove them entirely would not only rob us of a valuable cultural resource, but also do a shameful disservice to the species.

Even researchers who study wild horses with scientific objectivity find themselves fascinated and profoundly moved by these inspiring animals. Kirkpatrick (1994, p. 160) wrote:

If we are to appreciate these animals at all, we must appreciate them simply because of their aesthetic value as wild creatures. A glimpse of a wild horse has no promise of meat, of the thrill of the hunt, of license sales, or of market values. This glimpse offers only whatever value can be obtained from the view itself.

In writing this book, I have attempted to approach my subject with objectivity—the horses, their history, the management conflicts, the environmental concerns, and the sentiments flowing for and against them. As I worked, however, I came to realize that objectivity is an unattainable ideal. To be entirely objective, one must be detached, and I care too much to remain indifferent. My passion for wild horses has been a part of my emotional makeup since I was a small child. Perhaps I was born with it. By the time I completed the second draft of this book, I became aware of how my own sentiments were thwarting my efforts at neutrality.

I care about wild horses. They pull me out of a mind that thinks too much and allow me to experience the world as awe-inspiring. There are no words to describe the tranquility of soul I experienced standing knee-deep in the bay while the horses of Cedar Island foraged nearby, ominous storm clouds reflecting in the glassy water. I have camped alone in the emptiness of Shackleford Banks, where horses grazed the nearby dunes in the vermilion glow of the sinking sun. I have beheld playful bachelor stallions among the panoramic vistas of McCullough Peaks in Wyoming, watched foals cavort in the wildflowers at Stewart Creek, also in Wyoming, and seen patriarchs clash on the beach at Corolla. There are no words for the lightness of spirit I found on Montana's Pryor Mountain Wild Horse Range overlooking Bighorn Canyon, where a battle-scarred apricot dun stallion tenderly nuzzled his grulla colt on a rocky escarpment. These encounters have been some of the best moments of my life. The memories they have forged are indelible.

Wild horses should be honored and cherished because of what they are—tenacious, beautiful survivors who excite and inspire our domesticated souls. When we watch them grazing in the marshes or galloping like free spirits from the beach to the dunes, those of us with intact imaginations can feel the spirit of wildness well up in our own souls. For as long as they remain free, so do we.

References

American Wild Horse Preservation Campaign. (2011, November 23). *NYU legal forum on Wild Horse and Burro Act a success.* Retrieved from http://wildhorsepreservation.org/media/nyu-legal-forum-wild-horse-and-burro-act-success

Assateague Island National Seashore. [2008]. Feral horse management at Assateague Island National Seashore. Retrieved from http://parkplanning.nps.gov/showFile.cfm?projectID=17228&MIMEType=application/pdf&filename=Horse EA Open House Info Panels1.pdf&sfid=0&ei=JzvUU87aBNGzyAT6g4DoAg&usg=AFQjCNF47mw_b5h_h7Cas0woBqwoJJOcXQ

Höglund, D. (2010, March 23). *Management of the free-roaming horse* [Online forum comment]. Retrieved from http://pryorwild.wordpress.com/2010/02/14/february-15-2010-pzps-reversibility/

Hornaday, W.T. (1913). *Our vanishing wild life: Its extermination and preservation.* New York, NY: Charles Scribner's Sons.

Kirkpatrick, J. (1994). *Into the wind: Wild horses of North America*. Minocqua, WI: North-word Press.

Kirkpatrick, J., & Fazio, P. (2010, January). *Wild horses as native North American wildlife*. Retrieved from http://awionline.org/content/wild-horses-native-north-american-wildlife

Zimmerman, C., Sturm, M., Ballou, J., & Traylor-Holzer, K. (Eds.). (2006). *Horses of Assateague Island population and habitat viability assessment workshop: Final report*. Apple Valley, MN: IUCN/SSC Conservation Breeding Specialist Group.

Appendix

Equine Evolution

Fossil horses laid out in chronological order in obsolete museum exhibits create the impression that earlier species gave rise to later ones; that *Equus caballus*, the modern horse, was the "goal" of this evolutionary line; and that success meant gaining the features found in *Equus* and leaving others behind. But evolution has no goal, plan, or schedule. In truth, the equine family tree has many branches of various sizes, most of which died long ago. Many now-extinct branches were just as successful in their day as the modern horse is in ours, or more successful if time on earth is the benchmark.

Evolutionary trends are simply adaptations to environmental pressures. For a species to survive, individuals must have food, water, sound health, protection from predators and environmental extremes, and a chance to reproduce. Animals that control prime resources are more likely to survive to produce offspring, which may also survive and produce more offspring. Each species competes on a mutable game board. A novel disease, prolonged drought, or new predator can doom a flourishing species to seemingly unlikely extinction. Climates change, and in response habitats may shift from rainforest to grassland or deciduous woodland, dealing their denizens a new hand of cards. Rival species grab territory and vie for resources. Populations ebb and flow, shaped by abundance and shortage, boom and bust. Some lines thrive, and others decline or die off.

Although paleontology is a complicated science, writers and exhibit designers often depict orthogenesis (straight-line evolution) to create a simplified model more easily understood by nonscientists (MacFadden, 1992). One of the most popular models is the horse. As Switek writes (2010, October 10), linear evolution is "more than scientific shorthand for museum exhibits and books." Given the fragmentary nature of many dawn horse skeletons, the anatomy of modern horses was often used to fill in the missing bits and pieces. As an illustration, he compares a recently discovered skeleton of *Hyracotherium*, a very early ancestor of modern horses, nearly complete, with a 50% reconstruction from 1884. The reconstruction gave the animal the defined withers and rigid back of *Equus*; whereas the skeleton demonstrated that *Hyracotherium*'s spine was more flexible, like that of a dog (Wood, Bebej, Manz, Begun, & Gingerich, 2011).

Scientific inferences are not always correct. The best minds in paleontology once saw dinosaurs as universally cold-blooded and leathery and they greatly overestimated the body weight of many species. Now it appears that some dinosaurs were warm-blooded and active, and many had feathers (*Newly Discovered Dinosaur*, 2012).

Equus, which includes modern horses, zebras, and asses, is the only surviving genus of a once diverse family that included 27 genera (Kirkpatrick & Fazio, 2010). Although fossils of early equids are plentiful, and their chronology is well established, it is no easy task for paleontologists to unravel the enigmas presented by these extinct creatures. Scientists must often reconstruct an entire dynamic animal from a few fragmentary remains and determine relationships among ancient species without the benefit of genetic evidence (Parker et al., 2004).

Our understanding of equine evolution is progressing. In 1832, British naturalist Richard Owen (who coined the term *dinosaur*) examined pieces of fossilized mammal skull and jawbone found along the Thames. He initially thought the fragments had belonged to a monkey, but then concluded that they more closely resembled the bones of the modern hyrax, a small Afro-Asian herbivore related to elephants and manatees. He named the species *Hyracotherium leporinum*—"harelike hyrax-beast" (Owen, 1841; Switek, 2010). In 1876 the American O.C. Marsh uncovered the complete skeleton of a similar animal in the western United States and assigned it to another new genus, *Eohippus* ("dawn horse"). Over the following decades, paleontologists merged these discoveries with other proposed species (eventually more than 50) under the generic name *Hyracotherium*, and *Eohippus* fell into disfavor among academics. For a time *Hyracotherium* was widely considered the earliest horse.

Recent scholarship has partially reversed these developments. Following the lead of David Froehlich (2002), many paleontologists now classify two former members of *Hyracotherium* with tapirs and rhinoceroses and group Owen's *H. leporinum* with paleotheres ("ancient beasts"). Paleotheres, a now-extinct branch of mostly Old World perissodactyls (odd-toed hoofed mammals), were closely related to equids, but not horses exactly.

Emerging opinion also divides the remaining equids from *Hyracotherium* among seven new genera, six of them associated with North America (MacFadden, 2004). One new genus bears the revived name *Eohippus*. But as Rose (2006, p. 247) says, "Regardless of this proliferation of names, the anatomical differences among these early Eocene equids, and between them and some other basal perissodactyls, are so minor that even experts have difficulty distinguishing them." A number of researchers (e.g., Wood et al., 2011) have decided to stay with the old terminology for now.

Despite many advances in knowledge, the prehistory of the horse remains unsettled. Still, parts of the picture are coming into clearer focus. By current reckoning, the oldest equid is *Sifrhippus* \sif-′rip-əs\—"zero horse," from the Arabic *sifr*, "nothing," the root of the English words *zero* and *cipher*. It roamed the primeval forests of North America subsisting on leaves, buds, and low-hanging fruit and seeds about 55–56 million years ago, only 9–10 million years after dinosaurs had died out. This adaptable creature evidently shrank from the size of a small dog (12.3 lb/ 5.6 kg) to that of a house cat (8.6 lb/3.9 kg) in response to rising temperatures during the Paleocene-Eocene Thermal Maximum, then rebounded to about 15.4 lb/7 kg as the climate cooled, all over a span of less than 200,000 years (Secord et al., 2012).

Sifrhippus had a younger contemporary and neighbor. *Arenahippus* ("sand horse") was a bit larger, 17.6–22 lb/ 8–10 kg, but resembled *Sifrhippus* in that it had relatively long, slender legs, front feet that ended in four hoofed toes, and hind feet that had three (Franzen, 2010; Vaughan, Ryan, & Czaplewski, 2011)—14 hooves total! Its legs evidently had a greater range of rotational and side-to-side motion than those of modern horses. This would have been advantageous on uneven ground or in dense undergrowth. *Arenahippus* was probably about as fast as a modern horse over the limited open spaces that its habitat afforded (Franzen, 2010). Though stouter musculature helped stabilize its head as it ran (Wood et al., 2011), it retained sufficient mobility for browsing. Like *Sifrhippus*, it had 44

Sifrhippus, the earliest horse, was about the size of a cat. It had hooves on each of its 14 toes and a flexible backbone like a dog's.

teeth resembling those of primates (hence Owen's initial confusion), but its molars had changed slightly. The seemingly quick appearance of *Arenahippus* may suggest that *Sifrhippus* is older than its known fossils.

The preferred food of *Sifrhippus* and *Arenahippus* seems to have become scarcer by the mid–late Eocene as North America became cooler and drier, but there are earlier signs of climate change. *Orohippus* came along only 2 million years after *Sifrhippus* and several subsequent species. It was about the size of *Arenahippus*, but it had larger middle toes on all four feet, and it had lost the vestigial first and fifth toes on its hind feet. Its hind legs were proportionally longer. These changes suggest adaptation for running and jumping in more open country. Its head was longer, and changes in its teeth may indicate a diet higher in grass. Early equids and their kin migrated between Europe and North America and back again, probably over the land bridges that connected Great Britain, Iceland, and Greenland (Rose, 2006).

When more than 60 virtually intact skeletons of primitive equids were excavated from the Messel Pit in Germany, paleontologists gained new understanding of their behavior and physiology (Franzen, 2010). Surprisingly, Eocene equids, so different from what we call horses, had many of the physical features that characterize their descendants. Paleontologists unearthed a number of pregnant *Eurohippus messelensis* mares with well-preserved fetuses (and one with an identifiable placenta), contours of bodily soft tissue, and digestive organs containing chewed laurel leaves and grape seeds. These ancient animals had the same gender-specific pelvic architecture as modern horses, gestated single foals that were positioned in the uterus like modern horse fetuses, and had perky little horselike ears. Adults had short, fluffy tails similar to those of modern foals (Franzen, 2010). Even the skulls showed the elongated triangular shape typical of modern horses. Although Eurohippus was still obviously a browser, it had a cecum for digesting cellulose like modern horses.

Over millions of years, expanding prairies supplanted woodlands. A promising new ecological niche opened for species that could use it. Without sophisticated adaptations, a forest animal would starve on a diet of grass. Grass contains silica, the main component of sand, and is very abrasive to teeth. Many species grow close to the ground and tend to be gritty and dirty. Most species ancestral to modern horses developed characteristics that increased their success in the grasslands, such as molars durable enough for a diet that would have worn *Sifrhippus* teeth down to the gums. Faces grew longer and jaws gradually deepened to accommodate these teeth.

Grass is mostly cellulose, which vertebrates and even termites cannot use directly. Most grazing mammals have evolved fermentation chambers in their digestive tracts where symbiotic bacteria break cellulose down into usable molecules. Sheep, cattle, deer, hippopotami, and other ruminants evolved four stomachs; the first two are fermentation vats, the last two are chemical stomachs. Ruminant digestion is very efficient: the animal moves food through his system slowly and extracts most of the usable nutrients from every mouthful.

In contrast, horses ferment grass in a cecum, a large pouch between the small and large intestines. (Humans also have a small, vestigial cecum in roughly the same spot, terminating in the appendix.) Whereas a cow or a deer consumes a large amount of food and then lies down to ruminate and extract additional nutrition from it, a horse eats on the move, putting large quantities of plant matter through its digestive tract and taking whatever nutrients are easily available. This is not as efficient as a ruminant's strategy, but it does not have to be. A horse's digestive system processes poor-quality grasses quickly, allowing him to remain well nourished, as long as he spends most of his waking hours grazing on a low-protein, high-fiber diet that would not support a ruminant. The difference in digestion explains why horse manure is fibrous, stemmy briquettes, and cows produce pudding-like patties.

The cecum strategy for cellulose digestion increases fleetness of foot by allowing the animal to carry a greater proportion of body weight as muscle. The gut comprises 40% of a ruminant's body weight, but only 15% in cecum-fermenting species (Franzen, 2010).

Increased body bulk is crucial to the cecum strategy for grass eating. Small animals have greater ratio of surface area to body mass and lose heat proportionally faster. Their metabolic furnaces work furiously to achieve homeostasis. A large, bulky animal loses heat at a slower rate and requires fewer calories per unit of mass for homeostasis. Horses became large, well-insulated animals with slow metabolisms, an adaptation that allowed them to subsist on poor-quality forage and, later, to survive Ice Age winters. Horses maintained the mass needed for heat retention, but gained fleetness by trading a large, efficient digestive system for additional muscle needed for flight.

Because animals of the grasslands must feed in the open, survival involves keeping distance between themselves and predators. The horse evolved the ability to see distant objects in reasonably good detail within a nearly 360° visual field to spot approaching predators. To allow rapid escape, some of the leg bones of early grazing equids fused, eliminating most of the lateral movement of the legs, but granting greater straight-line speed. A "springing foot" arrangement of tendons and ligaments functioned as ropes and pulleys, propelling the horse with greater speed and efficiency.

Three-toed horses carried most of their weight on the larger middle toe, but side toes added stability in slippery footing or when rapidly changing direction. Tridactyly conferred advantages for tens of millions of years, but in some groups ligaments developed around the fetlock to give greater stability to a single hoof, and at least two lines of equids independently lost their side toes. One of them was ancestral to modern horses; only single-hoofed equids survive today.

Horses possess a unique system for circulating blood through their hooves. There are no large muscle masses in the lower legs of the horse to aid in the return of venous blood to the heart; the distal limbs are moved by tendons, as are human fingers and toes. Within each hoof is a digital cushion, a tough, elastic, fatty, spongy structure extending into the bulbs of the heels. As the horse takes a step, he applies pressure to the frog, a rubbery shock absorber at the bottom of the hoof, and to the phalanges, hoof wall, and other structures. The frog compresses the digital cushion, which flattens and presses outward against the lateral cartilages, forcing blood out of a large plexus of veins within the foot. When the horse raises his foot, he relieves the pressure on the hoof and the arterial pulse refills the hoof. A system of valves holds blood in the vessels of the hoof below the plexuses, producing a hydraulic cushion that reduces concussion and protects the delicate structures within the hoof. Each time the foot bears weight, the veins are compressed. This system also helps protect the feet from freezing in the winter.

By the Miocene, horses were plentiful and had split into many diverse species. The main evolutionary line split into two major branches, plus a less successful offshoot of "pygmy horses." One of these prominent branches was that of the Anchitheres, "near beasts," a lineage of three-toed forest horses that continued to browse like deer. They thrived for tens of millions of years, then died out, leaving no modern survivors.

Equus arose from the other branch, the line that took to the grasslands and evolved as grazers. The erroneously named *Merychippus* ("ruminant horse"), a three-toed grazer, appeared about 15 million years ago and became abundant. *Merychippus* was very horsy in appearance and stood 40 in. (1 m) at the shoulder, or 10 hands—about the height of a Falabella or a small Shetland Pony. As in today's equids, proportions varied from stocky to svelte. *Merychippus* had an unmistakably equine brain and, judging from its modern-looking fissuration pattern, greater intelligence than earlier species. The fossil record suggests that his behavior was similar to that of modern equids. *Merychippus* traveled primarily on his middle toe; the two toes on either side were still fully formed, but becoming increasingly unimportant. Springy ligaments in the lower legs served to keep his weight on the center toe, giving *Merychippus* speed and agility as it grazed on the open savanna.

During the reign of *Merychippus*, the horse family produced its greatest variety. There were browsers and grazers of every description, and most were highly successful, more so than contemporary artiodactyls, the even-toed hoofed mammals that dominate today. Scientists have long puzzled over the presence of fossae, or hollows, in the facial bones of horses of this period. Because soft tissue does not usually fossilize, and modern horses do not have fossae, it is a matter of speculation just what purpose they served. They must have been important, as nearly every species had some form of them. But if they were necessary for survival, modern horses would have them.

The mysterious fossae varied between species and between individuals of a species. Did the hollows house glands? Sensory organs? Attachments for structures that made an individual appealing or intimidating to others of the species? Muscles to move a flexible trunk? We can only guess why they evolved and what purpose they served.

It is unclear why three-toed horses died out. During their ascendancy, they were abundant and prolific. Perhaps someday the fossil record will reveal why tridactyl equids became extinct while their one-toed descendants lived on. Evidence of prior evolutionary stages can be observed not only in fossil collections, but also in modern horses. Every horse has splint bones in his legs—vestiges of his ancestors' three toes. Behind every fetlock is a small, horny growth, the remainder of what used to function as doglike pads on the feet of ancient relatives.

Pliohippus (Pliocene—"more recent"—horse) was one of the descendants of *Merychippus* that closely resembled *Equus* in size and conformation, traveling across the grasslands on four well-formed hooves. *Pliohippus*, though thoroughly horselike, was not ancestral to *Equus*, and it retained several characteristics of earlier species, such as pronounced facial fossae and curved cheek teeth.

Most scientists now believe that it was *Dinohippus*, "powerful horse," that gave rise to *Equus*. *Dinohippus* survived as a genus for nearly 7 million years and was abundant on the grasslands. Its fossils are found in the Upper Miocene and Pliocene of North America and date from 10.3–3.6 million years ago. *Dinohippus* had straight, *Equus*-like teeth, and its facial fossae were less pronounced than those of its ancestors. This genus also possessed a rudimentary version of the equine passive "stay apparatus," an anatomical adaptation that allows a horse to stand for long periods without muscular fatigue. *Dinohippus* seems to be the intermediate form between three-toed and single-toed horses—most of the *Dinohippus* skeletons excavated are monodactyl, but some possess three toes.

Modern horses, zebras, asses, and onagers belong to the genus *Equus*, the only surviving genus of the ancient family of horses. Although zebras and asses appear quite different from the horse, their anatomy is very similar, and the bones of any equid species can be difficult to distinguish from those of any other. All members of the horse family can interbreed, and when crossed they usually produce sterile offspring that are larger and hardier than either parent (Kelekna, 2009).

The first true horses were the size of a large pony, about 13.3–14 hands (55–56 in./1.40–1.42 m) in height, roughly 800 lb/363 kg, heavy-boned, and built similarly to the Przewalski Horse. *Equus* appeared in North America nearly 4 million years ago, the product of more than 50 million years of evolution, most of which took place on this continent. *E. simplicidens* (also known as the Hagerman horse or Hagerman zebra) appears to be the oldest member of the genus (McDonald, 1996). The fossil record indicates that the ancestors of all extant horses, asses, and zebras once lived in North America (U.S. Bureau of Land Management, 2005).

As the Pliocene drew to a close, the horse family took advantage of a land bridge that linked North America to Asia, across what is now the Bering Strait. The late Pliocene *E. idahoensis*—the progenitor of the first modern horses in North America—migrated to Asia and split into several subspecies (Kirkpatrick & Fazio, 2010).

Horse dispersal was bidirectional; *Equus* emigrated to Eurasia 2–3 million years ago, died out in North America, returned via the Bering Strait land bridge, continued to ebb and flow through multiple waves of migration and extinction, and eventually spread to all other continents except Australia and Antarctica (Kirkpatrick & Fazio, 2010). Asses and onagers settled in Asian desert regions while zebras and quaggas found a home in Africa. During ice ages, when glaciers seized the northern latitudes, caballoid horses thrived in the milder climate of the southern Ukraine and Turkestan. Similarly, the Pyrenees Mountains divided the Iberian Peninsula from the rest of Eurasia, creating a "glacial refuge for many species, including humans" (Achilli et al., 2012, p. 2452).

Some species of *Equus* migrated to South America and became common there. South America was subsequently detached from North America for tens of millions of years, during which equine species on the two continents diverged. When the volcanic Isthmus of Panama emerged from the sea floor, animals migrated freely between the continents in an event termed the Great American Interchange, which peaked about 3 million years ago.

At one point during the Pleistocene, the path to modern *Equus* narrowed. A team led by Italian geneticist Alessandro Achilli determined that all modern horses descend from a common ancestor, a mare that lived about 130,000–160,000 years ago (Achilli et al., 2012). There is no physical evidence of this "Equine Eve," but the team found genetic evidence of her presence by following the mitochondrial genomes of 83 dissimilar horses to their root at the Ancestral Mare Mitogenome (Achilli, et al., 2012). There were undoubtedly many other bloodlines present at the time, but if Achilli is correct, only one lineage ultimately survived.

The horse family had experienced astonishing success in the past, but these early days of *Equus* were perhaps the most successful. After bison and mammoths, horses were the most abundant large animals in North America. Some scientists think that they outnumbered bison. Around the globe, multitudes of horses grazed the steppes that fringed the great northern glaciers and straddled the Arctic Circle. Although they preferred cool, damp grassland and open forest, they adapted to oppressive heat, arid deserts, frigid periglacial tundra, insect-ridden tropical swamps, and exposed flanks of steep mountains.

Then around 13,000 years ago, North and South American horses began to die off, and then vanished within the span of a few thousand years. After numbering in the tens of millions for much of 55 million years, North American equids became extinct about 8,000 years ago (Franzen, 2010).

References

Achilli, A., Olivieri, A., Soares, P., Lancioni, H., Kashani, B.H., Perego, U.A., Torroni, A. (2012). Mitochondrial genomes from modern horses reveal the major haplogroups that underwent domestication. *Proceedings of the National Academy of Sciences of the United States of America, 109*(7), 2449–2454. doi: 10.1073/pnas.1111637109

Franzen, J.L. (2010). *The rise of horses: 55 million years of evolution.* Baltimore, MD: Johns Hopkins University Press.

Froehlich, D.J. (2002, February). Quo vadis eohippus? The systematics and taxonomy of the early Eocene equids (Perissodactyla). *Zoological Journal of the Linnean Society, 134*(2), 141–256. doi: 10.1046/j.1096-3642.2002.00005.x

Kelekna, P. (2009). *The horse in human history.* Cambridge, United Kingdom: Cambridge University Press.

Kirkpatrick, J., & Fazio, P. (2010, January). *Wild horses as native North American wildlife.* Retrieved from http://awionline.org/content/wild-horses-native-north-american-wildlife

MacFadden, B.J. (1992). *Fossil horses: Systematics, paleobiology, and evolution of the family Equidae*. Cambridge, United Kingdom: Cambridge University Press.

MacFadden, B.J. (2004). The last decade (more or less) of equid paleobiology. *Pony Express, 13*(2). Retrieved from http://www.flmnh.ufl.edu/ponyexpress/pony13_2/Pe132.html#paleobiology

McDonald, H.G. (1996). Population structure of the late Pliocene (Blancan) zebra *Equus simplicidens* (Perissodactyla: Equidae) from the Hagerman Horse Quarry, Idaho. In K.M. Stewart and K.L. Seymour (Eds.), *Palaeoecology and palaeoenvironments of late cenozoic mammals* (pp. 134–155). Toronto, Ontario, Canada: University of Toronto Press.

Newly discovered dinosaur implies greater prevalence of feathers: Megalosaur fossil represents 1st feathered dinosaur not closely related to birds. (2012, July 2). Retrieved from http://www.amnh.org/science/papers/feathers.php

Owen, R. (1841). Description of the fossil remains of a mammal (*Hyracotherium leporinum*) and of a bird (*Lithornis vulturinus*) from the London Clay. *Transactions of the Geological Society of London, Series 2*(6), 203–208.

Parker, H.G., Kim, L.V., Sutter, N.B., Carlson, S., Lorentzen, T.D., Malek, T.B., . . . Kruglyak, L. (2004). Genetic structure of the purebred domestic dog. *Science, 304*(5674), 1160–1164. doi: 10.1126/science.1097406

Rose, K.D. (2006). *The beginning of the age of mammals*. Baltimore, MD: Johns Hopkins University Press.

Secord, R., Bloch, J.I., Chester, S.G.B., Boyer, D.M., Wood, A.R., Wing, S.L., . . . Krigbaum, J. (2012, February 24). Evolution of the earliest horses driven by climate change in the Paleocene-Eocene Thermal Maximum. *Science, 335*(6071), 959–962. doi: 10.1126/science.1213859

Switek, B. (2010). *Written in stone: The hidden secrets of fossils and the story of life on earth*. New York, NY: Bellevue Literary Press.

Switek, B. (2010, October 10). *Running with* Arenahippus [Web log post]. Retrieved from http://www.wired.com/wiredscience/2010/10/running-with-arenahippus/

U.S. Bureau of Land Management, Wild Horse and Burro Program. (2005, March). *Strategic research plan: Wild horse and burro management*. Retrieved from http://www.blm.gov/pgdata/etc/medialib/blm/wo/Planning_and_Renewable_Resources/wild_horses_and_burros.Par.91906.File.dat/Strategic Research Plan.pdf

Vaughan, T.A., Ryan, J.M., & Czaplewski, N.J. (2011). *Mammalogy* (5th ed.). Sudbury, MA: Jones & Bartlett.

Wood, A.R., Bebej, R.M., Manz, C.L., Begun, D.L., & Gingerich, P.D. (2011). Postcranial functional morphology of *Hyracotherium* (Equidae, Perissodactyla) and locomotion in the earliest horses. *Journal of Mammalian Evolution, 18*(1), 1–32. doi: 10.1007/s10914-010-9145-7

Acknowledgments

About 20 years ago, I began to research my first book, *Hoofprints in the Sand: Wild Horses of the Atlantic Coast*. The experience enriched my life in many ways. I learned a tremendous amount about how the natural behavior of these horses differs from that of their domestic counterparts. Following herds through marsh and dune, I gained a greater appreciation of the complexity of their apparently simple lives in the wild. Similarly, with the writing of *Wild Horse Dilemma*, many helpful and extraordinary people have earned my gratitude, and many have become friends.

Dr. Jay F. Kirkpatrick, senior scientist at the Science and Conservation Center at ZooMontana in Billings and author of *Into the Wind: Wild Horses of North America*, has devoted much of his career to preserving the health, rights, and dignity of wild horses. He developed and implemented the immunocontraceptive program in use with many species of wildlife, including horses. He answered my questions, forwarded useful documents, crafted an insightful preface, and generously reviewed the manuscript before publication.

Don Höglund, DVM, is an internationally esteemed leader in horse training and management and the author of *Nobody's Horses*, a riveting book about the rescue of wild horses from the White Sands Missile Range. He has implemented numerous large-scale equine programs, including the Department of the Interior's Wild Horse Prison Inmate Training Program, which teaches prisoners to gentle horses while providing training for adoptable mustangs. His love and admiration for horses is evident in all that he does. When I approached him with a few questions, he responded enthusiastically and sent me a number of articles that shaped the backbone of my manuscript. I am grateful for his support and encouragement along the way and fortunate that he agreed to review the manuscript.

Dr. E. Gus Cothran, Texas A&M University's renowned expert on the genetics of wild and domestic horses, helped me to understand the significance of the Q-ac variant in certain wild horse herds and the concept of minimum viable population. He also found time in his busy schedule to review the manuscript. His research is the cornerstone of wild horse management, and I have cited it extensively.

D. Phillip Sponenberg, DVM, PhD, helped me to understand the genetic underpinnings of coat color and its implications for the free-roaming Banker horses. In his review of the manuscript, he offered great insights on Spanish horse origins and genetics, and his comprehensive articles on that topic were a valuable resource.

Dr. Sue Stuska, the wildlife biologist at Cape Lookout National Seashore who oversees the Shackleford Banks herd, has corresponded regularly about the status of the horses. When I visited, she taught me how to identify individuals, and showed me her dynamic census chart that tracks the members of each band and where they were last sighted. She explained how the current management plan makes optimal use of the existing gene pool by monitoring family lineage and contracepting certain mares. Sue also gave of her valuable time to review my manuscript for accuracy.

Karen McCalpin, director of the nonprofit Corolla Wild Horse Fund, Inc., found time in her impossibly busy schedule to meet me and discuss the genetic crisis facing the herd. She also reviewed the manuscript. Karen and the other members of the organization—mostly volunteers—have upended their lives to secure protection for these horses. Karen produced a beautifully written blog highlighting the triumphs and tragedies of the herd. It can be accessed at www.corollawildhorses.com

Dr. Ronald Keiper, Distinguished Professor of Biology (emeritus) at Penn State University, was one of the first scientists to study the behavior of horses in the wild, answered my questions about the foaling rate of lactating mares and shared his groundbreaking research detailing the behavior of the Assateague horses.

Wesley Stallings, former manager of the Corolla herd, took me in his truck several times as he patrolled the Outer Banks north of Corolla, following the movements of wild bands and logging herd data in his notebook. Sometimes we climbed on the roof of the truck to scout for horses. Sometimes Wes climbed a tree for a better view. At one point we encountered a flooded hollow and were forced to don hip boots and slog through surging currents occupied by cottonmouth snakes to evaluate the health of a newborn colt. I am grateful he allowed me to participate in his daily adventures.

Steve Edwards, by day an attorney for Isle of Wight County, Virginia, works magic in rehabilitating injured Corolla and Shackleford horses. At his farm, Mill Swamp Indian Horses in Smithfield, Va., he teaches children how to train wild horses with natural horsemanship techniques. He also established an off-site breeding program to preserve the herd's rare genes in case of disaster in the wild. Steve has been extremely supportive and helpful throughout the writing of this book, and has brought his expertise to the task of reviewing this book for accuracy before publication.

In 2012, I visited Mill Swamp and was captivated by the sight of children working in the round pen with young horses, many of them recently brought in from the wild. To this point, I had great esteem for the wild horses living on North Carolina's Outer Banks, but had never ridden one. I found these Colonial Spanish Horses astonishingly surefooted, brave, rugged, and smooth-gaited. The climax of my visit was a ride through the inky forest astride Manteo, a wild-born black stallion. He never missed a step despite exposed roots, steep embankments, deep pools, and deer crashing gracelessly through the underbrush. For the better part of an hour, we trotted and cantered through darkness so complete, I could not see my hand in front of my face. I had recently been injured in a riding accident, and I was working through many horse-related fears. It was a profound and humbling experience to trust a once-wild stallion to find his way through darkness that left me blind.

Carolyn Mason, president of the Foundation for Shackleford Horses, Inc., accompanied me to Shackleford Banks and generously shared her wealth of knowledge about the horses. She introduced me to the Banker Horses grazing in her yard, gentled animals awaiting adoption. My heart melted when a young gelding named Adagio followed me like a puppy and courted hugs and scratching.

Woody and Nena Hancock loaded me and my cameras into their boat and searched island and marsh for members of the Cedar Island herd. They introduced me to Bucky, the most genetically valuable horse on Cedar Island, and her 2-week-old look-alike filly, Gay; a mare who prefers the company of three burly wild bulls; and Shack, the robust sorrel patriarch whose photograph graces the front cover of this book. It was a profound, almost holy experience to stand calf-deep in warm estuarine waters under a moody sky, surrounded by peaceful wild horses, splashing pelicans, and wind-licked marsh grass.

Laura Michaels, the Park Service ranger in charge of pony care, took me behind the scenes to meet the Ocracoke horses. I also met Wenzel, Doran, Sacajawea, and Jitterbug, the Shackleford horses who will revitalize the Ocracoke herd. I even scratched the neck of the lovely black-and-white mare Easter Lady after admiring her from afar for years.

Roe Terry, former public relations specialist of the Chincoteague Volunteer Fire Company, invited me to the workshop where he carves graceful wooden waterfowl and discussed the challenges faced by the hardworking firefighters. Besides managing the herd of free-roaming ponies, these dedicated people donate their time to provide tax-free fire suppression, search and rescue, and emergency medical services in a town of 4,400 permanent residents that receives roughly 1.5 million visitors a year. He also granted me access to the optimal vantage point for the world-famous Chincoteague Pony Swim: whereas most onlookers stood in a field behind an orange fence, out of harm's way, I was able to stand directly on the grassy landing where the horses regained solid ground after swimming the channel from Chincoteague National Wildlife Refuge. Ponies rose out of the water like mythical creatures of the sea, dripping wet and looking very pleased with themselves. My feet were in their hoofprints, and occasionally I dove for cover as a stallion thundered by in pursuit of a rival. It was a magical experience.

Denise Bowden, his successor at the fire company, cheerfully supplied me with useful information. Her passion for the horses and the refuge are evident, and her enthusiasm enhances the overall festivity of Pony Penning week.

Lou Hinds, former manager of the Chincoteague NWR, took me around the refuge to show me unequivocal signs of dramatic environmental change, such as tree trunks, light poles, and chunks of peat that had once been on the bay side of Assateague until island migration situated them squarely on the beach. Studying the dynamic nature of barrier islands is one thing; seeing the evidence of their migration is another thing entirely.

Pam Emge, co-author of *Chincoteague Ponies: Untold Tails*, can identify all of the Chincoteague wild ponies and knows the intimate details of their relationships and lineages. She reviewed part of the manuscript, corrected errors, and filled in details.

Anthropologist Karen Dalke of the University of Wisconsin-Green Bay shared her doctoral dissertation and other writings, which provide unique perspective on what we feel about wild horses and how we define them. She also reviewed the manuscript prior to publication.

Paula Gillikin, manager of the Rachel Carson North Carolina National Estuarine Research Reserve, assisted me in researching the horses of Carrot Island and vicinity.

Philip Howard, nephew of Marvin Howard (1897–1969), who led Ocracoke's mounted Boy Scout troop in the late 1950s, and grandson of the legendary horseman Homer Howard, allowed me to use excerpts from his Web site detailing his family history with the wild ponies.

Allison Turner, biological science technician at Assateague Island National Seashore, supplied excellent information about the Maryland herd and shared Park Service photographs that vividly show the bites and kicks that occur when people get too close to wild horses.

DeAnna Locke, administrator of the Ocracoke Preservation Society, let me pore over and digitize the organization's fascinating scrapbooks, which included many pictures of the island's mounted Boy Scout troop.

Tim Ferry and Flickr user rich701 allowed me to reprint some of their unique historical photographs of the Chincoteague herd.

Craig Downer, a wild-horse ecologist and activist on the board of directors of the Cloud Foundation, shared several of his writings with me on the subject of mustang management.

Jean Bonde of the Buy-Back Babes has a contagious enthusiasm for the Chincoteague Ponies and many tales to tell. These ponies have a large cult following, and her e-mail group recounted the details of their lives—celebrating the romance of Copper Moose and Scotty's ET, pondering Rip Tide's status within Surfer Dude's band, and speculating on why Queenie and Suzy Sweetheart were wandering around the wildlife loop.

Tabetha Fenton of Barefoot Minis helped with proofreading and offered enthusiastic support.

Special thanks to my mother, Joyce Urquhart, who is an exceptionally good proofreader. She read every word of the manuscript and discovered errors that other readers missed.

My husband, Alex, is committed to giving me space in which to create and assisting me wherever possible in my creative pursuits. Besides proof-reading my manuscripts, he is the behind-the-scenes man who maintains the household, runs to the post office, and brings in the bird feeders at night so the bears don't destroy them. He is the love of my life and I am thankful every day that we are together.

Index

Aborigines, Australian   28

abortion, spontaneous   11, 102, 148, 178, 189–190, 321, 520–521

Acadia National Park (Maine)   206

Acadia (Canada)   196

Accomac, Va.   205, 214; Accomack County, 194

Accomack (Native American)   204

accretion, shoreline   197, 534

Ackerman, Leon   128–130

Adagio (horse)   481, 483, 566

Adobe Town Herd Management Area (Wyo.)   102

adoption   551; Assateague horses, 126; Carrot Island horses, 456; Chincoteague Ponies, ; Corolla horses, 300, 325, 327; Cumberland horses, 537; mustangs, 79, 84–85, 88, 91–93, 103, 243, 315, 423, 565–566; Shackleford horses, 460, 483

Africa   32–33, 35–38, 40, 52, 69, 137, 159, 274, 561

Afro-Turkic horse   42

alarm behavior   393, 432, 442

Alaska   32, 40, 310, 453

Albemarle Sound, N.C.   263, 270–271, 274, 320, 351, 360

Alberta, Canada   88, 90, 104, 424

Algonquian (Native American)   201, 268–270, 278–279

alleles   42, 297, 302, 375, 429

Allen, Hervey   364–365

Al-Marah Sunny Jim (horse)   244

Alonzo (horse)   404–405

Amadas, Philip   264

amble (gait)   214, 361–362

Amelia Island, Fla.   497

American beachgrass   138, 152, 156, 165, 470

American Indian Horse Registry   373

Amish, genetic disorders of   299–300

Amrhein, John, Jr.   215, 218, 220

anchithere   559

Andalusia   36, 360

Andalusian (breed)   37, 54, 360, 371, 374, 375, 378, 392, 403

anestrus, lactational   189

animal unit month (AUM)   99

Antelope Hills Herd Management Area (Wyo.)   88

Antelope Valley Herd Management Area (Nev.)   94, 96, 101

antelope, pronghorn   507

antibodies   80, 83, 237, 239, 392, 461, 463, 524

Antilles   37, 218, 362

Apache (Native American)   39

Appaloosa   42–46, 245–246, 370, 394, 495, 499–502; leopard, 42–43

appropriate management level (AML)   78–79, 86–87, 101, 300

aquifers   198, 536

Arabian (breed)   35–36, 42, 70, 183, 187, 206, 214, 219, 238, 243–247, 297, 358, 360, 366, 368, 371–372, 374, 376, 378, 384, 393, 499–502; added to Chincoteague herd, 243

Arenahippus   556–557

Argall, Samuel   206, 358, 368, 371

Argentina   44, 69, 104, 308, 314, 475

Arizona   55, 68, 177, 499

Armada, Spanish   54, 268

artiodactyls   559

Ash Wednesday Storm   129–130, 132, 178, 222, 235–236, 246, 285

Asia   29, 32–33, 42, 48–49, 158, 306, 556, 560–561

Askania Nova, Ukraine   50

ass   32, 34, 50, 52, 209, 555, 560; onager, 34, 50, 52, 560; Yukon wild ass, 32

Assateague (Native American)   202, 204, 209

Assateague herd   117–166 passim, 307, 399, 422, 426, 442, 445, 448, 455, 473–474, 500, 525, 532, 536, 550, 566

Assateague horses/ponies   117–166 passim,
    214, 220, 225, 236, 399, 500
Assateague Island, Md.–Va.   19, 21, 75,
    77, 84, 104, 117–166 passim, 173–248
    passim, 293, 298, 302, 307, 352, 358,
    399, 422, 425–426, 442, 445, 448, 455,
    465, 473–474, 500, 525, 532, 536, 550,
    566–567; development, 128–132
Assateague Island Alliance   422
Assateague Island National Seashore   21, 75,
    77, 84, 117–166 passim, 184, 218, 422,
    465, 474, 550, 567
Assateague Research Laboratory   239, 242
Assateague State Park (Md.)   117–166
    passim
Attwater's prairie chicken   96, 200, 315
auction   Carrot Island property, 453;
    domestic horses, 89; Chincoteague
    Ponies, 160, 173, 176–178, 180,
    182–185, 187–188, 191, 196, 207, 219,
    234, 237, 242, 244, 247–248, 275; Mt.
    Rogers/Grayson Highlands ponies, 315;
    U.S. Coast Guard Beach Patrol horses,
    429; wildfowl, 318
aurochs   54
Australia   28–29, 32, 44, 104, 159, 560
Avon, N.C.   274, 310, 384
Ayllón, Lucas Vázquez de   264, 266, 344–
    346
Aztec (Native American)   38–39, 45

bachelor horses   122, 138, 141–146, 160,
    287–288, 442, 456, 458, 480, 525, 528,
    552
Back Bay   318, 320, 404
Back Bay National Wildlife Refuge   285,
    290–293
Baden-Powell, Baron Robert   220
Bahamas   17, 44, 104, 302, 345
Baltimore Blvd. (Assateague)   128–129, 131
Baltimore, Md.   181, 215–216, 232, 505
Banker Horse   186, 263–264, 269, 271–274,
    284, 289, 300–301, 321, 327–328, 344,
    348, 350, 355, 357–359, 362–363, 367,
    373–375, 377–378, 380, 382, 384–390,
    392–393, 398, 402–405, 427, 430, 432,
    436, 448–450, 452–453, 469, 482, 565–
    566; see also individual herds
Bankers (human)   272–273, 276, 348, 382,
    384, 430, 450
Barb (breed)   35–38, 42, 206, 242, 325, 360,

371–374, 500
Barbados   210, 357
Barden Inlet, N.C.   355, 430, 432, 438
Barlowe, Arthur   264–265, 268
Barrier islands   54, 56, 104, 122, 129,
    132–133, 137–138, 145, 148, 150–158,
    164, 168–169, 180, 189, 194, 197–198,
    200–202, 210–211, 216–218, 242–243,
    245, 264, 269–271, 278–280, 285, 290–
    292, 294, 298, 307, 309–310, 321, 331,
    343–344, 346, 348–350, 358, 364, 373,
    378, 381–382, 395, 402, 427, 431–432,
    436, 439, 446, 458, 464–465, 470–471,
    478, 496, 507, 512, 522, 529, 531, 534,
    550; see also particular islands
bass   230, 448
bats   76, 240, 312
Bath, N.C.   122, 381
Baton Rouge (horse)   327
Baum, Washington   280
bay (coat)   38, 42, 66, 92, 177–178, 214,
    218, 243–247, 264, 297, 343, 376–377,
    383, 415–417, 421, 423, 440, 465, 495,
    499–500, 552, 567
Bay Girl (horse)   286
Beach Patrol   see U.S. Coast Guard
beach renourishment   151, 158
Beach Walk (joining of the herds)   174
bears   32, 68, 75, 86, 290, 514, 568
Beaufort, N.C.   352, 355, 380–381, 388, 424,
    434, 437, 439, 446, 452–454, 460, 464,
    482
Becky (horse)   470
Beebe family   178, 225, 235, 246
Beeswax (horse)   387
Belair Stud (Md.)   371
Bering Strait land bridge   310, 560
Berlin, Md.   118, 158
Bermuda   218, 308, 357
Bialovesh Forest Preserve (Poland)   51
Bighorn Basin   40, 49, 72, 552
Billings, Mont.   44, 69–70, 76, 80, 84–85,
    104, 462, 565
bio-bullet   79, 84
biological weapons   238–240
birth control   see castration, contraception,
    immunocontraception, and sterilization
bison   32, 52, 68, 96, 105, 315, 561
bit (tack)   41, 52, 228,
black (coat)   177, 214, 243, 355, 440, 469,
    500

Black Warrior (horse)   235
Black Hills Wild Horse Sanctuary   72
Blackbeard (pirate)   216, 325, 371, 380–381
black-throated blue
Blaze (horse)   390
blood type, equine   392, 427–429
bluestem   457–458
boardwalk   228, 328, 393, 539
boars   515, 517
boats   175, 236, 317, 351, 354–355, 373,
    386, 508, 512–513
bobcat   29, 514
Bodie Island, N.C.   351, 391
body condition   95, 102, 161, 190, 314,
    475–476, 478, 530; Henneke body
    condition scoring system, 314, 462–463,
    466–467, 472, 475, 478–479; Lactational
    Condition Index, 478–479
Boer War   69
Bogue Banks, N.C.   438, 457
Bonde, Jean   183–184, 238, 568
Bonita Sopresa (horse)   404
Bookcliffs mustangs   55
Boston, Mass.   270, 357, 420
Botai Culture   40–41, 52
botulism   240, 319
bovids   267; see also bison and cattle
Bowden, Denise   173, 184, 196, 246, 567
Boy Scouts   158, 220; mounted (Ocracoke),
    370, 374–375, 377, 388–391, 566–568
Brazil   44, 69
breeching   220, 222
Breton (breed)   206
bridle   50, 227, 361, 382, 450
British Columbia, Canada   90, 104
Broken Jaw (horse)   243
bronco   53, 178
browsing   31, 148, 159, 306, 313, 452, 458,
    531, 556, 558, 559
Brumby   44,
Brunswick   N.C., 380; Ga., 536
Bruselas, Geronimo de   38
buckskin (coat)   66, 173, 177, 197, 221, 245,
    377, 440, 457, 469–470
Bucky (horse)   457, 470, 566
Bulle Rock (horse)   368, 376
bullwhip   Chincoteague Pony Penning, 173,
    182; Florida Cracker Cowboys, 402
bunchgrasses   66, 97
bunting   John, 230; Harry, 242–243
Burns Amendment   88, 91

burro   51, 65, 68, 74–75, 85–86, 79–82, 88,
    91–93, 96, 98–100, 102–103, 105, 239,
    423, 536
Bush, President George W.   65, 88
Butterscotch (horse)   290
Buxton, N.C.   278, 310, 389, 435; Buxton
    Woods, 278
buy-back   176, 178, 182–185, 245; Buy-Back
    Babes, 178, 182–185, 568
Byerley Turk (horse)   360, 368, 372, 376

caballoid horses   40, 561
Cadiz, Spain   37, 218
Caffeys Inlet, N.C.   285, 316
calcium   135, 191, 319
Calico Mountains Complex (Nev.)   82, 102,
California   30, 70, 80, 86, 161, 183, 243,
    319, 358
Camargue region (France)   44, 140, 146,
    200, 314
camel   32, 36, 49, 52, 105, 310
Canaan (river)   346
Canada goose   29, 193, 316
Canada   17, 54, 89–90, 206, 240, 424; see
    also individual locations
Canadian Horse   206
Canary Islands   37, 356, 517
Candler family   506, 512
canter   327, 361–362, 491, 480, 566
Cape Canaveral, Fla.   511
Cape Charles, Va.   201
Cape Cod, Mass.   130, 527
Cape Fear, N.C.   346
Cape Fear River, N.C.   346, 264, 346, 380
Cape Hatteras, N.C.   17, 130, 198, 272, 279,
    309, 348, 352–353, 356, 381, 384, 434,
    436, 438, 510
Cape Hatteras National Seashore   205, 307,
    343–405 passim
Cape Henry, Va.   348
Cape Lookout, N.C.   309, 350–351, 355,
    416, 424–425, 430–431, 434, 436, 438–
    439, 444–445; Lookout Woods, 278
Cape Lookout National Seashore   30, 84,
    150, 304, 307, 309, 343, 356, 415–483
    passim, 531, 550, 565; Cape Lookout NS
    Free-Roaming Horse Law Amendment,
    68
Caribbean Sea and region   17, 37, 216, 269,
    356, 358, 362, 365, 426, 428
Carnegie   Andrew, 317, 496, 506, 508;

Andrew II, 508; family, 317, 496, 498–
500, 506–513; Lucy, 496, 506; Margaret
(Retta), 508; Thomas, 496, 506
Carova, N.C. 320–321, 323, 325, 384
Carrot Island, N.C. 84, 104, 307, 348, 352,
376–377, 403–404, 415–483 passim, 567
carrying capacity 72, 76–77, 82, 150–153,
313–314, 320, 453, 466, 469, 478–479,
496, 509, 517
Carson, Rachel 198, 234, 304, 307, 446,
448, 452–453, 455–458, 537, 567;
see also *Rachel Carson N.C. National
Estuarine Research Reserve*
cartilage 72, 559
Castillo, Bernal Díaz del 38
castration 80, 91, 209, 225, 388, 460, 466,
499
cats 27, 32, 44, 70, 86, 241, 304, 394, 438,
556–557
cattle 38, 40, 52, 54, 57, 65–66, 68, 72, 90,
96–99, 101–102, 105, 117, 140, 206–207,
209–211, 223–224, 233–234, 241,
264, 266–267, 269, 272, 274, 278–279,
282, 304, 308, 310, 312, 314–315, 344,
346–348, 355, 357–358, 380, 382–384,
389–391, 402, 430, 432, 435–436, 438,
450–452, 457, 469–470, 472, 498–499,
504, 506, 508–509, 521, 558; cows, 72,
97, 99, 105, 206, 266, 283, 382, 384, 397,
402, 436, 450, 518, 558
causeway Assateague, 124, 129;
Chincoteague, 233; Nags Head–Roanoke
Island, 280; Hatteras Island, 353–354
cavalry 35, 37, 40, 52, 55, 220, 387; see also
U.S. Army
cecum 558
cedar 155, 211, 275, 385, 431–432, 495,
500, 504
Cedar Island, N.C. 104, 300–302, 307, 348,
377, 382, 415–483 passim, 537, 552, 566;
Cedar Island National Wildlife Refuge,
469
Celtic (horse type) 37
census horse, 75, 77, 102, 148, 150–151,
162–164, 207, 296, 298, 300, 302, 304,
307, 313, 320–321, 370, 405, 451,
453, 458, 466, 480, 506, 518, 520, 524,
528, 530–532, 550, 565; bird, 316,
533; human, 276, 355, 506; hog, 516,;
seabeach amaranth, 309,; sika, 159
Cerbat Mountain (Ariz.) mustangs 55, 87,
373–374
champing 395
cheetah 32
Chernobyl, Ukraine 50
Cherokee (breed) 55, 463
Chesapeake Bay 158, 204, 207, 209–210,
218, 229, 264, 266, 269, 319–320, 346,
348
chestnut (coat) 38, 42, 66, 136, 173, 177,
183, 244–246, 264, 297, 360, 372, 376–
377, 404–405, 440, 469, 500
Chicamacomico, N.C. 279, 348
Chickasaw (breed) 269, 347, 359–364, 366,
368, 370–371, 378, 426
chickens 70, 96, 135, 200, 206, 232, 236,
315, 346, 382
Chicora 345
China 42–43, 50, 159, 240
Chincoteague, VA. 117, 137, 152, 165,
173–248 passim, 460
Chincoteague Bay (Md.–Va.) 134, 137, 212
Chincoteague National Wildlife Refuge 78,
118–119, 126, 131, 135, 160, 173–248
passim, 304, 307, 315, 457, 465, 472,
537, 567
Chincoteague Ponies 143, 165, 173–248
passim, 275, 307, 498, 567–568;
"Yankee" stock added to, 178
Chincoteague Pony Penning see *horse/pony
penning*
Chincoteague Volunteer Fire
Company 117–118, 160, 173–248
passim, 307, 567
Choctaw (breed) 55, 359–360, 362, 373,
463, 482
cholesterol 91
Christopher Columbus (horse) 387
chromosomes 42, 48, 296–297
Civil War, U.S. 227, 230, 276–278, 285,
317, 361, 366, 384, 402, 498, 505–506
Civilian Conservation Corps 150, 282
climate 32–33, 38, 43, 45, 50, 52, 76, 121,
149, 158, 196, 198–199, 206, 282, 311–
312, 347, 400, 424, 460, 476, 478, 483,
530, 536, 555–557, 561
climate change 196, 198–199, 282, 483, 536
Clinton, President William 92, 465, 468
Cloud (horse) 21, 422, 568; Cloud
Foundation, 66, 102, 422, 568
club-foot 297
coat 269, 297; condition, 122, 178, 274,

432; curliness, 145, 182; density, 400; grooming, 347, 445, 522; juvenile, 177, 400, 526; shedding, 400, 526; summer, 400; winter 195, 400, 403

coat colors and patterns   base, 343; desirability, 177; developmental changes, 177, 345; manipulation, 214, 244, 246; prehistoric, 42–43; role in mating, 446; see also particular varieties, such as *dun* and *grulla*

Cobb, Collier   278, 280

cobby   372, 374

Coggins test   176, 185, 237, 456, 461, 464–465, 537

coins, Spanish, on Assateague   217–218

cold-blooded breeds   370

colic   37, 126, 138, 238, 289, 323

Colonial Spanish Horse   104, 300–303, 327–328, 360, 362, 369, 373, 376, 378, 401, 410, 426, 500; see also individual breeds

Colorado   55, 65, 92

colostrum   392, 463, 522, 524

colts   44, 73, 141, 144–146, 180, 182–183, 187, 206–207, 225–226, 235, 237, 245, 248, 275, 287–288, 290, 296, 300–301, 323, 328, 363, 370, 383–384, 388, 392, 404–405, 416, 433, 437, 449, 482, 497, 515, 517, 520, 527–528, 552, 566

Columbus, Christopher   36–37, 344; Christopher Columbus (horse), 387

Comanche (Native American)   39

Comma (horse)   422, 442

Confederate/s   351, 379, 506; see also *Civil War, U.S.*

conformation   52, 85, 93, 104, 147, 188, 246, 328, 364, 374, 376–377, 403–404, 427, 440, 448, 499–500, 560

Connecticut   183, 357–359, 536

Connemara Pony   44, 139

conquistadors   35, 38–39, 54, 427

Constitution, U.S.   370, 436

contraception   79–85, 101, 119, 121–122, 161–162, 164–165, 248, 300, 455, 458, 460, 470–471, 473–474, 482, 531; see also *immunocontraception*

Cooper, James Fenimore   360

Copper Moose (horse)   219, 245, 568

copulation   146–147, 212

cordgrass (Spartina spp.)   198, 291, 308, 312, 316, 459, 470, 498, 513, 536; saltmarsh, 128, 138, 148, 162, 310, 435,

440, 452, 495, 526–527; saltmeadow, 138, 440

Core Banks, N.C.   150, 309, 355, 388, 390, 424, 429–432, 435, 438, 460, 469

Core Sound, N.C.   152, 429, 465

Coree (Native American)   429

Corolla, N.C.   84, 87, 263–328 passim, 359, 374–378, 401, 403–404, 408, 426, 468–469, 531, 552, 566; see also *Currituck Banks*

Corolla Wild Horse Fund   7, 286–304 passim, 313, 320–328 passim, 469, 566

Coronado, Francisco   39, 365

corral   68, 91, 94, 174–175, 177, 182, 184, 186, 198, 219, 223, 228, 237, 245, 265, 269, 293, 327, 370, 384–385, 404, 444, 450, 464–465, 472

Cortés, Hernando   38–40, 43, 45, 54

Cothran, E. Gus   32, 35, 40, 146–147, 162, 164–165, 214, 271, 298, 300–301, 362, 375–376, 403–404, 426–427, 468, 499, 529, 565

cottages, vacation   228, 276, 286, 294, 323, 352, 390, 438, 508

cotton   218, 501, 505–506, 514

courtship   142, 226, 396, 522

cowbird   29, 140

cowboys   21, 66, 73–74, 205, 217, 228–229, 384–386, "Cowboy" Oliphant, 232; Florida Cracker Cowboys, 399, 402–403; Saltwater Cowboys (Chincoteague), 173–177, 180, 182–184,

coyote   29, 75, 521

crabs   151, 516, 536; blue, 276, 316, 430; mole, 157; xanthid, 151

cranes   75, 522

Creek (breed)   359, 362

Croatoan (horse)   325, 327, 362

crossbreeding   35, 301, 375, 377–378, 392, 426

Crow (Native American)   46, 79

crystal skipper (butterfly)   457–458

Cuba   33, 38, 266, 344, 347, 362

Cubanito (horse)   378, 381, 392–393, 403

Cumberland Island, Ga.   27, 104, 141, 164, 186, 236, 302, 307, 387, 426, 466, 475, 495–540 passim, 550

Cumberland Island National Seashore   21, 77–78, 200, 283, 302, 307, 495–540 passim

Cumberland Island Wilderness Boundary

Adjustment Act of 2004   538
curlew   203, 382
currents   132–133, 137, 155, 198, 216, 344, 350, 355–356, 450, 534, 566; littoral, 132; Assateague Channel, 175; Ocracoke Inlet, 355–356; Equatorial, 356; Gulf Stream, 217–218, 356, 434
Currituck Banks, N.C.   104, 154, 164, 263–328 passim, 348–349, 351, 363, 384, 405, 425, 473; see also Corolla
Currituck National Wildlife Refuge   263–328 passim
Currituck N.C. National Estuarine Research Reserve   287, 294, 297, 321, 452
Czar Nicholas II   499

Dakotas   70
Dale, Sir Thomas   206, 358
Dales Pony (breed)   428
Dalke, Karen   7, 65, 73–74, 82–83, 85, 92, 567
dapple/dapple-gray   92, 301, 384. 495
Dare County, N.C.   280, 282–283, 285, 384, 388
Dare, Virginia   266
Darley Arabian (horse)   368, 372, 376
dart/dart gun   79–81, 84, 161–162, 298, 300, 458, 471
Dartmoor Pony   44
Darwin, Charles   54
De Bry, Theodor   269, 275, 277, 425
De Soto, Hernando   347
deer, white-tailed   101, 158–160, 223, 502, 512, 514, 526; Sika, 124, 158–160, 193, 298, 453
Deerwood Ranch eco-sanctuary   94
defecation   136, 140–141, 308, 391, 397, 399–400, 513
deforestation   98, 282
Delaware   132–133, 211
Delmarva Peninsula   135, 193, 202, 204, 212, 215–216; see also Eastern Shore and particular locations
delta   Rhône, 44; flood-tide, 353
demoflush   134
Denisovans   32
density dependence (population growth)   77, 164
Dereivka, Ukraine   52
Devine, Betts   236, 246
Dews Island, N.C.   291

Diamond City, N.C.   430, 433
Diamond Hills South (Nev.) Herd Management Area   86
Diamond Shoals, N.C.   356
diebacks/die-offs   marsh, 316, 527–529, 536; waterfowl, 318–319; wild horses, 452–454, 458
dingo   28–29
Dinohippus   560
dinosaurs   555–556
dished face   36, 187, 369, 374
displacement behavior   394–395
DNA   32–34, 40, 42–43, 75, 118, 146–147, 165, 214, 298, 300–301, 310, 375, 426–427, 468, 474, 514, 516; mitochondrial, 34, 40, 42, 298, 474; quagga, 34; swine, 516–512
dogs   27–29, 43–44, 54, 39, 43–44, 56, 98, 122, 206, 210, 241, 266, 302, 394, 397, 424, 445, 450, 460, 555–557; dog food, 56, 70–71, 95
Dogskin Mountain (Nev.) mustangs   86
doldrums   105
dolphin   192, 276; see also porpoises
domestic horses   40–44, 46–51, 53–54, 56, 66, 68–70, 72, 84, 89, 92, 104–105, 117, 122, 124, 137, 142, 161, 178, 186, 189–190, 207, 209, 237, 239, 241, 263, 268, 275, 289, 295–296, 301, 310, 323–324, 328, 347, 358, 376, 391, 393, 396, 401–402, 444, 446, 448, 461, 465, 470, 498, 519, 522, 525, 529–530, 565
dominance/submission behavior   52–53, 126–127, 136, 138, 141, 143–147, 175, 183, 391, 394–398, 404, 415–423 passim, 442–445, 464, 513, 518–520, 525, 527; see also hierarchy, social
dominant/recessive traits   87, 104, 147, 297, 299, 367, 377–378
Dominican Republic   266, 435
donkeys   52, 320
Doran (horse)   198, 403–404, 567
Dowdy, Betsy   263, 327
Down-Easters   460, 464–465
Downer, Craig   85–86, 98–99, 101–102, 105, 536, 568
draft breeds   37, 42, 55, 206, 214, 358, 371, 369, 426, 444
Drake, Sir Francis   266
Dream Dancer (horse)   182
dreaming, horse   398; see also REM sleep

dredging   129–134, 155, 201, 351, 452, 454–455, 457

dressage   93, 475

drought   78, 94–95, 100, 102, 147–148, 163, 190, 206, 302, 314, 447, 454, 466, 495–496, 514, 516–517, 520–521, 530, 534, 555

drowning   susceptibility of horses to, 222, 224, 323, 358; incidents of, 129, 163–164, 178, 222, 225, 235–236, 302, 350, 352, 357, 382, 386, 435–436

ducks   29, 190, 203, 229, 247, 304, 312, 314, 317–319, 439

dun (coat)   36, 42–43, 48, 51, 55, 104, 177–178, 377, 440, 552

dunes   18, 117, 129, 132–133, 150–158, 165–166, 190–193, 217, 272–273, 276–280, 282–285, 291, 294–295, 309–310, 323, 328, 343–344, 349, 352–353, 372, 388, 399, 415–416, 424, 431–432, 439, 442, 449, 451, 457, 470, 496, 509, 518, 525–526, 531, 533–534, 536, 539, 552; see also *hills*

dung   136, 306, 308, 316, 526, 536

Dungeness   495, 498–499, 502–506, 508–509, 513, 515, 520, 526, 528, 534, 537, 539

Dutch horses   274, 358–361, 366, 370–371, 500

eagle   30, 76, 318, 376, 439, 514, 533

Easter Lady (horse)   567

Eastern Seaboard   54, 137, 165, 197, 218, 264, 269, 357, 378

Eastern Shore   202–205, 211–212, 216, 221, 246; Virginia, 173, 185, 210–211; tribes, 202–205; see also *Delmarva Peninsula*

ebb   134, 422, 555, 560

ebbed   132

ebbing   495, 498

ebb-tidal shoal   134

eco-preserve/eco-resort/eco-sanctuary   94–96

ecosystem   18, 23, 31, 105, 148, 151–152, 160, 162, 164, 199, 280, 282, 304–305, 310–312, 314, 316, 320, 322, 452, 458, 471, 496, 528, 530, 536–537, 539, 549–550

ecotourism   82, 94

Edward Teach (horse)   301, 325; for the pirate, see *Blackbeard*

Edwards, Steve   263, 301, 315, 325, 327–328, 362, 403, 448, 470, 480, 482–483, 566

eelgrass   231–232, 274, 276, 316

eggs   232, 513; addling, 193; egg cells (see also *oocytes* and *ova*), 80, 161; "egging," 230, 236; mosquito, 236; piping plover, 151; turtle, 513, 515; wild bird, 213, 318, 513, 515, 533; worm, 391

egrets   140, 148, 199, 228, 317–318, 439, 514, 533

equine infectious anemia(EIA)   237, 247, 439, 451, 455–456, 460–461, 463–466, 470, 537

Elizabeth I (England)   264

Elizabeth City, N.C.   263, 278, 280

elk   97, 105, 306, 507; sika, 124, 158–160, 193, 298, 453

Elko County, Nev.   94

Ellis-van Creveld Syndrome   299

Ely, Nev.   86, 101

Emge, Pam   174, 567

Empire, British   69, 217

encephalitis   124, 239–240, 242, 289; Eastern equine (EEE), 163, 185, 236, 239, 302, 390, 521; Western equine, 185

England   54, 104, 201–202, 205, 209–210, 239, 265–270, 274, 347, 356–361, 366, 368, 370, 372, 379, 426, 429–431, 497

English horses   358–359; see also particular breeds

Eocene Epoch   556–557

Eohippus   556

epidemics   147, 439

epizootics   237, 242, 319

equator   38

Equatorial Current   356

equids   23, 31–32, 34, 40, 42, 47, 49, 74–75, 77, 85–86, 96, 98, 102–103, 200, 453, 460, 465, 551, 555–561

"Equine Eve" (Ancestral Mare Mitogenome)   561; see also *DNA, mitochondrial*

equine infectious anemia   91, 237, 242, 247, 300, 377, 439–440, 451, 455–456, 460–461, 463–466, 470, 537

Equus   31, 41, 43, 300, 303, 310, 401, 555, 559–561

Eriskay Pony   44

erosion   19, 56, 129, 132–133, 150–151, 154, 156–159, 164, 198, 279, 282–283,

298–299, 305, 307, 310, 352–353, 388, 457, 469, 529, 533–534, 536

española (horse type) 37

estancias 38

estrogen 56, 83, 189

estrone 189

estrus 80, 141–142, 144–145, 148, 189, 400, 442, 519; postpartum, 148, 189–190, 520; lactational anestrus, 189

estuaries 132, 152, 192, 198–200, 231, 287, 291, 308, 310, 316, 346, 446, 452, 458, 495, 537, 567

Eurasia 29, 32, 40, 42, 51–53, 560–561

Eurohippus 557–558

Europe 17, 31–33, 35, 39, 41–42, 51, 55–56, 70, 90, 96, 104, 152, 193, 202, 204, 216, 218, 220, 264, 269–270, 272, 306, 310, 314–315, 347, 351, 356, 358, 362, 368, 378, 426, 502–503, 517, 557

euthanasia 89, 91, 93, 237, 323, 327, 424, 464–466, 469, 471, 537

Evans, Bob 242

evolution 29, 31, 52–53, 144, 148, 152, 196, 240, 290, 310–311, 401, 416, 438, 474, 517, 555–561; horse, 555–561

exclosure 153, 308–309, 313, 320–322, 324, 529

Exmoor Pony 44

exotic species/subspecies 27–28, 30–31, 49, 72, 88, 118–119, 150, 158–159, 193, 245, 304, 308, 310, 312, 318, 320, 456, 459, 504, 507, 518, 539, 549; see also *feral*, *invasive species*, and *reintroduction/ repatriation*

extinct species/subspecies/populations 33–35, 40–42, 44, 50, 53, 193, 240, 242, 347, 453, 466, 503, 555–561; extinct horse breeds, 214, 296, 300, 359, 361–362, 375,

extinction 23, 28–29, 32–33, 50–51, 54, 86, 147, 151, 163, 165, 196, 276, 282, 298, 300, 302, 311, 313, 318, 327–328, 343, 373, 391–392, 401–403, 453, 466, 507, 514, 555–561

Fabio (horse)

Falabella (breed) 559

falcon, peregrine 192, 439

False Cape State Park (Va.) 291–293

False Cape, Va. 264, 384

farms and farming 29, 34, 36, 38, 47, 54–56, 68, 73, 89, 92, 97–98, 135, 180, 182,

207–208, 210, 212, 230, 232, 242, 247, 269–272, 275–276, 285, 316, 325, 328, 358, 363–364, 372, 402, 428, 434, 481, 483, 499, 505, 539, 566

farrier 120, 456, 471

Feather Fund 176, 183,

feces 75, 96134, 165,, 189, 441, 474, 519; see also *defecation*

fecundity 87, 100, 299, 404. 479

Fell Pony 44

fences Assateague, 118, 120, 175, 184, 187, 200, 223, 225, 234–235, 237; Cedar Island, 470; Cumberland Island, 529; Currituck Banks, 270–272, 274, 276, 278, 290–291, 294, 306, 313, 321–322, 324, 327–328; legal requirements, 208–210, 270–271, 282–283, 390; Ocracoke, 307, 383, 390; sand, 156, 158, 277, 284, 388; Shackleford Banks, 430, 433; taxes on, 54, 270

Fenwick Island, Del.–Md. 132

feral 31, 44, 47, 314; cattle, 57, 384, 430, 451, 469; definitions, 30–31; dogs, 29, 43–44; goats, 430, 451; horses, 25–57 passim, 70, 72, 76–77, 89, 104, 117, 119, 128–129, 137, 146, 148, 160, 189, 198, 200, 248, 271, 304–305, 309, 320–321, 362, 375–376, 384, 404, 430, 441, 451–452, 457, 465, 469, 499, 528–529, 532–533, 549–550; quaggas, 34; sheep, 430, 451; sika deer/elk, 159; swine, 30, 309, 312, 320, 324, 430, 515–518, 539

Ferdinand I (Spain) 36, 38

Ferghana (breed) 43

Ferguson Janet "Gogo," 496; Lucy, 499, 517; Robert, 509

fermentation, cecum 558

Fernandina Beach, Fla. 536

ferries 27, 129, 230, 243–245, 263, 280, 343, 353–355, 382, 389, 469, 482, 495, 508, 512–513

Ferry, Tim 7, 243–245, 568

fetlock 210, 392, 415, 559–560

Field, Samuel 232, 244

fillies 93, 144, 146, 162, 182, 248, 445–446, 519–520, 527–528

Firemen's Carnival (Chincoteague, Va.) 173–176, 184, 188, 226, 234, 245

First African Baptist Church 496, 539

fish and fishing 75, 96, 119, 134, 152, 158, 160, 174, 188–189, 191, 193–196,

198–202, 204, 211, 222–224, 226, 230, 233–234, 236, 239, 242, 272, 274–276, 285, 287, 292–294, 297, 304–305, 307, 309, 313, 315–316, 320, 324, 355, 386, 430, 448, 451–452, 455, 457, 464, 506, 512, 516, 536, 550; "fishing horses," 448; see also particular species

fishermen   223, 230, 276, 317, 388, 432; see also *watermen*

flehmen   139–140, 142, 144, 396, 400

Flemish horses   358–359

flies   127, 134, 139–140, 176, 212, 237, 393, 397, 447, 463–465, 537; horseflies, 461–463, 465; fruit flies, 163,

floods and flooding   27, 129, 137, 155, 158, 184, 191, 198, 235, 279, 282, 286, 302, 312, 350, 382, 398, 436. 518, 549. 566

flood-tide delta   353

Flores Man (Hobbit)   32

Florida   54–55, 74, 130, 183, 245, 248, 264, 278, 304, 344, 346–347, 359, 362–363, 366, 373, 378, 399, 401–403, 426, 428, 434–435, 501–503, 509, 514

Florida Cracker (horse)   55, 373, 399, 402–403, 426; Florida Cracker Horse Association, 373

fly-shaker muscle   140, 237, 397, 461

flyways   Atlantic, 192, 317–318, 439; Central, 316; Mississippi, 316, 319

foals   33–34, 36, 43, 49, 54, 70, 74, 80–83, 90, 101–102, 104, 117, 120–121, 124, 135, 137–138, 140, 142, 144–146, 148–149, 160–162, 164, 175–185, 187–191, 195–197, 207–208, 213, 221, 234–237, 242–245, 247–248, 264, 274, 286–287, 291, 296–297, 300–303, 313–315, 321–322, 327, 343, 368, 378, 384, 387, 392–393, 395–400, 404, 416, 422, 424, 438–445, 449, 452, 455–456, 458–461, 463, 465, 470–476, 478, 480–483, 495, 498–499, 514516,, 518–528, 530, 552, 557; foaling, 36, 76, 80, 142, 148, 161–162, 189, 238, 248, 370, 392, 405, 473, 475, 519–520, 522, 566; foal mortality, 520–521

follicle-stimulating hormone   83

forbs   96–97, 306, 530

forefeet   386

forehead   187, 287, 343, 361, 422, 500

foreleg   72, 141, 394, 397, 522

forelock   228, 400

forts   265, 267, 348, 497; Detrick, 239–240; Granville, 354, 380; Macon, 430, 457; Prince William, 503; San Pedro, 502; St. Andrew, 503

fossae   559–560

fossils   32, 40, 49, 56, 306, 555–557, 559–560

Foundation for Shackleford Horses   303, 416, 424, 450, 455, 465–466, 468, 470, 479–480, 483, 566

founder (medical condition)   328

four-beat gait   214

fox   30, 43, 178, 193, 214, 424; red, 193,

France   44, 52, 89, 91, 104, 146, 159, 200–202, 206, 308, 314, 347, 370, 460

Francis I (France)   39, 201, 207, 265–266, 358

Franciscan   502

freedmen   498, 506

free range   97, 208–210, 234, 269–270, 348, 358, 390, 498

free-ranging/roaming   11, 13, 18–19, 22, 27–28, 30–31, 44, 46–48, 54, 56, 65–66, 69, 74, 76, 78, 83–85, 88–89, 91, 93, 96, 98, 102, 104–105, 118, 120, 135, 137, 142, 148, 150, 152, 158–162, 166, 173, 186, 208–210, 216, 223, 232–233, 236–237, 244, 248, 264, 274, 276, 278–280, 282, 284–286, 298, 305, 308, 310, 312, 320, 328, 343–344, 346, 358, 364, 370, 374, 382–384, 387, 390, 401,424, 429, 437, 453, 459, 468, 470, 472, 474, 476, 478, 498–499, 504, 506, 517–518, 528, 532, 537, 549–551, 565, 567; working definition, 44–47

freeze-branding   91,182, 416, 454, 465

French and Indian War   354, 503

French horses   206, 358, 366, 371, 500

French Norman (breed)   206

freshwater   129, 137, 190–192, 198, 306, 424, 436, 439, 443, 447, 454, 512, 514, 517; freshwater lens, 129, 436, 443, 446–447

Friesian (breed)   42, 147

frog (amphibian)   76, 512, 515; horse anatomy, 559

fungi   100, 148, 316, 536, 538

gaits   37, 212, 214, 359–361, 374, 386, 391, 480, 482

galleons   117, 215, 218, 220

gallop   36, 55, 122, 144, 208, 210, 226, 263, 287, 325, 361, 366, 388, 399, 416, 420, 527, 552

Galloway (breed)   360–361, 378

gathers and gathering   47, 68–69, 72–73, 90, 160, 174, 185–186, 217, 224–225, 227, 232, 236–237, 244, 315, 348, 384, 442, 449, 451, 460, 464, 466, 470–471, 498, 537; Bureau of Land Management, 75, 79–80, 82–85, 87–88, 91, 93, 95, 99–100, 102–103, 300, 423; see also *roundups*

Gay (horse)   457, 470, 566

geese   23, 29, 148, 193–194, 203, 228–229, 312, 316, 436

gelding and geldings   39, 44, 46, 92–94, 139, 210, 216, 288–289, 327, 383, 388, 401, 404, 429, 439, 470, 480–481, 566

genes   34, 42–44, 55–56, 82, 86–87, 121, 147, 162, 164, 186, 206, 214, 242, 244, 247, 274, 288, 296–297, 299, 301–302, 323, 328, 345, 350, 358, 367–368, 376–378, 387, 392–393, 401, 403–404, 425–426, 429, 431, 448, 458, 460–461, 468–469, 471, 479, 482, 499, 517, 529, 565–566

genetic bottleneck   32, 42, 100, 300, 381

genomes   40–41, 44, 297, 301, 343, 500, 561

genotype   75

geographic information system (GIS)   294

Georgia   17, 100, 104, 141, 186, 236, 264, 270, 346, 364, 402, 495–540 passim,

gestation   178, 189, 302, 314, 368, 478, 515, 519–522, 557

Ghyben-Herzberg principle   446

Gideon (horse)   183

Gidget (horse)   184

Gillikin, Paula   453, 455, 458, 567

giraffes   33

glaciers and glaciation   32, 154, 196–197, 311, 561; see also *climate change, ice ages*

Global Important Bird Area   192

goats   30, 40, 54, 66, 99, 117, 206, 209–211, 264, 266, 269, 279, 282, 308, 310, 346, 348, 390, 430, 432, 438, 450–452, 521

Gobi Desert   49–50, 302

Godolphin Arabian (horse)   360, 371–372, 376

Goldsmith, Oliver   366

GonaConTM   83

gonadotropin-releasing hormone

(GnRH)   83

Gordillo, Francisco   345

gorillas   33, 75

Goshute Herd Management Area (Nev.)   94, 96

grain   46, 182, 186, 212, 274, 324, 384, 387, 389, 396, 401, 434

graminoid species   308, 320

grapes   265, 557

Graveyard of the Atlantic   272, 356

gray (coat)   38, 51–52, 92, 177, 243–244, 297, 345, 367, 377–378, 384, 394, 403, 474, 495, 497, 499–500; graying gene, 297, 345, 377–378, 388; see also *dapple/dapple-gray*

Grayson Highlands (Va.)   55–56, 97–98, 200, 307, 314, 317

grazing   18, 23, 31, 33, 40, 52, 54, 72, 78, 94, 96–99, 101–102, 105, 117, 119, 122, 144, 147–148, 151–154, 156–157, 160, 162, 165, 189, 198–200, 209–211, 216, 220–221, 224, 234, 264, 270–271, 304–305, 308–316, 320–322, 346–348, 355, 357, 372, 381, 384, 389, 391, 393–394, 399, 425, 435, 440, 442–443, 446, 448–449, 451–452, 457–459, 466, 470, 472–473, 475, 478–480, 496, 498, 500, 508–509, 513, 520, 526–527, 529–531, 534, 536–537, 552, 558–559, 561, 566

Great Abaco, Bahamas   17, 44, 302

Great American Interchange   561

Great Basin   78, 521

Great Dismal Swamp   270, 316

Grecko (horse)   290

Greeks   53

Greenbackville, Va.   234

greenbrier   23, 138

Greene, Nathanael   504–505

Greenland   557

Grenada   357

Grenville, Sir Richard   265–268, 347–348

Greyfield   496, 508, 538

grulla (coat)   42, 51, 177, 552

Guadalquivir River (Spain)   37

Gualdape   346

Gulf of Mexico   96, 315–316, 430

Gulf Stream   217–218, 356, 434

gulls   380, 53

gunning   see *hunting*

Guthrie Sam (horse)   386

habitat   30, 40, 52, 76, 85, 96–97, 102, 105,
    120, 129, 131–132, 134, 147–148, 150–
    152, 159–160, 162, 190–191, 193–195,
    198–200, 212, 279, 282, 291, 294, 302–
    303, 305–306, 308–318, 320, 328, 380,
    441, 453, 457–459, 478, 513–514, 517,
    530–531, 537, 550, 556
hack   78, 141, 214
hackamore   52
HMS *Hady*   273
Hagerman horse   560
haircloth   56
Haiti   435
halter   52, 178, 186, 222, 225, 227–228, 327,
    433–434
hammock, horse   37–38, 220; see also *sling*
hammock/hummock (topographical
    feature)   343, 382, 385, 399, 465
Hancock, Nena and Woody   469–470, 566
harems   49, 80–82, 104, 136–138, 140, 142,
    144–147, 162, 183, 247–248, 288, 295,
    301, 321, 323, 348, 398–399, 415, 417,
    421, 423, 441–442, 444, 458, 474–475,
    480, 495, 497, 513, 518–519, 522, 528;
    harem fidelity, 474–475
hares   265
Harkers Island, N.C.   352, 436, 450, 464,
    482
harness   33, 50, 361, 382
Harriot, Thomas   267–269, 347
Hatteras   Inlet, 283, 309, 343, 356–357,
    383–385, 388–389, ; Island, 155, 158,
    274, 276–280, 343, 348, 350–353, 356–
    358, 378, 381, 383–392, 404, 434, 436,
    438, 510; Village, 155, 274, 353, 389; see
    also *Cape Hatteras*
Hawaii   29
hawk   228, 273, 277–278, 280, 312, 382
hay   46, 91, 148, 175, 184, 186, 200, 274,
    305, 308, 323, 384, 387, 401, 455, 465
hearing (sense)   393–394, 396
hearings   congressional, 88, 240, 304–305,
    312, 320,; public, 286, 531
helicopter   75, 87, 99, 235, 460
Henneke Body Condition Scoring
    System   314, 463, 466–467, 472–473,
    475, 478–479
Henning, Jim/Jeannetta   378, 380, 384,
    386–392
Henry IV (France)   206
Henry, Marguerite   173, 178–179, 235, 242

herbicides   200, 302
herons   318, 439, 514
heterogeneity   298
hibernation   91, 196, 236
hierarchy, social   45, 52, 138, 141, 144–145,
    396, 442–444, 527; see also *dominance/
    submission behavior*
Highland Pony   44
hills   132, 156, 186, 190, 193, 222, 273, 275,
    277–280, 309, 385, 432; see also *dunes*
    and particular hills
Hilton Head, S.C.   431, 510
hindparts/hindquarters   34, 142, 220, 323,
    361, 374, 397, 400, 501
Hinds, Lou   15, 78, 119, 158, 189, 193, 196,
    198–200, 315, 567
hips   138, 461, 463, 467, 475, 495, 522, 566
Hispaniola   38, 266, 344, 346, 348
Herd Management Area (HMA)   78–79, 81,
    86, 92, 96; see also individual HMAs
Hobbit (Flores Man)   32
hobbles   41, 210, 222
hocks   263, 361
Hoffman, Doug   35, 42, 76, 396–398, 400,
    442, 498, 510–511, 513–514, 516–518,
    521–522, 528, 530–534, 536–537
Höglund, Donald   7, 11–13, 28, 84–85, 89,
    91, 102, 549–551, 565
hogs   see *swine*
Holland (horse)   480, 482
holly   278, 495
Holmes, Thompson   212–213, 216, 225–226
Holocene Epoch   32, 40, 42
homeostasis   558
hominids   32
homogeneity   96, 120, 297, 306, 366, 500
homo-/heterozygous   297
hoofprints, Misty   178–179
Hoofprints in the Sand   4, 17, 565, 571
hooves   31, 49, 53, 56, 66, 73, 90, 93, 117,
    122, 124, 135, 141, 173, 175, 186, 199,
    231, 276, 300, 316, 323, 328, 363, 386,
    391–392, 394, 396–397, 416, 432, 442,
    446, 448, 501, 522, 533, 556–557, 559–
    560
Hope (horse)   325
hormones   56, 79–80, 83, 142, 392, 458
Horntown, Va.   227
horse communication   387, 391, 393–396,
    400; see also *scent marking, tail,
    vocalization*

Horse of the Americas registry   327, 373, 375, 483

horse/pony penning/s   Assateague, 117; Cape Lookout, 444–445; Cedar Island, 469; Chincoteague, 117–118, 120, 143, 173–248 passim, 567; Currituck Banks, 348; Hatteras Island, 384; Neuse River, N.C., 269; Ocracoke, 355, 370, 382, 384–385, 388–389; Outer Banks, 274, 350, 430, ; Shackleford Banks, 432–434, 450, 460; see also *roundups*

horse slaughter   65, 69, 72–73, 79, 88–92, 93, 105, 346, 424, 537

horseflesh/horsemeat   56, 70, 89–91

horseflies   461–463, 465

horsehair   41

horsehide   56

horsemanship   53, 92, 227, 315, 389, 566

horsemen   28, 36, 39, 41, 228, 266, 361, 365, 368, 372, 384, 386, 388, 401, 448, 567

horses on ships   37–38, 220–222, 224–225, 358

horses, single-hoofed/single-toed   559–560

horses, three-toed   559–560

horse-tamers   227

horse-training   94

horse-watching   296, 482

hot-blooded breeds   42

Howard   Homer, 384–386, 567; James, 378, 384; Lawton, 404; Marvin, 275–276, 374, 377, 385, 388–390, 405, 567; Phillip, 378, 383–386, 388–392, 404–405, 474, 567

Humane Society of the United States (HSUS)   126, 180, 182, 193, 303, 463,

hunting   30, 32, 50, 53, 68, 70, 72–73, 159–160, 201, 204, 209, 285, 317–319, 460, 503–504, 506, 509, 511, 516, 522; aurochs, 32; bears, 514; birds, 228–230, 235, 242, 285, 317; burros, 68; deer, white-tailed, 159–160, 193; deer/elk, sika, 159–160; dolphins, 276; horses, Abaco, 302; horses, Corolla, 323; horses, free-range, 210; horses, prehistoric, 32; horses, wild, 52; swine, 30, 511, 516–517; mustangs, 68–70, 72–73 (see also *mustangs, mustangers*); Outer Banks livestock, 283; panther, Florida, 514; quaggas, 33–34; Takhis, 49; turtles, 276, 513; whales, 276, 430

hunting clubs   282, 285–287, 291, 297,

317–318, 551

hurricanes   23, 27, 132, 150, 155, 186, 191, 193, 218, 220, 222, 236, 277, 285, 302, 350–353, 356–358, 379–380, 424, 432, 434–435, 450, 456; Bill, 352; Bonnie, 150; Donna, 450; Floyd, 302; Great/Chesapeake-Potomac Hurricane (1933), 132, 222, 432; Great September Gust (1821), 222; Ida, 193; Irene, 155, 193, 422, 424; Isabel, 155, 186, 310, 350, 352–353; Ophelia, 424; San Ciriaco, 380, 435; Sandy, 191

hybridization   29, 48, 52

hypocalcemia   191

hypothalamus   83

Hyracotherium   555–556

Indians, American   39, 54–55, 88, 152, 155, 200–204, 208–209, 211, 215, 264, 268, 279, 311–312, 345–347, 354, 362–366, 370, 373, 378, 429, 501–503; see also *Native Americans*

Iberian horses   373–374, 376, 392, 426,

Iberian Peninsula   33, 561

ice ages   31–33, 43, 53, 154, 311, 558, 561; Little Ice Age, 397; see also *glaciers and glaciation*

Iceland   557

Icelandic (breed)   445, 480

Icy (horse)   248

Idaho   54, 86

immunity   39, 86, 295

immunocontraception   80, 82–83, 149, 162, 165, 189–190, 300, 303, 247, 456, 459, 466, 471, 474–475, 524, 531–532, 550, 565

immunodeficiency   297, 299

immunoglobulins   524

Ina May (horse)   470

inbreeding   36, 50, 86–87, 104, 146–147, 162–163, 165, 242, 246, 287, 297–300, 376, 392, 403, 468, 529–531

Inca   29, 39

Indians   39, 54–55, 152, 200–204, 208–209, 215, 264, 268, 270, 311–312, 345–346, 362–363, 365–366, 373, 429, 501–502; see also *Native Americans* and particular groups

Indies   54, 274, 344, 356–358, 362, 378, 426, 505

Indonesia   32

infants   424, 518, 524

infection   100, 140, 238–242, 325, 461, 464, 470, 495, 524, 536, 538

infestation   124. 278, 314, 495

influenza   39, 204, 238–241

injection   80, 84, 89, 161, 471

inlets   129, 132–134, 155, 158, 192–193, 198, 201, 216–218, 228, 231, 270, 272–274, 276, 278, 283, 285, 309–310, 316, 343, 348–356, 358, 381, 384–385, 388–389, 424, 430, 432, 436, 438–439, 533

insecticides   117, 140, 389

insects   117, 135, 139–140, 151, 173, 222, 235–237, 243, 264, 278, 328, 286, 373, 393, 397, 399, 426, 442, 457, 463, 518, 525, 561; see also particular kinds

insulin   30

interglacial   32, 154; see also *glaciers and glaciation*

invasive species   30, 105, 200, 305–306, 308, 319–320, 456, 549; see also *exotic species/subspecies*

invertebrates   135, 151–152, 199, 306, 316, 319, 439; see also *insects*, *crabs*, et al.

Ireland   159, 359

Irish Hobby (breed)   359–361, 378

Irish horses   358–361, 378

Iroquoian (Native American)   268

Isabela, Hispaniola   266

Islam   35–36

Island Breeze (horse)   185

Island Roxy Theater   178–179

islands   17, 19, 21, 23, 30, 33, 35, 37–38, 44, 54, 56, 104, 117, 121, 132–133, 137, 145, 148, 150–152, 154–158, 180, 194, 197–198, 200–202, 205, 210–212, 214, 216, 218, 226, 230, 232, 242, 264, 270–274, 276, 278–279, 282–283, 298, 309–310, 315, 343–344, 346, 348–350, 353–354, 356–357, 360, 364–365, 376, 381–382, 401–402, 424–427, 429–430, 432, 436–439, 446, 448, 451, 453–454, 456–457, 460, 464, 469–470, 495–498, 507, 516–517, 526, 531, 534, 566–567; see also particular islands

Isle Royale, Mich.   454

isoerythrolysis   392

intrauterine devices   80; see also *contraception*

jaca serrana (horse type)   37, 216

Jackson Mountains (Nev.)   87

Jamaica   38, 357

Jamestown, Va.   202–204, 206, 209, 268, 346, 358

Janus (horse)   372, 376

Japan   44, 89–91, 104, 159, 239, 453

jaundice   461

jaws, horse   394–395, 400, 440, 415

Jaycees   118

Jenifer, Daniel   210

Jennet   36–38, 43–44, 53–54, 214, 274, 347, 358, 362, 374, 376, 392, 403, 427, 497

Jester, Kendall   210–211

Jesuits   264

jetties   132–134, 231, 283, 351, 424, 439, 533

Jim (horse)   391–392, 404

Jitterbug (horse)   404–405, 567

Jockeys Ridge (N.C.)   272, 284

John's Island Stud (S.C.)   371

Johnston, Velma "Wild Horse Annie"   73

Jones, U.S. Rep. Walter   302–303, 463, 468

Jordan (river)   346

Juno (horse)   218

Justin Morgan (horse)   360

Kachina Grand Star (horse)   221

Kaimanawa horses (New Zealand)   44, 520

Kansas   68, 92

Kassie (horse)   470

Kathrens, Ginger   66, 74, 99, 101, 422

Kazakhstan   40–41

Keiper, Ronald   30, 118, 121, 137–138, 141–142, 144, 146–148, 152–153, 162, 180, 189, 243–244, 305, 395, 399, 422, 443–445, 519, 549, 566

Ken-L Ration   70

Kennedy, John F., Jr.   496

keratin   522

Kickotank (Native American)   202, 205

kicks and kicking   49, 119, 122–123, 126, 141–142, 144, 146, 160, 176, 208, 228, 289–290, 323, 325, 385–387, 389, 394–395, 397–398, 442–443, 482, 525, 568

kidneys   137

Kiger mustangs (Ore.)   55, 87

Kill Devil Hill, N.C.   277, 280

Kimball's Rainbow Delight (horse)   185

King   36, 38, 68–69, 176, 201, 206, 471, 503

Kingdom   56, 70, 91

Kingston, U.S. Rep. Jack   532, 538

Kirkpatrick, Jay   21, 28, 40, 44–45, 76, 78–
    80, 82–86, 88, 95, 99–100, 103–105, 117,
    124, 137–138, 140, 142, 148, 160–161,
    180, 189–190, 236, 298–299, 310, 394,
    471, 474–475, 520, 525, 549–551, 555,
    560, 565
Kitty Hawk, N.C.   273, 277–278, 280
knackers   70
koumiss   41, 56
K-selected   76

La Galga (ship)   218, 220
Lacey Act   318
lactation   56, 77, 137, 148, 180, 182, 189–
    190, 248, 276, 311, 314, 323, 415, 463,
    466–467, 472–474, 476, 478–479, 495–
    496, 499–500, 516, 520–521, 530, 535,
    537, 566; lactational anestrus, 189
Lactational Condition Index   314, 478–479
Lahontan mustangs (Nev.)   86
lameness   89, 91, 140, 295, 497
laminitis   302
land bridge   Atlantic, 557; Bering Strait, 310,
    560
Lander, Wyo.   69
Lane, Ralph   233, 266–267, 347–348
Laramie, Wyo.   94
laurel   97, 557
laws   11, 21, 23, 28, 36, 56, 69, 72–74, 85,
    87–88, 91, 98, 102–104, 168, 203–204,
    209–210, 218, 237, 270, 282, 304,
    318, 390, 464–465, 468, 470, 496–97,
    537–538, 551; Act Relating to Fences,
    and for the Protection of Crops (N.C.,
    1873), 270; Act . . . To Allow for an
    Adjustment in the Number of Free
    Roaming Horses Permitted in Cape
    Lookout National Seashore (2005), 468;
    Act To Ensure Maintenance of a Herd of
    Wild Horses in Cape Lookout National
    Seashore (1998, 2010), 468; Act To
    Place Certain Portions of Currituck
    County under the State-Wide Stock Law
    (N.C., 1937), 282; Act To Place Certain
    Portions of Dare County under the State-
    Wide Stock Law (N.C., 1935), 282; Act
    To Provide for the Removal of Cattle
    Remaining on Core Banks (N.C., 1959),
    390; Act To Restore the Prohibition on
    the Commercial Sale and Slaughter of
    Wild Free-Roaming Horses and Burros
(2007), 91; Bayne law (N.Y., 1911),
    318; Burns Amendment (Consolidated
    Appropriations Act, 2005), 88, 91; Clean
    Water Act (1972), 133; colonial, 203–
    204, 206–207, 209–210; Corolla Wild
    Horses Protection Act (2013), 291, 295,
    302–305, 312, 320; Lacey Act (1900),
    318; Maryland Wetlands Act (1970),
    133; Migratory Bird Treaty Act (1918),
    229, 318; National Historic Preservation
    Act (1966), 497; Restore Our American
    Mustangs Act (2009), 82; Safeguard
    American Food Exports Act (2013),
    90; Shackleford Banks Wild Horses
    Protection Act (1997), 468; stock law,
    209–210, 270, 282, 390; Wild and Free-
    Roaming Horses and Burros Act (1971),
    21, 74, 85, 88; Wild Horse Annie Act
    (1959), 73; Wilderness Act (1964), 496,
    538; see also ordinances
Lawson, John   54, 268, 271
Lawton Howard (horse)   383, 391, 404
Leah's Bayside Angel (horse)   184
Lee, Gen. Henry ("Light Horse
    Harry")   505
leg   90, 124, 248, 298, 323, 386–387, 398,
    417, 523, 558
Leicestershire, horses from   358–359
lentivirus   237, 460
Leonard   Donald, 224, 234, 248; Arthur, 248
leopards   32, 163
lethal white syndrome   297
leukemia   460
Libyan horses   36
lice   140
ligaments   72, 558–559
lighthouses   211, 216, 222–223, 232, 273,
    279, 281, 285, 287, 351–352, 354, 379,
    430–431, 433
Lindessa (horse)   404
lineage   horse, 42, 55–56, 104, 118, 147,
    162, 183, 187, 214, 247, 328, 359, 364,
    372, 376, 378, 405, 422, 427, 429, 471,
    477, 482, 500, 559, 561, 566–567
lions   32, 75, 521
lip   139–140, 396–397, 400
lipids   41
Lipizzaners   147, 374
liposomes   83
Little Cumberland Island, Ga.   512, 532, 534
Little Icicle (horse)   248

Little Red Man (horse)   291
Little Teach (horse)   389
Liu, Irwin   80, 83, 161
livestock   28, 30, 38, 40, 48, 50, 54–56, 65–66, 72–73, 79, 88, 92, 96, 98–99, 101–102, 105, 117, 157, 180, 200, 202, 206–212, 216, 218, 220, 223–224, 233–234, 236, 239, 265–271, 274, 276, 278–280, 282–284, 310, 343–344, 347–348, 350, 354, 356–358, 373, 382, 384, 386–387, 390, 402, 424–426, 432, 437, 450–451, 459, 464, 497–498, 502, 506, 517; see also particular kinds
London, England   41, 318, 503
Long Island, N.Y.   270, 428
long-term pasture (LTP)   93–94
Lost Colony   266–268
Louisiana   66, 96, 315–316, 527, 536
Lp (gene)   42
Luna (horse)   403–404
Lundy Pony   44
Lusitano (breed)   374, 392
luteinizing hormone   83
Lyme disease   124, 289

Mackay Island National Wildlife Refuge   293, 309, 318
magnesium   135, 319
mammal   32–33, 137, 320, 556
mammalian   23, 161
manatee   192, 512–513, 556
mane   48, 52, 120, 123, 140, 213, 226, 297, 385–386, 397, 400, 404, 432, 440457, 459, 463, 501, 526
Manteo (horse)   263, 325, 566
Manteo, N.C.   280, 325
mares   36–39, 41–43, 49, 51, 56, 66, 68, 70, 77, 79–84, 86, 89, 93, 101–102, 104, 118–119, 122–123, 127–128, 137–138, 140–149, 160–162, 164, 175, 178, 180, 182, 182–185, 188–191, 206–209, 213, 216, 221, 235, 238, 241, 243–248, 263, 274, 266, 286–288, 290–291, 296, 300–303, 311, 313–315, 321, 323–324, 343, 348–349, 358–360, 362–363, 368, 370, 372, 377–378, 382, 386–387, 392–400, 4024–405, 415–419, 421–422, 429, 437–446, 449–450, 452, 456–461, 463, 466–467, 469–479, 482, 495–500, 502, 513, 516–531, 535, 537, 550, 557, 561, 566

Marismeño (breed)   37
maritime forest   132, 190, 192, 272, 276–279, 285, 291, 328, 424, 430–431, 439, 452, 496, 526, 535
Marsh Tackies   364–365, 378, 402–403, 426, 499, 506; "Marsh Tackies" (poem), 364–365
marsh   17, 21, 23, 37, 54, 128–129, 132–133, 135,, 148, 151–152, 154–155, 160, 162, 173, 185, 190, 192, 198, 200, 205, 207, 212, 220–222, 225, 229–230, 232, 235–236, 245, 247–248, 264, 270–271, 277, 282, 284, 287, 291, 305–310, 314, 316, 318, 327–328, 346–348, 364, 373, 382, 384–385, 388–392, 399, 402–403, 424, 426, 430, 432, 434–436, 438–440, 447–448, 452–454, 457, 465–466, 469–470, 482, 495–496, 498–499, 506–507, 509, 512–515, 522, 527–531, 533–534, 536–537, 552, 556, 566–567
Maryland   17, 21, 37, 104, 117–166 passim, 177–178, 184, 186, 189–191, 193, 195, 197, 200–201, 203–204, 210, 214, 218, 222, 224–226, 230, 236–239, 241–242, 247–248, 293, 357–358, 360, 362–363, 368, 370–371, 422, 429, 453, 465, 473, 567
Maryland Department of Agriculture   237
marsh periwinkle   316, 527, 536
Mason, Carolyn   450, 465, 481, 483, 566
Massachusetts   85, 198, 272, 357–359, 420, 536
Massachusetts Bay   357, 359
maternal behavior   118, 522, 524–525, 527
maternal lineage and matrilines   42, 118, 147, 298, 301, 422, 471, 477; see also DNA, mitochondrial, and lineage
maximum sustainable/permitted herd size   76–77, 82, 147–148, 163, 189, 300, 453, 466, 480, 530; see also carrying capacity
Maya (horse)   404
McCalpin, Karen   292–293, 295, 298, 300, 302, 304, 312, 320–321, 323, 327–328, 469, 566
meadows   23, 176, 209, 213, 215, 280, 312, 496, 512
mean age at death (MAD)   161
megafauna   311–312
melanocytes   465
Menéndez de Aviles, Pedro   347

menhaden 230

Mennonites, genetic disorders of 299–300

Merychippus 559–560

Messel Pit (Germany) 557

mesteño 54

metabolism 122, 189, 299, 558

metapopulation 87, 300

Metompkin Island, Va. 210

Mexico 38–39, 89–90, 92, 218, 242, 347, 356, 362, 373

mice 76, 124, 400, 514

Michaels, Laura 186, 404, 567

Midnight (horse) 287

Mill Swamp Indian Horses 7, 263, 301, 315, 325, 328, 566

minimum viable population (MVP) 86, 163, 165, 298–300, 302, 565

Miocene Epoch 559–560

miscarriage see *abortion, spontaneous*

missionaries and missions Chesapeake Bay, 264, 266; Cumberland Island, 501–503, 517

Mississippi Flyway, 316, 319; River, 160, 347, 359, 362; state, 347, 516

Missouri Fox Trotter (breed) 214

Missouri 104, 316

Misty (horse) 173, 178–179, 183, 235–236, 244, 246, 551

Misty II (horse) 244

Mongolia 29, 49–50, 302

Montana 40, 44–46, 55, 68, 72, 76, 80, 82, 84, 86, 88, 104, 186, 196, 422, 427, 429, 476, 552

Montezuma 39, 45

moose 32, 52, 75, 290, 454

Morehead City, N.C. 436

Moreno, Pedro 38

Morgan Horse 54–56, 360, 376, 428; added to Chincoteague herd, 214, 243, 246

morphology 34, 51, 87, 284, 534

Mosbach Horse 42

mosquitoes 76, 124, 128, 139–140, 148, 163, 176, 236, 239, 302, 306, 389–390, 463; larvae, 236, 457; mosquito-borne diseases, 163, 302, 390

moss, Spanish 202, 495, 500

Mount Rogers and Mt. Rogers National Recreation Area 55, 97–98, 200, 307, 314–315,

mountains 21, 33, 39–40, 45–47, 49, 51, 53, 55, 72, 75–76, 79, 82, 87–87, 96–98, 101–104, 143, 146, 200, 214, 278, 314, 362, 365, 373–374, 403, 422–423, 427, 429, 476, 521, 552; see also particular mountains and ranges

mountain lion 75, 521

Mr. Bob(by) (horse) 392

mules 36, 68, 70, 266, 504

mullet 430, 438, 447

Munich, Germany 50

muscles 38, 276, 369, 395, 467, 475, 522, 559–560; "fly-shaker, 140, 237, 397, 461; muscle mass of ruminant/nonruminant herbivores, 558; musculature, 463, 556; relaxation in sleep, 398

museums and exhibits 34, 130, 178–179, 215–216, 267, 294, 460, 549, 555

musk ox 32

mussels 29, 137

mustangs 40, 49, 53–56, 65–105 passim, 121, 137–138, 148, 178, 214, 242–247, 294, 296, 300, 303, 307, 327–328, 358, 373–375, 378, 392, 422, 427–429, 431, 482, 499–500, 549, 565, 568; added to Chincoteague herd, 214, 242–247; mustang hunting, 66–70; mustangers, 66, 70, 72–73; see also particular populations

mutations 241, 297

muzzles 122, 136, 141–142, 160, 187, 395–396, 400, 415–416, 440, 523, 526

myrtle 138, 345, 385, 431, 495–496

Mystery (horse) 183, 241

Nags Head, N.C. 272–273, 276–278, 280, 351–352; Nags Head Woods, 430

Namibia 34, 104; Namib Desert, 137

naming of horses 420, 422, 424

Narragansett Pacer (breed) 54, 214, 263, 357, 360–362, 364, 368, 371, 376, 378, 426

Narvaez, Pánfilo de 346

National Aeronautics and Space Administration 194, 244, 511

National Audubon Society 192, 318

National Park Service see *U.S. National Park Service*

National Oceanic and Atmospheric Administration 196

Native Americans 39–40, 54, 56, 94, 202–204, 211, 266, 268–270, 272–273, 362, 428–429, 501; see also *Indians, American,* and individual tribes

Neanderthal 32

Neapolitan (breed) 36

neck (anatomy) 17, 36, 52, 69, 72, 90, 136, 140–142, 145, 210, 228, 238, 245, 270, 272, 287, 301, 325–326, 361, 374, 385, 391, 394–397, 400, 445, 461–463, 467, 475, 477, 513, 522, 525–526, 567

necks (topography) 54, 117, 141, 210, 216, 232, 270–271, 364; see also *peninsulas*

neigh 226, 396

Neptune, King/Queen 176

Netherlands 97, 274, 358, 370; see also *Dutch horses*

Neuse River, N.C. 269, 346

Nevada (state) 55, 66, 69, 73–74, 78, 82–83, 86, 92, 94–95, 97, 99–102, 140, 144, 146, 186, 214, 243–245, 474, 521

Nevada (horse) 404–405

New Bern, N.C. 354, 452, 464

New England 429–431; horse exports, 205, 356–357, 360–361, 378–379, 426; horse imports, 210, 274, 359–360, 366

New England Pacing Horse see *Narragansett Pacer*

New Forest Pony 44

New Jersey 182, 193, 237, 358, 366

New Mexico 39, 90, 92, 242, 373

New Orleans, La. 69, 214

New World 31, 35, 37–38, 54–55, 206, 214, 216, 218, 264–266, 270, 274–275, 310, 328, 344, 346–347, 356, 374, 378, 392, 425

New Zealand 44, 75, 104, 159, 520

newborn 175, 180, 182, 241, 248, 286, 296, 303, 323, 392, 399, 439, 476, 514, 520, 522, 524–526, 566

Nicuesa, Diego de 37

nicker 142, 173, 396, 522

Nixon, President Richard 240, 512

nor'easters 129, 193–194, 198, 222, 277–278, 285, 352, 358, 456

Norfolk, Va. 178, 280, 363

Norman horses 206

Norse Dun (breed) 36

North Carolina 17, 43, 46, 87, 104, 150–151, 154–155, 157–158, 180, 186, 198, 215, 218, 222, 263–328 passim, 343–405 passim, 415–483 passim, 500, 536–537, 566–567; see also particular locations

North Carolina Department of Agriculture 460, 465, 469–470

North Dakota 404

North Star (horse) 183, 185, 245

Northampton County, Va. 194

Norwood, Henry 202–203, 205

nostrils 72, 142, 222, 386, 393, 395–396, 400, 415, 522

Nova Scotia, Canada 17, 44, 78, 104, 121, 190, 201, 206, 264, 358

National Parks Conservation Association 533

Numidian horse 36

Nunki (horse) 302

nursing 144, 148, 178, 180, 189, 237, 248, 314, 395, 461, 467, 472–473, 478, 495, 498–499, 517, 519–521, 523–525; see also *lactation*

oaks 275, 278, 285, 311, 356, 399, 431, 435, 495, 500, 504, 512

oats 276, 278–279, 440, 516, 533, 536

Oceania 33

Ocracoke, N.C. 46, 104, 158, 186, 218, 275–277, 282, 298, 307–309, 343–405 passim, 425–426, 435–436, 438–439, 469, 567–568; beach patrol (Park Service), 392

odd-toed hoofed mammals 556

odor 139, 396, 522; see also *scent* and *smell*

Oglethorpe, James 497, 502–504

Ojeda, Alonso 37

Okies Rainbow (horse) 405

Oklahoma 92, 463

Old Jerry (horse) 355

Old Paint (horse) 381, 392

Old Widdie (horse) 386

Old Wildy (horse) 385

Oliphant, Cooper H. "Cowboy," 232

onager see *ass*

oocytes 83–84

Opelousas (breed) 370

open-woodlands husbandry 208

orangutans 33

orcas 420, 422

ordinances 287, 289, 296; see also *laws*

Oregon 54–55

Oregon Inlet, N.C. 273, 343, 348–349, 351, 356, 384

organs 276, 396

Oriental, N.C. 269, 444

Oriental breeds 42

Orohippus 557

orthogenesis 555

Ossabaw Island hog   516

otters   439, 512

outcrossing   54, 121, 147, 164–165, 245, 367, 375, 378, 393, 499

Outer Banks   43, 104, 144, 150–152, 154, 158, 164, 198, 201, 215, 218, 263–328 passim, 343–405 passim, 415–483 passim, 500, 531, 537, 552, 565–566; see also particular locations

ova   80, 302; see also *eggs* and, *oocytes*

ovaries   80, 83–84

overgrazing   78, 96, 148, 150, 160, 282, 309, 312–315, 320, 322, 449, 452–453, 459, 470, 473, 495, 498, 517, 534, 536

overharvesting   231, 316

overhunting   311

overo (coat)   42, 376

overpopulation   84, 99, 101–102, 305, 314, 453, 459, 464, 531, 535

overpredation   33

overwash   129, 132, 150–158, 193, 197–198, 278–279, 283, 285, 309–310, 343, 349, 352–354, 424, 470

owls   29, 324

oystercatcher, American   514, 533

oysters   173, 202–203, 224, 229–232, 235, 276, 316, 354, 430, 534

pacers and pacing (gait)   214, 358–363, 366, 522; see also *Narragansett Pacer*

Pacific Ocean   203, 264, 428

paddocks   46, 307, 344, 381, 401, 446

paint (coat)   147, 246, 376, 381, 392

Paleocene-Eocene Thermal Maximum   556

Paleo-Indians   152, 200, 311, 501

Paleolithic Period   32, 44

paleotheres   556

palmetto   495, 512, 515, 531, 535

Paloma (horse)   343–345, 404–405

palomino (coat)   173, 177, 183, 248, 377, 440

Pamlico, County, N.C.   269

Pamlico River, N.C.   343, 346, 381

Pamlico, Sound, N.C.   274, 354, 385, 470

Pampas   308

Panama   37, 218, 561

Paragard®   80

parakeet   33, 347

parasites   43, 77, 152, 186, 232, 314, 391, 455, 463, 475, 478–479, 495, 521, 535

parks   118–120, 137, 140, 151, 165, 195,

198, 200, 312, 533, 550

Parliament, British   54, 270

Parramore Island, Va.   210

Parris Island, S.C.   347

parrot-mouth   297

parturition   522

Paso Fino (breed)   214, 374, 426–428, 500–501

patriarchs   142, 245, 287, 367, 441, 552, 567; see also *stallions*

p-cresol   400

Pea Island, N.C.   158, 310, 351, 353; horses, 276, 378; National Wildlife Refuge, 349

pedigrees   118, 366, 376, 422, 471, 474, 482

peninsulas   35–36, 42, 54, 132, 135, 193, 204, 215, 225–226, 270, 351, 382, 446, 561; see also *necks*

Pennsylvania   19, 31, 93, 183, 193, 211, 228, 285, 299, 358

Penny (Lewark) Hill   278–279, 285, 328, 363

pennywort   440

perissodactyls   556

permafrost   32–33, 310

Peru   38, 184

Peruvian Paso (breed)   214, 374

pesticides   302

pests   19, 96, 140, 237, 320

phalanges   559

Philadelphia, Pa.   230, 285

phlebotomy   426

Phragmites   119, 138, 200, 306–308, 310, 312

Pickens, T. Boone   94

piebald (coat)   370

pigeon   33, 318, 394

pigs   see *swine*

Pike, Zebulon   66

pilots, maritime   266, 276, 354, 356

pine   23, 75, 215, 222, 225–226, 244, 278, 311, 399, 495, 510, 512, 536; loblolly, 23, 278

Pine Mountain (Va.)   97

pinto (coat)   38, 42, 66, 117, 121, 136, 173, 177, 183–184, 188, 207, 214, 238, 244, 246–248, 297, 343, 345, 367, 377, 383, 392, 394, 403–405, 440, 500

pinworms   397

pirates and piracy   205, 216–218, 243, 272–273, 371, 380–382, 389, 502–503; see also *privateers*

Pittsburgh, Pa. 138
pituitary gland 83
Pi-W allele 427
Pizarro, Francisco 39
placenta 523, 557
plague, bubonic 39, 240
plantations 209, 212, 271, 302, 356, 499, 501, 504–506, 514, 536, 539
Pleistocene Epoch 32, 34–35, 40, 42–44, 46, 56, 104, 152, 310–312, 507, 561
plexus 559
Pliocene Epoch 560
Pliohippus 560
plover mountain, 96; piping, 151, 174, 190–191, 193, 309–310, 312, 318, 439, 514; Wilson's, 514
plows and plowing 208, 386
Plum Orchard 498, 500, 508, 523, 538–539
Plymouth, Mass. 268
Pocahontas 204, 206
Pocomoke City, Md. 178; Pocomoke River, 202; Pocomoke Tribe, 202, 204
Poland 51, 104, 159
Pomiac (horse) 325
Ponce de Léon, Juan 344, 364
ponds 29, 129, 137, 188, 191–192, 248, 286, 291, 430, 438, 447, 454
ponies 33, 37, 42, 44, 54–56, 70, 78, 89, 97–98, 117–128, 134, 135, 137–138, 140, 143, 147–148, 150, 152–153, 160, 162–163, 165, 173–248 passim, 272–276, 278, 282–283, 295, 307–308, 314–317, 327, 344, 355, 357, 359, 363, 370–375, 377–378, 380–382, 384–393, 398–405, 422, 426, 428, 430, 434, 436–438, 444, 447–450, 453, 457–458, 469, 483, 498, 500, 506, 559–560, 567–568; see also particular breeds and populations
pony patrol (Assateague) 128
porcine zona pellucida (PZP) 80–81, 83–84, 101, 161, 300, 458, 471, 474
porpoises 276, 355, 430, 435; see also *dolphins*
Port Royal, N.S., Canada 206,
Portsmouth, N.C. 276, 343, 348, 354–356, 380, 382, 424, 436, 469
postpartum estrus 148, 189–190, 520; see also *estrus*
potassium 135
potato 127, 230, 272
Potomac River 202, 222

prairie 21, 45, 83, 96, 200, 311, 315, 402, 558
Preakness Stakes 483
pregnancy 56, 77, 79–81, 83, 89, 122, 137, 144, 148, 161–162, 165, 175, 178, 180, 182, 185, 189, 221, 238, 276, 321, 404, 458, 471, 473–476, 478, 482, 499, 519–521, 524, 557
PremarinTM 56
Premierre (horse) 245
presidents, U.S. 65, 88, 188, 217, 240, 318, 468, 512
Pribilof Islands 453
Primavera (horse) 300
Prince 183–184, 248, 503
privateers 266, 354, 380–381
progesterone 189
proteins 80, 524
proto-Arabians 376
Pryor Mountain mustangs Mont.–Wyo.) 40, 45–47, 49, 53, 55, 72, 76, 79, 82–83, 87, 98, 101, 103–104, 143, 146, 365, 373–374, 422–423, 427, 429, 476, 522; Pryor Mountain Wild Horse Range, 76, 82, 552
Przewalski Horse (Takhi) 29, 34, 42, 44, 48–51, 137, 302, 442, 560
Pueblo revolt 39, 54
Puerto Rico 38, 44, 104, 265–267, 344, 347, 426–428, 435
pulse 559
purebred horses 36, 244–245, 375, 378, 499
Pyle, Howard 215, 222, 228–230
Pyrenees Mountains 561
PZP see *porcine zona pellucida*

Q-ac allele 365, 427–429, 565
quagga 33–34, 41, 50, 561
Quakers 270, 360
Quarter Horse 54–56, 147, 246, 276, 366, 368–369, 372, 376–378, 389, 402, 428, 500, 501; in Chincoteague herd, 246;
Quebec, Canada 91, 206
Queen Anne's War 381
Queenie (horse) 568
Quejo, Pedro d 345

rabbits 76, 265, 424, 439, 454, 514
rabicano (coat) 299, 377
rabies 185, 240, 293
raccoons 193, 312, 439, 510, 514, 533
racehorses 142, 344, 366, 372, 426, 483

Rachel Carson N.C. National Estuarine
    Research Reserve   198, 307, 452–458,
    446, 448, 537, 567
racing   365–366, 368, 370, 372, 389, 426,
    482, 483; quarter-racing, 366, 368, 370,
    372, 389
radiocarbon-dating   278
raffle, pony   176
Raffles (horse)   244
rain   76, 78, 132, 137, 236, 309, 352, 400,
    436, 446–447, 454, 495–496, 517, 521,
    526, 530, 534, 536
Rainbow (horse)   325
Raleigh, Sir Walter   264–265, 275, 279, 347,
    351
rangers, park   123–127, 186, 391–392, 404,
    482, 567
rats   29, 43, 514
Rayo (horse)   405
Reconstruction   430, 555
red (coat)   52, 66, 377, 415
Red (horse)   246
Red Feather (horse)   327, 362
re-domesticated   402
Reed, Thomas   235, 242
reeds   138, 200, 202, 232, 235, 242, 306,
    308, 369, 376, 392
reefs   194, 355, 432–434, 447, 534
registries   373
registry   245, 327, 373–375, 392, 483
rein   120
reindeer   48, 52, 453
reintroduction/repatriation   23, 28, 34–35,
    40, 44, 50, 56–57, 83, 88, 103, 193, 303,
    310, 312, 344, 390, 456, 470, 507, 514,
    549; see also rewilding
REM (rapid eye movement) sleep   398, 518
Representative   130, 206, 302, 355, 452,
    455, 460, 463, 468, 471, 532
Restore Our American Mustangs Act   82
Retuertas horses (Spain)   86, 298
Revolution   49, 213–214, 363, 430, 504–505
Revolutionary   217, 452, 505
rewilding   312
rhinoceros
rhinoceroses   33, 75, 556
Rhode Island (breed)   94, 270, 357, 359–
    362; see also Narragansett Pacer
Ribault, Jean   502
rice   72, 133, 197, 279, 349–350, 504
Ricketson family   496, 507

ridges (topography)   97, 156, 190, 225, 232,
    272, 284, 310, 314, 343, 465, 534
Rio Grande   39, 54
rivers   37, 39, 51, 160, 202–203, 206, 264,
    269–270, 272, 316, 320, 343, 345–346,
    357, 359, 362, 380–381, 439, 498, 512
roan (coat)   38, 66, 177, 500, 502
Roanoke Inlet, N.C.   272, 285, 316, 347,
    351, 354
Roanoke Island, N.C.   215, 266–268, 272,
    280, 316, 347, 351
Roanoke River (Va.–N.C.)   370
Robins, John   211
Rock Springs Herd Management Area
    (Wyo.)   104
Rocky Mountain Horse   214
Rocky Mountains   96
Rodanthe, N.C.   348, 352
rodeo   53, 73, 88, 124, 182, 229, 391, 404,
    499
Ronald (horse)
Roosevelt, Theodore   68; National Park,
    200, 404
ropes   38, 41, 52, 72, 220, 558
roping   402
roundups   54, 72, 81, 85, 99, 102, 117, 160,
    173–174, 177, 180, 185–186, 208, 225,
    234, 238, 247, 264, 269, 274, 290, 355,
    384, 387–388, 444, 448, 450–451, 460,
    465–466, 469, 472, 479, 498–499, 531;
    see also horse/pony penning
r-selected   76
Ruffin, Edmund   355, 358, 425–426, 432,
    434, 447
ruminants   159, 558–559
Russia   33, 41, 91, 159

sabino (coat)   299, 377
Sable Island, N.S., Canada   17, 44, 78, 104,
    121
Sacajawea (horse)   402, 404, 567
saddle   50–52, 55, 66, 69–70, 73, 89, 212,
    214, 274, 295, 325, 357, 359–364, 366,
    368, 375, 378, 384–386, 392, 426, 429,
    450, 499, 551; McClellan, 384
Saddlebred   376, 428
Saiga   33
Salazar, Ken   65, 82, 88
saliva   522
Salt Wells Creek Herd Management Area
    (Wyo.)   102

saltgrass   162

saltwater   23, 198, 274, 316, 435, 439

Saltwater Cowboys   173–175, 177, 180, 205, 217

Salvo, N.C.   348

San Clemente Island, Cal.   314

San Felipe   502

San Juan Bautista   345–346

San Lorenzo (ship)   215–216

San Mateo, Fla.   347

San Miguel de Gualdape   346

San Pedro   island, 502; missions, 502; ship, 218

sandbars   105, 216, 222, 358, 290, 424

Sandbridge, Va.   292–293

sandhills   see *hills*

Sandwash Basin Herd Management Area (Colo.)   92

Santa Cruz mustangs (Cal.)   373

Santiago (horse)   367, 383, 404

Santo Domingo   218, 357

Saskatchewan, Canada   90

savanna   311, 559

Savannah, Ga.   504, 516

Saving America's Mustangs Foundation   94, 303

scallops   232, 276

Scarborough/Scarbrough, Edmund   204

scent   17, 75, 139–142, 393, 400, 415, 522; scent-marking, 142, 397, 400, 441; see also *smell*

Schiltberger, Johannes   49

Science and Conservation Center   7, 21, 44, 76, 80, 84–85, 565

scleras   394, 501

Scottish horses   216, 360–361

Scythians   43, 53

Sea Camp   505–506, 531

Sea Feather (horse)   176

sea level   18, 132, 284, 432, 447; sea-level rise, 18, 154, 156–157, 191, 196–200, 309–310, 349, 432, 439, 457, 536

sea oats   276, 278, 440, 516, 533, 536,

seabeach amaranth   150–151, 191, 193, 309

seagull   117, 278, 285

sea oats   276, 278–279, 440, 449, 516, 533, 536

seawater   129, 137, 158, 223, 447, 536

Seminole (breed)   269, 347, 359, 362, 366, 378, 402

senses   216, 396, 400, 523, 559; see also

hearing, odor, scent, sight, and *smell*

Seville, Spain   36–37, 218

Shackleford (horse)   483

Shackleford Banks, N.C.   43, 104, 144, 150–152, 158, 164, 264, 279, 300–304, 307–310, 315, 343, 348, 352, 355, 364, 374–378, 384, 388, 390, 397, 401–405, 415–483 passim, 531–532, 537, 552, 565–567

sheep   40, 49, 54, 72, 97–99, 105, 117, 206–211, 223–224, 228, 264, 266, 269, 272, 274, 278–279, 282, 308, 310, 314, 348, 355, 382, 385, 390, 430, 432, 435–436, 438, 450–452, 521, 558; sheep penning (Assateague), 224, 228

Shell Castle, N.C.   354–355

Shetland Pony   44, 55–56, 97, 121, 178, 200, 214, 216, 244, 246, 274, 314, 317, 444, 500, 559

shipwrecks   27, 117, 129, 188, 205, 212, 215–218, 220–222, 224, 264, 272–273, 276–277, 353, 356–358, 376, 378, 380–381, 425–426, 430–431, 435, 438

Shire (horse)   44

shorebirds   151, 191–192, 308–309, 316, 514, 533

Siberia   32–33, 42, 48, 159

Sifrhippus   556–558

sight   216, 394

sika elk/deer   124, 158–160, 193, 298, 453

Silver Lake, N.C.   354, 373

Sinepuxent Bay (Md.)   129, 132, 134, 137, 230

singlefoot (gait)   35, 214

sinuses   400

Siouan (Native American)   268

Sioux (Native American)   88

skeleton   555–557, 560

skewbald (coat)   370

Skowreym (horse)   244–245

Slash (horse)   117

sleep   398–399, 518, 526; REM, 398, 518

sling   37, 220, 222, 248

smallpox   39, 204, 241

smell   139–140, 396, 522

Smith, John   204, 206, 347

Smithfield, Va.   263, 301, 325, 327, 566

snaffle   52

snails   316, 527, 536

Snip (horse)   360, 376

snow goose   148, 316, 533

Sockett To Me (horse)   184

Solutré   44, 52

sorrel (coat)   38, 66, 92, 177–178, 184, 214, 264, 360–361, 363, 377, 392, 404, 415, 417, 440, 471, 500, 567

sound   391, 394, 396, 400,; see also *hearing*

sounds (topography)   117, 152, 225–226, 263, 270–271, 273–278, 285, 290–291, 316–320, 324, 349–351, 354, 356, 360, 384–386, 389–391, 420, 429–430, 432, 436, 442, 446, 449–450, 465, 469–470, 536; see also particular sounds

South Carolina   201, 264, 270, 345, 347, 360, 364, 371, 382, 402–403, 431, 510, 536; see also particular locations

South Carolina Jockey Club   371

South Dakota   69

Soviet Union   239–240

Spain   29, 36–38, 86, 201, 215–218, 220, 264, 298, 344, 346–347, 356, 362, 370, 374, 497

Spanish horses   36–39, 43–44, 54–56, 70, 87, 90, 104, 121, 206, 214, 216, 242, 264–269, 271, 274, 276, 328, 344, 346, 358, 360, 362–366, 370–376, 565; see also *Colonial Spanish Horse* and individual breeds

Spanish Barb Breeders Association, Southwest   373

Spanish Mustang Association   373

Spanish Mustang Registry   373, 375, 392,

SpayVacTM   83–84

Spirit (horse)   183, 343, 404

Sponenberg, D. Philip   37, 42, 44, 51, 55–56, 70, 104, 216, 302, 362, 369, 373–374, 376, 378, 402, 426, 469, 565

Spruce Mountain (Nev.)   94; Spruce/ Pequop Herd Management Area, 94, 96

squeal   136, 141, 144, 175, 228, 386, 396

squirrel   193, 514

St. Augustine, Fla.   266, 347, 502

St. Marys, Ga.   512, 533, 536

St. Matthew Island, Alaska   453,

St. Paul Island, Russia   32

Stafford   Robert, 499, 505; Robert, Jr., 505–506; mansion, 508, 539

Stallings, Wesley   294, 296, 299–300, 321–323, 325, 327–328, 566

stallions   21, 36–37, 39, 42, 50, 66, 68, 70, 79–82, 93, 103–104, 117–118, 122, 127–128, 133, 136–138, 140–147, 162, 164, 175, 183–184, 206, 209–210, 213, 216, 219, 235, 237, 242–248, 263–264, 266, 287–291, 294–296, 299, 301–302, 306, 314–315, 321–323, 325, 327, 343, 348, 357–358, 360, 362, 364, 367–368, 372, 375, 377–378, 382, 385–389, 391–393, 396–401, 403–405, 415–423, 429, 437, 439–443, 445–446, 449, 451, 456, 460, 464, 470–471, 474–475, 477, 482, 495, 497, 499–500, 513, 518–520, 522, 525, 528–529, 551–552, 566–567; multi-stallion bands, 519

Standardbred   214, 376

Star (horse)   287

Station   180, 182, 248, 285, 293, 351, 356, 379–380, 384, 421, 430–431

status, social   145, 397, 438, 442–444, 446, 513, 532, 568; see also *dominance/ submission behavior*

steppe   33, 40–42, 48–50, 561

sterilization   80, 297

steroids   79, 83

stock-dipping   384

stockmen   33, 50, 54, 69, 72, 88, 95, 117, 208, 216, 223, 225, 232, 270–271, 282, 355, 357, 364, 378, 380, 383, 385, 402, 429, 448, 450–451, 498

Stormy (horse)   178, 235, 242, 244

stud pile   140–141

studs   68, 140–142, 145, 287, 315, 343, 358, 360, 363, 368, 371–372, 387, 442, 446, 497, 499

Stuska, Sue   81, 150, 343, 416, 425, 440, 458, 470–471, 474–475, 479–480, 482, 565–566

submission   see *dominance/submission behavior*

Suerte (horse)   325

Sulphur mustangs (Utah)   55, 87, 373

Sundance (horse)   482

Sunny (horse)   325

Surfer Dude (horse)   184, 188, 243, 568

Suzy Q (horse)   184

Suzy Sweetheart (horse)   568

swamp fever   see *equine infectious anemia*

swamps   91, 105, 237, 263, 270, 301, 311,315–316, 325, 328, 432, 460, 537, 561, 566

Swan Cove, Va.   174, 192

swans   174, 192–193, 203, 293, 321, 324

Swimmer (horse)   327

swimming   27, 152, 173–248 passim, 290, 324, 327, 355, 357–358, 373, 378, 385, 432, 437, 449–450, 453, 498, 567; see also *drowning*

swine/hogs/pigs   29–30, 40, 80, 83, 117, 161, 205–208, 209–211, 223, 233, 241, 264, 266, 269, 272, 274, 276, 278–279, 282–283, 304, 309–310, 312, 320, 324, 346–348, 355, 390, 430, 438, 448, 452, 499, 504, 506, 508, 510–511, 514–518, 528, 539; Ossabaw Island hog, 516; see also *porcine zona pellucida*

Tacatacuru   502

tail   36–37, 46, 69, 72, 140–142, 144, 176, 213, 226, 228, 297, 361, 374, 376, 385–387, 393–394, 397, 400, 404, 415–416, 445, 457, 462–463, 475, 501, 522; carriage, 36, 141,142, 226, 361, 385, 393–394, 397, 415–416; flattening, 394; swishing, 140, 236, 393–394, 397, 463

Takhi   see *Przewalski Horse*

Tallaboa Bay, Puerto Rico   265, 267

Tampa Bay   346–347

Tarpan   34, 37, 42, 51

Taylors Creek, N.C.   434, 437–438

Teach, Edward   pirate, see *Blackbeard*; Edward Teach (horse), 301, 325; Little Teach (horse), 389

tendons   300, 392, 558–559

Tennessee Walking Horse   54, 214, 376, 499–501

Terry, Roe   185, 188, 567

testosterone   80, 83, 144

Texas   39–40, 66, 89, 92, 96, 99, 146, 159, 315–316, 327, 375, 432, 565

thongs, prehistoric   40–41

Thoroughbred   55, 70, 93, 142, 147, 214, 246, 360–361, 365–366, 368, 370–372, 376–377, 387, 483, 499–501

ticks   124, 140, 135, 383–384

tides   132, 134, 137–138, 154, 160, 173, 175, 183, 196, 201, 212–213, 236, 248, 285, 306, 309, 320, 344, 350, 355,353, 382, 424, 430, 439, 446–448, 453, 470, 495, 498, 513, 516, 530, 533, 568, 537

Tiger (ship)   267, 348

Timucua   347, 502–503

titanium   509

tobacco   208–210, 218, 368

tobiano (coat)   42, 367, 377–378

Tom Creef (horse)   380

Tornado's Prince of Tides   183

Tomochichi   503

Toms Cove, Va.   158, 174, 193, 225, 232, 235

Toonahowie   503

tourism   94, 104, 123, 140, 195, 242, 277, 280, 286, 288–290, 295, 353, 355, 393, 450, 460

trade   Africa, 52; wool, 207; colonial, 270, 276, 285, 354–356; horse, ancient, 35; colonial, 273–274, 344, 356–378 passim; Grenville expedition, 266, 344, 348; Native American, 40, 54; New England, 356–360; wildfowl, 318; 52, 130, 184, 186–187, 203, 207, 216–217, 266, 270, 273, 276, 318, 344, 352, 354, 356–360, 425

Trade Winds   356

Tradewind (horse)   315, 325, 327

translocation   86–87, 298, 300, 469

travois   52

treasure fleets, Spanish   218, 273

Tresie (horse)   325

T-Rex (horse)   323

tridactyly   559–560

Turkestan   561

Turkoman (breed)   372

turn-back   see *buy-back*

Turner, Allison   126, 165, 184, 567

turtles   192, 276, 439, 510–511, 513–515, 532–533; loggerhead, 192, 276, 439, 514, 533

Tuscarora (Native American)   268, 270, 429

U.S. Army   39, 69–70, 75, 132, 155, 230, 238–240, 242, 263, 353, 384, 429, 431, 452, 454–455; biological weapons, 238–240; Cavalry, 55, 387; Corps of Engineers, 132–134, 150, 155, 353, 452, 454–455; Remount Service, 69–70

U.S. Bureau of Land Management   46, 65–105 passim, 119, 148, 160, 164, 242–243, 245, 300, 422–423, 479, 550; see also *U.S. Department of the Interior*

U.S. Coast Guard   175, 212, 222, 232, 266, 280, 285–286, 351, 387, 389, 400, 416, 428–429, 431, 440; Beach Patrol, 428–429, 431; see also *U.S. Lifesaving Service*

U.S. Congress   21, 28, 74, 82, 90–91,

285, 302–304, 317, 352, 439, 468, 483, 501, 538; see also U.S. House of Representatives and U.S. Senate

U.S. Department of Agriculture   89–90, 100, 384–384

U.S. Department of the Interior   65, 74, 82, 88, 102, 131, 303, 319, 468–469, 51, 550, 565; see also *U.S. Bureau of Land Management, U.S. Fish and Wildlife Service,* and *U.S. National Park Service*

U.S. Fish and Wildlife Service   96, 119, 158, 160, 173–248 passim, 285–324 passim, 457, 550; see also *U.S. Department of the Interior*

U.S. Forest Service   55–56, 69–70, 85, 97–98, 200, 278, 314, 317

U.S. House of Representatives   82, 91, 194, 303, 468

U.S. Lifesaving Service   188, 216, 221, 223, 285, 356, 379–380, 384; see also *U.S. Coast Guard*

U.S. National Park Service   23, 27, 30, 46, 77, 117–166 passim, 186, 188, 191, 193–194, 198, 200, 218, 237, 242, 282, 284, 303–304, 307–309, 343–405 passim, 415–483 passim, 495–540 passim, 550–551, 567; see also *U.S. Dept. of the Interior* and particular parks and seashores

udder   142, 522–523

Ukraine   33, 41, 50–52, 561

Ulva (horse)   138, 470

Union   230, 356, 379; see also *Civil War, U.S.*

Uno (horse)   325

urak/arkan   49

urination   127, 144, 397, 399

urine   56, 96, 135, 137, 142, 144, 189, 399–400, 441

USS *Despatch*   217

USS *Monitor*   353, 356

Utah   55, 86, 92, 195, 242, 319

uterus   84, 476, 522, 557

uveitis   43

vaccination   46, 79–81, 83–84, 91, 93, 120, 149, 161–162, 176, 185, 197, 236–237, 239–240, 248, 295, 300, 391, 456, 458, 460, 471, 473–474, 482, 531, 550

Valor (horse)   325

van der Waals forces   156–157

Venezuelan Equine Encephalitis   239–240

Vermont   85, 180, 182, 290

Verrazzano, Giovanni da   201–202, 264, 272

vertebrae   376, 462, 495; lumbar, 376, 462

Virginia   17, 37, 55–56, 73, 97–98, 104, 117–118, 120–121, 128–129, 143, 148, 150, 158, 173–248 passim, 263–264, 266, 269–272, 274–275, 280, 285, 287, 290–293, 307, 309, 317, 346–348, 350, 357–358, 360–363, 366, 368, 370, 372, 381, 384, 402, 404, 425, 429, 470, 536, 566

vocalization   396, 398; see also *neigh, whinny,* and *squeal*

vomeronasal organ   139–140, 396

vomit   323

Voodoo (horse)   142

vulva   36, 142, 393, 522

walk (gait)   36, 214, 399

Wallops Island, Va.   194, 210, 234, 239, 244

warbler   192, 533

Ward, Rick   127–128

warm-blooded breeds   42

Wash Woods, N.C.–Va.   285, 348

waterfowl   151, 191–192, 195, 199–200, 223, 228–230, 234–235, 242, 265, 276, 285–286, 291, 304, 310, 314, 317–319, 457, 567

waterholes   50, 66, 313, 416, 418–419, 421, 423, 446, 448

watermen   195, 200, 223, 230, 235, 276, 343, 430

watershed   135, 203, 536

weaning   178, 180, 182, 185, 189–191, 213, 247, 433, 463, 525

weanling   176, 184, 242

weapons   35, 45, 318; biological, 238–242

wells   94, 102, 204, 228

Welsh Pony   44, 246, 404

Wenzel (horse)   343, 404, 567

West Nile virus   185, 521

Westerlies   217–218, 356

wetland   133, 152, 191–192, 197–198, 203, 234, 280, 305–306, 312, 314, 319, 353, 514, 517, 536

Whalehead Club   285, 287,

whales and whaling   192, 276, 285, 422, 430, 438, 512

whinny   175, 178, 226, 396

White Dandy (horse)   378, 384–385

wigeon 200, 314, 317
Wilbur-Cruce mustangs 373
Wilburn Ridge Pony Association 97, 314
wild 27–56 passim; see also *exotic species/ subspecies, feral, free-ranging/roaming, invasive species,* and *reintroduction/ repatriation*
"Wild Horse Annie" (Velma Johnston) 73
Wild Horse Prison Inmate Training Program 565; comparable state programs, 92
wild horse sanctuaries, private and quasiprivate 72, 94–96; Cedar Island, 456, 469; Currituck Banks, 287, 295, 328,
wilderness area, Cumberland Island
willet 306, 308, 514
windmills 276, 354
Winyah Bay, S.C. 345–346
Witch Doctor (horse) 219, 247
withers 33–34, 72, 120, 386, 393, 445, 475, 495, 538, 555
wolves 29, 32–33, 43–44, 54, 57, 75, 86, 203, 207, 424, 432; gray, 29; red, 57
wool 207, 223, 228
woolly mammoth 32, 105, 310, 312, 561
World War I 70
World War II 50, 70, 73, 91, 173, 239, 241,

276, 370, 428, 431
worms, parasitic 91, 93, 120, 176, 185, 316, 391, 396–397, 456
Wrangel Island, Russia 32
wranglers 87, 91–92, 405, 498
Wright brothers 273, 277, 280; Wright Brothers National Memorial, 280
Wyoming 40, 45, 49, 55, 72, 74, 82, 88, 92, 102, 104, 232, 311, 427, 429, 552

yawning 395
yearlings 68, 79, 92–93, 238, 248, 290, 325, 400, 452, 455, 475–476, 481, 500, 525, 530
Yellowstone National Park (Wyo., Mont., Idaho) 81, 290
Yeomans, Steve 448, 450

zebra 29, 33–34, 41, 48, 50, 424, 555, 560–561
zinc cadmium sulfide 240
Zion National Park (Utah) 195
zona pellucida see *porcine zona pellucida*
ZonaStat-H 85
zoos 19, 29–30, 34, 41, 44, 49–51, 158, 164, 302

About the Author

Bonnie Urquhart Gruenberg is a multifaceted person who wishes that sleep were optional. She is the author of the award-winning textbook *Birth Emergency Skills Training* (Birth Guru/Birth Muse, 2008); *Essentials of Prehospital Maternity Care* (Prentice Hall, 2005); and *Hoofprints in the Sand: Wild Horses of the Atlantic Coast* (as Bonnie S. Urquhart; Eclipse, 2002), as well as articles in publications as dissimilar as *Equus* and the *American Journal of Nursing*. She is an artist and photographer and has illustrated all her own books.

By profession, she is a Certified Nurse-Midwife and Women's Health Nurse Practitioner who welcomes babies into the world at a freestanding birth center in Lancaster County, Pa. She obtained her MSN from the University of Pennsylvania after completing her BSN at Southern Vermont College, and she spent 10 years attending births in tertiary-care hospitals before returning to out-of-hospital practice. Prior to her career in obstetrics, she worked as an urban paramedic in Connecticut.

Horses have been her passion from infancy. For nearly two decades, she has spent countless hours researching and photographing the private lives of wild horses in both Western and Eastern habitats. She has been riding, training, teaching, and learning since her early teens, from rehabilitating hard-luck horses to wrangling trail rides in Vermont and Connecticut. In her vanishing spare time, she explores the hills and hollows of Lancaster County astride her horses Andante and Sonata.

More information and a collection of her photographs can be found at her Web site, www.BonnieGruenberg.com Additional information about the Atlantic Coast horse herds is on the Web at www.WildHorseIslands.com

www.ingramcontent.com/pod-product-compliance
Lightning Source LLC
Chambersburg PA
CBHW050617290326
41929CB00062B/2935